VOLUME SIX HUNDRED AND FOUR

METHODS IN ENZYMOLOGY

Marine Enzymes and Specialized Metabolism
Part A

METHODS IN ENZYMOLOGY

Editors-in-Chief

ANNA MARIE PYLE
Departments of Molecular, Cellular and Developmental Biology and Department of Chemistry
Investigator, Howard Hughes Medical Institute
Yale University

DAVID W. CHRISTIANSON
Roy and Diana Vagelos Laboratories
Department of Chemistry
University of Pennsylvania
Philadelphia, PA

Founding Editors

SIDNEY P. COLOWICK and NATHAN O. KAPLAN

VOLUME SIX HUNDRED AND FOUR

METHODS IN ENZYMOLOGY

Marine Enzymes and Specialized Metabolism Part A

Edited by

BRADLEY S. MOORE
Center for Marine Biotechnology and Biomedicine, Scripps Institution of Oceanography; Skaggs School of Pharmacy and Pharmaceutical Sciences, University of California at San Diego, La Jolla, CA, United States

ELSEVIER
ACADEMIC PRESS
An imprint of Elsevier

Academic Press is an imprint of Elsevier
50 Hampshire Street, 5th Floor, Cambridge, MA 02139, United States
525 B Street, Suite 1800, San Diego, CA 92101–4495, United States
The Boulevard, Langford Lane, Kidlington, Oxford OX5 1GB, United Kingdom
125 London Wall, London, EC2Y 5AS, United Kingdom

First edition 2018

ISBN: 978-0-12-813959-2
ISSN: 0076-6879

Publisher: Zoe Kruze
Acquisition Editor: Zoe Kruze
Editorial Project Manager: Alina Cleju
Production Project Manager: Divya Krishna Kumar
Cover Designer: Miles Hitchen

Typeset by SPi Global, India

CONTENTS

Section II
Biosynthesis by Bacterial Symbionts of Marine Invertebrate Animals

Section III
Natural Product Biosynthetic Enzymology of Marine Bacteria and Fungi

CONTRIBUTORS

Vinayak Agarwal
School of Chemistry and Biochemistry; School of Biological Sciences, Georgia Institute of Technology, Atlanta, GA, United States

Courtney C. Aldrich
University of Minnesota, Minneapolis, MN, United States

Silvia C. Bobeica
Howard Hughes Medical Institute, University of Illinois at Urbana–Champaign, Urbana, IL, United States

W. Clay Brown
University of Michigan, Ann Arbor, MI, United States

Michael D. Burkart
University of California San Diego, San Diego, CA, United States

Qingyun Dan
University of Michigan, Ann Arbor, MI, United States

Tony D. Davis
University of California San Diego, San Diego, CA, United States

Shi-Hui Dong
Department of Biochemistry, University of Illinois at Urbana-Champaign, Urbana, IL, United States

Alessandra S. Eustaquio
University of Illinois at Chicago, College of Pharmacy, and Center for Biomolecular Sciences, Chicago, IL, United States

Amy E. Fraley
University of Michigan, Ann Arbor, MI, United States

Michael F. Freeman
Department of Biochemistry, Molecular Biology, and Biophysics & BioTechnology Institute, University of Minnesota-Twin Cities, St. Paul, MN, United States

Neha Garg
School of Chemistry and Biochemistry, Georgia Institute of Technology, Atlanta, GA, United States

Lena Gerwick
Scripps Institution of Oceanography, University of California, San Diego, CA, United States

William H. Gerwick
Scripps Institution of Oceanography; Skaggs School of Pharmacy and Pharmaceutical Sciences, University of California, San Diego, CA, United States

Evgenia Glukhov
Scripps Institution of Oceanography, University of California, San Diego, CA, United States

Wenjia Gu
Department of Medicinal Chemistry, University of Utah, Salt Lake City, UT, United States

Jingjing Guo
Key Laboratory of Combinatorial Biosynthesis and Drug Discovery, Ministry of Education, and Wuhan University School of Pharmaceutical Sciences, Wuhan University, Wuhan, China

Kui Hong
Key Laboratory of Combinatorial Biosynthesis and Drug Discovery, Ministry of Education, and Wuhan University School of Pharmaceutical Sciences, Wuhan University, Wuhan, China

Sylvia Kunakom
University of Illinois at Chicago, College of Pharmacy, and Center for Biomolecular Sciences, Chicago, IL, United States

Andrew J. Lail
School of Chemistry and Biochemistry, Georgia Institute of Technology, Atlanta, GA, United States

Tiago Leao
Scripps Institution of Oceanography, University of California, San Diego, CA, United States

Xinyu Liu
University of Pittsburgh, Pittsburgh, PA, United States

Andrew N. Lowell
Life Sciences Institute, University of Michigan, Ann Arbor, MI, United States

Wanying Mai
Key Laboratory of Combinatorial Biosynthesis and Drug Discovery, Ministry of Education, and Wuhan University School of Pharmaceutical Sciences, Wuhan University, Wuhan, China

Finn P. Maloney
University of Michigan, Ann Arbor, MI, United States

Shaun M.K. McKinnie
Center for Marine Biotechnology and Biomedicine, Scripps Institution of Oceanography, University of California at San Diego, La Jolla, CA, United States

Zachary D. Miles
Center for Marine Biotechnology and Biomedicine, Scripps Institution of Oceanography, University of California at San Diego, La Jolla, CA, United States

Bradley S. Moore
Center for Marine Biotechnology and Biomedicine, Scripps Institution of Oceanography; Skaggs School of Pharmacy and Pharmaceutical Sciences, University of California at San Diego, La Jolla, CA, United States

Brandon I. Morinaka
Institute of Microbiology, Eidgenössische Technische Hochschule (ETH) Zurich, Zurich, Switzerland; Department of Pharmacy, National University of Singapore, Singapore, Singapore

Nathan A. Moss
Scripps Institution of Oceanography, University of California, San Diego, CA, United States

Satish K. Nair
Department of Biochemistry, University of Illinois at Urbana-Champaign, Urbana, IL, United States

Jörn Piel
Institute of Microbiology, Eidgenössische Technische Hochschule (ETH) Zurich, Zurich, Switzerland

Snigdha Sarkar
Department of Medicinal Chemistry, University of Utah, Salt Lake City, UT, United States

Eric W. Schmidt
Department of Medicinal Chemistry, University of Utah, Salt Lake City, UT, United States

Vikram V. Shende
Life Sciences Institute, University of Michigan, Ann Arbor, MI, United States

David H. Sherman
Life Sciences Institute, Department of Medicinal Chemistry; Department of Chemistry; Department of Microbiology and Immunology, University of Michigan, Ann Arbor, MI, United States

Meredith A. Skiba
University of Michigan, Ann Arbor, MI, United States

Samuel T. Slocum
Life Sciences Institute, Department of Biological Chemistry, University of Michigan, Ann Arbor, MI, United States

Janet L. Smith
Life Sciences Institute, Department of Biological Chemistry, University of Michigan, Ann Arbor, MI, United States

Robin Teufel
ZBSA, Center for Biological Systems Analysis; Faculty of Biology, University of Freiburg, Freiburg, Germany

Hem R. Thapa
School of Chemistry and Biochemistry, Georgia Institute of Technology, Atlanta, GA, United States

Ashootosh N. Tripathi
Life Sciences Institute, University of Michigan, Ann Arbor, MI, United States

Agustinus R. Uria
Faculty of Pharmaceutical Sciences, Hokkaido University, Sapporo, Japan

Anna L. Vagstad
Institute of Microbiology, Eidgenössische Technische Hochschule (ETH) Zurich, Zurich, Switzerland

Wilfred A. van der Donk
Howard Hughes Medical Institute, University of Illinois at Urbana–Champaign, Urbana, IL, United States

Toshiyuki Wakimoto
Faculty of Pharmaceutical Sciences, Hokkaido University, Sapporo, Japan

Jingjing Yan
Key Laboratory of Combinatorial Biosynthesis and Drug Discovery, Ministry of Education, and Wuhan University School of Pharmaceutical Sciences, Wuhan University, Wuhan, China

Wei Yuan
Key Laboratory of Combinatorial Biosynthesis and Drug Discovery, Ministry of Education, and Wuhan University School of Pharmaceutical Sciences, Wuhan University, Wuhan, China

Wenjun Zhang
University of California, Berkeley; Chan Zuckerberg Biohub, San Francisco, CA, United States

Xuejun Zhu
University of California, Berkeley, CA, United States

PREFACE

The ocean and the organisms that call it home continue to inspire and fascinate us for their beauty and mystery. Covering over 70% of the surface of planet Earth and containing roughly 97% of all Earth's water, the ocean supports unique marine species that have evolved to flourish in all corners of the globe. To thrive in seawater and navigate physical challenges such as extreme pressure, salinity, low temperature, and the continual darkness of the deep as well as oceanic nutrient minimum "dead" zones, marine organisms have developed specialized features that are often without parallel in their terrestrial counterparts (if a counterpart even exists!). Often these adaptations are metabolic and involve specialized biochemical pathways to small molecules and proteins that sustain life or facilitate communication. One only has to think of bioluminescence and the discovery of the green fluorescent protein from jellyfish or oceanic harmful algal blooms and the characterization of massive polyether toxins as prime examples of nature's incredible engineering feats in the ocean.

To highlight these unique features, I am pleased to present two volumes of *Methods in Enzymology* that focus on Marine Enzymes and Specialized Metabolism. This two-part series in Volumes 604 and 605 encompasses 34 chapters written by many of the recognized experts in the field who study biochemical processes in marine systems. Their contributions provide illuminating accounts on how to investigate and apply a wide range of marine enzymes and products from diverse organisms.

In *Marine Enzymes and Specialized Metabolism, Part A*, the 17 chapters are largely focused on marine enzymes and pathways associated with microbial specialized pathways to biologically active small molecules or "natural products." Natural product compounds produced by marine organisms often have fascinating chemical traits not featured in terrestrial molecules, and as such, modern genomic sequencing efforts have illuminated a plethora of new enzymes and metabolic pathways from diverse organisms in the marine environment that expand our understanding of nature's biochemical capacity.

The first section of five chapters in Volume 604 highlights marine cyanobacteria, photosynthetic organisms with widespread distribution in the marine environment from the nearshore coast to the open ocean and that live in the water column, in microbial mats, or in close association with animal hosts. These chapters highlight some of the enzymatic novelty of marine cyanobacteria from assembly line biosynthesis to posttranslational peptide biochemistry and also offer guidance in their collection, culturing, and genomics.

The second section of four chapters focuses on symbiotic bacteria associated with marine invertebrate hosts, namely bryozoans and sponges. Recent advances in metagenomics have revealed that the remarkable chemistry of marine invertebrates is often the result of their choreographed microbiomes. Without the ability to directly culture and interrogate symbiotic microbes outside their host, the assembly and mining of invertebrate-associated metagenomes have transformed the study of exotic enzymatic reactions through heterologous biochemistry and promise to help solve supply issues of invertebrate natural products that have hindered the application of many marine "animal" natural products.

The third and last section of eight chapters in Volume 604 chronicles a wide variety of biosynthetic enzymes from marine bacteria and fungi involved in diverse halogenation, terpene synthesis, and redox reactions. In many of the presented cases, marine biosynthetic enzymes are being explored for their bioengineering or synthetic biology potential in making designer molecules and pharmaceutical agents.

In closing, I wish to thank all of the authors who contributed insightful and descriptive articles to this volume set. I hope that you enjoy reading and learning more about some of the amazing things that marine enzymes can do.

Bradley S. Moore
Scripps Institution of Oceanography & Skaggs School of Pharmacy and Pharmaceutical Sciences, University of California at San Diego, La Jolla, CA, United States

SECTION I

Marine Cyanobacteria, Enzymes and Pathways

CHAPTER ONE

Collection, Culturing, and Genome Analyses of Tropical Marine Filamentous Benthic Cyanobacteria

Nathan A. Moss*, Tiago Leao*, Evgenia Glukhov*, Lena Gerwick*, William H. Gerwick*,†,1

*Scripps Institution of Oceanography, University of California, San Diego, CA, United States
†Skaggs School of Pharmacy and Pharmaceutical Sciences, University of California, San Diego, CA, United States
[1]Corresponding author: e-mail address: wgerwick@ucsd.edu

Contents

Methods in Enzymology, Volume 604
ISSN 0076-6879
https://doi.org/10.1016/bs.mie.2018.02.014

Abstract

Decreasing sequencing costs has sparked widespread investigation of the use of microbial genomics to accelerate the discovery and development of natural products for therapeutic uses. Tropical marine filamentous cyanobacteria have historically produced many structurally novel natural products, and therefore present an excellent opportunity for the systematic discovery of new metabolites via the information derived from genomics and molecular genetics. Adequate knowledge transfer and institutional know-how are important to maintain the capability for studying filamentous cyanobacteria due to their unusual microbial morphology and characteristics. Here, we describe workflows, procedures, and commentary on sample collection, cultivation, genomic DNA generation, bioinformatics tools, and biosynthetic pathway analysis concerning filamentous cyanobacteria.

1. INTRODUCTION

Cyanobacteria are carbon-fixing, oxygenic photosynthetic prokaryotes that play essential roles in nearly every biotic environment. Moreover, the development of oxygenic photosynthesis in cyanobacteria was largely responsible for creating Earth's original oxygen-rich atmosphere, thereby stimulating the evolution of the extraordinary species diversity currently present (Flores, López-Lozano, & Herrero, 2015; Shih et al., 2013). In the open ocean, nitrogen-fixing (N_2-fixing) cyanobacteria are the major source of biological nitrogen, and this can be a limiting factor for the productivity of these oligotrophic environments (Zehr, 2011). Apart from their importance in biogeochemical cycles as a result of their primary metabolism, cyanobacteria are also a prolific and sometimes distinctive source of secondary metabolites known as natural products (NPs). NPs have been a major inspiration for the development of many of the pharmaceutical drugs currently available (Newman & Cragg, 2016). Despite the diversity of cyanobacteria and their importance for both primary and secondary metabolism, genomic investigations are unevenly distributed through this phylum (Keller, Leão, & Gerwick, 2017).

It was recently observed that the number of natural products discovered per year have remained relatively constant over the past several years, and that the number of truly novel structures have also stayed constant during this period (Pye, Bertin, Lokey, Gerwick, & Linington, 2017). To this point, the hope is that genome mining can direct isolation efforts toward underexplored microbial sources, such as tropical filamentous marine cyanobacteria, the subject of this review. This anticipation is predicated upon past traditional investigations of the structurally unique secondary metabolites of these organisms, which were developed through the process of chemical extraction, chromatography, spectroscopic determination of structure, and biological evaluations. However, the natural products chemistry dimension of these studies is not the focus of this chapter, and other reviews have summarized various methods in the area of natural products chemistry and drug discovery (Queiroz, Hostettmann, & Wolfender, 2012). This chapter describes our experiences with the collection of these tropical marine filamentous cyanobacteria over the past 35 years as part of a natural product drug discovery program, and our ensuing investigation of their laboratory culture, extraction of DNA, sequencing and bioinformatic analysis, and feeding studies to confirm aspects of the biosynthesis of their unique secondary metabolites.

2. FIELD COLLECTION AND PROCESSING OF CYANOBACTERIAL BIOMASS FOR LABORATORY ANALYSES

2.1 Biogeography and Morphology of Cyanobacteria Collections

A general guideline in the field of natural products discovery is that structurally novel metabolites, deriving from equally new and unique biosynthetic pathways, can be encountered by sampling previously understudied locations on the planet. This hypothesis is widely held to be true and is supported by strong foundational knowledge in the area of biogeography (Leal et al., 2013); however, it is clearly based on an incomplete sampling in any given location, both in terms of the range of species or the range of potential metabolites within a given species. Indeed, some have argued that "everything can be found everywhere" (Li & Ma, 2016), and there are some notable examples of elaborate secondary metabolites being found in distant parts of the globe and often in taxonomically distinct organisms. In our experience, the biogeographical diversity concept is generally useful for the discovery of novel chemical entities; new metabolites can be discovered by sampling species or strains emanating from unstudied geographical locations.

For the case of marine cyanobacteria, while examples can be found in all the world's oceans, both benthically and pelagically, they are especially prevalent as filaments growing on the benthos in tropical waters. There one finds an abundance of macroscopic tufts, mats, sheets, and other growth morphologies which emanate from three principle taxonomic groups of filamentous cyanobacteria; the *Synechococcales, Nostocales,* and the *Oscillatoriales* (Komárek, Kaštovský, Mareš, & Johansen, 2014).

2.2 Legal, Ethical, and Logistical Considerations

In 1993, a convention of the world's governments was held in Rio de Janeiro that resulted in the treaty on biodiversity conservation (TBC), governing and regulating access to genetic resources from participatory nations. There is no clause that specifies that samples smaller than a given size are exempt; all genetic resources must be dealt with in the same fashion and according to the same rules. Many academic societies, as well as any downstream pharmaceutical company working in the international arena, require that leads coming into their company be obtained in full compliance with the TBC. Generally, this involves signing a memorandum of understanding (MOU) with the designated national ministry in charge of such genetic resources in the foreign country that addresses the various areas of responsibility listed earlier.

Once MOUs have been duly signed by both parties, and permits applied for and received, a collection expedition can be planned. The geography, logistics, and laws in the target country largely dictate the basic nature of the expedition. For example, most of the southern Caribbean island country of Curaçao can be effectively sampled by car as the road system follows the perimeter of the island, and many public access entry points can be found to the surrounding waters. In Papua New Guinea, no such roadway system exists for the majority of the country, which is spread out over numerous islands, and thus a liveaboard dive boat-based expedition is necessary. For some tropical country locations, such as Panama, a mixture of land-based access complemented by small boats has worked well.

2.3 Cyanobacteria Habitat Considerations

Cyanobacteria are prevalent in almost all photic zone habitats in the tropics, from the supratidal and shallow subtidal, extending down to significant depths of 30 m or more. All of these habitats should be examined to the extent possible given infrastructure and safety considerations. The supratidal can be accessed by shore, but care must be taken to not crush delicate marine life and corals, and at the same time avoid the spines of sea urchins and venomous fish. The shallow subtidal can be conveniently sampled by snorkeling, and fairly large areas can often be reconnoitered, possibly in advance of making collections, as sometimes a cyanobacterial strain appearing as occasional tufts in one section of a lagoon or back reef bay can be found in abundance nearby. Caution should be taken to avoid venomous animals and sunburn. Exploring deeper water habitats requires the use of SCUBA. Most scientific diving activities conducted by US-based scientists should be performed in accordance with the American Academy of Underwater Science (AAUS) Standards for Scientific Diving (http://aaus.org/diving_standards). While liveaboard dive vessels will have the appropriate infrastructure for SCUBA, land-based SCUBA shops can be difficult to find in some remote locations, and in other localities the use of SCUBA is unlawful.

2.3.1 SCUBA Safety Considerations

As SCUBA diving in remote tropical locations poses inherent dangers, a few safety recommendations are provided here. First, Divers Alert Network (or equivalent) supplemental dive accident and emergency evacuation insurance is a necessity, as this provides for emergency infrastructure and evacuation should a diving accident occur. It is wise to evaluate the quality of the dive operation providing SCUBA cylinders and breathing gases, considering the overall setup, valid date stamps on tanks, location of air intake for filling tanks, and the availability of emergency oxygen kits for taking into the field. The location and contact information for the nearest recompression chamber should be predetermined as part of the predive plans. The current dive and health status of all divers should be discussed, and clear and professional communication about specific dive plans must occur. Small boat safety should also be considered, including general seaworthiness of the vessel for the conditions at hand, adequate fuel, available safety and first-aid equipment, radio contact infrastructure, and provision of boat plans to shore-based colleagues. These are all part of the American Academy of Underwater Sciences (AAUS) safety program for scientific diving.

2.4 Biomass Collection

The protocols for collection of cyanobacterial samples are essentially the same regardless of the depth of collection. First, it is valuable and highly recommended to photograph the specimen to be collected before collections begin. Then, collection may occur in three scales for three different purposes as described later: acquisition of biomass for chemistry samples, preservation of DNA, and live cultures for laboratory growth.

2.4.1 Materials

- Fine mesh bag with drawstring. This allows escape of water through pores in the bag, and thus the cyanobacterial filaments are captured in the bag and do not float out with each new sample added
- High-density polyethylene (HDPE) bottles of varying size (50-mL to 2-L)
- T25-mL plastic culture flasks with gas permeable lid (BD Biosciences, San Diego, California)
- Isopropanol (IPA)
- 1 Quart zip lock bags with plastic slider
- RNAlater solution (ThermoFisher Scientific, Waltham, MA)
- 15-mL Falcon centrifuge tubes or similar

2.4.2 Procedure for Collection of Biomass for Chemistry Samples

1. Gather biomass by hand and place it in mesh bag underwater, closing drawstring after addition. Ensure that individual collections of different species are kept separately.
2. Bring the mesh bag with the collected materials to shore and tightly press to remove surface water, and place in a wide-mouth HDPE jar.
3. Add isopropanol or isopropanol:seawater solution to a final ratio of 1:1 IPA:seawater, and then pickle for 48–72h in this 50:50 mixture of seawater and isopropanol.

Note, for downstream chemistry efforts, it is most convenient to collect approximately 2L of compressed cyanobacterial biomass, although smaller collections are certainly of value. Isopropanol is generally available from pharmacies in the tropics, but the label should be carefully inspected for percentage of isopropanol such that the final concentration of alcohol is 50%, and that no adulterants are present. After a few days, the excess fluid can be decanted and the sample jar tightly closed for transport inside of sealed plastic bags.

2.4.3 Procedure for Collection of Biomass for DNA Preservation

1. Gather biomass by hand and place it in zip locking bag underwater, ensuring at least half the bag's volume contains seawater
2. Immediately following collection, place samples in shade or otherwise ensure native temperature until processing. Process as soon as possible following collection
3. To process, fill 15-mL centrifuge tube halfway with RNAlater solution
4. Add approximately 1 g of tissue, close the tube, and gently invert several times to spread out the cyanobacterial filaments
5. Fill the rest of the 15-mL centrifuge tube with RNAlater, seal tightly
6. Store in the cool and in the dark until such time as it can be confidently kept under continuously frozen conditions. Freeze–thaw is to be avoided

2.4.4 Collection of Biomass for Laboratory Cultures

The same materials segregated for the RNAlater sample can also be used for establishing cultures. A small tuft of healthy filaments, roughly the size of a pencil eraser, can be pinched off underwater in the bag and, on land, rapidly transferred to a 25-mL culture bottle (T25) with gas permeable lid and containing 30% of its volume of seawater from the location of collection (Fig. 1). This latter aspect is important because of regional differences in pH, salinity, and nutrients.

2.4.5 Notes

Labeling of all bottles, tubes, vials, and cultures should provide an unequivocal code for the sample, and records made that specify the types and number of sample bottles for each material collected. This code number and record needs to be correlated with collection location, depth, temperature, substrate, physical description, field identification, and in situ photography. Subsequently in the laboratory, all deriving materials including extracts, fractions, pure compounds, cultures, DNA samples, gene sequences, and biological assay results need to be unequivocally coded in a fashion that leads

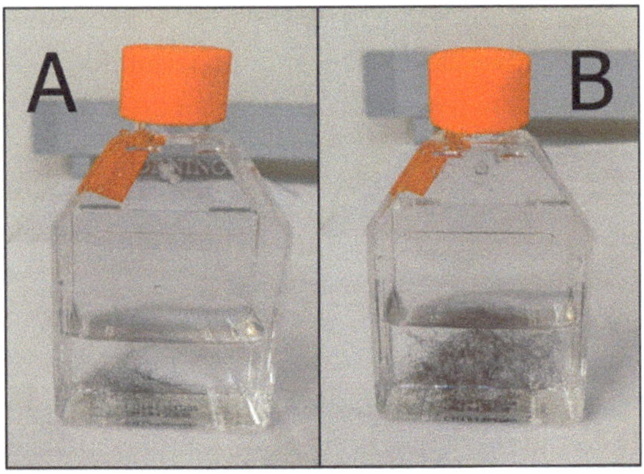

Fig. 1 (A) Freshly collected tuft of filamentous cyanobacteria *Moorea producens.* (B) Collected tuft of *M. producens* after 10 days of growth.

back to this specific sample. Field identifications can be assisted using taxonomic works on the cyanobacteria (Littler & Littler, 2003; Littler, Littler, Bucher, & Norris, 1989), along with microscopic examinations using a field microscope.

2.5 Sample Transport

Samples must be transported back to the laboratory safely. Double-bag chemistry bulk samples and place with padding in checked luggage. A small amount of liquid that could potentially leak is nonflammable at these dilute concentrations of isopropanol. RNAlater-preserved samples can be similarly placed in checked luggage with padding and double bagging. Culture samples should be brought in hand luggage to insure minimal temperature variations, and given current airline rules, is limited to that which can be placed in a 1 quart zip lock bag per person. All documents confirming their legal collection and export should be easily on hand when going through airport security. This documentation may also need to be shown to agricultural inspectors upon arriving at your home country.

3. NATURAL PRODUCTS CHEMISTRY FROM MARINE FILAMENTOUS CYANOBACTERIA

Increased interest in the genomics of cyanobacteria is prompted in part by the wide diversity of natural product structures that emanate from different characteristic biosynthetic pathways. For example, they produce unique hydrocarbons as well as oxidized fatty acids, known as oxylipins, which include such molecules as grenadamide (Sitachitta & Gerwick, 1998) and other oxidized 18-carbon polyunsaturated acids (Lang & Feussner, 2007). Cyanobacteria elaborate notable terpenoid compounds such as a portion of the lyngbyatoxin A structure (Cardellina, Marner, & Moore, 1979). The alkaloidal UV sunscreen pigment scytonemin (Proteau, Gerwick, Garcia-Pichel, & Castenholz, 1993) and the saccharide containing toxin polycavernoside A (Navarro et al., 2015; Yotsu-Yamashita, Haddock, & Yasumoto, 1993) are also well-established products of cyanobacteria. But the dominant pathways we and others have encountered from these organisms are products of the polyketide synthase and nonribosomal peptide synthetase pathways, as well as hybrids of these two (Tan, 2013). Two striking examples of polyketides are debromoaplysiatoxin, a protein kinase C activator (Hanaki et al., 2018) and phormidolide, a cytotoxic natural product of very unusual structure and one of the few trans-AT PKS pathways characterized to date in cyanobacteria (Bertin et al., 2016; Williamson, Boulanger, Vulpanovici, Roberts, & Gerwick, 2002). In the pure peptide class, a notable example is provided by ribosomally encoded natural product patellamide (Donia et al., 2006). Illustrations of hybrid NRPS/PKS products, the most common type of cyanobacterial secondary metabolite, are given by apratoxin A (Luesch, Yoshida, Moore, Paul, & Corbett, 2001), a promising anticancer lead acting through a novel mechanism of action, and hoiamide A, a voltage-gated sodium channel activator (Pereira, Cao, Murray, & Gerwick, 2009).

3.1 Separation and Refinement of Chemical Extracts for Natural Products Discovery

For the top-down approach to drug discovery from marine cyanobacteria, a voucher specimen of the sample biomass is first taken (stored frozen), and then the remainder of the material is soaked multiple times in warm 2:1 dichloromethane:methanol. The filtered extracts are combined, solvents reduced, and the extract fractionated by any number of chromatographic methods, such as normal phase vacuum liquid chromatography (Ebada, Edrada, Lin, & Proksch, 2008). It is useful to perform a prefractionation prior to biological assay, as this improves the success rate of detecting useful biological activities, in part through the increase in titer of minor constituents in the fractions, and in part through the segregation of nuisance compounds that can be discounted from further pursuit. The fractions are then tested in either mechanism-based or phenotypic screening campaigns, and active fractions additionally purified by HPLC. These subfractions are then reevaluated in the same bioassay, and the process of purification and biological evaluation is repeated until a pure and active compound is obtained. This is then subject to structure determination as briefly described later.

3.2 Genome-Guided Natural Products Discovery From Cyanobacteria

As a complementary approach, potentially unique natural products can be located through the process of DNA isolation, genome sequencing, and bioinformatics analysis; this is described in some detail in subsequent sections. Nevertheless, the extract and fractions as made above can be queried for the presence of such a deduced compound, using mass spectrometry (MS) in concert with rough molecular size, presence of heteroatoms or halogens, or anticipated UV chromophores, to help locate the substance. The presumed biosynthetic product of the biosynthetic gene cluster can then be targeted for isolation and structure elucidation; an example of this approach is given by our recent discovery of the columbamides (Kleigrewe et al., 2015). There are a number of variants of the "bottom-up" genome-guided approach to natural products discovery, and this is an area of active pursuit with much innovation being developed (Blin, Kim, Medema, & Weber, 2017).

3.3 Dereplication of Analytical Information

It is highly recommended to dereplicate the compounds being targeted for isolation as early in the process as possible, and many approaches have been noted in the literature using spectroscopic or genomic information (Hubert, Nuzillard, & Renault, 2017). A relatively new approach that is finding broad application in natural products research is termed Molecular Networking, and is based on the ability to identify and link a metabolite to a compound family on the basis of its MS^2 spectrum. These families of molecules are depicted in Cytoscape and can be compared with an ever-increasing database of MS^2 information on the Global Natural Products Social Molecular Networking platform (Wang et al., 2016) (https://gnps.ucsd.edu/ProteoSAFe/static/gnps-splash.jsp). One extremely powerful aspect of this approach is that known compounds, or compounds closely related to known compounds, can be quickly recognized by analysis of the extract or fraction using LC-MS^2 data. Depending on the questions being considered in the project, these data can be used to quickly prioritize or deprioritize a particular extract or project for further study.

Another recently appearing dereplication approach involves comparison of the ^{1}H–^{13}C HSQC spectrum with a database of such spectra. This spectrum results from an NMR experiment which depicts on a 2D graph both the proton and carbon chemical shifts of protonated carbon atoms in a molecule. The method of making the comparison is via an algorithm learned by applying a deep convolutional neural network to the database; in effect, this uses facial recognition artificial intelligence to "recognize" a spectrum, or to "recognize" that a given spectrum is related to a known one. This algorithm has been termed "Small Molecular Automatic Recognition Technology", or SMART (Zhang et al., 2017).

3.4 Molecular Structure Determination

Once a pure compound is obtained by either the top-down or bottom-up approach, and it is determined to be new substance, the next task is to rigorously determine its molecular structure. It is important to accumulate a nearly complete data set before setting off on the process of building a structure hypothesis, as what is cryptic by NMR may be very apparent by IR or UV, and vice versa. An accurate mass is obtained to determine molecular formula, UV to deduce chromophores, IR for certain types of functional groups such as ketones, and then a series of 2D NMR spectra are recorded. The basic set of NMR spectra for structure elucidation is comprised of the following: ^{1}H and ^{13}C 1D NMR, ^{1}H–^{1}H COSY, ^{1}H–^{13}C HSQC, and ^{1}H–^{13}C HMBC; however, there are many dozens of other specialized experiments which selectively examine particular relationships between atoms in the molecule. The general idea is to use these data to produce a series of well-defined partial structures, and then "stitch" these together using either the existing data set or employing more specialized experiments selected for the specific question at hand. Once a planar structure is formulated and confirmed by additional analysis of the MS and NMR data sets, stereochemical features in the molecule must be resolved. Again, much has been written on this subject, but the major methodologies involve analysis of proton–proton coupling values, proton–proton through space interactions, optical rotary properties, circular dichroism, derivatization with chiral reagents and NMR analysis (Mosher ester method), and partial degradation with comparison by a chiral chromatographic procedure to authentic standards (e.g., for the chirality of amino acids in peptides,

Marfey's analysis is very effective) (Almaliti et al., 2017). The desired outcome of this investigation is a rigorously established molecular framework that includes the 3D or chiral features of the new natural product.

4. CULTURING FILAMENTOUS CYANOBACTERIA

4.1 Identification of Cultures by Microscopy and Morphology

Preliminary taxonomic identification of field collections can be made by microscopic analysis of the cyanobacterial filaments. Characteristics, such as unicellular vs filamentous, presence of a sheath, filament thickness as well as cell shape and dimension, can give information for preliminary species identifications (Komárek, 2006; Komárek et al., 2014). However, a phylogenetic analysis is always needed for final taxonomic determination. One fairly straightforward method for phylogenetic determination of bacteria is 16S rRNA sequence analysis, for discerning species of relatively distant bacterial lineages (Clarridge, 2004; Engene, Coates, & Gerwick, 2010), or multilocus sequence analysis for parsing very closely related species or even strains of the same species (Calteau et al., 2014; Engene et al., 2012). These methods can be carried out on extracted genomic DNA (gDNA) samples described in Section 5, by PCR amplification using cyanobacterial-specific primers used in the Gerwick laboratory (Table 1) or from Nübel, Garcia-Pichel, & Muyzer, 1997, followed by Sanger sequencing, or alternately with full genome sequence analysis.

4.2 Conditions for Growth of Cyanobacteria

The culture protocols provided later were adapted to study the filamentous cyanobacterial genera *Moorea*, *Okeania*, and *Leptolyngbya*, but they also may work efficiently with other filamentous and unicellular species of cyanobacteria. These bacteria are best grown under photoautotrophic conditions with 16/8h light–dark cycles in an essentially inorganic/mineral medium (apart from vitamins). These conditions favor growth of photosynthetic organisms over that of heterotrophic bacteria.

4.2.1 Media Preparation

As many cyanobacteria are unable to fix nitrogen, a good general growth media is BG-11 (Castenholz, 1988); this media contains exogenously added nitrate. However, the recipe given here is slightly modified to decrease final nitrate concentration and other additives in order to suppress growth of heterotrophic bacteria (Tables 2–4), and to accommodate additional salts that

Table 1 Cyanobacterial Primers That Can Be Used for Initial PCR Amplification of the 16S rRNA Gene

	Lineage Specificity	Sequence (5′-3′)
CYA106F	General cyano	CGG ACG GGT GAG TAA CGC GTG A
1509R	General	GGT TAC CTT GTT ACG ACT T
CSL128F	*Symploca/Moorea*	GGT CGG GGA CAA CCA
CSL1438R	*Symploca/Moorea*	CTT CGG GCG TGG CCA G
CSL220F	*Symploca/Moorea*	CTC GCG TCT GAT TAG CTA
CSL1489R	*Symploca/Moorea*	AAC CCT GCC TTC GGC
CSL1445R	*Symploca/Moorea*	GGT AAC GAC TTC GGG CG
CSL140F	*Symploca/Moorea*	GGT CGG GGA CAA CCA C
52F	Unicell cyano	TAA CAC ATG CAA GTC GAA CG
1486R	Unicell cyano	CGG TAC GGC TAC CTT GTT A
809R	General cyano	GCT TCG GCA CGG CTC GGG TCG ATA
740F	General cyano	GGC YRW AWC TGA CAC TSA GGG A
1494R	General cyano	TAC GGC TAC CTT GTT ACG AC
27F1	General cyano	TAGTGTAAAACGACGGCCAGTAGAGTITGATCCTGGCTCA (*note*: has inosine in one position)
1494R	General cyano	TAGCAGGAAACAGCTATGACCTACGGCTACCTTGTTACGA

Table 2 Seawater BG-11 Stock Solution Components, Comparing Modified Low-Nitrate Formula With Castenholz Original 1988 Recipe

Compound	Modified BG-11	BG-11 (Castenholz, 1988)
Na_2EDTA	0.2g	0.2g
Citric acid	1.2g	1.2g
$NaNO_3$	**100g**	300g
$K_2HPO_4 \times 3H_2O$	8.0g	8.0g
$MgSO_4 \times 7H_2O$	15.0g	15.0g
$CaCl_2 \times 2H_2O$	**7.0g**	7.2g
$Na_2CO_3 \times H_2O$	4.0g	4.0g
Ferric ammonium citrate	1.2g	1.2g
$NiSO_4(NH_4)_2SO_4 \times 6H_2O$ (1m*M* stock solution)	**50mL**	30mL
Na_2SeO_4 (1m*M* stock solution)	**20mL**	40mL
Microelements mix (see Table 3)	200mL	200mL
Deionized water	To 10L	To 10L

Changes are bolded

Table 3 Microelements Mix Components, Comparing Modified Low-Nitrate With Castenholz Original 1988 Recipe

Compound	Modified BG-11	BG-11 (Castenholz, 1988)
H_2SO_4, 98% (w/v)	0.5mL	0.5mL
$ZnSO_4 \times 7H_2O$	0.5g	0.5g
H_3BO_3	0.5g	0.5g
$CuSO_4 \times 5H_2O$	0.025g	0.025g
$NaMoO_4 \times 2H_2O$	0.025g	0.025g
$MnSO_4 \times H_2O$	**0.00228g**[a]	2.28g
$CoCl_2 \times 6H_2O$	**0.135g**	0.045g
Deionized water	To 1L	To 1L

[a]In our experience, filamentous marine cyanobacteria grow well in media containing from 1.2μ*M* (our modified SWBG-11 media with 1.2μ*M* manganese contributed from Instant Ocean) to 13.5μ*M* (as in Castenholz media).
Changes are bolded

Table 4 Vitamin Mix Components (Nelson, Waterbury, & Jannasch, 1982)

Compound	Mass
Nicotinic acid	100 mg
4-Aminobenzoic acid (PABA)	10 mg
Biotin	1 mg
Thiamine	200 mg
B_{12}	1 mg
Folic acid	1 mg
i-Inositol	1 mg
Ca-pantothenate	100 mg
Deionized water	To 100 mL

are present in the Instant Ocean® salt mixture as opposed to pure NaCl. Volumes may be scaled as appropriate.

4.2.2 Media Components

- Instant Ocean® salt mixture (Spectrum, Blacksburg, VA)
- Deionized water
- Standard 0.22-μ*M* sterile filtration apparatus
- Noble agar or agarose
- Plastic Petri dishes
- 15-mL round-bottom Falcon tubes (BD)
- 50-mL centrifuge tubes (BD)

4.2.3 Liquid Media Preparation Procedure

1. Mix 1 L "Microelements" solution (Table 3), by adding all components, bringing volume to 1 L with deionized water, and sterile filter. Aliquot into 50 mL volumes. Store frozen at −20°C for up to 1 year. Defrost aliquot immediately prior to use.
2. Prepare BG-11 stock solution (Table 2) by adding each component (including "Microelements" solution from Table 3), then bring volume to 10 L with deionized water. Mix well and aliquot into 10 1-L bottles. Freeze aliquots for up to 1 year. Defrost aliquot immediately prior to use.
3. Prepare vitamin mix solution (Table 4), sterile filter. Aliquot into 1–10 mL volumes and store at 4°C for up to 6 months.

4. Prepare 10 L batches of seawater BG-11 media (SW BG-11) by adding 500 mL of BG-11 stock (Table 2) to 9.5 L of deionized water in a 10 L carboy. Add 330 g of Instant Ocean® and mix until dissolved. Freshwater BG-11 (FW BG-11) is prepared identically, but without adding the Instant Ocean® salt mix. Autoclave for 70 min.
5. Allow carboy to cool to room temperature overnight in laminar flow hood.
6. Add 1 mL of vitamin mix (Table 4), close cap tightly, and mix by shaking vigorously.
7. Prepared media is stored in a clean space at room temperature and used for up to 6 weeks. Quality control of the freshly prepared media is recommended by culturing the most sensitive cyanobacterium in the new batch and ensuring proper growth and morphology characteristics. The pH of SW BG-11 prepared by this recipe is 7.8 and in general marine media should be slightly alkaline (Rippka, Waterbury, & Stanier, 1981) to mimic current ocean pH of 8.1 (Jacobson, 2005). Some cultures, for example, the genus *Okeania*, require extra nitrogen to be present in the media. For this genus, the amount of $NaNO_3$ is tripled doubled (the final culture media has a $NaNO_3$ concentration of 1.5 g/L). Investigators are encouraged to systematically determine the optimal media concentrations for their particular strains and species of cyanobacteria.

4.2.4 Solid Media Preparation Procedure

Because vitamins will decompose if added to hot liquid agar, these are omitted in the preparation of solid media. The following preparation is for 1 L of solid media, to be aliquoted into culture tube slants or plastic petri dishes. The recipe may be scaled as needed.

1. Prepare 500 mL 2% (w/v) Noble agar or agarose in deionized water.
2. Separately, prepare 500 mL of media by adding 25 mL of BG-11 stock solution, Table 2, to 475 mL deionized water, followed by 16.6 g of Instant Ocean® salt mix (for FW BG-11 plates, omit Instant Ocean®).
3. Autoclave each bottle separately for 20 min and mix thoroughly after cooling to 40–45°C (Allen, 1968). The resulting 1% Noble agar or agarose mix is poured into 15-mL sterile Falcon culture tubes for creating agar slants (7–8 mL) and Petri dishes (20–30 mL). Culture tubes and Petri dishes may be stored in sealed bags at +4°C for up to 6 months.

Fig. 2 From *left* to *right*, microscopy images of cyanobacterial strains *Leptolyngbya* ISB3Nov94-8A at 200× magnification, *Okeania hirsuta* PAB10Feb10-1 at 1000× magnification, and *Moorea producens* 3L at 554× magnification.

4.3 Isolation of Monocultures of Cyanobacterial Strains From Field Collections

Under ideal circumstances, cyanobacterial research should be conducted using monocultures (Fig. 2) (Acreman, 1994). Although cyanobacterial monocultures can be achieved, axenic cultures can be very difficult to obtain (Waterbury, 2006). Many heterotrophic bacteria live within the sheath that surrounds filamentous cyanobacteria, and removal of these surface associated bacteria has proven very difficult. It is possible that some heterotrophic bacteria may be symbiotic (Cummings et al., 2016). Several techniques are used to produce cyanobacterial monocultures from environmental collections brought into a laboratory environment.

4.3.1 Materials and Media

- Seawater (SW) BG-11 liquid media and Noble agar plates (Castenholz, 1988)—recipe modified (Tables 2–4)
- Freshwater (FW) BG-11 liquid media and Noble agar plates (Castenholz, 1988)—recipe modified (Tables 2–4)
- Sterile culture vessels, Petri dishes, and forceps
- Microscope with 40× and 100× objectives

4.3.2 Isolation Procedure

1. Examine each collection using microscopy with 40× and 100× objectives. Take archival photographs immediately upon arrival.
2. Allow field collections to acclimate for a 3- to 7-day period in the laboratory environment. During this period, keep at a lower light intensity (150–220 lx or approximately 2–3 μmol photons m^{-2} s^{-1}) under fluorescent lamps using a 16/8 h light–dark regimen. The culture room is kept at a constant temperature of 27°C ± 1°C.

3. After the acclimatization period, subculturing individual strains from the collected samples may begin. Environmental samples often contain up to three or more different cyanobacteria, so multiple subcultures are created from almost every collection. Grasp one to three filaments and either place or embed them in 0.8%–1.0% Noble agar or agarose Petri dishes in the inverted position. Some strains may form well-separated monoculture filaments that are easily physically separated using forceps or a pipet. Alternatively, the partially cleaned tip from a trichome may be cut and subcultured in fresh media every 5–10 days. Sticky contaminants can be separated from the filaments by gripping the end of a filament with forceps and dragging it through agar. Single filaments can be placed in six-well plates that are filled with 5 mL of SW BG-11 or FW BG-11.
4. If present, diatoms can be eliminated by treatment with germanium dioxide (GeO_2) (Lewin, 1966). GeO_2 is added to the culture media from a sterile filtered 20 mg/L stock solution so as to obtain a final concentration of 0.4 mg/mL. This is continued until there is no visual signs of diatoms as determined by microscopy; this may take up to 4 weeks. Heterotrophic bacterial growth is retarded in these monocyanobacterial nonaxenic cultures by frequent subculturing and dragging of single filaments through agar. Axenic cultures of filamentous bacteria can be formed using antibiotics (Rippka et al., 1981); however, this is not always successful due to the adherence of heterotrophic bacteria deep inside the sheath material as well as the sensitivity of cyanobacteria to antibiotics. Microscopy records should be produced throughout the isolation process.

4.3.3 Subculturing

To propagate filamentous cyanobacterial cultures, 10%–20% of the parent biomass is transferred by sterile technique to fresh media in a new sterile vessel. In general, a small bundle of filaments is introduced into a 150- to 500-mL sterile vessel, either Erlenmeyer or single use plastic tissue culture flasks, each containing liquid media of approximately 15%–30% of the flask volume. It is important to have a high surface area to volume ratio in order to foster good gas exchange. Liquid cultures usually grow well for 1–4 months, while stock cultures can survive for longer periods on agar slants, up to 12 months; however, this is highly strain dependent. Therefore, the subculturing protocol should be repeated every 30 days for liquid cultures and every 6 months for solid media to assure that none of the essential nutrients is depleted. Moreover, each culture should be inspected at least once a week and subcultured if showing signs of distress. We recommend keeping duplicates of the most important cultures in a separate incubator to ensure backup in case of problems with lights or temperature in main culture room.

4.3.4 Growth in Solid Media

Solid media is equilibrated to 27°C prior to use. A small number of filaments are transferred, using sterile forceps or a 2- to 5-mL sterile tissue culture pipette, to the slants or plates, respectively. *Okeania* filaments prefer to be stabbed inside the solid media in order to grow, whereas other strains grow well when simply placed on the agar surface. Tubes and dishes are then sealed with parafilm, labeled with permanent ink, with the label covered by transparent tape. Agar slants are preferred for long-term storage, as in our experience, benthic filamentous cyanobacteria do not survive cryopreservation. Unfortunately, some cyanobacterial species, such as *Moorea bouillonii* and *Moorea producens*, only grow in liquid media; therefore, preserving these strains on agar plates has not yet been possible.

4.3.5 Morphology Considerations: Mat vs Solitary Filaments

Cultures that form a mat-like morphology in culture are transferred using a 2- to 5-mL tissue culture pipette. Those with solitary "hair-like" filaments are moved as an inoculum of filaments of approximately 4 cm × 0.5 cm using sterilized forceps. Forceps can be sterilized by spraying with 70% ethanol, wiping with Kimwipes, left in a hot bead sterilizer for 5 min, and then cooled to room temperature in a laminar flow hood. The inoculum of filaments is cut to approximately 2 cm in length with scissors that are sterilized by the same protocol, and then placed in the new growth media.

4.3.6 Culture Maintenance: Seed Cultures and Stepwise Scale-Up

If there is a need for larger biomass to be produced, such as for biosynthetic feeding experiments, compound isolation, or DNA or RNA isolation, cyanobacterial cultures are subcultured for a scale-up process using a stepwise increase of seed culture medium and flask volumes. Scale-up growth is initially performed using 1- to 2.8-L glass Fernbach flasks (Fig. 3A and B), followed by transfer to autoclaved 15-L polypropylene trays (43 cm × 36 cm × 13 cm) filled with 3–5 L of liquid media and sealed with air permeable polyvinyl chloride film (Fisher) for 4–6 weeks before harvest (Fig. 3C). The media in the vessels should be manually and gently agitated at least once a week to improve distribution of trace nutrients in the vessel. More vigorous agitation, or use of rotary shaker, results in filaments breaking apart or rolling into ropes that do not grow well.

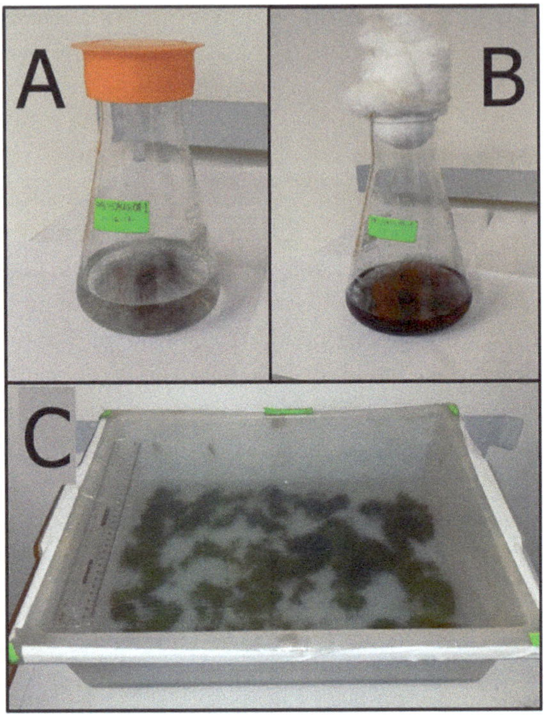

Fig. 3 Culture scale-up (A) 1-L Erlenmeyer flask with sparse culture of *Moorea producens*. (B) 1-L Erlenmeyer flask with dense culture of *M. producens*, ready for subsequent scale-up. (C) Culture pan containing *M. producens* with breathable cover of plastic film.

4.3.7 Culture Room Setup

Established cultures are grown under a 16/8 h light/dark cycle in a 26–28°C temperature-controlled room. The room is illuminated using 32 W 4100 K cool broad spectrum white fluorescent lights that provides each shelf containing cultures a light intensity ranging from 100 to 800 lx, or 1.35–10.8 μmol photons m^{-2} s^{-1}) depending on shelf location and light position. Flasks are placed according to their field depth at time of collection; those collected closer to the surface are placed on more intensively lit upper shelves. Cultures are checked for viability, and flask positions are periodically adjusted for optimal growth.

4.3.8 Glassware Maintenance

Algal biomass is considered a biohazardous waste and must be destroyed in a 10% bleach solution for at least 1 h before discarding. Following bleaching, glassware is cleaned thoroughly by washing flasks with a bottlebrush and warm tap water. If there is no remaining residue on flask surfaces afterward, it is rinsed well (including dishwasher cycle and with deionized water) and autoclaved. If the flask contains any adherent residue, this may be removed by soaking for 1 h to overnight in 0.1% HCl, followed by a thorough rinse in hot water prior to placing it in a dishwasher. After the dishwasher cycle is complete, the flasks are rinsed at least five times with hot tap water followed by one rinse with deionized water. The top is sealed with tight cotton plug or washable silicone stopper, and then autoclaved for 20 min.

4.4 Biomass Measurement Techniques

4.4.1 Measuring Cell Growth by Length

Since filament width remains constant for a given species, the growth rate for filamentous bacteria can be determined by measurement of their increased length. Using a subset of 20 individually grown filaments SW BG-11 media (Esquenazi, Jones, Byrum, Dorrestein, & Gerwick, 2011), the length can be measured from images taken every 48 h by a camera attached to a dissecting scope, followed by analysis in computational imaging software such as the publicly available software program ImageJ (Abràmoff, Magalhães, & Ram, 2004).

4.4.2 Dry Cell Weight and Storage

Biomass productivity of a cyanobacterial culture can be estimated by the weight of the freeze-dried filaments. Once fully grown, the cyanobacterial filaments are harvested using nonsterile vacuum filtration though Whatman paper filter No 4. The collected filaments are dislodged from the paper and placed into specimen bags. The biomass is then subjected to lyophilization, weighed, and stored frozen at −20°C.

4.5 Disposal of Biohazard Waste

Any remaining cyanobacterial biomass is rendered nonviable by treatment with 10% bleach, and then discarded into a solid biohazard waste container to prevent any possible contamination or human exposure.

4.6 Record Keeping

It is essential to establish a good record keeping system for the cyanobacterial culturing process. Culture health and absence of contaminants are monitored periodically with microscope observations. Pictures are recommended to be taken at least once a year and stored for future reference, as well as for measurement of physical parameters such as filament width and cell dimensions. Tables summarizing all established cultures, their recommended media, isotope feeding, DNA sequencing, detected natural products, established taxonomy, and parameters of cell dimensions should be created. Also, records of the last subculturing event, calculated date for the next subculturing, and current culture status (e.g., healthy, contaminated, or problematic) are recommended. It is advised to develop a standard operating procedure for each step in the culturing process and to maintain detailed records of the cultures as visualized by microscopy.

5. EXTRACTION OF GDNA FROM CRYOPRESERVED AND FRESH CYANOBACTERIAL SAMPLES

5.1 gDNA Extraction Considerations

Obtaining high-quality gDNA from tropical marine benthic filamentous cyanobacteria can be challenging. Both environmental samples and, to a lesser degree, cultures, can be highly heterogeneous, containing many different microbial species (Cummings et al., 2016). Additionally, in most benthic filamentous cyanobacteria, the cells are organized into stacks of discs that are surrounded by a rigid polysaccharide sheath (Hoiczyk, 1998; Jürgens & Weckesser, 1985; Rossi & De Philippis, 2015; Weckesser, Hofmann, Jürgens, Whitton, & Raffelsburger, 1988). The sheath imparts both structural rigidity as well as a mucilaginous capacity to the cyanobacterial biomass, thereby complicating standard bacterial DNA extraction procedures. Extraction therefore needs to be vigorous enough to release high-molecular weight gDNA from the confines of the inner and outer cell membranes, cell wall, and polysaccharide sheath without degrading the extracted DNA molecules. Here, we described the general procedure for obtaining high-molecular weight gDNA from field-preserved or -cultured specimens of marine benthic filamentous cyanobacteria.

5.2 Biomass Handling and gDNA Extraction

5.2.1 Equipment

- Buchner funnel and Whatman qualitative filter paper over a vacuum flask
- Benchtop centrifuge with 1.5- to 2.0-mL microcentrifuge tube rotor
- 1.5- and 2.0-mL microcentrifuge tubes
- Liquid nitrogen
- Small mortar and pestle
- G20 Genomic Tip (Qiagen)
- Standard agarose gel materials and apparatus

5.2.2 Reagents

- gDNA Buffer Set (Qiagen)
- RNAlater solution (ThermoFischer Scientific)
- Isopropanol
- 70% ethanol solution
- Sterile (autoclaved) milliQ water
- 1%–5% SDS (optional)

5.2.3 RNAlater Sample and Freezer Storage

1. Upon returning from making field collections, wrap the lids of the preserved biomass sample tubes in parafilm and store at −20°C until they are processed for gDNA extraction.
2. Avoid freeze–thaw cycling by ensuring that samples stay at −20°C during freezer cleanouts and transfers, because it may lyse cells and degrade DNA. Process the samples as soon as possible after returning from the field; preserved biomass samples as well as extracted gDNA degrades with time.

5.2.4 Biomass Rinsing, Pulverization, and gDNA Extraction Procedure

1. Immediately prior to extraction, a thumbnail-sized RNAlater-preserved sample or live culture specimen should be placed in a Buchner funnel with filter paper, followed by washing and cleaning with a gentle stream of deionized water.
2. Use small tweezers to clean away any detritus, dead filaments, or other field contaminants. Remove as much water as possible via vacuum, but do not lyophilize. Immediately proceed to biomass extraction.
3. (Optional) If samples are particularly dirty, or a very clean sample is required, the biomass can be placed in a 15-mL centrifuge tube containing a 1%–5% solution of SDS. The tube is gently rotated on a rotary shaker for 30 min, followed by several washes with deionized water using a Buchner funnel and vacuum. This is done to encourage the release of contaminants from the exterior of the cyanobacterial sheath.
4. After the biomass has been cleaned, fill a small mortar with liquid nitrogen. Transfer biomass to mortar and freeze for 10–15 s.
5. Gently crush the biomass as the liquid nitrogen evaporates. As the final few drops of liquid nitrogen evaporate, pulverize the biomass to a grainy powder.
6. Using a small spoon or spatula, transfer the biomass into a 2.0-mL microcentrifuge tube. Immediately proceed to extract the DNA from the pulverized biomass.

7. The preferred method for extraction of cyanobacterial DNA from liquid nitrogen-crushed marine cyanobacteria filaments is the Qiagen Bacterial gDNA isolation procedure (Qiagen Genomic DNA Buffer Set), followed by gDNA cleanup and purification using the G20 Genomic Tip protocol (Qiagen). Prior to transfer of gDNA from the extraction tube to the G20 Genomic Tip, spin down the biomass/extracted DNA mixture at 16,000 × RCF for 10 min. The supernatant contains the gDNA and should be carefully transferred with a wide mouth (or cut 1000 μL) tip, onto the G20 Genomic Tip, to avoid shearing of the DNA. Make sure that none of the pellet is accidentally transferred.
8. Follow the Qiagen G20 Genomic Tip protocol to clean and purify the gDNA. Summarizing the final steps of the protocol, elute the gDNA from the G20 Genomic Tip in 1 mL of prewarmed elution buffer QF, precipitate with 100% isopropanol, and wash with 1 mL of ice cold 70% ethanol. Resuspend final dried pellet in 200 μL TE buffer (a component of the Qiagen buffer kit) or sterile water, and dissolve gDNA overnight.

5.2.5 Alternative DNA Extraction Procedure

Some samples may respond differently to the liquid nitrogen pulverizing procedure, Qiagen buffers, or the G20 tip purification process. This results from cyanobacteria possessing a wide variety of differing morphologies and polysaccharide sheath materials, resulting in lower DNA yields or quality. Alternative methods for isolation of gDNA include using the cetyltrimethylammonium bromide (CTAB) method, followed by extraction with chloroform:isoamyl alcohol (24:1) and isolation of intact DNA (http://1ofdmq2n8tc36m6i46scovo2e-wpengine.netdna-ssl.com/wp-content/uploads/2014/02/JGI-Bacterial-DNA-isolation-CTAB-Protocol-2012.pdf) (Clarke, 2009). The CTAB method was developed for extraction of gDNA from plant cells, as CTAB binds polysaccharides that may interfere with the extraction process.

5.2.6 DNA Quality Assessment

DNA may be assessed by any number of tools to calculate DNA concentration and quality, such as by measuring the A260/280 ratio. This may include running a standard 0.6%–1.5% agarose analytical electrophoresis gel or use of instruments such as a Qubit (ThermoFischer Scientific), Nanodrop (ThermoFischer Scientific), or a Tapestation (Agilent, Santa Clara, CA).

5.2.7 Notes

The Qiagen Bacterial DNA isolation procedure followed by the Genomic Tip G20 protocol contains several timed incubation steps, enabling high enzymatic activity by lysozyme, RNAse A, and proteinase K. These timed steps may be lengthened, and buffer volumes may be appropriately scaled in order to achieve a more homogeneous solution and to reduce viscosity. Experimentation with individual strains to maximize the resulting DNA quality and yield is encouraged. For particularly viscous pulverized biomass, doubling of the volume of Bacterial DNA Isolation buffers is recommended.

6. GENOME MINING NATURAL PRODUCT RICH MARINE CYANOBACTERIA

6.1 Cyanobacterial Genomics Overview

Historically, cyanobacteria have been characterized into five morphology-based subsections (Rippka, Deruelles, Waterbury, Herdman, & Stanier, 1979), and later refined to eight via genomic analyses (Komárek et al., 2014). However, the current phylogenetic distribution of 425 publicly available cyanobacterial genomes in NCBI RefSeq database (September 2017) is heavily biased toward certain subsections. About half (212 out of 425) of the NCBI strains are from the genera *Prochlorococcus* and *Synechococcus* (subsection II). This is probably due to their importance in oceanic photosynthesis and nitrogen fixation (Flores et al., 2015). Moreover, organisms from those two genera usually have a smaller genome and contain fewer symbiotic heterotrophs, which has facilitated genome sequencing and assembly (Alvarenga, Fiore, & Varani, 2017; Leão et al., 2017). However, they tend to be less important for natural product genome mining (and drug discovery) due to the low quantity and diversity of natural product-encoding biosynthetic gene clusters (BGC's) (Dittmann, Gugger, Sivonen, & Fewer, 2015). Here, we focus on protocols for sequencing and obtaining high-quality genomes from marine filamentous cyanobacteria that are rich in natural product pathways. This is followed by a detailed view of genome-mining analysis in the search for novel natural products.

6.2 Obtaining High-Quality Draft Genomes From Complex Cyanobacterial Samples

Given the vast and underexplored diversity of natural product pathways in marine filamentous cyanobacteria (Kleigrewe, Gerwick, Sherman, & Gerwick, 2016; Leão et al., 2017), we have focused our collection, sequencing, and assembly efforts on this type of microbe. Environmental cyanobacterial assemblages have been collected worldwide from subtidal marine tropical ecosystems. Short read sequencing via Illumina HiSeq4000® has yielded excellent results when using high-performance library preparation kits such as the KAPA HyperPlus from Roche®. However, assembled genomes still tend to contain hundreds or even thousands of contiguous DNA sequence fragments (contigs) due to the lack of reference sequences and other challenges (Alvarenga et al., 2017). In order to obtain high-quality genomes from these complex, diverse environmental assemblages, it is necessary to modify the typical microbial assembly pipeline. The lack of an efficient and fully automated genome assembly pipeline is a major reason for the current scarceness of natural product rich cyanobacterial genomes. Our current genome assembly workflow is outlined in the sections that follow.

6.2.1 Procedure for Cyanobacterial Genome Assembly

1. De novo metagenomic assembly of short reads via metaSPAdes v3.11 (http://bioinf.spbau.ru/spades): this step uses paired end reads from our environmental samples to output a ".fasta" file (a common text-based format for representing either DNA or amino acid sequences) with the assembled contigs. SPAdes/metaSPADES creates the initial de Bruijn assembly graph according to the *k*-mer size selected (Bankevich et al., 2012). Standard settings for the current version of SPAdes uses *k*-mer values as 21, 33, 55, and only the results for the best iteration are kept at the end of the assembly (e.g., *k*-mer that generated the least number of fragments). We suggest the use of a wider range of *k*-mers (e.g., 21, 33, 55, 77, 99, and 127) for marine cyanobacteria, although it is important to keep in mind that the processing requirements and the error rates can vary according to the *k*-mer length that is selected.
2. Binning of cyanobacterial contigs: there are a plethora of recently reviewed binning tools available for metagenomic microbial assemblies (Kang, Rubin, & Wang, 2016). Unfortunately, most of them do not perform well with environmental natural product rich marine cyanobacterial assemblages. Based on our experience, binning via phylogenetic assignments from DarkHorse (http://darkhorse.ucsd.edu) (Podell & Gaasterland, 2007), GC content, and identified rRNA genes via Barrnap (https://github.com/tseemann/barrnap) has yielded much better results. Customized python scripts can extract the phylogenetic inferences for each gene found in a given assembled contig through

the DarkHorse workflow (with comparison to the most updated NCBI database). Then, the same scripts can extract the contigs that contain putative cyanobacterial genes. This customized binning can be refined by selecting %GC thresholds and rRNA hits, eliminating false-positive putative cyanobacterial genes that may be homologs horizontally transferred from the cyanobacterium host to associated microbiota. Once binned, these contigs can be used to extract the subset of reads from the total metagenomic assembly, followed by a reassembly. Alternatively, if the number of binned contigs are very low (less than two dozen), we suggest proceeding to scaffolding and gap filling, which may lead more quickly to a complete genome assembly.

3. Reference scaffolding: despite the scarceness of reference genomes for cyanobacteria in the public databases, reference scaffolding can sometimes significantly improve draft genomes and even generate complete genomes. Medusa (https://github.com/combogenomics/medusa) has been an excellent tool for ordering contigs from de novo draft genomes according to similarities in the reference genome of choice. Two major advantages of Medusa are that it can use multiple reference genomes at once and it uses contigs, rather than reads, for scaffolding.
4. Gap filling: this penultimate step in the workflow consists of closing the gaps and obtaining a circular chromosome (this is most common, although bacteria can also have linear chromosomes) and, if applicable, plasmids. For binned out cyanobacterial genomes (from step 2) that contain a few contigs and have a good reference for scaffolding, PCR followed by Sanger sequencing can be a robust and rapid choice for closing gaps. On the other hand, for complex samples with many contigs, this approach is not feasible due to the number of possible permutations. In the latter case, an in silico method is more suitable, such as using paired reads with high confidence alignments to the contig edges (e.g., using the tool GapFiller https://www.baseclear.com/genomics/bioinformatics/basetools/gapfiller) (Boetzer & Pirovano, 2012). This in silico approach requires a high read coverage because only a very small subset of the total reads can be used for those alignments.
5. Quality check can be performed using the automated quality assessment with QUAST (http://bioinf.spbau.ru/quast) (Gurevich, Saveliev, Vyahhi, & Tesler, 2013) and CheckM (https://github.com/Ecogenomics/CheckM) (Parks, Imelfort, Skennerton, Hugenholtz, & Tyson, 2015). This can aid in the choice of the best assembly parameters and binning

methods for your samples. Quality check can also help to identify assembly errors and incorrect scaffolding that can bias the algorithms in the programs listed in previous steps; such errors can be very detrimental to the downstream genome mining analyses.

6.3 Obtaining Complete Genomes With Additional Long-Read Sequencing

Long-read sequencing with PacBio RSII (Pacific Bioscience®) or MinION (Oxford Nanopore Technologies®) can facilitate obtaining complete genomes by either scaffolding the assembled (and binned) contigs from short-read generated assemblies, or by generating a nearly complete de novo draft that can be error corrected and gap filled by the short reads. Additionally, short-read sequencing can provide an initial assessment for the complexity of the analysis, guiding the choice of which long-read sequencing method to use. Drafts from short reads can be used to estimate the abundance of the cyanobacterium in the sample (vs contaminating heterotrophic bacteria), indicating the need for enrichment before the next gDNA extraction. The short-read *k*-mer signature from the binned contigs can estimate the expected genome size quite accurately using the program Jellyfish and custom scripts further explained in this tutorial: https://bioinformatics.uconn.edu/genome-size-estimation-tutorial/ (Marçais & Kingsford, 2011). Regarding the long-read sequencing output, PacBio technology tends to have higher variability within similar sequencing efforts (e.g., two different strains from same genus) and Nanopore has been problematic for multiplexing samples, limiting the number of samples and increasing the cost per sample. If initial sample assessment indicates that a large number of reads (for example, for eukarya) is a major requirement, Nanopore can be an excellent choice. If read throughput is not a concern, PacBio can be more cost effective. In either case, high molecular weight gDNA is crucial for the efficiency and success of the long-read sequencing.

6.4 Detection of BGC's From Genomic Assemblies

Identifying BGC's from the assembled data is mainly performed by two different types of algorithms: high-confidence/low-novelty or high-novelty/low-confidence (Medema & Fischbach, 2015). For the first type, the tool antiSMASH (Blin, Kim, et al., 2017; Blin, Wolf, et al., 2017) has been widely used and has, for the most part, yielded excellent results. However, some misidentification of certain domains may occur in comparison to the DELTA-BLASTP Domain-Enhanced Lookup Time Accelerated Basic

Local Alignment Search Tool (NCBI) analysis. Additionally, antiSMASH structural predictions need to be further improved, in order to better accommodate well-characterized scaffold modifications, such as beta-branching of polyketide chains, a common cyanobacterial biosynthetic gene cassette (Gu, Wang, Kulkarni, Geders, et al., 2009; Gu, Wang, Kulkarni, Gehret, et al., 2009). Clusterfinder (Cimermancic et al., 2014) is an example of a high-novelty/low-confidence algorithm and it is also available on the antiSMASH server (an option on the online antiSMASH input screen—https://antiSMASH.secondarymetabolites.org/#!/start). Normally, the use of antiSMASH is preferred for untargeted analyses because a low rate of false positives is essential for this kind of analysis. AntiSMASH has been recognized as one the most accurate tools for BGC identification, therefore, it is also commonly used as a proxy for drug discovery potential (e.g., the DNA percentage dedicated to biosynthesis of NPs) (Leão et al., 2017; Letzel et al., 2017). However, the combination of antiSMASH and manual inspection and border refinement does not significantly differ from crude ClusterFinder statistics regarding the percentage of the genome dedicated to the biosynthesis of secondary metabolites (Leão et al., 2017). On the other hand, since antiSMASH uses a high-confidence/low-novelty algorithm, it normally does not detect rare and unprecedented BGC arrangements (BGC's of unusual domain composition), such as in the *bmp* pathway found in marine nonphotosynthetic bacteria and *Hormoscilla spongeliae* (a cyanobacterial sponge symbiont) (Agarwal et al., 2017). In such cases, full genome comparison between fairly distinct organisms that produce the same natural product can indicate the sequence containing the rare BGC, as the cluster may be the most homologous DNA fragment between the two genomes. The cluster can then be confirmed via open reading frame translation and the predicted biosynthesis. Alternatively, genome comparison between a producer and very similar nonproducer can also help identify unique BGC's. Algorithms for genome comparison are still under construction; however, one can envision the use of machine learning to aid in this challenge, such as by defining the limits of uncharacterized BGC's. Such algorithms can also be used for better assembly of contigs containing BGC's. BGC-containing contigs tend to be the most challenging (and most interesting) part of the assembly graph, often leading to a very fragmented draft genome.

6.5 Genome Mining Strategies for Novel Gene Cluster Families

Certain types of cyanobacteria tend to harbor a prolific and unique metabolic potential (Kleigrewe et al., 2016; Leão et al., 2017), portending a promising future for genome mining from these strains. However, challenges in bioinformatics and high-throughput screening have hampered access to this NP discovery. The key challenges include: difficulty to obtain high quality de novo genomes for mining (explained in Section 6.2), lack of rapid and precise tools for distinguishing known vs new biosynthetic genes, difficulty to systematically connect biosynthetic genes to known structures, inability to systematically connect new biosynthetic genes to unknown molecules, and lack of efficient high-throughput heterologous expression systems to provide sources of novel compounds when they are not expressed in the native organism. For example, the discovery of the columbamides (Kleigrewe et al., 2015) and cryptomaldamide (Kinnel et al., 2017) identifies approaches that have yielded new NPs. Networking analysis tends to stand out due to the success of molecular networking while dereplicating MS data (Wang et al., 2016). Here, we briefly discuss gene cluster networking, whereby BGC's are grouped into families based on scaffolding and gene types. Gene cluster networking can be combined with additional approaches, such as MS and NMR, for better dereplication accuracy.

6.5.1 Gene Cluster Networking Workflow

1. Parse GenBank files for the BGC's to be networked from assemblies (detected via antiSMASH with the option ClusterFinder off) and from other databases, for example, the antiSMASH database (Blin, Medema, Kottmann, Lee, & Weber, 2017) and minimum information about a biosynthetic gene cluster (MIBiG) (Medema et al., 2015). BioCompass package (http://biocompass.net) also offers a code for downloading the 100 most similar BGC's from NCBI per queried BGC.
2. Once the BGC files are parsed into the same folder, use BiG-SCAPE (https://git.wageningenur.nl/medema-group/BiG-SCAPE) for networking BGC's. BiG-SCAPE calculates raw distances based in the domain content present in the BGC's. Correct and precise domain identification is key for this analysis. Alternatively, BioCompass calculates similarity scores via gene alignments, not being limited by domain detection. However, BiG-SCAPE tends to be very fast and memory efficient, therefore, it is the tool of choice for large untargeted gene cluster networks. After selecting a particularly interesting gene cluster, reanalyze the genomes containing the queried gene cluster family (GCF) via BioCompass (https://biocompass.net). This combination of both tools works very well to identify relevant GCFs.

3. Beta-diversity calculation: because the previous steps can sort BGC's into families, the similarity percentile can be used to calculate between a pair of samples based on the presence or absence of their GCFs. Beta-diversity via Brays–Curtis distance is a common metric for microbial diversity studies and it can be used to calculate BGC diversity. The average beta-diversity for the BGC's of one sample can indicate how novel that sample is compared to the already known data set. Finally, this novelty can be combined with dereplication of known families (e.g., GCF containing MiBIG BGC's) for directing chemical isolation efforts toward the most novel NP scaffolds.

6.6 Module Deduction

Individual biosynthetic gene cluster modules may be analyzed with additional fidelity by online bioinformatic tools. Two currently useful tools enumerated here are DELTA BLASTP (Boratyn et al., 2012) and antiSMASH (Medema et al., 2011), discussed partially in Section 6.4. Practically speaking, antiSMASH applies a visualization with each domain that is contingent on a structurally complete or near-complete domain, while NCBI BLASTP does not have such a contingency; therefore, when using DELTA BLASTP, modules may contain a number of truncated or incomplete domains which do not necessarily reflect the true function of the module. It may be helpful for the investigator to use both tools, focusing on antiSMASH to obtain a complete picture of a biosynthetic gene cluster, while using DELTA BLASTP to analyze individual domains, tailoring, and standalone genes. Additional clues to enzymatic function may be suggested by Phyre2, which generates a 3D homology model to known protein structures based on alignment from an amino acid query (Kelley, Mezulis, Yates, Wass, & Sternberg, 2015). An investigator may consider overall domain completeness, presence (or absence) of key active site residues, as well as overall gene context within a biosynthetic pathway in order to connect queried genes to known, putative, or unknown molecules.

7. BIOSYNTHETIC PATHWAY VALIDATION

As noted earlier, tropical marine filamentous cyanobacteria harbor heterotrophic bacteria among its sheath filaments, and while gDNA extraction, sequencing, and assembly strategies are targeted toward DNA extracted specifically from cyanobacteria, some DNA from heterotrophic bacteria is likely to be coprocessed with the cyanobacterial DNA. Therefore, biosynthetic pathways that are deduced from genomic data should be validated by laboratory techniques that specifically assign molecular structure or production in coordination with the presence of the biosynthetic pathway. These strategies may include classical stable isotope-labeled feeding studies, heterologous expression, and gene knockouts. The advent of CRISPR will likely enable more specific and targeted manipulation of pathways with this goal in mind.

7.1 Stable Isotope-Labeled Feeding to Verify Incorporation of Predicted Backbone

A classical method of interrogating the biosynthetic origin of isolated molecules may be accomplished by the feeding of stable isotope-labeled small molecules to growing cultures of the suspected producer, confirming the bioinformatically predicted incorporation of precursors into the molecule (Vederas, 1987). This lends itself particularly well to cultured cyanobacteria (Edwards et al., 2004; Garson, 1993) as many filamentous benthic cyanobacteria have proven difficult if not impossible to manipulate using molecular genetic techniques. Here is described a general procedure for carrying out such experiments with healthy cultures of a cyanobacterial strain. It should be noted that strains may have different rates of growth, preferred light intensities, and reactions to exogenous addition of nutrients or carbon sources to the growth media (Castenholz, 1988; Jones et al., 2011). Additionally, the rate of incorporation of labeled substrate may vary significantly between strains, experiments, and growth conditions (Esquenazi et al., 2011). Therefore, systematic testing of culture conditions should be carried out prior to stable isotope-labeled feeding experiments, particularly with concern to the concentration of the exogenous metabolite added to the culture. A recommended range of concentration to aim for is 0.5–1.0 m*M* of stable isotope-labeled metabolite in the growth media. Prior to using labeled substrate, it is recommended to test a wide range of concentrations of unlabeled substrate on small culture aliquots over a period of 1–4 weeks, from 0.05 to 5 m*M* final concentration. Given the heterotrophic bacterial community commonly present on and in cyanobacterial filaments, addition of exogenous carbon sources (acetate and amino acids, for example) may result in their rapid growth. In turn, this may have deleterious effects on the cultured cyanobacteria, or alternatively, may metabolize and therefore encourage scrambling of the stable isotope-labeled metabolite. Therefore, culture growth should be monitored and the culture harvested if heterotrophic bacteria appear to negatively affect the viability of cyanobacterial filaments.

7.1.1 Equipment

- Large autoclave pans with >5 L capacity, or similar vessel, sterilized
- 2.8-L glass Erlenmeyer flasks and caps, sterilized
- Gas-permeable plastic wrap (Fisher Scientific)
- Sterile metal forceps and shears

7.1.2 Reagents

- SW or FW BG-11 (Castenholz, 1988, media modified as described in Tables 2–4)
- Stable isotope-labeled metabolite, dissolved in sterile SW or FW BG-11 or H_2O

7.1.3 Stable Isotope Feeding Procedure

1. Under strain-specific ideal growth conditions, grow several grams (approximate wet weight) of biomass of the strain of interest, in sterile growth pans or 2.8-L Fernbach flasks, as described in the culturing section of this chapter. Most tropical marine filamentous cyanobacteria will require 4–12 weeks and may require several changes of media and subculturing into new flasks or pans.
2. After strains have reached sufficient density to result in 2–4 g of wet biomass weight, concentrate the biomass into the minimal volume possible that still ensures light can reach most of the filaments. Concentrating biomass to a smaller volume encourages metabolite uptake and ensures a higher final concentration of stable isotope-labeled metabolite and therefore may result in a higher percentage of labeled natural products for analysis. For example, a typical experimental setup may use the biomass grown in 10 L of media in a large culture pan and transfer this into 1 L of media in a 2.8-L Fernbach flask.
3. Allow the culture to acclimate to its new subculture for 24–48 h.
4. Add stable isotope-labeled precursor under investigation, to a concentration predetermined by previous tolerance testing for the molecule. As high concentrations of exogenous metabolite may be degraded and then metabolically scrambled by heterotrophic bacteria, it is recommended to add the metabolite of interest in two to four boluses of material over a period of 5–15 days, depending on the growth characteristics of the organism. If possible, it is recommended to run experiments in parallel, so as to capture the metabolite of interest at maximum labeled percentage and maximum production of compound prior to metabolic scrambling. If there is significant biomass present, a microchemistry extraction of a small sample followed by sensitive liquid chromatography–mass

spectrometry (LC–MS) to determine incorporation rate may be used to guide harvesting and feeding time.

5. Harvesting may take place under the following circumstances: the predetermined harvest date has been reached as per earlier unlabeled feeding studies, or the culture begins to suffer from either light attenuation, concentrated biomass, or heterotrophic bacterial growth.
6. After harvesting, extract the biomass as referenced in Section 3.1, and purify the molecule or molecules of interest using a combination of semipreparative or preparative HPLC and other standard analytical techniques, followed by the appropriate NMR or MS experiment to monitor incorporation.

7.2 Recombinant Protein Expression

Recombinant protein expression is an alternative technique to consider implementing in parallel with stable isotope-labeled feeding experiments. As many novel pathways contain interesting enzymes of unknown or poorly understood function (Jones, Gu, Sorrels, Sherman, & Gerwick, 2009), expression and activity assays of such enzymes may provide additional confirmation that the pathway of interest matches the bioinformatically predicted molecule. This technique has the dual functionality of clarifying and exploring unknown biochemistry often found in cyanobacterial biosynthetic gene clusters (Calteau et al., 2014). Protein expression coupled with enzymatic assays and MS was used to probe a novel HMG-CoA-synthase-like β-branching enzyme (Gu, Wang, Kulkarni, Geders, et al., 2009; Gu, Wang, Kulkarni, Gehret, et al., 2009), terminal alkene formation in curacin A, and the olefin synthase pathway (Gehret et al., 2011; Gu, Wang, Kulkarni, Geders, et al., 2009; Gu, Wang, Kulkarni, Gehret, et al., 2009), while a novel initiator alkyne-encoding cassette was described from the jamaicamide pathway (Zhu, Liu, & Zhang, 2015).

7.3 Heterologous Expression and Molecular Genetics

Marine filamentous cyanobacteria of the genus *Moorea* have not, thus far, been genetically tractable. It has proven difficult to transfer genetic materials via electroporation or conjugation into filamentous cyanobacterial cells. While stable isotope labeling and protein expression experiments to probe biosynthetic features have been carried out for many years, molecular genetic techniques have only recently enabled alternative methods for probing the biosynthetic gene clusters in cyanobacteria. To date, the heterologous expression of the lyngbyatoxin and barbamide BGC's has been carried out in *Escherichia coli* and *Streptomyces venezuelae*, respectively (Kim et al., 2012; Ongley et al., 2013). Recently, transformation-associated recombination cloning of large DNA molecules, as well as expression of lyngbyatoxin in *Anabaena* (Nostoc) PCC 7120, has enabled its consideration as a model strain for expressing BGC's (Videau, Wells, Singh, Gerwick, & Philmus, 2016). However, a comprehensive platform for successfully expressing large cyanobacterial biosynthetic gene clusters has not yet been successfully implemented. Microinjections as well as biolistic techniques may be helpful for the introduction of DNA into marine benthic filamentous cyanobacteria, speeding their use as platforms for expression. Development of CRISPR–Cas9 (Wendt, Ungerer, Cobb, Zhao, & Pakrasi, 2016) or CRISPR–Cpf1 (Ungerer & Pakrasi, 2016) machinery and guide RNAs may enable precise editing of cyanobacterial genomes for study of their natural products via markerless knock-ins, knockouts, and point mutations (especially for characterizing novel catalytic domains). However, tools and knowledge surrounding Cpf1 are still being developed, while Cas9 has shown significant off-target effects and toxicity in cyanobacteria (Cradick, Fine, Antico, & Bao, 2013). As such, heterologous expression of large cyanobacterial BGC's is still very much under development and the future is ripe for opportunity.

8. SUMMARY

Despite many recent advances in genome mining, the correlation between genomic data (genotype) and molecular structural data (phenotype) remains poorly understood. As recently stated (Palazzolo, Simons, & Burke, 2017), the boundedness of the natural product universe "reveals an extraordinary opportunity to leverage more systematized approaches to natural product research." In parallel with genomics, there has been a shift from studying one gene at a time to the comprehensive study of all genes simultaneously. When such a shift occurs in microbial genome mining, natural product research will experience a new level of interest and productivity, and thereby provide new avenues for small-molecule therapeutic discovery, characterization of protein–protein interaction, and new modes of gene regulation. Cyanobacteria have long been recognized as excellent sources of novel metabolites, but only recently have researchers delved into studies of cyanobacterial genomics, biosynthetic gene clusters, and details of their biosynthetic processes. It is our hope that this chapter may provide a foundation for the expanded collection, culture, and genomic analysis of filamentous marine cyanobacteria in the future, and thus help enable the next generation of therapeutic discoveries from these chemically prolific organisms.

REFERENCES

Abràmoff, M. D., Magalhães, P. J., & Ram, S. J. (2004). Image processing with ImageJ. *Biophotonics International*, *11*, 36–42.

Acreman, J. (1994). Algae and cyanobacteria: Isolation, culture and long-term maintenance. *Journal of Industrial Microbiology & Biotechnology*, *13*(3), 193–194.

Agarwal, V., Blanton, J. M., Podell, S., Taton, A., Schorn, M. A., Busch, J., et al. (2017). Metagenomic discovery of polybrominated diphenyl ether biosynthesis by marine sponges. *Nature Chemical Biology*, *13*, 537–543.

Allen, M. M. (1968). Simple conditions for growth of unicellular blue-green algae on plates. *Journal of Phycology*, *4*, 1–4.

Almaliti, J., Malloy, K. L., Glukhov, E., Spadafora, C., Gutiérrez, M., & Gerwick, W. H. (2017). Dudawalamides A–D, antiparasitic cyclic depsipeptides from the marine cyanobacterium *Moorea producens*. *Journal of Natural Products*, *80*, 1827–1836.

Alvarenga, D. O., Fiore, M. F., & Varani, A. M. (2017). A metagenomic approach to cyanobacterial genomics. *Frontiers in Microbiology*, *8*, 809.

Bankevich, A., Nurk, S., Antipov, D., Gurevich, A. A., Dvorkin, M., Kulikov, A. S., et al. (2012). SPAdes: A new genome assembly algorithm and its applications to single-cell sequencing. *Journal of Computational Biology*, *19*, 455–477.

Bertin, M. J., Vulpanovici, A., Monroe, E. A., Korobeynikov, A., Sherman, D. H., Gerwick, L., et al. (2016). The phormidolide biosynthetic gene cluster: A trans-AT PKS pathway encoding a toxic macrocyclic polyketide. *Chembiochem*, *17*, 164–173.

Blin, K., Kim, H. U., Medema, M. H., & Weber, T. (2017). Recent development of antiSMASH and other computational approaches to mine secondary metabolite biosynthetic gene clusters. *Briefings in Bioinformatics*, bbx146. https://doi.org/10.1093/bib/bbx146 (Epub ahead of print).

Blin, K., Medema, M. H., Kottmann, R., Lee, S. Y., & Weber, T. (2017). The antiSMASH database, a comprehensive database of microbial secondary metabolite biosynthetic gene clusters. *Nucleic Acids Research*, *45*(D1), D555–D559. gkw960.

Blin, K., Wolf, T., Chevrette, M. G., Lu, X., Schwalen, C. J., Kautsar, S. A., et al. (2017). antiSMASH 4.0—Improvements in chemistry prediction and gene cluster boundary identification. *Nucleic Acids Research*, *45*(W1), W36–W41.

Boetzer, M., & Pirovano, W. (2012). Toward almost closed genomes with GapFiller. *Genome Biology*, *13*(6), R56.

Boratyn, G. M., Schäffer, A. A., Agarwala, R., Altschul, S. F., Lipman, D. J., & Madden, T. L. (2012). Domain enhanced lookup time accelerated BLAST. *Biology Direct*, *7*, 12.

Calteau, A., Fewer, D. P., Latifi, A., Coursin, T., Laurent, T., Jokela, J., et al. (2014). Phylum-wide comparative genomics unravel the diversity of secondary metabolism in cyanobacteria. *BMC Genomics*, *15*, 977.

Cardellina, J. H., Marner, F. J., & Moore, R. E. (1979). Seaweed dermatitis: Structure of lyngbyatoxin A. *Science*, *204*, 193–195.

Castenholz, R. W. (1988). [3] Culturing methods for cyanobacteria. *Methods in Enzymology*, *167*, 68–93.

Cimermancic, P., Medema, M. H., Claesen, J., Kurita, K., Brown, L. C. W., Mavrommatis, K., et al. (2014). Insights into secondary metabolism from a global analysis of prokaryotic biosynthetic gene clusters. *Cell*, *158*, 412–421.

Clarke, J. D. (2009). Cetyltrimethyl ammonium bromide (CTAB) DNA miniprep for plant DNA isolation. *Cold Spring Harbor Protocols, 2009*, pdb-prot5177.

Clarridge, J. E. (2004). Impact of 16S rRNA gene sequence analysis for identification of bacteria on clinical microbiology and infectious diseases. *Clinical Microbiology Reviews, 17*, 840–862.

Cradick, T. J., Fine, E. J., Antico, C. J., & Bao, G. (2013). CRISPR/Cas9 systems targeting β-globin and CCR5 genes have substantial off-target activity. *Nucleic Acids Research, 41*, 9584–9592.

Cummings, S. L., Barbé, D., Leão, T. F., Korobeynikov, A., Engene, N., Glukhov, E., et al. (2016). A novel uncultured heterotrophic bacterial associate of the cyanobacterium Moorea producens JHB. *BMC Microbiology, 16*, 198.

Dittmann, E., Gugger, M., Sivonen, K., & Fewer, D. P. (2015). Natural product biosynthetic diversity and comparative genomics of the cyanobacteria. *Trends in Microbiology, 23*, 642–652.

Donia, M. S., Hathaway, B. J., Sudek, S., Haygood, M. G., Rosovitz, M. J., Ravel, J., et al. (2006). Natural combinatorial peptide libraries in cyanobacterial symbionts of marine ascidians. *Nature Chemical Biology, 2*, 729–735.

Ebada, S. S., Edrada, R. A., Lin, W., & Proksch, P. (2008). Methods for isolation, purification and structural elucidation of bioactive secondary metabolites from marine invertebrates. *Nature Protocols, 3*, 1820–1831.

Edwards, D. J., Marquez, B. L., Nogle, L. M., McPhail, K., Goeger, D. E., Roberts, M. A., et al. (2004). Structure and biosynthesis of the jamaicamides, new mixed polyketide-peptide neurotoxins from the marine cyanobacterium Lyngbya majuscula. *Chemistry & Biology, 11*, 817–833.

Engene, N., Coates, R. C., & Gerwick, W. H. (2010). 16S rRNA gene heterogeneity in the filamentous marine cyanobacterial genus Lyngbya. *Journal of Phycology, 46*, 591–601.

Engene, N., Rottacker, E. C., Kaštovský, J., Byrum, T., Choi, H., Ellisman, M. H., et al. (2012). Moorea producens gen. nov., sp. nov. and *Moorea bouillonii* comb. nov., tropical marine cyanobacteria rich in bioactive secondary metabolites. *International Journal of Systematic and Evolutionary Microbiology, 62*, 1171–1178.

Esquenazi, E., Jones, A. C., Byrum, T., Dorrestein, P. C., & Gerwick, W. H. (2011). Temporal dynamics of natural product biosynthesis in marine cyanobacteria. *Proceedings of the National Academy of Sciences of the United States of America, 108*, 5226–5231.

Flores, E., López-Lozano, A., & Herrero, A. (2015). Nitrogen fixation in the oxygenic phototrophic prokaryotes (cyanobacteria): The fight against oxygen. *Biological Nitrogen Fixation, 2*, 879–890.

Garson, M. J. (1993). The biosynthesis of marine natural products. *Chemical Reviews, 93*, 1699–1733.

Gehret, J. J., Gu, L., Gerwick, W. H., Wipf, P., Sherman, D. H., & Smith, J. L. (2011). Terminal alkene formation by the thioesterase of curacin a biosynthesis structure of a decarboxylating thioesterase. *Journal of Biological Chemistry, 286*, 14445–14454.

Gu, L., Wang, B., Kulkarni, A., Geders, T. W., Grindberg, R. V., Gerwick, L., et al. (2009). Metamorphic enzyme assembly in polyketide diversification. *Nature, 459*, 731–735.

Gu, L., Wang, B., Kulkarni, A., Gehret, J. J., Lloyd, K. R., Gerwick, L., et al. (2009). Polyketide decarboxylative chain termination preceded by *O*-sulfonation in curacin A biosynthesis. *Journal of the American Chemical Society, 131*, 16033–16035.

Gurevich, A., Saveliev, V., Vyahhi, N., & Tesler, G. (2013). QUAST: Quality assessment tool for genome assemblies. *Bioinformatics, 29*, 1072–1075.

Hanaki, Y., Shikata, Y., Kikumori, M., Hotta, N., Imoto, M., & Irie, K. (2018). Identification of protein kinase C isozymes involved in the anti-proliferative and pro-apoptotic activities of 10-methyl-aplog-1, a simplified analog of debromoaplysiatoxin, in several cancer cell lines. *Biochemical and Biophysical Research Communications, 495*, 438–445.

Hoiczyk, E. (1998). Structural and biochemical analysis of the sheath of Phormidium uncinatum. *Journal of Bacteriology*, *180*, 3923–3932.

Hubert, J., Nuzillard, J. M., & Renault, J. H. (2017). Dereplication strategies in natural product research: How many tools and methodologies behind the same concept? *Phytochemistry Reviews*, *16*, 55–95.

Jacobson, M. Z. (2005). Studying ocean acidification with conservative, stable numerical schemes for nonequilibrium air-ocean exchange and ocean equilibrium chemistry. *Journal of Geophysical Research: Atmospheres*, *110*(D7), D07302. https://doi.org/10.1029/2004JD005220.

Jones, A. C., Gu, L., Sorrels, C. M., Sherman, D. H., & Gerwick, W. H. (2009). New tricks from ancient algae: Natural products biosynthesis in marine cyanobacteria. *Current Opinion in Chemical Biology*, *13*, 216–223.

Jones, A. C., Monroe, E. A., Podell, S., Hess, W. R., Klages, S., Esquenazi, E., et al. (2011). Genomic insights into the physiology and ecology of the marine filamentous cyanobacterium Lyngbya majuscula. *Proceedings of the National Academy of Sciences of the United States of America*, *108*, 8815–8820.

Jürgens, U. J., & Weckesser, J. (1985). The fine structure and chemical composition of the cell wall and sheath layers of cyanobacteria. *Annales de l'Institut Pasteur Microbiologie*, *136*, 41–44.

Kang, D. D., Rubin, E. M., & Wang, Z. (2016). Reconstructing single genomes from complex microbial communities. *IT-Information Technology*, *58*, 133–139.

Keller, L., Leão, T., & Gerwick, W. H. (2017). Chemical biology of marine cyanobacteria. In D. J. Newman, G. M. Cragg, & P. Grothaus (Eds.), *Chemical biology of natural products*. Boca Raton, FL, USA: CRC Press.

Kelley, L. A., Mezulis, S., Yates, C. M., Wass, M. N., & Sternberg, M. J. (2015). The Phyre2 web portal for protein modeling, prediction and analysis. *Nature Protocols*, *10*, 845–858.

Kim, E. J., Lee, J. H., Choi, H., Pereira, A. R., Ban, Y. H., Yoo, Y. J., et al. (2012). Heterologous production of 4-O-demethylbarbamide, a marine cyanobacterial natural product. *Organic Letters*, *14*, 5824–5827.

Kinnel, R. B., Esquenazi, E., Leao, T., Moss, N., Mevers, E., Pereira, A. R., et al. (2017). A Maldiisotopic approach to discover natural products: Cryptomaldamide, a hybrid tripeptide from the marine cyanobacterium *Moorea producens*. *Journal of Natural Products*, *80*, 1514–1521.

Kleigrewe, K., Almaliti, J., Tian, I. Y., Kinnel, R. B., Korobeynikov, A., Monroe, E. A., et al. (2015). Combining mass spectrometric metabolic profiling with genomic analysis: A powerful approach for discovering natural products from cyanobacteria. *Journal of Natural Products*, *78*, 1671–1682.

Kleigrewe, K., Gerwick, L., Sherman, D. H., & Gerwick, W. H. (2016). Unique marine derived cyanobacterial biosynthetic genes for chemical diversity. *Natural Product Reports*, *33*, 348–364.

Komárek, J. (2006). Cyanobacterial taxonomy: Current problems and prospects for the integration of traditional and molecular approaches. *Algae*, *21*, 349–375.

Komárek, J., Kaštovský, J., Mareš, J., & Johansen, J. R. (2014). Taxonomic classification of cyanoprokaryotes (cyanobacterial genera) 2014, using a polyphasic approach. *Preslia*, *86*, 295–335.

Lang, I., & Feussner, I. (2007). Oxylipin formation in Nostoc punctiforme (PCC73102). *Phytochemistry*, *68*, 1120–1127.

Leal, M. C., Munro, M. H., Blunt, J. W., Puga, J., Jesus, B., Calado, R., et al. (2013). Biogeography and biodiscovery hotspots of macroalgal marine natural products. *Natural Product Reports*, *30*, 1380–1390.

Leão, T., Castelão, G., Korobeynikov, A., Monroe, E. A., Podell, S., Glukhov, E., et al. (2017). Comparative genomics uncovers the prolific and distinctive metabolic potential of the cyanobacterial genus Moorea. *Proceedings of the National Academy of Sciences of the United States of America, 114*, 3198–3203.

Letzel, A. C., Jing, L., Amos, G., Millán-Aguiñaga, N., Ginigini, J., Abdelmohsen, U. R., et al. (2017). Genomic insights into specialized metabolism in the marine actinomycete Salinispora. *Environmental Microbiology, 19*, 3660–3673.

Lewin, J. (1966). Silicon metabolism in diatoms. V. Germanium dioxide, a specific inhibitor of diatom growth. *Phycologia, 6*, 1–12.

Li, L., & Ma, Z. S. (2016). Testing the neutral theory of biodiversity with human microbiome datasets. *Scientific Reports, 6*, 31448.

Littler, D. S., & Littler, M. M. (2003). *South Pacific reef plants*. Offshore Graphics, Inc.

Littler, D. S., Littler, M. M., Bucher, K. E., & Norris, J. N. (1989). *Marine plants of the Caribbean; a field guide from Florida to Brazil*. Washington DC: Smithsonian Institution Press.

Luesch, H., Yoshida, W. Y., Moore, R. E., Paul, V. J., & Corbett, T. H. (2001). Total structure determination of Apratoxin a, a potent novel Cytotoxin from the marine cyanobacterium *Lyngbya majuscula*. *Journal of the American Chemical Society, 123*, 5418–5423.

Marçais, G., & Kingsford, C. (2011). A fast, lock-free approach for efficient parallel counting of occurrences of k-mers. *Bioinformatics, 27*, 764–770.

Medema, M. H., Blin, K., Cimermancic, P., de Jager, V., Zakrzewski, P., Fischbach, M. A., et al. (2011). antiSMASH: Rapid identification, annotation and analysis of secondary metabolite biosynthesis gene clusters in bacterial and fungal genome sequences. *Nucleic Acids Research, 39*, W339–W346.

Medema, M. H., & Fischbach, M. A. (2015). Computational approaches to natural product discovery. *Nature Chemical Biology, 11*, 639–648.

Medema, M. H., Kottmann, R., Yilmaz, P., Cummings, M., Biggins, J. B., Blin, K., et al. (2015). Minimum information about a biosynthetic gene cluster. *Nature Chemical Biology, 11*, 625–631.

Navarro, G., Cummings, S., Lee, J., Moss, N., Glukhov, E., Valeriote, F. A., et al. (2015). Isolation of polycavernoside D from a marine cyanobacterium. *Environmental Science & Technology Letters, 2*, 166–170.

Nelson, D. C., Waterbury, J. B., & Jannasch, H. W. (1982). Nitrogen fixation and nitrate utilization by marine and freshwater Beggiatoa. *Archives of Microbiology, 133*, 172–177.

Newman, D. J., & Cragg, G. M. (2016). Natural products as sources of new drugs from 1981 to 2014. *Journal of Natural Products, 79*, 629–661.

Nübel, U., Garcia-Pichel, F., & Muyzer, G. (1997). PCR primers to amplify 16S rRNA genes from cyanobacteria. *Applied and Environmental Microbiology, 63*, 3327–3332.

Ongley, S. E., Bian, X., Zhang, Y., Chau, R., Gerwick, W. H., Müller, R., et al. (2013). High-titer heterologous production in E. coli of lyngbyatoxin, a protein kinase C activator from an uncultured marine cyanobacterium. *ACS Chemical Biology, 8*, 1888–1893.

Palazzolo, A. M., Simons, C. L., & Burke, M. D. (2017). The natural productome. *Proceedings of the National Academy of Sciences of the United States of America, 114*, 5544–5546.

Parks, D. H., Imelfort, M., Skennerton, C. T., Hugenholtz, P., & Tyson, G. W. (2015). CheckM: Assessing the quality of microbial genomes recovered from isolates, single cells, and metagenomes. *Genome Research, 25*, 1043–1055.

Pereira, A., Cao, Z., Murray, T. F., & Gerwick, W. H. (2009). Hoiamide a, a sodium channel activator of unusual architecture from a consortium of two Papua New Guinea cyanobacteria. *Chemistry & Biology, 16*, 893–906.

Podell, S., & Gaasterland, T. (2007). DarkHorse: A method for genome-wide prediction of horizontal gene transfer. *Genome Biology, 8*, R16.

Proteau, P. J., Gerwick, W. H., Garcia-Pichel, F., & Castenholz, R. (1993). The structure of scytonemin, an ultraviolet sunscreen pigment from the sheaths of cyanobacteria. *Experientia*, *49*, 825–829.

Pye, C. R., Bertin, M. J., Lokey, R. S., Gerwick, W. H., & Linington, R. G. (2017). Retrospective analysis of natural products provides insights for future discovery trends. *Proceedings of the National Academy of Sciences of the United States of America*, *114*, 5601–5606.

Queiroz, E. F., Hostettmann, K., & Wolfender, J. (2012). Modern approaches in the search for new active compounds from crude extracts of natural sources. In V. Cechinel-Filho (Series Ed.) & B. Wang (Vol. Ed.), *Series in drug discovery and development: 17. Plant bioactives and drug discovery: Principles, practice, and perspectives* (pp. 43–79). 17. Hoboken, NJ, USA: Wiley & Sons.

Rippka, R., Deruelles, J., Waterbury, J. B., Herdman, M., & Stanier, R. Y. (1979). Generic assignments, strain histories and properties of pure cultures of cyanobacteria. *Microbiology*, *111*, 1–61.

Rippka, R., Waterbury, J. B., & Stanier, R. Y. (1981). Isolation and purification of cyanobacteria: Some general principles. In M. P. Starr, H. Stolp, H. G. Trüper, A. Balows, & H. G. Schlegel (Eds.), *The prokaryotes* (pp. 212–220). Berlin, Heidelberg: Springer.

Rossi, F., & De Philippis, R. (2015). Role of cyanobacterial exopolysaccharides in phototrophic biofilms and in complex microbial mats. *Life*, *5*, 1218–1238.

Shih, P. M., Wu, D., Latifi, A., Axen, S. D., Fewer, D. P., Talla, E., et al. (2013). Improving the coverage of the cyanobacterial phylum using diversity-driven genome sequencing. *Proceedings of the National Academy of Sciences of the United States of America*, *110*, 1053–1058.

Sitachitta, N., & Gerwick, W. H. (1998). Grenadadiene and grenadamide, cyclopropyl-containing fatty acid metabolites from the marine cyanobacterium Lyngbya majuscula. *Journal of Natural Products*, *61*, 681–684.

Tan, L. T. (2013). Marine cyanobacteria: A prolific source of bioactive natural products as drug leads. In S. K. Kim (Ed.), *Marine microbiology: Bioactive compounds and biotechnological applications* (pp. 59–81). Hoboken, NJ, USA: Wiley & Sons.

Ungerer, J., & Pakrasi, H. B. (2016). Cpf1 is a versatile tool for CRISPR genome editing across diverse species of cyanobacteria. *Scientific Reports*, *6*, 39681.

Vederas, J. C. (1987). The use of stable isotopes in biosynthetic studies. *Natural Product Reports*, *4*, 277–337.

Videau, P., Wells, K. N., Singh, A. J., Gerwick, W. H., & Philmus, B. (2016). Assessment of Anabaena sp. strain PCC 7120 as a heterologous expression host for cyanobacterial natural products: Production of lyngbyatoxin A. *ACS Synthetic Biology*, *5*, 978–988.

Wang, M., Carver, J. J., Phelan, V. V., Sanchez, L. M., Garg, N., Peng, Y., et al. (2016). Sharing and community curation of mass spectrometry data with global natural products social molecular networking. *Nature Biotechnology*, *34*, 828–837.

Waterbury, J. B. (2006). The cyanobacteria—Isolation, purification and identification. In M. Dworkin, S. Falkow, E. Rosenberg, K. H. Schleifer, & E. Stackebrandt (Eds.), *The prokaryotes* (pp. 1053–1073). New York, NY, USA: Springer.

Weckesser, J., Hofmann, K., Jürgens, U. J., Whitton, B. A., & Raffelsburger, B. (1988). Isolation and chemical analysis of the sheaths of the filamentous cyanobacteria Calothrix parietina and C. scopulorum. *Microbiology*, *134*, 629–634.

Wendt, K. E., Ungerer, J., Cobb, R. E., Zhao, H., & Pakrasi, H. B. (2016). CRISPR/Cas9 mediated targeted mutagenesis of the fast growing cyanobacterium Synechococcus elongatus UTEX 2973. *Microbial Cell Factories*, *15*, 115.

Williamson, R. T., Boulanger, A., Vulpanovici, A., Roberts, M. A., & Gerwick, W. H. (2002). Structure and absolute stereochemistry of phormidolide, a new toxic metabolite from the marine cyanobacterium Phormidium sp. *The Journal of Organic Chemistry*, *67*, 7927–7936.

Yotsu-Yamashita, M., Haddock, R. L., & Yasumoto, T. (1993). Polycavernoside A: A novel glycosidic macrolide from the red alga Polycavernosa tsudai (Gracilaria edulis). *Journal of the American Chemical Society*, *115*, 1147–1148.

Zehr, J. P. (2011). Nitrogen fixation by marine cyanobacteria. *Trends in Microbiology*, *19*, 162–173.

Zhang, C., Idelbayev, Y., Roberts, N., Tao, Y., Nannapaneni, Y., Duggan, B. M., et al. (2017). Small molecule accurate recognition technology (SMART) to enhance natural products research. *Scientific Reports*, 7, 14243.

Zhu, X., Liu, J., & Zhang, W. (2015). De novo biosynthesis of terminal alkyne-labeled natural products. *Nature Chemical Biology*, *11*, 115–120.

CHAPTER TWO

PKS–NRPS Enzymology and Structural Biology: Considerations in Protein Production

Meredith A. Skiba*, Finn P. Maloney*, Qingyun Dan*, Amy E. Fraley*, Courtney C. Aldrich†, Janet L. Smith*,[1], W. Clay Brown*,[1]
*University of Michigan, Ann Arbor, MI, United States
†University of Minnesota, Minneapolis, MN, United States
[1]Corresponding authors: e-mail address: janetsmith@umich.edu; wclayb@umich.edu

Contents

Methods in Enzymology, Volume 604
ISSN 0076-6879
https://doi.org/10.1016/bs.mie.2018.01.035

Abstract

The structural diversity and complexity of marine natural products have made them a rich and productive source of new bioactive molecules for drug development. The identification of these new compounds has led to extensive study of the protein constituents of the biosynthetic pathways from the producing microbes. Essential processes in the dissection of biosynthesis have been the elucidation of catalytic functions and the determination of 3D structures for enzymes of the polyketide synthases and nonribosomal peptide synthetases that carry out individual reactions. The size and complexity of these proteins present numerous difficulties in the process of going from gene to structure. Here, we review the problems that may be encountered at the various steps of this process and discuss some of the solutions devised in our and other labs for the cloning, production, purification, and structure solution of complex proteins using *Escherichia coli* as a heterologous host.

1. INTRODUCTION

Natural product megasynthases, including polyketide synthases (PKS) and nonribosomal peptide synthetases (NRPS), encompass a wealth of fascinating enzymes that catalyze a vast array of chemical transformations and thus are prime candidates for development as synthetic tools (Akey, Gehret, Khare, & Smith, 2012; Keatinge-Clay, 2017; Payne, Millar, Jackson, & Ochs, 2017; Walsh, 2016; Winn, Fyans, Zhuo, & Micklefield, 2016; Yuzawa, Keasling, & Katz, 2016). Purified, recombinant proteins are key reagents in the discovery of the catalytic activity, chemical mechanism, substrate scope, and 3D structure of PKS and NRPS enzymes and in the manipulation of their activity. Recombinant production of individual enzymes in a heterologous host presents several challenges, including coding sequences adapted for proteins of low natural abundance, uncertain boundaries within multidomain polypeptides, unusual G+C content, and poor solubility when excised from the full-length natural protein. Multidomain proteins present additional challenges for folding to a native state in a heterologous host cell.

Fortunately an abundance of annotated and unannotated sequences for PKS and NRPS gene clusters provides a rich database of target proteins and, in combination with 3D structures from the growing number of homologs in the structure database (Keatinge-Clay, 2012; Miller & Gulick, 2016), can inform experiment design. Many reagents and tools are available to aid in production of active recombinant proteins; however, each protein behaves in a unique manner when produced in a heterologous host cell. Thus, the success or failure of an expression and purification method for a particular protein can be poorly predictive for the applicability of that strategy to a homologous protein. A certain amount of trial and error is generally needed. Any one of several problems can be vexing, leading us to adapt the Anna Karenina principle: "All happy recombinant proteins are alike, each unhappy recombinant protein is unhappy in its own way." The challenge is to identify a range of experimental possibilities by applying all knowledge about a target protein and to investigate these possibilities rapidly and economically.

Here we describe the approaches developed in our laboratory for gene expression in an *Escherichia coli* host, for purification of soluble recombinant PKS and NRPS proteins, and for addressing the most common challenges of these target proteins, with examples from our and others research. We have applied these methods widely to proteins of marine cyanobacterial origin, but they are equally applicable to many other systems. Although the host expression system is bacterial, we have used it for production of eukaryotic proteins. While smaller quantities, reduced purity, and soluble aggregates are generally well tolerated in enzyme assays, our approach is directed to producing milligram quantities of highly pure, monodisperse protein because solving crystal structures is a major activity in our lab. Our method to produce recombinant proteins is based on using all available sequence and 3D-structure information to design an experiment, and then applying the fundamentals of *E. coli* physiology to survey several conditions in parallel on a small scale.

2. cDNA SELECTION

2.1 Amino Acid Sequence Selection

The design of a construct encoding a protein of interest generally starts with a multiple sequence alignment of the target and several related proteins, which can be found with tools such as BLAST (Johnson et al., 2008). The sequence database is deep for PKS and NRPS enzymes, and generally a subset of hits must be selected as a workable number (10–20) for alignment. Selection criteria may favor sequences of proteins that are characterized biochemically or are from the biosynthetic pathways of characterized natural products. Sequence identities with the target protein sequence are an important consideration; those in a range of 20%–80% should yield an alignment with enough information to identify approximate domain boundaries, conserved hotspots, insertion/deletion sites, and intrinsically disordered regions. A search of the structure database (PDB, http://www.rcsb.org/pdb/) (Berman et al., 2000) with the sequence of the target protein can aid in understanding the multiple sequence alignment and inform decisions about domain boundaries. The structures of any related proteins (>20% identity) can be powerful tools to generate hypotheses about function. We generally search the PDB with the BLAST server, using "Protein Data Bank proteins (PDB)" as the search set database.

The rapidly decreasing cost of synthetic DNA makes the entire sequence database accessible for protein production and has the added advantage of codon optimization for gene expression in a heterologous host. This can be extremely helpful when a target protein proves to be recalcitrant to purification in a stable form, or when other complications arise.

2.2 Translation Start Sites

The amino acid sequences for most PKS and NRPS proteins are based on annotation of genome or gene cluster sequences; thus they should be regarded as working hypotheses for the translation start and stop sites. Incorrect annotation of start sites can impede biochemical and structural studies of recombinant proteins. For example, the SpnF Diels–Alderase was initially characterized in a recombinant protein that could be produced in soluble form only with chaperones (Kim, Ruszczycky, Choi, Liu, & Liu, 2011). Researchers later identified a start codon upstream and in-frame with the annotated *spnF*, resulting in stable protein amenable to crystallization (Fage et al., 2015). Start codon identification can be challenging for polycistronic, overlapping genes, which are common in prokaryotes (Kozak, 1983; Osterman, Evfratov, Sergiev, & Dontsova, 2013). This can be assisted by locating the Shine–Dalgarno (SD) ribosome-binding site; SD sites 4–12 nucleotides upstream of the start codon have optimal function. For example, a polycistronic gene encodes the CurD hydroxymethylglutaryl synthase (HMGS) of PKS β-branching in the gene cluster for curacin A, and the *curD* start codon was originally annotated at a site 59 nucleotides upstream of the stop codon for the preceding gene *curC*. The correct start codon, four nucleotides downstream of the *curC* stop codon, was recognized by finding a consensus SD sequence immediately upstream of the *curC* stop (Maloney, Gerwick, Gerwick, Sherman, & Smith, 2016). A multiple sequence alignment of homologs can also aid in detection of incorrectly annotated translation start site, as for the marine haloalkane dehalogenase DmmA (Gehret et al., 2012).

It is also important to consider the start codon identity of the producing organism. For example, a GTG codon is frequently used for initiation in G+C-rich *Streptomyces* species (Villegas & Kropinski, 2008). Consideration of several potential GTG start codons for the LipPKS1 loading module from the lipomycin biosynthetic pathway led to the identification of alternative start sites, some of which were not predicted by gene-finding tools (Yuzawa et al., 2017). In some cases, use of a different start site altered the production level and catalytic activity of recombinant protein from *E. coli* and a heterologous *Streptomyces* host. Alternative start codons have also been identified in cyanobacterial species (Sazuka & Ohara, 1996). Additionally, identification of translation start sites can be complicated in the case of "leaderless" genes, which exist in many bacterial genomes (Nakagawa, Niimura, Miura, & Gojobori, 2010). Leaderless genes lack the SD sequence, but may encode a "TA-like signal" ~12 nucleotides upstream of the translation start codon (Zheng, Hu, She, & Zhu, 2011). Cyanobacterial genomes also include a substantial number of "atypical" leaderless genes that lack both an SD sequence and a TA-like signal, although some of these may have a dipyrimidine immediately upstream of the translation start site (Nakagawa et al., 2010; Zheng et al., 2011).

2.3 Domain Boundaries

If the goal is to produce the target protein in its full natural length, then the critical decision is a straightforward choice of biological source. However, if the target protein consists of one or more domains from a megasynthase, then excision sites must be chosen within interdomain linker peptides, typically of length 5–40 amino acids. The 3D structures of homologs can be extremely helpful in this task, as can the established boundaries of adjacent domains. The limits of the well-conserved region in a multiple sequence alignment generally define the folded core of a protein domain, but this may not include all regions essential to function or stability. It is often necessary to investigate multiple constructs to identify the most stable and/or active fragment, bearing in mind that "linker" regions longer than ~100 amino acids may be folded domains, and if so, should be included or excluded in their entirety.

Domain boundaries within many PKS megasynthases were determined originally through proteolysis and later rationalized when crystal structures revealed important conserved details that could be applied to other pathways. For example, the sequence for the first ketoreductase (KR) crystal structure from the erythromycin PKS (Fig. 1A) (Keatinge-Clay & Stroud, 2006) was based on fragments from limited proteolysis (Aparicio, Caffrey, Marsden, Staunton, & Leadlay, 1994). The structure revealed that the KR consists of tandem structural (KR_S) and catalytic (KR_C) domains and that an enoylreductase (ER) domain, when present, is inserted between KR_S and KR_C (Fig. 1B and C). A more subtle feature was a conserved β-strand that precedes KR_S in the amino acid sequence, but serves as a critical element of the NADPH-binding core of KR_C.

Much later, this structural detail was critical to understanding how *C*-methyltransferases (CMT) fit into PKS modules (Skiba et al., 2016). In the absence of a 3D structure for any close relative, it was impossible to ascertain the CMT domain boundaries beyond the easily recognized methyltransferase core motifs (Kozbial & Mushegian, 2005), so we determined the boundaries experimentally by constructing 12 plasmids for the CurJ CMT with N-termini spanning 30 residues from the end of the preceding dehydratase (DH) domain and three different C-termini. Nine constructs yielded soluble protein, and one yielded a crystal structure, which defined the true interdomain linkers (Skiba et al., 2016). The structure led to the unexpected finding that the KR N-terminal β-strand precedes the CMT (Fig. 1D), and the conclusion that this critical element is essential to producing soluble KR domains from homologous PKS modules.

A similar approach was used to identify soluble, active MT domains in "GNAT" loading modules, which contain a GCN5-related *N*-acetyltransferase (GNAT) domain (Skiba et al., 2017). In this context, the MT is located between an "adaptor region" (AR) (Gu et al., 2007) and a GNAT domain. We screened constructs encoding five different N- and C- termini for the production of a soluble MT domain from two biosynthetic pathways (Grindberg et al., 2011; Young et al., 2013). Only those constructs encoding an AR-MT-GNAT tridomain yielded a stable, active protein. The crystal structure of the AprA AR-MT-GNAT tridomain from apratoxin A revealed that the AR is a large MT lid and the GNAT is closely associated with the MT core, explaining the insolubility of shorter constructs (Skiba et al., 2017).

Crystallization behavior is not predictable and can be highly sensitive to the termini of excised domains. Thus, several termini within the putative interdomain linker regions should be tested for crystallization behavior.

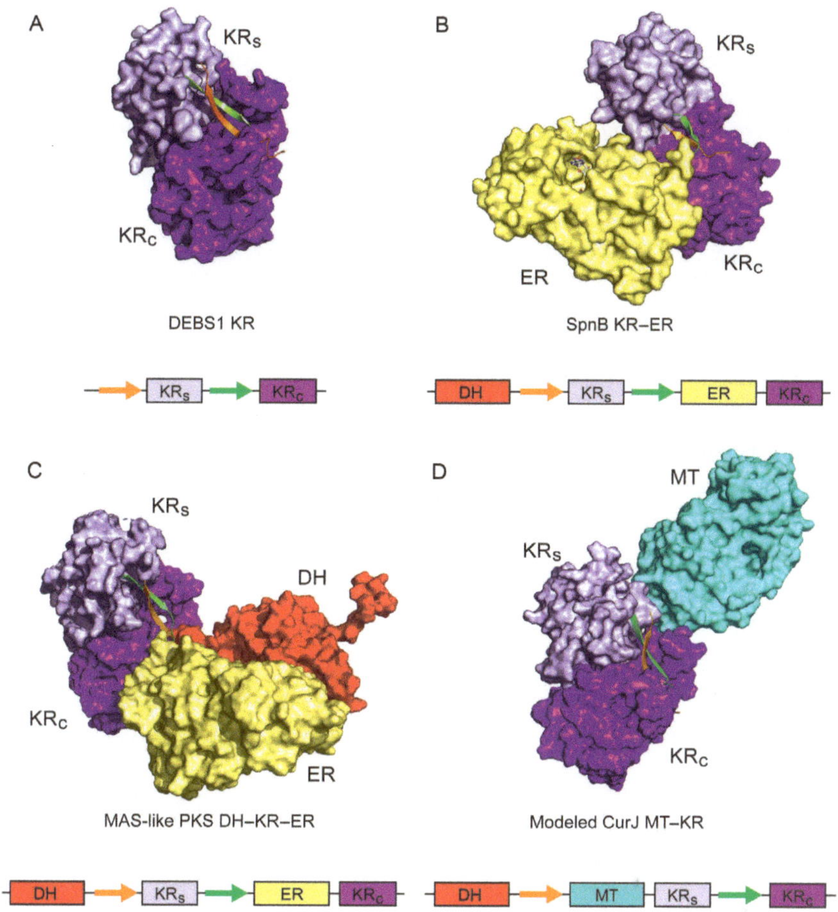

Fig. 1 Architectures of PKS modification domains: DH (*red*), MT (*cyan*), KR (*light* and *dark purple* KR_S *and* KR_C *subdomains*), ER (*yellow*). A two-stranded β-ribbon (*orange* and *green β-strands*) is an integral part of the KR regardless of the domain context. The presence of other modifying domains such as ER or MT can complicate identification of the full KR (*β-orange*—KR_S—*β-green*—KR_C). (A) DEBS1 KR (PDB 2FR0): A β-strand (*orange*) at the KR_S N-terminus pairs with a β-strand (*green*) between the KR_S and KR_C subdomains (Keatinge-Clay & Stroud, 2006). (B) SpnB KR–ER (PDB 3SLK): The ER domain is located after the *green β-strand* following KR_S but prior to the KR_C (Zheng, Gay, Demeler, White, & Keatinge-Clay, 2012). (C) MAS-like PKS DH–KR–ER (PDB 5BP4): The *orange β-strand* preceding KR_S follows the DH domain (Herbst, Jakob, Zahringer, & Maier, 2016). (D) Modeled CurJ MT–KR (MT PDB 5THY, KR homology model): CMT domains are inserted between the *orange β-strand* and KR_S in *cis*-AT PKS (Skiba et al., 2016).

Domain termini were screened (three N-termini and two C-termini) to optimize DHs from the curacin A pathway, resulting in crystal structures of all four DHs (Akey et al., 2010). Crystal quality was improved by inclusion of five C-terminal amino acids that are not highly conserved but form a short 3_{10} helix.

3. SELECTION OF EXPRESSION PLASMID AND BACTERIAL STRAIN

3.1 Cloning Method

The production of recombinant protein generally begins with the cloning of the gene encoding the protein(s) of interest. The first choice in a cloning strategy is the method of cloning itself. Cloning was originally done using restriction digestion to generate complimentary ends on the insert and vector. A large number of commercially available vectors have multiple-cloning sequences (MCSs) to accommodate many combinations of restriction sites to facilitate gene cloning for expression. The probability of a conflict arising from sequences in the gene matching restriction sites in the MCSs increases with larger inserts. Over the last 25 years, many ligation-independent cloning (LIC) systems have been developed to address this and other disadvantages of restriction cloning. LIC systems may be used universally; they function independently of gene sequence.

The Gateway system (Invitrogen) relies on enzymatic recombination to create the final expression clone. The large number of Gateway-adapted vectors makes the system flexible, and the high efficiency makes it easy for first-time users. However, the requirement for proprietary enzymes may be cost prohibitive when making more than a few clones. The Gateway system also adds up to eight amino acids to the target protein through translation of the recombination sites that are added to the gene fragment for cloning. The In-Fusion system (Clontech) relies on the use of complimentary sequences in the vector and insert and employs a proprietary enzyme to render these sequences single-stranded and then join them. Advantages include no addition of amino acids and applicability in any vector (Berrow et al., 2007). Our primary cloning method relies on vectors containing the LIC region developed at the Midwest Center for Structural Genomics (MCSG) (Stols et al., 2002). Complimentary sequences in the vector and insert are rendered single stranded through the exonuclease activity of T4 DNA polymerase, a less expensive method than the

In-Fusion or Gateway reagents. The MCSG-LIC and In-Fusion systems have 15-nucleotide single-stranded tails and thus comparable cloning efficiencies. However, the MCSG method adds an alanine codon at the beginning of the gene sequence to provide a guanine required for stopping the T4 exonuclease activity.

Numerous variations of LIC integrate a PCR step into the process, including polymerase incomplete primer extension (Klock, Koesema, Knuth, & Lesley, 2008) and overlap extension cloning (Unger, Jacobovitch, Dantes, Bernheim, & Peleg, 2010). The popular Gibson assembly system (Gibson et al., 2009) uses the exonuclease step to generate single-stranded overlaps in combination with PCR to amplify the gene and vector fragments. The reaction mix is transformed directly into bacteria where positive clones are identified by the power of antibiotic selection. Commercially available enzyme kits can be used at lower volumes than suggested to reduce the cost. We use Gibson assembly with the MCSG vectors to increase cloning efficiency while keeping costs low. Although the Gibson method was initially developed for the construction of very large pathway assemblies, it has the disadvantage of increased mutation rates with larger fragments due to the reliance on a second round of PCR for native templates. Use of a synthetic template reduces this problem if the cloning-complimentary sequences are synthesized into the fragment, which can then go into the Gibson reaction directly without prior PCR amplification. Gibson assembly is the method of choice for generating chimeric proteins and multienzyme polypeptides (Hagen et al., 2016).

3.2 Base Content

Actinobacteria, such as the pikromycin producer *Streptomyces venezuelae* (He et al., 2016), are rich sources of PKS and NRPS gene clusters, but have upward of 70% G+C content and present challenges for cloning of large multidomain constructs especially from cosmid or genomic DNA. The use of hot-start high-fidelity *Thermococcus kodakaraensis* (KOD) DNA polymerases, such as KOD Xtreme (Novagen), can greatly improve PCR efficiency and product purity, expediting the cloning process. In contrast, cyanobacteria, such as the marine *Moorea* species with 40%–50% G+C content (Leao et al., 2017), often contain long runs of A or T nucleotides, which are challenging for primer design due to the likelihood of false priming or primer dimerization. In such cases, the introduction of silent mutations into primers can assist in successful amplification.

3.3 Promoter Choice

The choice of promoter to drive transcription is a major consideration in vector selection. The bacteriophage T7 RNA polymerase–promoter system, the most widely used in bacterial production, is one of the strongest available for *E. coli* due to its polymerase binding and transcription initiation and is an ideal first-choice system. T7 polymerase travels over DNA at eightfold the rate of *E. coli* RNA polymerase (Iost, Guillerez, & Dreyfus, 1992), thus uncoupling transcription and translation. This uncoupling is not detrimental for soluble production of many proteins, but for others may lead to insolubility due to misfolding and aggregation. Slowing the rate of production can increase the solubility of these proteins. This is commonly done by lowering the growth temperature to slow the rates of transcription and translation (Section 6.2). However, if the protein of interest is primarily insoluble at lower temperature, a change to a weaker promoter may be needed. A number of commercially available vectors with native *E. coli* promoters can be used for tight coupling of transcription and translation. Coupling allows for the coordinated sensitivity of RNA polymerase and ribosomes to naturally programed pausing, which may be required for some difficult folds. Improved folding often increases solubility. The *trp-lac* fusion promoter *trc* (Amann, Brosius, & Ptashne, 1983) is widely used as a native promoter alternative to the T7 system.

The choice of promoter influences the choice of host strain. If a T7 promoter plasmid is chosen, the host strain must contain the gene for T7 RNA polymerase unless it is delivered exogenously. The most common host strain is BL21(DE3) (Studier, Rosenberg, Dunn, & Dubendorff, 1990), a lambda lysogen of *E. coli* strain BL21 carrying a copy of the gene for T7 RNA polymerase under the control of the inducible *lac*UV5 promoter. Several features have been incorporated to reduce the problems caused by leakage transcription from the *lac*UV5 promoter, for example, insertion of lac repressor sites downstream of the T7 promoter in the expression vector to block progression of T7 RNA polymerase prior to induction. When used in conjunction with a (DE3) host, the use of IPTG or lactose as an inducer both stimulates transcription of the T7 RNA polymerase gene and relieves transcriptional repression of the target gene. Another level of repression can be achieved with a secondary plasmid (pLysS or E) (Studier et al., 1990) carrying a constitutively expressed copy of the T7 gene for lysozyme, which binds T7 RNA polymerase and inhibits transcription. The lysozyme can assist in cell lysis prior to protein purification, but may also cause unexpected cell lysis. The presence of pLysS also limits the ability to employ other secondary plasmids to deliver accessory functions to enhance soluble protein production.

An alternative host strain for T7 vectors is BL21 AI (Invitrogen), which carries the gene for T7 RNA polymerase under the control of the promoter *araBAD*, derived from the arabinose metabolic pathway (Guzman, Belin, Carson, & Beckwith, 1995). The *araBAD* promoter is more tightly controlled than *lac*UV5 with a significantly lower level of leakage transcription but still capable of high-level expression from a T7 target plasmid when induced by the addition of arabinose (Chao, Chiang, & Hung, 2002). Since the target vectors rely on the *lac* repressor to reduce leakage from the T7 promoter, the AI host requires arabinose in addition to IPTG or lactose for full induction. We have found the BL21 AI strain to be preferable to pLys-containing strains when working with genes encoding proteins toxic to *E. coli*.

When shifting from a T7 promoter plasmid to a weaker promoter such as *trc*, it is prudent to change the host strain as well. If a T7-compatible strain is used with a lactose-inducible promoter plasmid, the T7 RNA polymerase will be induced in competition with the target gene. The T7 RNA polymerase message is stable and well translated to a stable protein that may accumulate to fairly high levels. It is best to remove this as a competitor, especially if the target protein is not made well and/or folds with difficulty. When using *trc* promoter plasmids, we have used BL21 as the production host. Since this is the parent strain of the (DE3) and AI variants used with the T7 system, expression results with these strains may be directly compared. Other hosts commonly used with native promoter containing plasmids include DH5α and JM109.

4. AFFINITY HANDLES AND FUSION PARTNERS

The most widely used affinity handles are some variation of the polyhistidine tag (His-tag), usually differing by the number and placement of histidine residues. The advantages of His-tags are the relatively small size of the tag and the low cost of metal-affinity chromatography. All the vectors we use incorporate a His_6-tag so they may be used in parallel with standardized conditions for expression and purification. Small-scale expression trials including an immobilized-metal-affinity chromatography (IMAC) step conveniently determine the target solubility and ease of purification.

Although most vectors have an N-terminal His-tag, this placement is not always compatible with the encoded protein. In some cases, a C-terminal tag may avoid tag interference with enzymatic activities or protein–protein interactions. For protein targets prone to truncation due to proteolysis or premature termination of translation, a C-terminal tag may allow for purification of full-length protein away from truncated fragments. A series of MCSG vectors are available for LIC construction with C-terminal His-tags (Eschenfeldt et al., 2010).

One of the most common issues in heterologous protein production in *E. coli* is insolubility of the product. The His-tag is an effective affinity handle, but it does not generally function as a solubility fusion partner. At the cloning step, insolubility problems can be addressed through the use of solubility fusion partners that are naturally soluble and produced to high level. Numerous proteins have been developed as fusion partners with varying levels of success (Costa, Almeida, Castro, & Domingues, 2014). It is unclear what properties of a protein improve the solubility of a passenger protein although high charge and an acidic p*I* appear to correlate with effectiveness (Fox, Routzahn, Bucher, & Waugh, 2003; Su, Zou, Feng, Zhou, & Cao, 2007). For example, the commonly used maltose-binding protein (MBP) is a stable, acidic protein that also has tight specific binding to amylose, making it an effective affinity handle. When combined with a His-tag, MBP allows for multistep purification to a high level of purity (Donnelly et al., 2006; Pryor & Leiting, 1997).

Problems associated with the use of MBP as a solubility partner arise from its ability to solubilize protein aggregates and from difficulties in purifying the passenger protein away from MBP after the fusion is cleaved. These problems led us to develop an additional fusion partner to complement MBP, the Mocr fusion (DelProposto, Majmudar, Smith, & Brown, 2009). Mocr is a highly charged, highly acidic ($pI = 3.8$) protein one-third the molecular weight of MBP. This compactness makes Mocr less likely to interfere with passenger protein activity while they are still fused. We and others have shown that the Mocr fusion has effective solubilizing activity (Banda, Tiwari, Darici, & Tse-Dinh, 2016; Buchholz et al., 2009; Gehret et al., 2011; Key, Dydio, Clark, & Hartwig, 2016), and have rarely observed difficulty in removing Mocr from the passenger protein.

To maximize the chance of producing soluble protein, we routinely test His-tag, His-MBP, and His-Mocr fusions in parallel. All N-terminal tags encoded in the MCSG vectors in our lab are cleavable with tobacco etch virus (TEV) protease (Raran-Kurussi, Cherry, Zhang, & Waugh, 2017), which has high specificity and only rarely detected cleavage at other sites in the protein. TEV protease leaves three amino acids at the N-terminus (Ser-Asn-Ala), which have no impact on activity or crystallization properties of most targets. Cleavage of the fusion is not always necessary, as it may not interfere with biochemical activity, particularly the smaller Mocr-tag and His-tag. For crystallography, it is usually desirable to remove the solubility fusions (MBP and Mocr), but the His-tag alone often does not interfere with crystallization (Akey et al., 2010; Maloney et al., 2016; Skiba et al., 2017).

Many other fusion proteins are available with varying levels of effectiveness or utility. Although glutathione-*S*-transferase (GST) is viewed as a solubility partner (Smith & Johnson, 1988; Waugh, 2005), we have found it to be ineffective presumably because it forms homodimers. If the passenger protein is prone to form oligomers or aggregate, then the GST dimer may nucleate extensive aggregation. We have found effective solubilizing activity in the SUMO fusion partner, which is available in LIC-compatible vectors (Weeks, Drinker, & Loll, 2007). More importantly, SUMO-specific protease cleaves the link between the SUMO C-terminus and the target protein N-terminus to yield a target protein with its native N-terminus (Malakhov et al., 2004). The SUMO fusion is an excellent option for proteins requiring a solubilizing fusion partner and a native N-terminus for biochemical activity or partner interactions (Whicher et al., 2013).

5. HELPER PLASMIDS

Production of protein is an integral function of the cell. It is a very complex series of closely coordinated processes requiring assorted cellular machinery and pools of resources. Occasionally the normal limitations of these resources are reached when using a cell to produce recombinant protein, requiring the provision of additional resources within the cell.

5.1 Codon Usage

Discussions of codon bias during heterologous expression usually revolve around the differences in codon usage between the host species, often *E. coli*, and the species of the gene being used for expression. For eukaryotic genes, there is a clear codon bias in *E. coli* that can reduce both the quantity and the quality of the protein product. When such a bias exists, it is widely accepted that supplementing the cells with additional copies of the genes for the tRNAs of low-use codons can have a significant impact on heterologous protein yield. This supplementation approach can also aid in expressing genes of bacterial origin. The tables of codon usage for a given species are constructed from averages across the genome, but this presentation obscures the heterogeneity in codon usage among genes within a genome (Sharp & Li, 1986). Although the patterns of usage in related organisms are similar, differences exist and are generally correlated with the evolutionary distance of the organisms (Sharp et al., 1988). When expressing genes from bacteria not closely related to *E. coli*, low-use codons can be a complication. Additionally, highly expressed genes tend to have high-use codons, whereas genes expressed at low levels display little or no bias for codon usage; they have a mix of high- and low-use codons (Sharp et al., 1988). The gene clusters for many interesting natural products are unlikely to be highly expressed in the native organism and are likely to contain low-use codons in greater number. When working in *E. coli* with a gene for a low-expressing protein from a distantly related bacterium, supplementing the cell with additional copies of the genes for the tRNAs of low-use codons can have a significant impact on production of the heterologous protein. This is accomplished by the introduction of a plasmid carrying anywhere from 2 to 12 tRNA genes for low-use codons.

Codon usage issues may also be addressed with synthetic DNA. The cost of DNA synthesis has dropped dramatically over the last few years, making synthetic gene fragments with sequences optimized for common heterologous hosts an attractive option. If fragments are made synthetically using predominantly high-use codons, then a production strain carrying a tRNA supplementation plasmid may not be needed. Fragments longer than several thousand nucleotides may be cost prohibitive at present, but the economics are changing rapidly. Constructs from synthetic fragments with optimized sequences are becoming the norm for ever larger recombinant proteins.

5.2 Chaperone Assistance

For many years, it was thought that folding was a purely thermodynamic process and newly synthesized proteins could fold to their native state without assistance. As the functions of heat-shock proteins and chaperones were elucidated, it became clear that not all proteins spontaneously fold to their native state, but require chaperone assistance. Chaperones mediate folding by binding unstable conformers of client proteins, followed by a series of binding and release steps associated with the hydrolysis of ATP. Possible outcomes include proper folding, assembly of homo- and hetero-oligomers, trafficking to and from subcellular compartments, or proteolytic degradation (Hartl, 1996).

This understanding of chaperone function led researchers studying heterologous protein production to postulate that insufficient chaperone function in the cell can lead to target protein misfolding and aggregation. As with low-use tRNAs, supplementation could solve this problem. Two major sets of chaperones play different but complimentary roles in protein folding in *E. coli*. The DnaK–DnaJ–GrpE set binds and stabilizes nascent polypeptides very early in translation, often when still bound to the ribosome. The GroEL–GroES set acts at a later stage when a protein has been released from the ribosome with extensive secondary structure but has not reached its final tertiary structure (Hartl, 1996). A series of plasmids were constructed carrying various combinations of the genes for these chaperone sets under several different promoters (Nishihara, Kanemori, Kitagawa, Yanagi, & Yura, 1998). In small-scale coexpression tests, problematic recombinant proteins can be tested with the various combinations of chaperones to find a combination that maximizes production of soluble protein. Takara has made a subset of these plasmids commercially available.

Chaperone assistance was essential to our recent structural characterization of the PKS HMGS enzyme (Maloney et al., 2016). Several HMGS proteins were incorrigibly insoluble when produced in *E. coli*, even with use of culture additives, fusion partners, and native promoters. Coexpression of *curD*, the curacin A HMGS gene, with the Takara plasmids encoding either the GroEL/ES (Takara pGro7) or DnaK/DnaJ (Takara pKJE7) chaperones (Nishihara et al., 1998) resulted in production of soluble protein. Furthermore, this strategy could be applied to solubilize CurD homologs from several other biosynthetic pathways. Thus, when chaperone coexpression is successful for stabilizing a protein of interest, one may consider applying the strategy to homologs from other biosynthetic pathways during structural studies. Similarly, coexpression of *malA* and a Takara plasmid encoding GroEL/GroES was essential to production of the flavin-dependent halogenase from the malbrancheamide pathway (Fraley et al., 2017).

5.3 Simultaneous tRNA Augmentation and Chaperone Assistance

Most of the plasmids encoding low-use-codon tRNAs have a pACYC backbone and rely on chloramphenicol selection (Rosetta from Novagen, Codon Plus from Agilent). The Takara chaperone sets are also cloned in pACYC backbones and some of these also rely on chloramphenicol selection. Plasmids with a common origin of replication are thought to be incompatible (Novick, 1987), such that a second plasmid introduced into a cell causes instability of a resident plasmid. This incompatibility may not be as stringent as once thought; two plasmids from the same compatibility group may be retained over short periods through antibiotic selection (Velappan, Sblattero, Chasteen, Pavlik, & Bradbury, 2007). For cell lines maintained for longer than several overnight growths, however, it is advisable to use plasmids from different compatibility groups.

Over the years we encountered several protein production projects having problems with both codon usage and proper folding, and thus we created a supplemental tRNA plasmid (pRARE2-CDF) (Lopanik et al., 2008) compatible with the Takara chaperone plasmids. To ensure long-term stability of both a tRNA plasmid and a chaperone plasmid in the cell, we needed to move one of the components to a different plasmid backbone. The five plasmids in the Takara chaperone set are pACYC-based, so we moved the tRNA genes into a plasmid from a different compatibility group (Novagen pCDF-1). The secondary plasmid pRare2 (Novagen), which carries tRNA genes corresponding to low-use codons in *E. coli* and has a copy number of 10, and the plasmid pCDF-1b (Novagen), which has a copy number of 20, were each digested with restriction endonucleases *Drd*I and *Xba*I. The larger fragment from pRare2 carrying the tRNA genes and the smaller fragment from pCDF-1b, which carries the origin of replication and the spectinomycin resistance marker gene, were each purified from a low-melting agarose gel. The two fragments were ligated to create pRARE2-CDF. This plasmid allows us to use chaperone supplementation at the same time as tRNA supplementation. Another advantage of the system is the copy number increase from 10 to 20 for the plasmid bearing tRNA genes. Increasing the level of the tRNAs has been beneficial and does not compete at all with target protein production, as might occur if the level of chaperones were increased.

5.4 Coexpression of NRPS A Domains With MtbH-Like Proteins

A strategy analogous to the coexpression of chaperones has been used to improve production of soluble NRPSs. MbtH-like proteins (MLPs) comprise a family of small proteins whose coding sequences are found in many NRPS gene clusters. Deletions of MLP-encoding genes provided genetic evidence that MLPs are essential for the production of nonribosomal peptide natural products (Wolpert, Gust, Kammerer, & Heide, 2007), while coexpression and purification identified them as integral components of NRPSs (Felnagle et al., 2010; Imker, Krahn, Clerc, Kaiser, & Walsh,

2010; Miller, Drake, Shi, Aldrich, & Gulick, 2016). It has since been shown that NRPSs from heterologous production found to be inactive or unobtainable due to insolubility could be rescued through coexpression of the cognate MLP-encoding gene (McMahon, Rush, & Thomas, 2012). Cross talk between MLPs and NRPSs from different pathways (Lautru, Oves-Costales, Pernodet, & Challis, 2007) suggests that a similar effect may be obtained with noncognate MLPs. Coexpression may be achieved by cloning a bicistronic insert of the NRPS and MLP genes into a standard expression vector such as pET (Imker et al., 2010) or through addition of a second compatible vector, such as pACYC Duet (McMahon et al., 2012) carrying the MLP gene. This coexpression strategy has become a common means to obtain soluble NRPS proteins where the use of fusions, low temperature, and variations in induction conditions has failed (Zolova & Garneau-Tsodikova, 2012).

6. GROWTH CONDITIONS

6.1 Media and pH

E. coli employed for heterologous protein production has been likened to a factory. As with factories, the level of production is dependent on the efficiency of operation. The best results are usually obtained by tailoring expression conditions to maintain bacterial cell growth within physiologically optimal conditions and by avoiding triggers of stress responses that divert energy and resources away from production of the target protein(s). The long-standing use of *E. coli* as a model organism provides a large body of literature to shape protocols and numerous well-characterized vectors from multiple complementation groups to use for expression and coexpression experiments.

Although most heterologous production of PKS and NRPS protein constituents is carried out in *E. coli*, alternative systems have been developed. A recently developed synthetic biology platform allows for the modular assembly of plasmid elements into cyanobacterial host-vector systems (Taton et al., 2014). A shuttle vector was derived from the broad-host-range plasmid RSF1010, making it compatible with a broad range of cyanobacterial hosts. This system was employed to express a cyanobacterial P450 protein that could not be obtained from *E. coli* (Agarwal et al., 2017). Alternatively, the pSyn_6 vector (Thermo Fisher Scientific) can be used to insert a gene of interest into a neutral site of the *Synechococcus elongatus* PCC 7942 genome via homologous recombination (Clerico, Ditty, & Golden, 2007). *Rhizobium leguminosarum* has also been used as a heterologous host for expression of cyanobacterial proteins (Freeman, Helf, Bhushan, Morinaka, & Piel, 2017).

E. coli grows over a pH range from 5.5 to 9.0 (Padan, Zilberstein, & Schuldiner, 1981). The metabolites from *E. coli* in culture can cause rapid and significant changes in the pH of an unbuffered growth medium. The carbon source affects the metabolites produced and their impact on pH (Losen, Frolich, Pohl, & Buchs, 2004). Incorporating a buffer into the medium will prevent alkaline or acid stress responses from growth-induced shifts above pH 9.0 or below pH 5.5. We compared the growth of *E. coli* BL21(DE3) cultures in 96-well deep-well blocks in three media: unbuffered lysogeny broth (LB) (Bertani, 1951), buffered ZYM 5052 (Studier, 2005), and buffered Terrific Broth (TB) with glycerol (Tartoff & Hobbs, 1987). The 1.0-mL cultures grew to strikingly different final cell densities and pHs (Table 1). Cells in unbuffered LB went into alkaline stress and stopped growing at a fairly early timepoint with the growth curve flattening at ~10 h. For ZYM 5052 medium, the growth curve flattened at ~22 h, and for TB the growth rate had begun to slow when the experiment was stopped at 25 h. Following this test and further optimization experiments, we settled on TB as our default medium for expression testing and larger scale production. TB is a fermentation-type medium that avoids or reduces cellular stress from triggers for the general stress response (Storz & Hengee-Aronis, 2000) including carbon starvation, fluctuations in pH, low osmolarity, and slow growth in late log phase approaching stationary phase. TB is a rich, complex medium containing relatively high amounts of yeast extract and tryptone, which ensures against carbon starvation and micronutrient deprivation at high density. TB is also a buffered medium, which reduces pH fluctuations. In our overnight growth studies, we found that cultures in TB stay in log phase for over 20 h. We usually harvest induction cultures at about 18 h.

The standard recipe for TB contains phosphate buffer, but if a protein requires other additives that are not compatible with phosphate, which

Table 1 Comparisons of Growth Parameters for Different Media

Medium	Time to Exit From Log Phase (h)	Culture Density (OD_{600})	pH
LB	10	4.0–4.7	9.5
ZYM 5052	22	12–16	6.7–7.3
TB	>25	18–20	7.3

Culture density and pH were measured when the cells began to emerge from log-phase growth. Data points represent ranges or averages of 96 cultures for each medium, grown in 96-well blocks.

may cause precipitation, then HEPES, MOPS or another buffer with a pK_a near pH 7.4 can be used. TB recipes all include glycerol, generally at 0.4% but up to 4.0% for large-scale production or fermentation. We also use the higher 4.0% glycerol in our TB formulation (Table 2) in small-scale expression tests, so the results will parallel large-scale production up to 20 L fermentation. Glycerol acts as an alternative carbon source, reduces pH fluctuations by altering the metabolite profile during growth (Losen et al., 2004), and may be a chemical chaperone promoting proper protein folding (Brown, Hong-Brown, Biwersi, Verkman, & Welch, 1996; Rariy & Klibanov, 1997). We compared several target protein constructs for soluble production in three media and found that insoluble protein was produced in LB and ZYM 5052, whereas soluble protein was produced in TB (Buchholz et al., 2009; Lopanik et al., 2008).

6.2 Temperature

The 4–10-fold higher rates of transcription and translation in *E. coli* compared to eukaryotic cells may cause folding problems for eukaryotic proteins in *E. coli* (Widmann & Christen, 2000), leading to aggregation and/or proteolysis. Similarly, rapid production in an *E. coli* overexpression system can also cause folding problems for more complex bacterial proteins, particularly those that are not normally expressed to very high levels. This is a common problem for the multidomain proteins of PKS and NRPS systems. Slowing the rate of production through reduced temperature is a widely used technique to limit aggregation and promote solubility during heterologous protein production (Schein & Noteborn, 1988).

E. coli grows over a temperature range from ~8°C (Broeze, Solomon, & Pope, 1978) to ~49°C (Herendeen, VanBogelen, & Neidhardt, 1979).

Table 2 Recipe and Protocol for Phosphate-Buffered TB Medium

	Nutrient Solution	**Buffer Solution**
Components	12 g tryptone	23.1 g KH_2PO_4
	24 g yeast extract	125.4 g K_2HPO_4
	40 mL of glycerol	
Volume	900 mL in deionized H_2O	1 L in deionized H_2O
Treatment	Autoclave separately	Autoclave separately
TB (mix immediately before use)	900 mL	100 mL

Maintenance cultures and starter cultures are usually grown at 37°C. Production cultures are generally started at 37°C and grown to a specific density (OD_{600}), and then the temperature may be downshifted. Literature reports have a very wide variety of temperatures, ranging from 12°C to 37°C, for induction of protein production. *E. coli* has a linear growth range from 21°C to 37°C (Herendeen et al., 1979) within which the steady-state levels of protein do not vary. Outside this range, growth rates decrease sharply, and the relationship of temperature and growth rate is nonlinear (Herendeen et al., 1979). We generally induce cultures after they have reached 20°C when downshifted from 37°C.

For small-scale (0.5 mL) cultures, we allow an hour to equilibrate at 20°C before induction. The 17°C downshift in temperature is large enough to trigger cold shock in these small cultures, which reach the target temperature very quickly. They emerge from cold shock after an hour (Jones, Cashel, Glaser, & Neidhardt, 1992; Thieringer, Jones, & Inouye, 1998). Cold shock is not a significant problem with larger cultures (500 mL to 1 L) since the culture itself does not change temperature as quickly as the incubator. Often, a 1-L culture may not have fully reached 20°C after 1 h from the time of the downshift. It is essential that the culture itself has reached 20°C before induction, so we monitor the temperature of larger cultures, beginning at 1 h postdownshift. We have observed that proteins prone to insolubility above the reduced growth temperature tend to go into the insoluble fraction if cultures are induced before reaching the reduced temperature. This suggests that the first protein made sets the pattern for protein made subsequently. Once the protein starts to misfold and aggregate at the higher temperature, subsequent protein made after the culture has reached the target temperature continues to misfold and aggregate even though it would fold and be soluble if it had not been induced until the culture was at the lower temperature.

Growth temperatures below 20°C may yield soluble protein, but as these temperatures are outside the linear range for *E. coli* growth, the results may not be reproducible. If the protein of interest is insoluble when produced at 20°C, we suggest shifting to an alternative weaker promoter (Section 6.4) to obtain a more extensive decrease in the rates of transcription and translation rather than shifting to even lower temperature. A weaker promoter combined with 20°C temperature will result in dramatic decreases in the speed and level of transcription as well as the rate of translation.

Temperature shifts are sometimes achieved by removing flasks from a shaker and placing in a cold room or on ice for rapid cooling, with the great disadvantage that the cultures will become hypoxic, a stress trigger detrimental to producing soluble protein. Rapid temperature reduction also may place the cultures at temperatures below the growth range for *E. coli* and present risk of some cell death. Too rapid cooling may trigger cold shock, and the cells will be unresponsive to inducer until they acclimate to the final induction temperature, which for larger cultures may be longer than the time to cool naturally. In the absence of refrigerated incubators, cultures can be shifted from a 37°C incubator to one at room temperature to maintain optimal aeration. The cells may require more than 1 h to reach the target temperature, but they should grow well and remain in log phase without significant stress.

6.3 Aeration

Aeration is another factor in obtaining optimal *E. coli* culture growth. To maximize the surface-to-volume ratio, the vessel should be wide and the volume of medium relatively shallow. For small-scale production to parallel large-scale production, shaking 50-mL conical tubes with 5 mL of medium produces well aerated cultures. When using 96-well blocks, a culture volume of 0.5 mL and a shake speed of at least 400 RPM will maximize aeration. For large-scale cultures in shake flasks, the surface of the medium should not go above the point of maximum width of the flask (one-quarter volume of an Erlenmeyer flask, i.e., 1 L of medium in a 4-L flask; 800–1000 mL for a 2.8-L Fernbach flask), and the shake speed should be at least 300 RPM. When using TB, optimal aeration is important to take full advantage of the achievable high cell densities.

6.4 Induction

Considerations for inducing heterologous gene expression include choice of inducer, inducer concentration, and time of induction. Lactose, the native inducer for the most widely used expression systems, is much less expensive than isopropyl β-D-1-thiogalactopyranoside (IPTG). Lactose can be present in the medium at the time the culture is inoculated, but induction is delayed until other carbon sources have been exhausted (Studier, 2005). Autoinduction requires less handling, an advantage for high-throughput expression trials. The timing of induction in autoinduction medium is dependent on a number of factors including plasmid composition, arrangement of promoter and repressor elements, culture aeration during growth, and autoinduction medium composition (Blommel, Becker, Duvnjak, & Fox, 2007).

Proper coordination of the timing of induction with the change in growth temperature is critical to produce soluble protein and requires significant effort to perfect. For this reason, we induce with added IPTG, which enters the cell through passive diffusion, resulting in nearly immediate induction after addition.

We measured protein production as a function of inducer concentration for various promoters at various temperatures (Table 3). At levels of inducer commonly reported in the literature (400 μ*M* up to 1 m*M*), protein production was lower than with the optimal concentrations. Titrating the inducer concentration is another way to reduce the rate of transcription, especially with a native *E. coli* promoter where limiting the inducer has a continuing impact on the native promoter. With a T7 system, limiting the inducer will slow the production of only the T7 RNA polymerase; once T7 RNA polymerase saturates the available plasmid, the system will be insensitive to the inducer limitation.

We studied the effect of culture optical density (OD_{600}) at the time of induction on protein yield. Literature reports suggest OD_{600} values for induction ranging from 0.6 to 1.0. We grew parallel cultures in TB, induced production beginning at an OD_{600} of 0.6 then at OD_{600} of 1.0 and at intervals of 1.0 OD_{600} unit up to an OD_{600} of 7.0, grew the cultures overnight, purified the target protein by metal affinity, and compared yields by SDS-PAGE analysis. We observed no significant difference in yield between the cultures induced at the lowest ODs and those induced at the highest. These experiments were run with MBP as the target protein. As MBP is made well, these results may not extrapolate completely to a more problematic protein, but they suggest that the culture density at induction is not a critical factor in the success of the production. We generally induce at an OD_{600} of roughly 1.0–2.0.

Based on the considerations summarized in Sections 6.1–6.4, we developed an optimized set of conditions for initial production of PKS and NRPS proteins (Table 4). Variations of these conditions are used as needed.

Table 3 Optimal Inducer Concentrations for Different Promoters and Temperatures

Promoter	Temperature (°C)	Optimal Inducer Concentration (μ*M*)
T7	20	150
	37	50
trc	20	300
	37	125

Table 4 Default Bacterial Production Conditions

Parameter	Condition
Cloning	pMCSG vectors (LIC variants of pET) T7 promoter with fusions (His-tag, MBP, Mocr)
Expression cell line	*E. coli* BL21AI/pRare2CDF
Medium	Terrific Broth w/ 4% glycerol
Aeration	Optimal surface-to-volume ratio and shaker speed for optimal aeration
Induction	200 μ*M* IPTG, 0.2% arabinose (300 μ*M* IPTG with *trc* promoter)
Temperature	Grow at 37°C, shift to 20°C gradually (and confirm), induce, continue growth at 20°C for ~18 h, spin, and freeze cell pellet at −80°C
Extras	Protein and/or chemical chaperones as needed

6.5 Additives

When produced alone in an overexpression system, some proteins do not fold quickly or completely enough for purification from the soluble fraction of a cell lysate. For such proteins, proper folding may be stimulated by coexpression with a natural partner protein. The production of soluble protein for the dehalogenase Bmp8 from *Marinomonas mediterranea* MMB1 required coexpression of the thioesterase Bmp1 from the same pathway (El Gamal, Agarwal, Rahman, & Moore, 2016). It was reasoned that, as Bmp1 is the substrate for Bmp8, the two proteins form a functional complex at some point and interaction with Bmp1 could help Bmp8 fold to its native stable form in an overexpression system. Similarly, production of soluble halogenase CylC from the cylindrocyclophane biosynthetic gene cluster required coexpression with the partner fatty acid-ACP ligase (CylA) and ACP (CylB) (Nakamura, Schultz, & Balskus, 2017). Untagged CylC was copurified with tagged CylB, indicating formation of a halogenase–ACP complex, which required prior formation of fatty acyl-CylB by CylA.

Other proteins may be stabilized in vivo by the addition of a ligand, either a cofactor or an agonist or antagonist of activity. A group of small organic compounds have been identified as "chemical chaperones" based on their reversal of protein mislocalization or aggregation (Brown et al., 1996; Perlmutter, 2002). They have since been adapted to improve recombinant protein folding (Prasad, Khadatare, & Roy, 2011). The group includes polyols such as glycerol (discussed earlier), triethylamines such as triethylamine-*N*-oxide (TMAO), some amino acid derivatives, and sugars such as trehalose and sorbitol. Chemical chaperones may be employed to enhance folding in vivo through addition to the growth medium (100 m*M*–1 *M*) or in vitro by addition to the lysis and purification buffers. When used at lysis, the chemical chaperones or cosolvents, such as kosmotropic salts or detergents, may rescue partially folded proteins, prevent aggregation and partitioning into the insoluble fraction, and assist in achieving a native fold (Bondos & Bicknell, 2003; Lindwall, Chau, Gardner, & Kohlstaedt, 2000). The presence of the osmolytes is thought to stabilize the properly folded regions of the protein and alter solvent interactions in unfolded regions to favor a more compact and folded structure. Kosmotropic salts are thought to work in a similar fashion by modulating the interactions of the protein surface with solvent to favor more compact and folded structures.

Proteins requiring cofactors or ligands for stability may benefit from the addition of the cofactor, a ligand, or a precursor at the time of induction. For example, cultures producing flavin-dependent enzymes can be stabilized by addition of 50 μ*M* riboflavin, a precursor to the cofactors FAD and FMN. Similarly, cultures coexpressing *cylC* (halogenase), *cylA* (fatty acyl-ACP ligase), and *cylB* (ACP) required supplementation with the CylA substrate decanoic acid in order to produce stable CylC as a complex with CylB, which was produced as the decanoyl-ACP (Nakamura et al., 2017).

Occasionally components of PKS or NRPS pathways are membrane associated, such as fatty acid desaturases (Zhu, Liu, & Zhang, 2015; Zhu, Su, Manickam, & Zhang, 2015) and ε-poly-L-lysine synthetase (Yamanaka, Maruyama, Takagi, & Hamano, 2008). Such proteins are assayed in the membrane fraction or extracted with detergents.

7. PRODUCTION OF ACPs IN DEFINED STATE

For functional and structural studies, it is usually necessary to produce acyl carrier proteins (ACPs) or peptidyl carrier proteins (PCPs) in a defined state such as the inactive apo-ACP or the native activated ACP with either a 4′-phosphopantetheine (Ppant) arm (holo-ACP) or a modified 4′-phosphopantetheine arm (crypto-ACP). The latter crypto-ACPs containing unnatural acyl substrates on the Ppant arm of the ACP have been widely used for functional studies to elucidate the substrate specificity of PKS and NRPS domains (Belshaw, Walsh, & Stachelhaus, 1999; El Gamal et al., 2016; Fridman et al., 2007; Sieber, Walsh, & Marahiel, 2003; Vitali, Zerbe, & Robinson, 2003). Modification of carrier proteins with reactive acyl groups incorporating epoxides, chloroacetates, imidazoles, vinylsulfonamides, and 3-alkynyl sulfones has facilitated cross-linking of ACPs with other domains for structural studies (Drake et al., 2016; Haslinger et al., 2014; Haushalter, Worthington, Hur, & Burkart, 2008; Hur et al., 2009; Ishikawa, Haushalter, Lee, Finzel, & Burkart, 2013; Kapur et al., 2008; Liu & Bruner, 2007; Liu, Zheng, & Bruner, 2011; Mitchell, Shi, Aldrich, & Gulick, 2012; Sundlov, Shi, Wilson, Aldrich, & Gulick, 2012; Worthington, Porter, & Burkart, 2010; Worthington, Rivera, Torpey, Alexander, & Burkart, 2006). Lastly, modification of ACPs with bioorthogonal handles, fluorophores, and biotin have been explored for a variety of chemical biology applications (Clarke, Mercer, La Clair, & Burkart, 2005; La Clair, Foley, Schegg, Regan, & Burkart, 2004; Meier, Mercer, Rivera, & Burkart, 2006).

ACPs are typically isolated as a mixture of apo-ACP and holo-ACP during expression in *E. coli*. To ensure complete conversion to the holo form, we coexpress with a phosphopantetheinyl transferase from a low-copy plasmid following the general protocol described in *Methods in Enzymology* (Sunbul, Zhang, & Yin, 2009) or in the Bap1 cell line, in which a phosphopantetheinyl transferase is integrated into the *E. coli* genome (Pfeifer & Khosla, 2001). The promiscuous Sfp from *Bacillus subtilis* has been the most widely employed because of its high catalytic activity and relaxed substrate specificity (Lambalot et al., 1996), but *Streptomyces verticillus* Svp has proven to be equally versatile (Sánchez, Du, Edwards, Toney, & Shen, 2001). Additionally, an Sfp mutant (R4-4 Sfp) will load the simpler and more synthetically accessible 3-dephospho-CoA analogues (Kittilä & Cryle, 2017; Zou & Yin, 2009).

Preparation of apo-ACPs required for synthesis of crypto-ACPs on the other hand can be more challenging due to the presence of the endogenous phosphopantetheinyl transferase EntD, which we have found in some cases results in partial phosphopantetheinylation of recombinantly expressed ACPs. The *E. coli* gene *entD* is part of an operon that encodes for enterobactin, an NRPS-derived siderophore (small-molecule iron chelator), whose biosynthetic genes are induced under iron-restricted conditions (Lambalot et al., 1996). The LB and TB media commonly used for recombinant protein production do not contain added iron and may become iron depleted as the medium is exhausted during growth, causing induction of *entD* and other genes required for enterobactin biosynthesis and transport.

Indeed, the production of enterobactin in LB medium has been observed, indicating expression of *entD* (Goetz et al., 2002).

We have found that apo-ACPs can be reproducibly obtained by addition of iron to the medium along with other trace metals to repress induction of *entD*. Our general protocol involves supplementation of 1 L of medium containing the appropriate antibiotic with 1 mL of a 1000 × trace metals mix to provide final concentrations of the following trace metals: iron (50 μ*M*), calcium (20 μ*M*), manganese (10 μ*M*), zinc (10 μ*M*), cobalt (2 μ*M*), copper (2 μ*M*), nickel (2 μ*M*), molybdenum (2 μ*M*), selenium (2 μ*M*), and boron (2 μ*M*) (see Table 5). This method is noted for its experimental simplicity and effectiveness with a wide variety of ACPs from PKS and NRPS systems (Maloney et al., 2016; Skiba et al., 2016, 2017).

A complimentary approach to generate apo-ACP employs an ACP hydrolase (AcpH) to cleave the Ppant arm of holo-ACPs (Kosa, Haushalter, Smith, & Burkart, 2012). AcpH orthologs from *E. coli*, *Pseudomonas aeruginosa*, *Shewanella oneidensis*, and *Pseudomonas fluorescens* have been investigated and shown to catalyze phosphopantetheine hydrolysis of holo-ACPs from PKS, NRPS, and fatty acid synthase (FAS) proteins. However,

Table 5 1000 × Trace Metals Mix (100 mL)[a]

Chemical[b]	**Volume**	**MW**	**[Final]**
0.1 *M* $FeCl_3 \cdot 6H_2O$[c]	50 mL	270.3	50 μ*M* iron
1 *M* $CaCl_2$	2 mL	111.0	20 μ*M* calcium
1 *M* $MnCl_2 \cdot 4H_2O$	1 mL	197.9	10 μ*M* manganese
1 *M* $ZnSO_4 \cdot 7H_2O$	1 mL	287.6	10 μ*M* zinc
0.2 *M* $CoCl_2 \cdot 6H_2O$	1 mL	237.9	2 μ*M* cobalt
0.1 *M* $CuCl_2 \cdot 2H_2O$	2 mL	170.5	2 μ*M* copper
0.2 *M* $NiCl_2 \cdot 6H_2O$	1 mL	237.7	2 μ*M* nickel
0.1 *M* $Na_2MoO_4 \cdot 2H_2O$	2 mL	242.0	2 μ*M* molybdenum
0.1 *M* $Na_2SeO_3 \cdot 5H_2O$[d]	2 mL	263.0	2 μ*M* selenium
0.1 *M* H_3BO_3	2 mL	61.8	2 μ*M* boron
H_2O	36 mL	18.0	—

[a]Add all solutions consecutively to 36 mL water and then filter sterilize; when preparing phosphate-buffered growth media, add trace metals prior to buffer to minimize precipitation.
[b]All stock solutions of metals except $FeCl_3 \cdot 6H_2O$ are autoclaved and then stored at room temperature.
[c]Dissolved in 0.1 *M* HCl.
[d]A brief precipitation appears upon addition of Na_2SeO_3, which rapidly redissolves.

these AcpHs are marred by variable activity toward PKS and NRPS proteins (Kosa, Pham, & Burkart, 2014). Despite these clear limitations, AcpH provides an interesting alternative method for phosphopantetheine removal.

8. PRODUCTION OF MULTIDOMAIN PROTEINS

Multidomain proteins pose special problems for overexpression systems because domains generally fold independently and may do so at very different rates, posing a risk of aggregation before all domains are folded. The risk of aggregation is increased if any of the domains are discontinuous in the amino acid sequence (Fig. 1), as more polypeptide must emerge from the ribosome before folding is complete.

A deficiency in production of recombinant cyanobacterial PKS and NRPS multidomain proteins in *E. coli* compared to those from other bacterial species (Dutta et al., 2014; Edwards, Matsui, Weiss, & Khosla, 2014; Lowell et al., 2017; Miller et al., 2016; Roberts, Staunton, & Leadlay, 1993; Tarry, Haque, Bui, & Schmeing, 2017; Whicher et al., 2014; Yuzawa, Eng, Katz, & Keasling, 2013) indicates that the cyanobacterial targets may require additional strategies. We encountered problems with several cyanobacterial multidomain proteins when using T7-based expression systems that had been used successfully for actinobacterial multidomain proteins. Slowing the rate of protein production by use of LIC-compatible expression plasmids containing the *trc* promoter (Smith, Brown, Harms, & Smith, 2015), and induction with 200 μ*M* IPTG at 20°C as described in Section 3.3, led to robust expression levels for several previously problematic proteins. Coupled rates of transcription and translation provided by the native *E. coli* RNA polymerase rather than the more rapid T7 RNA polymerase may have enhanced production of the large multidomain proteins in soluble form. Previously, the *trc* promoter was used successfully for the expression of the NRPS SrfA-C (Tanovic, Samel, Essen, & Marahiel, 2008).

Some recombinant multidomain proteins were insoluble even with use of the *trc* promoter and an MBP solubility tag. In some cases, much improved levels of soluble protein were achieved by supplementing the growth medium with 250–500 m*M* of the osmolyte TMAO, which promotes protein folding (Bandyopadhyay et al., 2012) and is found in marine organisms (Yancey, Clark, Hand, Bowlus, & Somero, 1982). Ni-affinity chromatography, followed by cleavage of the fused solubility tag and size exclusion chromatography, yielded multiple milligrams of >90% pure protein per liter of culture.

Additional strategies have been applied to successfully express constructs encoding multidomain PKS and NRPS proteins. In the case of HMWP1 and HMWP2 from the yersiniabactin synthetase, reducing the rate of protein production by lowering growth temperatures to 16–18° C and reducing inducer concentration (50–75 μM IPTG) yielded soluble protein (Keating, Miller, & Walsh, 2000; Suo, Tseng, & Walsh, 2001). Alternatively, coexpression with *Streptomyces* chaperones has also facilitated expression (Herbst et al., 2016) and improved activity (Betancor, Fernandez, Weissman, & Leadlay, 2008).

9. CELL LYSIS

To achieve high yields of active recombinant protein, it is critical to achieve full cell lysis under conditions appropriate for stabilizing the protein of interest. Incomplete lysis can be a problem with heterologous expression in *E. coli*, due to presence of a rigid peptidoglycan layer. (For heterologous expression in insect cells, incomplete lysis is rarely a problem.) Resuspension of cells in 4 mL of lysis buffer containing 0.05–0.2 mg/mL lysozyme per gram of cell paste assists in degrading the *E. coli* peptidoglycan layer. To achieve complete lysis, physical methods such as homogenization or sonication are often used. However, temperature should be monitored during lysis to prevent overheating of the lysate, which may damage the target protein. In the cell lysis process, we add Benzonase nuclease (MilliporeSigma) or DNase to degrade nucleic acid, reduce viscosity, and facilitate later purification steps. Cofactor-dependent proteins can be provided with the cofactor at the cell lysis step if this helps stabilize the protein (Fraley et al., 2017).

Cell lysis is a drastic change from the cellular environment to a rather simple buffer, which can stress a delicate target protein causing denaturation, proteolysis, and/or aggregation. Protease inhibitors added during cell lysis and purification can inhibit proteolysis. For problems with protein stability and aggregation, we have had success with a systematic screening process to identify conditions that promote protein stabilization.

The perception that protein aggregation during overproduction in *E. coli* is driven by formation of a "glob" of unfolded polypeptides has given way to the understanding that aggregates may be highly ordered and relatively close to their native state but have become associated through nonnative β-sheet structures (Ventura & Villaverde, 2006). The current view suggests that aggregated proteins could be dissociated and encouraged to complete folding to their native state. An indicator that a protein may have rescue potential is the partitioning of the majority of the protein to the insoluble fraction of the cell lysate with a small portion in the soluble fraction. Proteins with this profile may be candidates for a lysis buffer screen in which a variety of chemical cosolvents are added to the buffer to disaggregate the protein and promote greater solubility.

Based on literature reports (Bondos & Bicknell, 2003; Lindwall et al., 2000), we designed a 96-condition screen and tested it on proteins from several natural product biosynthetic pathways. While the more extreme conditions displayed negative effects, a number of conditions appeared to have positive effects in a variety of different combinations. We identified a small set of components for use in focused combinatorial screening rather than a large set of standard conditions (Table 6). The elements that have shown the most success in increasing soluble yield include pH, ionic salts, kosmotropic salts, organic cosolvents, and detergents.

Components of the lysis buffer can have a large impact on protein stability by affecting physical and chemical properties of the buffer, by interfering with interactions between solvent molecules and the target protein, or by direct interactions with the protein. As pH affects protein charge state, finding a suitable pH is of vital importance for many proteins. We use pH 7.0 and pH 8.5 in initial trials. Ionic salts interact with protein surface charges and reduce electrostatic interactions, while kosmotropic salts impact protein–solvent interactions in favor of the native state

Table 6 Parameters of a Lysis Buffer Screen

Parameter	Condition
pH	7.4, 8.5
Ionic salt	None; 50 m*M* or 100 m*M* NaCl
Kosmotropic salt	50 m*M* $(NH_4)_2SO_4$; 50 m*M* Na_2SO_4
Organic cosolvent	None; 2.5% or 5% DMSO; 5% or 10% glycerol; 2.5% or 5% ethanol
Detergent	None; 0.5% or 1.0% CHAPS; 0.5% or 1.0% Triton; 1.0% octylglucoside
Other	None; 100 m*M* urea; 50 m*M* arginine; 50 m*M* glutamate

(Bondos & Bicknell, 2003). Organic cosolvents weaken hydrophobic interactions and may disaggregate proteins or cause localized unfolding. Detergents disrupt hydrophobic interactions and provide interactions with the solvent. Arginine has been shown to have a stabilizing effect on several proteins and is proposed to reduce hydrophobicity by interacting with aromatic amino acids on the protein surface (Ito et al., 2011).

In order to screen lysis conditions for improved protein solubility, we developed a 14-condition small-scale lysis screen (Table 7; Fig. 2). To carry out the buffer screen, a His-tagged target protein is overexpressed in *E. coli* cell culture. The culture is split into 14 equal-volume aliquots (1 mL), and cells in each aliquot are resuspended in one of the test buffers and lysed by sonication. After separation, the soluble and insoluble fractions are assayed by SDS-PAGE. An optimal buffer is selected for use in large-scale purification. The lysis buffer screen not only benefits protein purification but also provides useful information on buffer choice for biochemical assays (Fiers et al., 2015; Fraley et al., 2017).

Table 7 Lysis Buffer Screen Formulation

Condition	Buffer	Additives
1	50 m*M* Tricine pH 8.5	500 m*M* NaCl, 10% glycerol
2	50 m*M* Tricine pH 8.5	50 m*M* $(NH_4)_2SO_4$, 10% glycerol
3	50 m*M* Tricine pH 8.5	50 m*M* NaCl, 5% DMSO
4	50 m*M* Tricine pH 8.5	50 m*M* NaCl, 5% ethanol
5	50 m*M* Tricine pH 8.5	50 m*M* NaCl, 1% Triton X-100
6	50 m*M* Tricine pH 8.5	50 m*M* NaCl, 100 m*M* urea
7	50 m*M* Tricine pH 8.5	50 m*M* NaCl, 50 m*M* Arg, 50 m*M* Glu
8	50 m*M* HEPES pH 7.0	50 m*M* $(NH_4)_2SO_4$, 10% glycerol
9	50 m*M* HEPES pH 7.0	50 m*M* NaCl, 5% DMSO
10	50 m*M* HEPES pH 7.0	50 m*M* NaCl, 5% ethanol
11	50 m*M* HEPES pH 7.0	50 m*M* NaCl, 1% Triton X-100
12	50 m*M* HEPES pH 7.0	50 m*M* NaCl, 100 m*M* urea
13	50 m*M* HEPES pH 7.0	50 m*M* NaCl, 50 m*M* Arg, 50 m*M* Glu
14	50 m*M* HEPES pH 7.0	500 m*M* NaCl, 10% glycerol

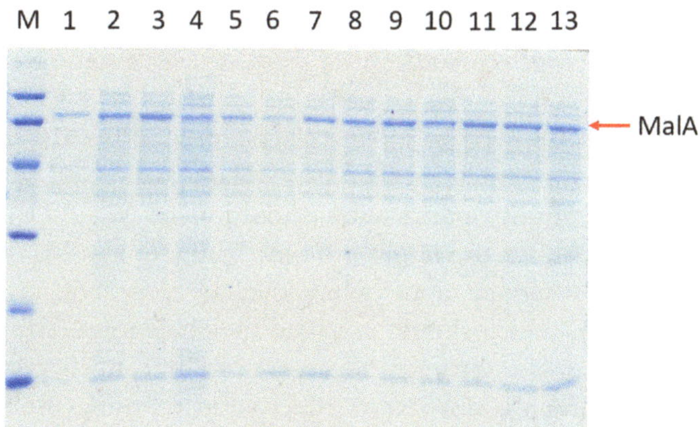

Fig. 2 Lysis buffer screen used in developing a purification protocol for the MalA halogenase (Fraley et al., 2017). Cells from equal-volume aliquots of an *E. coli* culture were lysed by sonication into 13 different lysis solutions (Table 6), and quickly purified by Ni-affinity chromatography, and analyzed by SDS-PAGE. Lanes in the gel are molecular weight markers (M: 97.4, 66.2, 45.0, 31.0, 21.5, 14.4 kDa) and lysis solutions 1–13. MalA molecular weight is 73 kDa.

10. PROTEIN PURIFICATION

Once appropriate conditions for cell lysis have been identified, protein purification can proceed. In general, the first purification step of a His-tagged protein involves affinity chromatography, such as metal-affinity chromatography for a His-tagged protein. The inclusion of a low level of imidazole (20 m*M*) in buffers used during loading and wash steps can inhibit nonspecific binding of *E. coli* proteins to the resin and improve the purity of the target protein during the first purification step. Use of an imidazole gradient or stepwise elution with increasing imidazole concentrations can improve protein purity over the simpler wash-and-elute protocol. If the target protein is sensitive to imidazole, the use of cobalt resin rather than nickel resin allows elution of the protein at lower imidazole concentrations. Despite the near universal application of metal-affinity resins to His-tagged proteins, complications can ensue. Care must be taken to avoid a protein buffer containing strong chelators that could strip metal from the resin. Similarly, weakly bound native metals can be stripped from metalloproteins by the high imidazole concentrations used to elute proteins from the resin. Disulfide-reducing agents may be essential to the stability of some proteins, but can be deleterious to metal-affinity resins; if a reducing agent is needed, the resin manufacturer's instructions should be consulted for compatibility.

In cases where the gene of interest is coexpressed with a chaperone-bearing plasmid, the target protein may retain significant amounts of chaperone during affinity purification. Addition of Mg-ATP can cause the chaperone to release the target protein (Fraley et al., 2017; Maloney et al., 2016). This can be done while the target protein is bound to metal-affinity resin, which should not bind the chaperone. Size exclusion chromatography can also separate the target protein from chaperones. The addition of ATP alone is not always effective in releasing the chaperone from the target protein. Another release method exploits the affinity of GroEL for denatured protein, which acts as a competitor of the target protein (Rohman & Harrison-Lavoie, 2000). The addition of urea- or heat-denatured protein with ATP was found to be necessary for the successful purification of the A-domain of MtbA away from bound chaperones (Somu et al., 2006).

A solubility fusion partner or N-terminal His-tag can be removed from the target by incubation with TEV protease in dialysis to remove imidazole. The tag-free target protein can be separated from uncleaved protein and the fusion partner by a second pass over metal-affinity resin. As TEV protease is a cysteine protease, a low level of reducing agent is added to the cleavage reaction to maintain protease activity, but not interfere with any additional metal-affinity steps. If the protein of interest is sensitive to the presence of strong reducing agents, a glutathione redox buffer can be substituted during the cleavage step to maintain protease activity (Saez, Cristofori-Armstrong, Anangi, & King, 2017).

A single affinity chromatography step can yield protein of sufficient purity for many applications. However, greater purity is usually required for crystallization. We generally include a second orthogonal step of size exclusion chromatography, which provides information about protein oligomer state and monodispersity and also eliminates soluble aggregates. Some biochemical assays are susceptible to interference from contaminating proteins such as ATPases (Drake, Duckworth, Neres, Aldrich, & Gulick, 2010) or acyltransferases (Florova, Kazanina, & Reynolds, 2002), particularly when carrier proteins are used as substrates or in single-turnover experiments. For such applications, most remaining trace contaminants can be eliminated by a third purification step, for example, ion-exchange chromatography.

11. REFINEMENTS FOR CRYSTALLIZATION

Even after conditions to produce stable recombinant protein have been established, an ensuing crystallization campaign may require additional refinements. For crystallization of enzymes from natural product biosynthetic pathways, we have found that successful strategies include screening multiple homologs of a target, altering the N- or C-terminus of the protein, cocrystallization with substrates/cofactors, and targeted mutagenesis.

The addition of cofactor or removal of a buffer component, such as glycerol, may make a protein more amenable to crystallization in sparse matrix screening. In cases where poorly diffracting crystals have been obtained, additional components identified through additive screening can greatly improve crystal quality (Bergfors, 2007). For example, initial CurJ CMT crystals were multicrystal spherulites. The addition of acetone, identified in a round of additive screening, produced poorly diffracting crystals (8 Å). Further addition of glutathione, identified in a second round of additive screening, yielded single crystals that diffracted to 2 Å (Skiba et al., 2016).

Mutagenesis may be considered in cases where crystallization trials fail, when poor crystal quality precludes structure solution, or where crystallization is too unreliable for a study requiring structures in several states of catalysis or substrate binding. For several projects in our lab, changes in the boundaries of an expression construct were crucial to producing protein amenable to crystallization. Exploration of polypeptide chain termini is strongly recommended for proteins excised from a PKS or NRPS megasynthase and intended for crystallization. As described in Section 2.3, exploration of domain boundaries was necessary to produce soluble CurJ CMT, however, among several soluble CMT variants, only one crystallized (Skiba et al., 2016). Similarly, the GNAT-MT2 didomain from the apratoxin loading module was produced in soluble form from a construct that, upon comparison to the CurJ CMT structure, included a large C-terminal linker region. Deletion of the linker led to successful crystallization. Finally, a change of even a single amino acid at the terminus may affect crystallization. Initial crystallization trials of the CurF enoylreductase failed, but removal of the C-terminal amino acid yielded crystals diffracting to 0.96 Å (Khare et al., 2015).

A strategy known as surface entropy reduction (SER) has been effective in cases where crystals are not obtained. SER, developed for proteins that cannot be induced to leave solution, aims to replace flexible amino acid side chains on the protein surface with alanine to reduce the entropic penalty of forming a crystal contact at that surface (Goldschmidt, Cooper, Derewenda, & Eisenberg, 2007). Predicted SER substitutions may be compared to a homolog of known structure (if available) to eliminate candidate substitutions that are near the active site, internal, or at an oligomer interface. SER is typically applied in cases where a target is exceedingly soluble, as exemplified by the CurM sulfotransferase (McCarthy et al., 2012). The CurD HMGS is a more unorthodox example of SER-assisted crystallization (Maloney et al., 2016). CurD is marginally stable and utterly insoluble unless it is produced with chaperones. Extensive screening of HMGS proteins from four biosynthetic pathways yielded no crystals. Despite the apparent instability of these proteins, an SER variant of CurD was stable, yielded diffraction-quality crystals in hours, and had activity comparable to the wild type. The SER substitutions are near a crystal contact but do not directly participate, contrary to the usual logic for SER. Occasionally, mutagenesis can be serendipitous. Initial crystallization conditions for AprA MT1-Ψ GNAT were obtained from a protein containing two PCR-induced substitutions (S274I and Q528P) (Skiba et al., 2017). The substitutions are near crystal contacts in two different crystal forms, and the wild-type protein did not crystallize. Similarly, MalA from the terrestrial fungus *Malbranchea aurantiaca* RRC1813 was recalcitrant to crystallization. However, the 99% identical homolog MalA′ from the marine fungus *Malbranchea graminicola* (086937A), which differs at two amino acid positions (L276P and R428P), yielded crystals (Fraley et al., 2017). The Pro428 occurs in a crystal contact that would be incompatible with an Arg side chain.

In some cases, mutagenesis to repair poor crystal lattice contacts can solve a problem of poor crystal reproducibility. This may be considered when the scope of a study extends beyond a single structure, requiring the capture of multiple states or substrate/cofactor-bound complex. This was the case for the MycF *O*-methyltransferase (Bernard et al., 2015), for which crystals grew in multiple space groups and typically had poor-quality diffraction. In the initial MycF structure, we discovered that repulsion of two glutamate side chains in a crystal contact was likely the culprit for this behavior. A double glutamine/alanine substitution of these residues reproducibly yielded crystals that diffracted beyond 2 Å, allowing three new substrate complexes to be captured. Targeted mutagenesis may also be used to introduce methionine substitutions in proteins where they are lacking and are needed for crystallographic phasing (Newmister et al., 2018).

12. SUMMARY

The production of recombinant proteins is essential for elucidating the biochemical steps of PKS and NRPS pathways, and for solving structures of the component enzymes. However, poor gene expression, protein solubility, or protein stability is often an impediment. We and others have identified several strategies to overcome such obstacles during the cloning, expression, and purification processes. After initial protein production, additional steps can improve efforts toward structural characterization. Here we have described many strategies that led to the successful production of protein targets from PKS and NRPS pathways, but this is not an exhaustive list. Challenging protein targets require careful screening of multiple constructs and conditions for expression and purification, along with an ounce of creativity.

ACKNOWLEDGMENTS

This work has been supported by the NIH (R01-DK42303, R01-GM076477, R01-CA108874 to J.L.S. and R01-AI070219 to C.C.A.) and by the Margaret J. Hunter Professorship to J.L.S. M.A.S. has been supported by predoctoral fellowships from an NIH Cellular Biotechnology Training Program (T32-GM008353) and the University of Michigan Rackham Graduate School.

REFERENCES

Agarwal, V., Blanton, J. M., Podell, S., Taton, A., Schorn, M. A., Busch, J., et al. (2017). Metagenomic discovery of polybrominated diphenyl ether biosynthesis by marine sponges. *Nature Chemical Biology*, *13*(5), 537–543.

Akey, D. L., Gehret, J. J., Khare, D., & Smith, J. L. (2012). Insights from the sea: Structural biology of marine polyketide synthases. *Natural Product Reports*, *29*(10), 1038–1049.

Akey, D. L., Razelun, J. R., Tehranisa, J., Sherman, D. H., Gerwick, W. H., & Smith, J. L. (2010). Crystal structures of dehydratase domains from the curacin polyketide biosynthetic pathway. *Structure*, *18*(1), 94–105.

Amann, E., Brosius, J., & Ptashne, M. (1983). Vectors bearing a hybrid trp-lac promoter useful for regulated expression of cloned genes in *Escherichia coli*. *Gene*, *25*(2–3), 167–178.

Aparicio, J. F., Caffrey, P., Marsden, A. F., Staunton, J., & Leadlay, P. F. (1994). Limited proteolysis and active-site studies of the first multienzyme component of the erythromycin-producing polyketide synthase. *The Journal of Biological Chemistry*, *269*(11), 8524–8528.

Banda, S., Tiwari, P. B., Darici, Y., & Tse-Dinh, Y. C. (2016). Investigating direct interaction between *Escherichia coli* topoisomerase I and RecA. *Gene*, *585*(1), 65–70.

Bandyopadhyay, A., Saxena, K., Kasturia, N., Dalal, V., Bhatt, N., Rajkumar, A., et al. (2012). Chemical chaperones assist intracellular folding to buffer mutational variations. *Nature Chemical Biology*, *8*(3), 238–245.

Belshaw, P. J., Walsh, C. T., & Stachelhaus, T. (1999). Aminoacyl-CoAs as probes of condensation domain selectivity in nonribosomal peptide synthesis. *Science*, *284*(5413), 486–489.

Bergfors, T. (2007). Screening and optimization methods for nonautomated crystallization laboratories. *Methods in Molecular Biology*, *363*, 131–151.

Berman, H. M., Westbrook, J., Feng, Z., Gilliland, G., Bhat, T. N., Weissig, H., et al. (2000). The protein data bank. *Nucleic Acids Research*, *28*(1), 235–242.

Bernard, S. M., Akey, D. L., Tripathi, A., Park, S. R., Konwerski, J. R., Anzai, Y., et al. (2015). Structural basis of substrate specificity and regiochemistry in the MycF/TylF family of sugar *O*-methyltransferases. *ACS Chemical Biology*, *10*(5), 1340–1351.

Berrow, N. S., Alderton, D., Sainsbury, S., Nettleship, J., Assenberg, R., Rahman, N., et al. (2007). A versatile ligation-independent cloning method suitable for high-throughput expression screening applications. *Nucleic Acids Research*, *35*(6), e45.

Bertani, G. (1951). Studies on lysogenesis. I. The mode of phage liberation by lysogenic *Escherichia coli*. *Journal of Bacteriology*, *62*(3), 293–300.

Betancor, L., Fernandez, M. J., Weissman, K. J., & Leadlay, P. F. (2008). Improved catalytic activity of a purified multienzyme from a modular polyketide synthase after coexpression with *Streptomyces* chaperonins in *Escherichia coli*. *Chembiochem*, *9*(18), 2962–2966.

Blommel, P. G., Becker, K. J., Duvnjak, P., & Fox, B. G. (2007). Enhanced bacterial protein expression during auto-induction obtained by alteration of lac repressor dosage and medium composition. *Biotechnology Progress*, *23*(3), 585–598.

Bondos, S. E., & Bicknell, A. (2003). Detection and prevention of protein aggregation before, during, and after purification. *Analytical Biochemistry*, *316*(2), 223–231.

Broeze, R. J., Solomon, C. J., & Pope, D. H. (1978). Effects of low temperature on *in vivo* and *in vitro* protein synthesis in *Escherichia coli* and *Pseudomonas fluorescens*. *Journal of Bacteriology*, *134*(3), 861–874.

Brown, C. R., Hong-Brown, L. Q., Biwersi, J., Verkman, A. S., & Welch, W. J. (1996). Chemical chaperones correct the mutant phenotype of the delta F508 cystic fibrosis transmembrane conductance regulator protein. *Cell Stress & Chaperones*, *1*(2), 117–125.

Buchholz, T. J., Geders, T. W., Bartley, F. E., 3rd, Reynolds, K. A., Smith, J. L., & Sherman, D. H. (2009). Structural basis for binding specificity between subclasses of modular polyketide synthase docking domains. *ACS Chemical Biology*, *4*(1), 41–52.

Chao, Y. P., Chiang, C. J., & Hung, W. B. (2002). Stringent regulation and high-level expression of heterologous genes in *Escherichia coli* using T7 system controllable by the araBAD promoter. *Biotechnology Progress*, *18*(2), 394–400.

Clarke, K. M., Mercer, A. C., La Clair, J. J., & Burkart, M. D. (2005). In vivo reporter labeling of proteins via metabolic delivery of coenzyme A analogues. *Journal of the American Chemical Society*, *127*(32), 11234–11235.

Clerico, E. M., Ditty, J. L., & Golden, S. S. (2007). Specialized techniques for site-directed mutagenesis in cyanobacteria. *Methods in Molecular Biology*, *362*, 155–171.

Costa, S., Almeida, A., Castro, A., & Domingues, L. (2014). Fusion tags for protein solubility, purification and immunogenicity in *Escherichia coli*: The novel Fh8 system. *Frontiers in Microbiology*, *5*, 63.

DelProposto, J., Majmudar, C. Y., Smith, J. L., & Brown, W. C. (2009). Mocr: A novel fusion tag for enhancing solubility that is compatible with structural biology applications. *Protein Expression and Purification*, *63*(1), 40–49.

Donnelly, M. I., Zhou, M., Millard, C. S., Clancy, S., Stols, L., Eschenfeldt, W. H., et al. (2006). An expression vector tailored for large-scale, high-throughput purification of recombinant proteins. *Protein Expression and Purification*, *47*(2), 446–454.

Drake, E. J., Duckworth, B. P., Neres, J., Aldrich, C. C., & Gulick, A. M. (2010). Biochemical and structural characterization of bisubstrate inhibitors of BasE, the self-standing nonribosomal peptide synthetase adenylate-forming enzyme of acinetobactin synthesis. *Biochemistry*, *49*(43), 9292–9305.

Drake, E. J., Miller, B. R., Shi, C., Tarrasch, J. T., Sundlov, J. A., Allen, C. L., et al. (2016). Structures of two distinct conformations of holo-non-ribosomal peptide synthetases. *Nature, 529*(7585), 235–238.

Dutta, S., Whicher, J. R., Hansen, D. A., Hale, W. A., Chemler, J. A., Congdon, G. R., et al. (2014). Structure of a modular polyketide synthase. *Nature, 510*(7506), 512–517.

Edwards, A. L., Matsui, T., Weiss, T. M., & Khosla, C. (2014). Architectures of whole-module and bimodular proteins from the 6-deoxyerythronolide B synthase. *Journal of Molecular Biology, 426*(11), 2229–2245.

El Gamal, A., Agarwal, V., Diethelm, S., Rahman, I., Schorn, M. A., Sneed, J. M., et al. (2016). Biosynthesis of coral settlement cue tetrabromopyrrole in marine bacteria by a uniquely adapted brominase-thioesterase enzyme pair. *Proceedings of the National Academy of Sciences of the United States of America, 113*(14), 3797–3802.

El Gamal, A., Agarwal, V., Rahman, I., & Moore, B. S. (2016). Enzymatic reductive dehalogenation controls the biosynthesis of marine bacterial pyrroles. *Journal of the American Chemical Society, 138*(40), 13167–13170.

Eschenfeldt, W. H., Maltseva, N., Stols, L., Donnelly, M. I., Gu, M., Nocek, B., et al. (2010). Cleavable C-terminal His-tag vectors for structure determination. *Journal of Structural and Functional Genomics, 11*(1), 31–39.

Fage, C. D., Isiorho, E. A., Liu, Y., Wagner, D. T., Liu, H. W., & Keatinge-Clay, A. T. (2015). The structure of SpnF, a standalone enzyme that catalyzes [4 + 2] cycloaddition. *Nature Chemical Biology, 11*(4), 256–258.

Felnagle, E. A., Barkei, J. J., Park, H., Podevels, A. M., McMahon, M. D., Drott, D. W., et al. (2010). MbtH-like proteins as integral components of bacterial nonribosomal peptide synthetases. *Biochemistry, 49*(41), 8815–8817.

Fiers, W. D., Dodge, G. J., Li, Y., Smith, J. L., Fecik, R. A., & Aldrich, C. C. (2015). Tylosin polyketide synthase module 3: Stereospecificity, stereoselectivity and steady-state kinetic analysis of beta-processing domains via diffusible, synthetic substrates. *Chemical Science, 6*(8), 5027–5033.

Florova, G., Kazanina, G., & Reynolds, K. A. (2002). Enzymes involved in fatty acid and polyketide biosynthesis in *Streptomyces glaucescens*: Role of FabH and FabD and their acyl carrier protein specificity. *Biochemistry, 41*(33), 10462–10471.

Fox, J. D., Routzahn, K. M., Bucher, M. H., & Waugh, D. S. (2003). Maltodextrin-binding proteins from diverse bacteria and archaea are potent solubility enhancers. *FEBS Letters, 537*(1–3), 53–57.

Fraley, A. E., Garcia-Borràs, M., Tripathi, A., Khare, D., Mercado-Marin, E. V., Tran, H., et al. (2017). Function and structure of MalA/MalA', iterative halogenases for late-stage C-H functionalization of indole alkaloids. *Journal of the American Chemical Society, 139*(34), 12060–12068.

Freeman, M. F., Helf, M. J., Bhushan, A., Morinaka, B. I., & Piel, J. (2017). Seven enzymes create extraordinary molecular complexity in an uncultivated bacterium. *Nature Chemistry, 9*(4), 387–395.

Fridman, M., Balibar, C. J., Lupoli, T., Kahne, D., Walsh, C. T., & Garneau-Tsodikova, S. (2007). Chemoenzymatic formation of novel aminocoumarin antibiotics by the enzymes CouN1 and CouN7. *Biochemistry, 46*(28), 8462–8471.

Gehret, J. J., Gu, L., Geders, T. W., Brown, W. C., Gerwick, L., Gerwick, W. H., et al. (2012). Structure and activity of DmmA, a marine haloalkane dehalogenase. *Protein Science, 21*(2), 239–248.

Gehret, J. J., Gu, L., Gerwick, W. H., Wipf, P., Sherman, D. H., & Smith, J. L. (2011). Terminal alkene formation by the thioesterase of curacin A biosynthesis: Structure of a decarboxylating thioesterase. *The Journal of Biological Chemistry, 286*(16), 14445–14454.

Gibson, D. G., Young, L., Chuang, R. Y., Venter, J. C., Hutchison, C. A., 3rd, & Smith, H. O. (2009). Enzymatic assembly of DNA molecules up to several hundred kilobases. *Nature Methods*, *6*(5), 343–345.

Goetz, D. H., Holmes, M. A., Borregaard, N., Bluhm, M. E., Raymond, K. N., & Strong, R. K. (2002). The neutrophil lipocalin NGAL is a bacteriostatic agent that interferes with siderophore-mediated iron acquisition. *Molecular Cell*, *10*(5), 1033–1043.

Goldschmidt, L., Cooper, D. R., Derewenda, Z. S., & Eisenberg, D. (2007). Toward rational protein crystallization: A web server for the design of crystallizable protein variants. *Protein Science*, *16*(8), 1569–1576.

Grindberg, R. V., Ishoey, T., Brinza, D., Esquenazi, E., Coates, R. C., Liu, W. T., et al. (2011). Single cell genome amplification accelerates identification of the apratoxin biosynthetic pathway from a complex microbial assemblage. *PLoS One*, *6*(4), e18565.

Gu, L., Geders, T. W., Wang, B., Gerwick, W. H., Hakansson, K., Smith, J. L., et al. (2007). GNAT-like strategy for polyketide chain initiation. *Science*, *318*(5852), 970–974.

Guzman, L. M., Belin, D., Carson, M. J., & Beckwith, J. (1995). Tight regulation, modulation, and high-level expression by vectors containing the arabinose PBAD promoter. *Journal of Bacteriology*, *177*(14), 4121–4130.

Hagen, A., Poust, S., Rond, T., Fortman, J. L., Katz, L., Petzold, C. J., et al. (2016). Engineering a polyketide synthase for *in vitro* production of adipic acid. *ACS Synthetic Biology*, *5*(1), 21–27.

Hartl, F. U. (1996). Molecular chaperones in cellular protein folding. *Nature*, *381*(6583), 571–579.

Haslinger, K., Brieke, C., Uhlmann, S., Sieverling, L., Süssmuth, R. D., & Cryle, M. J. (2014). The structure of a transient complex of a nonribosomal peptide synthetase and a cytochrome P450 monooxygenase. *Angewandte Chemie (International Ed. in English)*, *53*(32), 8518–8522.

Haushalter, R. W., Worthington, A. S., Hur, G. H., & Burkart, M. D. (2008). An orthogonal purification strategy for isolating crosslinked domains of modular synthases. *Bioorganic & Medicinal Chemistry Letters*, *18*(10), 3039–3042.

He, J., Sundararajan, A., Devitt, N. P., Schilkey, F. D., Ramaraj, T., & Melancon, C. E., 3rd. (2016). Complete genome sequence of *Streptomyces venezuelae* ATCC 15439, producer of the methymycin/pikromycin family of macrolide antibiotics, using PacBio technology. *Genome Announcements*, *4*(3).

Herbst, D. A., Jakob, R. P., Zahringer, F., & Maier, T. (2016). Mycocerosic acid synthase exemplifies the architecture of reducing polyketide synthases. *Nature*, *531*(7595), 533–537.

Herendeen, S. L., VanBogelen, R. A., & Neidhardt, F. C. (1979). Levels of major proteins of *Escherichia coli* during growth at different temperatures. *Journal of Bacteriology*, *139*(1), 185–194.

Hur, G. H., Meier, J. L., Baskin, J., Codelli, J. A., Bertozzi, C. R., Marahiel, M. A., et al. (2009). Crosslinking studies of protein-protein interactions in nonribosomal peptide biosynthesis. *Chemistry & Biology*, *16*(4), 372–381.

Imker, H. J., Krahn, D., Clerc, J., Kaiser, M., & Walsh, C. T. (2010). N-acylation during glidobactin biosynthesis by the tridomain nonribosomal peptide synthetase module GlbF. *Chemistry & Biology*, *17*(10), 1077–1083.

Iost, I., Guillerez, J., & Dreyfus, M. (1992). Bacteriophage T7 RNA polymerase travels far ahead of ribosomes in vivo. *Journal of Bacteriology*, *174*(2), 619–622.

Ishikawa, F., Haushalter, R. W., Lee, D. J., Finzel, K., & Burkart, M. D. (2013). Sulfonyl 3-alkynyl pantetheinamides as mechanism-based cross-linkers of acyl carrier protein dehydratase. *Journal of the American Chemical Society*, *135*(24), 8846–8849.

Ito, L., Shiraki, K., Matsuura, T., Okumura, M., Hasegawa, K., Baba, S., et al. (2011). High-resolution X-ray analysis reveals binding of arginine to aromatic residues of lysozyme surface: Implication of suppression of protein aggregation by arginine. *Protein Engineering, Design & Selection*, *24*(3), 269–274.

Johnson, M., Zaretskaya, I., Raytselis, Y., Merezhuk, Y., McGinnis, S., & Madden, T. L. (2008). NCBI BLAST: A better web interface. *Nucleic Acids Research*, *36*(Web server issue), W5–9.

Jones, P. G., Cashel, M., Glaser, G., & Neidhardt, F. C. (1992). Function of a relaxed-like state following temperature downshifts in *Escherichia coli*. *Journal of Bacteriology*, *174*(12), 3903–3914.

Kapur, S., Worthington, A., Tang, Y., Cane, D. E., Burkart, M. D., & Khosla, C. (2008). Mechanism based protein crosslinking of domains from the 6-deoxyerythronolide B synthase. *Bioorganic & Medicinal Chemistry Letters*, *18*(10), 3034–3038.

Keating, T. A., Miller, D. A., & Walsh, C. T. (2000). Expression, purification, and characterization of HMWP2, a 229 kDa, six domain protein subunit of Yersiniabactin synthetase. *Biochemistry*, *39*(16), 4729–4739.

Keatinge-Clay, A. T. (2012). The structures of type I polyketide synthases. *Natural Product Reports*, *29*(10), 1050–1073.

Keatinge-Clay, A. T. (2017). The uncommon enzymology of cis-acyltransferase assembly lines. *Chemical Reviews*, *117*(8), 5334–5366.

Keatinge-Clay, A. T., & Stroud, R. M. (2006). The structure of a ketoreductase determines the organization of the beta-carbon processing enzymes of modular polyketide synthases. *Structure*, *14*(4), 737–748.

Key, H. M., Dydio, P., Clark, D. S., & Hartwig, J. F. (2016). Abiological catalysis by artificial haem proteins containing noble metals in place of iron. *Nature*, *534*(7608), 534–537.

Khare, D., Hale, W. A., Tripathi, A., Gu, L., Sherman, D. H., Gerwick, W. H., et al. (2015). Structural basis for cyclopropanation by a unique enoyl-acyl carrier protein reductase. *Structure*, *23*(12), 2213–2223.

Kim, H. J., Ruszczycky, M. W., Choi, S. H., Liu, Y. N., & Liu, H. W. (2011). Enzyme-catalysed [4+2] cycloaddition is a key step in the biosynthesis of spinosyn A. *Nature*, *473*(7345), 109–112.

Kittilä, T., & Cryle, M. J. (2017). An enhanced chemo-enzymatic method for loading substrates onto carrier protein domains. *Biochemistry and Cell Biology*. https://doi.org/10.1139/bcb-2017-0275.

Klock, H. E., Koesema, E. J., Knuth, M. W., & Lesley, S. A. (2008). Combining the polymerase incomplete primer extension method for cloning and mutagenesis with microscreening to accelerate structural genomics efforts. *Proteins*, *71*(2), 982–994.

Kosa, N. M., Haushalter, R. W., Smith, A. R., & Burkart, M. D. (2012). Reversible labeling of native and fusion-protein motifs. *Nature Methods*, *9*(10), 981–984.

Kosa, N. M., Pham, K. M., & Burkart, M. D. (2014). Chemoenzymatic exchange of phosphopantetheine on protein and peptide. *Chemical Science*, *5*(3), 1179–1186.

Kozak, M. (1983). Comparison of initiation of protein synthesis in procaryotes, eucaryotes, and organelles. *Microbiological Reviews*, *47*(1), 1–45.

Kozbial, P. Z., & Mushegian, A. R. (2005). Natural history of S-adenosylmethionine-binding proteins. *BMC Structural Biology*, *5*(1), 19.

La Clair, J. J., Foley, T. L., Schegg, T. R., Regan, C. M., & Burkart, M. D. (2004). Manipulation of carrier proteins in antibiotic biosynthesis. *Chemistry & Biology*, *11*(2), 195–201.

Lambalot, R. H., Gehring, A. M., Flugel, R. S., Zuber, P., LaCelle, M., Marahiel, M. A., et al. (1996). A new enzyme superfamily—The phosphopantetheinyl transferases. *Chemistry & Biology*, *3*(11), 923–936.

Lautru, S., Oves-Costales, D., Pernodet, J. L., & Challis, G. L. (2007). MbtH-like protein-mediated cross-talk between non-ribosomal peptide antibiotic and siderophore biosynthetic pathways in *Streptomyces coelicolor* M145. *Microbiology*, *153*(Pt. 5), 1405–1412.

Leao, T., Castelao, G., Korobeynikov, A., Monroe, E. A., Podell, S., Glukhov, E., et al. (2017). Comparative genomics uncovers the prolific and distinctive metabolic potential of the cyanobacterial genus *Moorea*. *Proceedings of the National Academy of Sciences of the United States of America*, *114*(12), 3198–3203.

Lindwall, G., Chau, M., Gardner, S. R., & Kohlstaedt, L. A. (2000). A sparse matrix approach to the solubilization of overexpressed proteins. *Protein Engineering*, *13*(1), 67–71.

Liu, Y., & Bruner, S. D. (2007). Rational manipulation of carrier-domain geometry in nonribosomal peptide synthetases. *Chembiochem*, *8*(6), 617–621.

Liu, Y., Zheng, T., & Bruner, S. D. (2011). Structural basis for phosphopantetheinyl carrier domain interactions in the terminal module of nonribosomal peptide synthetases. *Chemistry & Biology*, *18*(11), 1482–1488.

Lopanik, N. B., Shields, J. A., Buchholz, T. J., Rath, C. M., Hothersall, J., Haygood, M. G., et al. (2008). *In vivo* and *in vitro* trans-acylation by BryP, the putative bryostatin pathway acyltransferase derived from an uncultured marine symbiont. *Chemistry & Biology*, *15*(11), 1175–1186.

Losen, M., Frolich, B., Pohl, M., & Buchs, J. (2004). Effect of oxygen limitation and medium composition on *Escherichia coli* fermentation in shake-flask cultures. *Biotechnology Progress*, *20*(4), 1062–1068.

Lowell, A. N., DeMars, M. D., 2nd, Slocum, S. T., Yu, F., Anand, K., Chemler, J. A., et al. (2017). Chemoenzymatic total synthesis and structural diversification of tylactone-based macrolide antibiotics through late-stage polyketide assembly, tailoring, and C-H functionalization. *Journal of the American Chemical Society*, *139*(23), 7913–7920.

Malakhov, M. P., Mattern, M. R., Malakhova, O. A., Drinker, M., Weeks, S. D., & Butt, T. R. (2004). SUMO fusions and SUMO-specific protease for efficient expression and purification of proteins. *Journal of Structural and Functional Genomics*, *5*(1–2), 75–86.

Maloney, F. P., Gerwick, L., Gerwick, W. H., Sherman, D. H., & Smith, J. L. (2016). Anatomy of the β-branching enzyme of polyketide biosynthesis and its interaction with an acyl-ACP substrate. *Proceedings of the National Academy of Sciences of the United States of America*, *113*(37), 10316–10321.

McCarthy, J. G., Eisman, E. B., Kulkarni, S., Gerwick, L., Gerwick, W. H., Wipf, P., et al. (2012). Structural basis of functional group activation by sulfotransferases in complex metabolic pathways. *ACS Chemical Biology*, 7(12), 1994–2003.

McMahon, M. D., Rush, J. S., & Thomas, M. G. (2012). Analyses of MbtB, MbtE, and MbtF suggest revisions to the mycobactin biosynthesis pathway in *Mycobacterium tuberculosis*. *Journal of Bacteriology*, *194*(11), 2809–2818.

Meier, J. L., Mercer, A. C., Rivera, H., Jr., & Burkart, M. D. (2006). Synthesis and evaluation of bioorthogonal pantetheine analogues for in vivo protein modification. *Journal of the American Chemical Society*, *128*(37), 12174–12184.

Miller, B. R., Drake, E. J., Shi, C., Aldrich, C. C., & Gulick, A. M. (2016). Structures of a nonribosomal peptide synthetase module bound to MbtH-like proteins support a highly dynamic domain architecture. *The Journal of Biological Chemistry*, *291*(43), 22559–22571.

Miller, B. R., & Gulick, A. M. (2016). Structural biology of nonribosomal peptide synthetases. *Methods in Molecular Biology*, *1401*, 3–29.

Mitchell, C. A., Shi, C., Aldrich, C. C., & Gulick, A. M. (2012). Structure of PA1221, a nonribosomal peptide synthetase containing adenylation and peptidyl carrier protein domains. *Biochemistry*, *51*(15), 3252–3263.

Nakagawa, S., Niimura, Y., Miura, K., & Gojobori, T. (2010). Dynamic evolution of translation initiation mechanisms in prokaryotes. *Proceedings of the National Academy of Sciences of the United States of America*, *107*(14), 6382–6387.

Nakamura, H., Schultz, E. E., & Balskus, E. P. (2017). A new strategy for aromatic ring alkylation in cylindrocyclophane biosynthesis. *Nature Chemical Biology*, *13*(8), 916–921.

Newmister, S. A., Li, S., Garcia-Borràs, M., Sanders, J. N., Yang, S., Lowell, A. N., et al. (2018). Structural basis of the Cope rearrangement and cyclization in hapalindole biogenesis. *Nature Chemical Biology*. (in press).

Nishihara, K., Kanemori, M., Kitagawa, M., Yanagi, H., & Yura, T. (1998). Chaperone coexpression plasmids: Differential and synergistic roles of DnaK-DnaJ-GrpE and GroEL-GroES in assisting folding of an allergen of Japanese cedar pollen, Cryj2, in *Escherichia coli*. *Applied and Environmental Microbiology*, *64*(5), 1694–1699.

Novick, R. P. (1987). Plasmid incompatibility. *Microbiological Reviews*, *51*(4), 381–395.

Osterman, I. A., Evfratov, S. A., Sergiev, P. V., & Dontsova, O. A. (2013). Comparison of mRNA features affecting translation initiation and reinitiation. *Nucleic Acids Research*, *41*(1), 474–486.

Padan, E., Zilberstein, D., & Schuldiner, S. (1981). pH homeostasis in bacteria. *Biochimica et Biophysica Acta*, *650*(2–3), 151–166.

Payne, J. T., Millar, J. J., Jackson, C. R., & Ochs, C. A. (2017). Patterns of variation in diversity of the Mississippi river microbiome over 1,300 kilometers. *PLoS One*, *12*(3), e0174890.

Perlmutter, D. H. (2002). Chemical chaperones: A pharmacological strategy for disorders of protein folding and trafficking. *Pediatric Research*, *52*(6), 832–836.

Pfeifer, B. A., & Khosla, C. (2001). Biosynthesis of polyketides in heterologous hosts. *Microbiology and Molecular Biology Reviews*, *65*(1), 106–118.

Prasad, S., Khadatare, P. B., & Roy, I. (2011). Effect of chemical chaperones in improving the solubility of recombinant proteins in *Escherichia coli*. *Applied and Environmental Microbiology*, 77(13), 4603–4609.

Pryor, K. D., & Leiting, B. (1997). High-level expression of soluble protein in *Escherichia coli* using a His6-tag and maltose-binding-protein double-affinity fusion system. *Protein Expression and Purification*, *10*(3), 309–319.

Raran-Kurussi, S., Cherry, S., Zhang, D., & Waugh, D. S. (2017). Removal of affinity tags with TEV protease. *Methods in Molecular Biology*, *1586*, 221–230.

Rariy, R. V., & Klibanov, A. M. (1997). Correct protein folding in glycerol. *Proceedings of the National Academy of Sciences of the United States of America*, *94*(25), 13520–13523.

Roberts, G. A., Staunton, J., & Leadlay, P. F. (1993). Heterologous expression in *Escherichia coli* of an intact multienzyme component of the erythromycin-producing polyketide synthase. *European Journal of Biochemistry*, *214*(1), 305–311.

Rohman, M., & Harrison-Lavoie, K. J. (2000). Separation of copurifying GroEL from glutathione-S-transferase fusion proteins. *Protein Expression and Purification*, *20*(1), 45–47.

Saez, N. J., Cristofori-Armstrong, B., Anangi, R., & King, G. F. (2017). A strategy for production of correctly folded disulfide-rich peptides in the periplasm of *E. coli*. *Methods in Molecular Biology*, *1586*, 155–180.

Sánchez, C., Du, L., Edwards, D. J., Toney, M. D., & Shen, B. (2001). Cloning and characterization of a phosphopantetheinyl transferase from *Streptomyces verticillus* ATCC15003, the producer of the hybrid peptide-polyketide antitumor drug bleomycin. *Chemistry & Biology*, *8*(7), 725–738.

Sazuka, T., & Ohara, O. (1996). Sequence features surrounding the translation initiation sites assigned on the genome sequence of *Synechocystis* sp. strain PCC6803 by amino-terminal protein sequencing. *DNA Research*, *3*(4), 225–232.

Schein, C. H., & Noteborn, M. H. M. (1988). Formation of soluble recombinant proteins in *Escherichia coli* is favored by lower growth temperatures. *Nature Biotechnology*, *6*, 291–294.

Sharp, P. M., Cowe, E., Higgins, D. G., Shields, D. C., Wolfe, K. H., & Wright, F. (1988). Codon usage patterns in *Escherichia coli*, *Bacillus subtilis*, *Saccharomyces cerevisiae*, *Schizosaccharomyces pombe*, *Drosophila melanogaster* and *Homo sapiens*; a review of the considerable within-species diversity. *Nucleic Acids Research*, *16*(17), 8207–8211.

Sharp, P. M., & Li, W. H. (1986). Codon usage in regulatory genes in *Escherichia coli* does not reflect selection for 'rare' codons. *Nucleic Acids Research*, *14*(19), 7737–7749.

Sieber, S. A., Walsh, C. T., & Marahiel, M. A. (2003). Loading peptidyl-coenzyme A onto peptidyl carrier proteins: A novel approach in characterizing macrocyclization by thioesterase domains. *Journal of the American Chemical Society*, *125*(36), 10862–10866.

Skiba, M. A., Sikkema, A. P., Fiers, W. D., Gerwick, W. H., Sherman, D. H., Aldrich, C. C., et al. (2016). Domain organization and active site architecture of a polyketide synthase *C*-methyltransferase. *ACS Chemical Biology*, *11*(12), 3319–3327.

Skiba, M. A., Sikkema, A. P., Moss, N. A., Tran, C. L., Sturgis, R. M., Gerwick, L., et al. (2017). A mononuclear iron-dependent methyltransferase catalyzes initial steps in assembly of the apratoxin A polyketide starter unit. *ACS Chemical Biology*, *12*(12), 3039–3048.

Smith, A. M., Brown, W. C., Harms, E., & Smith, J. L. (2015). Crystal structures capture three states in the catalytic cycle of a pyridoxal phosphate (PLP) synthase. *The Journal of Biological Chemistry*, *290*(9), 5226–5239.

Smith, D. B., & Johnson, K. S. (1988). Single-step purification of polypeptides expressed in *Escherichia coli* as fusions with glutathione *S*-transferase. *Gene*, *67*(1), 31–40.

Somu, R. V., Wilson, D. J., Bennett, E. M., Boshoff, H. I., Celia, L., Beck, B. J., et al. (2006). Antitubercular nucleosides that inhibit siderophore biosynthesis: SAR of the glycosyl domain. *Journal of Medicinal Chemistry*, *49*(26), 7623–7635.

Stols, L., Gu, M., Dieckman, L., Raffen, R., Collart, F. R., & Donnelly, M. I. (2002). A new vector for high-throughput, ligation-independent cloning encoding a tobacco etch virus protease cleavage site. *Protein Expression and Purification*, *25*(1), 8–15.

Storz, G., & Hengee-Aronis, R. (2000). *Bacterial stress responses*. Washington, DC: ASM Press.

Studier, F. W. (2005). Protein production by auto-induction in high density shaking cultures. *Protein Expression and Purification*, *41*(1), 207–234.

Studier, F. W., Rosenberg, A. H., Dunn, J. J., & Dubendorff, J. W. (1990). Use of T7 RNA polymerase to direct expression of cloned genes. *Methods in Enzymology*, *185*, 60–89.

Su, Y., Zou, Z., Feng, S., Zhou, P., & Cao, L. (2007). The acidity of protein fusion partners predominantly determines the efficacy to improve the solubility of the target proteins expressed in *Escherichia coli*. *Journal of Biotechnology*, *129*(3), 373–382.

Sunbul, M., Zhang, K., & Yin, J. (2009). Chapter 10 using phosphopantetheinyl transferases for enzyme posttranslational activation, site specific protein labeling and identification of natural product biosynthetic gene clusters from bacterial genomes. *Methods in Enzymology*, *458*, 255–275.

Sundlov, J. A., Shi, C., Wilson, D. J., Aldrich, C. C., & Gulick, A. M. (2012). Structural and functional investigation of the intermolecular interaction between NRPS adenylation and carrier protein domains. *Chemistry & Biology*, *19*(2), 188–198.

Suo, Z., Tseng, C. C., & Walsh, C. T. (2001). Purification, priming, and catalytic acylation of carrier protein domains in the polyketide synthase and nonribosomal peptidyl synthetase modules of the HMWP1 subunit of yersiniabactin synthetase. *Proceedings of the National Academy of Sciences of the United States of America*, *98*(1), 99–104.

Tanovic, A., Samel, S. A., Essen, L. O., & Marahiel, M. A. (2008). Crystal structure of the termination module of a nonribosomal peptide synthetase. *Science*, *321*(5889), 659–663.

Tarry, M. J., Haque, A. S., Bui, K. H., & Schmeing, T. M. (2017). X-ray crystallography and electron microscopy of cross- and multi-module nonribosomal peptide synthetase proteins reveal a flexible architecture. *Structure*, *25*(5), 783–793. e784.

Tartoff, K. D., & Hobbs, C. A. (1987). Improved media for growing plasmid and cosmid clones. *Bethesda Research Lab Focus*, *9*, 12.

Taton, A., Unglaub, F., Wright, N. E., Zeng, W. Y., Paz-Yepes, J., Brahamsha, B., et al. (2014). Broad-host-range vector system for synthetic biology and biotechnology in cyanobacteria. *Nucleic Acids Research*, *42*(17), e136.

Thieringer, H. A., Jones, P. G., & Inouye, M. (1998). Cold shock and adaptation. *BioEssays*, *20*(1), 49–57.

Unger, T., Jacobovitch, Y., Dantes, A., Bernheim, R., & Peleg, Y. (2010). Applications of the restriction free (RF) cloning procedure for molecular manipulations and protein expression. *Journal of Structural Biology*, *172*(1), 34–44.

Velappan, N., Sblattero, D., Chasteen, L., Pavlik, P., & Bradbury, A. R. (2007). Plasmid incompatibility: More compatible than previously thought? *Protein Engineering, Design & Selection*, *20*(7), 309–313.

Ventura, S., & Villaverde, A. (2006). Protein quality in bacterial inclusion bodies. *Trends in Biotechnology*, *24*(4), 179–185.

Villegas, A., & Kropinski, A. M. (2008). An analysis of initiation codon utilization in the domain bacteria—Concerns about the quality of bacterial genome annotation. *Microbiology*, *154*(Pt. 9), 2559–2661.

Vitali, F., Zerbe, K., & Robinson, J. A. (2003). Production of vancomycin aglycone conjugated to a peptide carrier domain derived from a biosynthetic non-ribosomal peptide synthetase. *Chemical Communications (Cambridge)*, *21*, 2718–2719.

Walsh, C. T. (2016). Insights into the chemical logic and enzymatic machinery of NRPS assembly lines. *Natural Product Reports*, *33*(2), 127–135.

Waugh, D. S. (2005). Making the most of affinity tags. *Trends in Biotechnology*, *23*(6), 316–320.

Weeks, S. D., Drinker, M., & Loll, P. J. (2007). Ligation independent cloning vectors for expression of SUMO fusions. *Protein Expression and Purification*, *53*(1), 40–50.

Whicher, J. R., Dutta, S., Hansen, D. A., Hale, W. A., Chemler, J. A., Dosey, A. M., et al. (2014). Structural rearrangements of a polyketide synthase module during its catalytic cycle. *Nature*, *510*(7506), 560–564.

Whicher, J. R., Smaga, S. S., Hansen, D. A., Brown, W. C., Gerwick, W. H., Sherman, D. H., et al. (2013). Cyanobacterial polyketide synthase docking domains: A tool for engineering natural product biosynthesis. *Chemistry & Biology*, *20*(11), 1340–1351.

Widmann, M., & Christen, P. (2000). Comparison of folding rates of homologous prokaryotic and eukaryotic proteins. *The Journal of Biological Chemistry*, *275*(25), 18619–18622.

Winn, M., Fyans, J. K., Zhuo, Y., & Micklefield, J. (2016). Recent advances in engineering nonribosomal peptide assembly lines. *Natural Product Reports*, *33*(2), 317–347.

Wolpert, M., Gust, B., Kammerer, B., & Heide, L. (2007). Effects of deletions of mbtH-like genes on clorobiocin biosynthesis in *Streptomyces coelicolor*. *Microbiology*, *153*(Pt 5), 1413–1423.

Worthington, A. S., Porter, D. F., & Burkart, M. D. (2010). Mechanism-based crosslinking as a gauge for functional interaction of modular synthases. *Organic & Biomolecular Chemistry*, *8*(8), 1769–1772.

Worthington, A. S., Rivera, H., Torpey, J. W., Alexander, M. D., & Burkart, M. D. (2006). Mechanism-based protein cross-linking probes to investigate carrier protein-mediated biosynthesis. *ACS Chemical Biology*, *1*(11), 687–691.

Yamanaka, K., Maruyama, C., Takagi, H., & Hamano, Y. (2008). Epsilon-poly-L-lysine dispersity is controlled by a highly unusual nonribosomal peptide synthetase. *Nature Chemical Biology*, *4*(12), 766–772.

Yancey, P. H., Clark, M. E., Hand, S. C., Bowlus, R. D., & Somero, G. N. (1982). Living with water stress: Evolution of osmolyte systems. *Science*, *217*(4566), 1214–1222.

Young, J., Stevens, D. C., Carmichael, R., Tan, J., Rachid, S., Boddy, C. N., et al. (2013). Elucidation of gephyronic acid biosynthetic pathway revealed unexpected SAM-dependent methylations. *Journal of Natural Products*, *76*(12), 2269–2276.

Yuzawa, S., Bailey, C. B., Fujii, T., Jocic, R., Barajas, J. F., Benites, V. T., et al. (2017). Heterologous gene expression of N-terminally truncated variants of LipPks1 suggests a functionally critical structural motif in the N-terminus of modular polyketide synthase. *ACS Chemical Biology*, *12*(11), 2725–2729.

Yuzawa, S., Eng, C. H., Katz, L., & Keasling, J. D. (2013). Broad substrate specificity of the loading didomain of the lipomycin polyketide synthase. *Biochemistry*, *52*(22), 3791–3793.

Yuzawa, S., Keasling, J. D., & Katz, L. (2016). Insights into polyketide biosynthesis gained from repurposing antibiotic-producing polyketide synthases to produce fuels and chemicals. *Journal of Antibiotics (Tokyo)*, *69*(7), 494–499.

Zheng, J., Gay, D. C., Demeler, B., White, M. A., & Keatinge-Clay, A. T. (2012). Divergence of multimodular polyketide synthases revealed by a didomain structure. *Nature Chemical Biology*, *8*(7), 615–621.

Zheng, X., Hu, G. Q., She, Z. S., & Zhu, H. (2011). Leaderless genes in bacteria: Clue to the evolution of translation initiation mechanisms in prokaryotes. *BMC Genomics*, *12*, 361.

Zhu, X., Liu, J., & Zhang, W. (2015). De novo biosynthesis of terminal alkyne-labeled natural products. *Nature Chemical Biology*, *11*(2), 115–120.

Zhu, X., Su, M., Manickam, K., & Zhang, W. (2015). Bacterial genome mining of enzymatic tools for alkyne biosynthesis. *ACS Chemical Biology*, *10*(12), 2785–2793.

Zolova, O. E., & Garneau-Tsodikova, S. (2012). Importance of the MbtH-like protein TioT for production and activation of the thiocoraline adenylation domain of TioK. *Medicinal Chemistry Communications*, *3*, 950–955.

Zou, Y., & Yin, J. (2009). Phosphopantetheinyl transferase catalyzed site-specific protein labeling with ADP conjugated chemical probes. *Journal of the American Chemical Society*, *131*(22), 7548–7549.

CHAPTER THREE

Terminal Alkyne Biosynthesis in Marine Microbes

Xuejun Zhu*, Wenjun Zhang*,†,1
*University of California, Berkeley, CA, United States
†Chan Zuckerberg Biohub, San Francisco, CA, United States
[1]Corresponding author: e-mail address: wjzhang@berkeley.edu

Contents

Abstract

The terminal alkyne is a readily derivatized functionality valued for its diverse applications in material synthesis, pharmaceutical science, and chemical biology. The synthetic biology routes to terminal alkynes are highly desired and yet underexplored. Some marine natural products contain a terminal alkyne functionality, and the discovery of the biosynthetic gene clusters for jamaicamide B and carmabin A marked the beginning of a new era in the understanding and engineering of terminal alkyne biosynthesis. In this chapter, we will overview recent advances in understanding the biosynthetic machinery for terminal alkyne synthesis. We will first describe how to elucidate terminal alkyne biosynthetic mechanism through heterologous expression, purification, and in vitro biochemical assays of individual pathway proteins. This will be followed by the description of an in vivo reporting system for the characterization of a membrane-bound bifunctional desaturase/acetylenase involved in terminal alkyne

Methods in Enzymology, Volume 604
ISSN 0076-6879
https://doi.org/10.1016/bs.mie.2018.01.040

formation. The chapter will also cover the strategies for discovering additional protein homologs for terminal alkyne synthesis from microbes as well as the applications of click chemistry to identify and quantify terminal alkyne-bearing metabolites from microbial cultures. We will conclude this chapter with current challenges and future directions in this field.

1. INTRODUCTION

The alkyne is a chemical functionality that has been prevalently used in modern chemistry and biology (Moses & Moorhouse, 2007). It is a particularly useful moiety in alkyne–azide cycloaddition reaction (known as a click reaction), which has allowed synthesis of complex molecular scaffolds for applications in pharmaceutical and material sciences (Jiang, Chen, Deng, Suuronen, & Zhong, 2014; Kolb & Sharpless, 2003; Lutz, 2007). In addition, the small size and bioorthogonal features of the alkyne have also enabled its diverse applications in chemical biology. The alkyne installed on medicinally and biologically important molecules either through metabolic or chemical synthesis can be directly visualized using stimulated Raman scattering microscopy (Wei et al., 2014), or bonds with azido-probes for a diverse of applications such as quantification and mode of action studies (Sletten & Bertozzi, 2009; Thirumurugan, Matosiuk, & Jozwiak, 2013; Zhu, Shieh, Su, Bertozzi, & Zhang, 2016). The importance of alkynes in both chemistry and biology motivates the discovery of a synthetic biology route to alkynes.

Marine microbes are known for producing diverse terminal alkyne-bearing natural products, including jamaicamide B, carmabin A, guineamide G, georgamide, and dragomabin, among others (Fig. 1), and many of them display bioactivities such as antimalarial, antitumor, anti-HIV, and antiparasitic (Chai, Yang, Lin, & Han, 2016). However, the biosynthetic route to the terminal alkyne functionality residing in these natural products was not well understood until recently. Biosynthetic gene clusters of jamaicamide B (Edwards et al., 2004) and carmabin A (Jones et al., 2011) have been identified from marine cyanobacteria *Moorea producens* (originally assigned as *Lyngbya majuscula*), and bioinformatics analysis indicated a conserved trigene cassette, as exemplified by *jamABC and camABC* in jamaicamide and carmabin gene clusters, respectively, to be responsible for alkyne formation (Fig. 2A). The trigene operon encodes for an acyl-CoA ligase homolog, a membrane-bound desaturase homolog, and

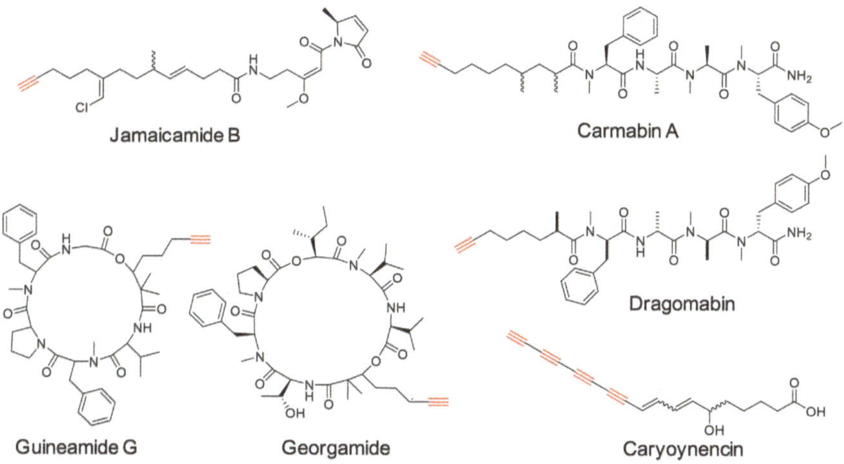

Fig. 1 Examples of alkyne-bearing natural products. Jamaicamide B, carmabin A, dragomabin, guineamide G, and georgamide are isolated from marine microbes. Caryoynencin is isolated from the carnation pathogen *Burkholderia caryophylli*.

an acyl carrier protein (ACP). Experimental studies further demonstrated an unprecedented ACP-dependent terminal alkyne biosynthetic pathway employed by the three proteins (Fig. 2B): a fatty acyl-ACP ligase (e.g., JamA) activates a fatty acid using ATP and further loads it onto a cognate ACP (e.g., JamC), and a membrane-bound bifunctional desaturase/acetylenase (e.g., JamB) subsequently performs the sequential terminal dehydrogenation of the fatty acyl moiety to generate a terminal alkyne functionality (Dorrestein et al., 2006; Edwards et al., 2004; Zhu, Liu, & Zhang, 2015).

Since the discovery of the first ACP-dependent terminal alkyne biosynthetic machinery, progress has been made toward additional discovery, characterization, and application of this unique biosynthetic machinery and the methodology thereof. In this chapter, we will first describe how to elucidate the terminal alkyne biosynthetic mechanism through the expression, purification, and in vitro biochemical assays of individual pathway proteins. This will be followed by the description of an in vivo reporting system for the characterization of a membrane-bound bifunctional desaturase/acetylenase involved in terminal alkyne formation. The chapter will also cover the strategies for discovering additional protein homologs for terminal alkyne synthesis from microbes as well as the application of click chemistry to identify and quantify terminal alkyne-bearing metabolites from microbial cultures. We will conclude this chapter with current challenges and future directions in this field.

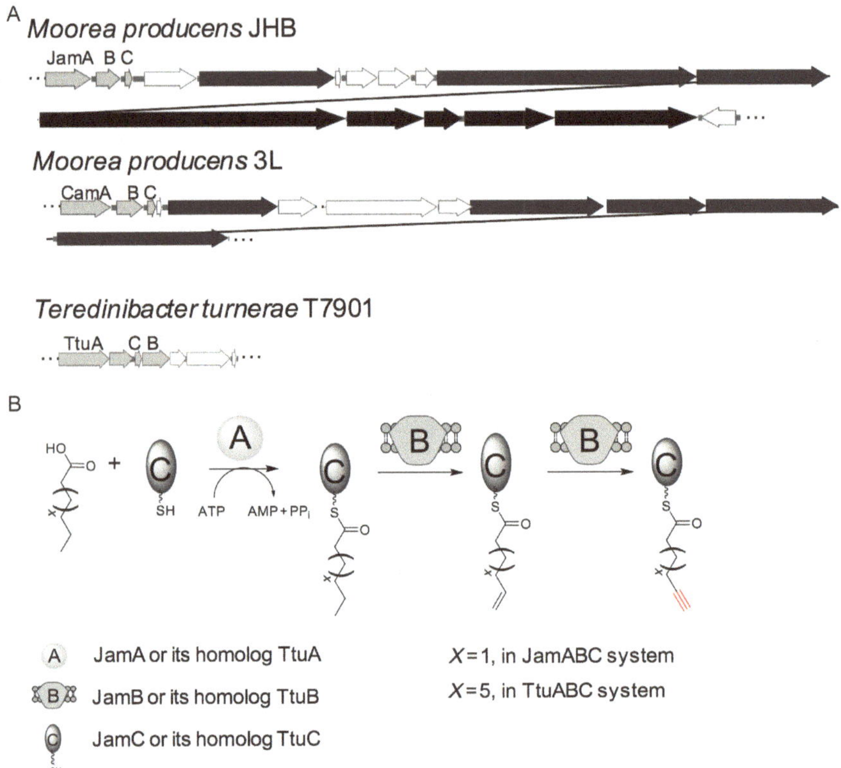

Fig. 2 Terminal alkyne biosynthesis in marine microbes. (A) JamABC are encoded in jamaicamide B biosynthetic gene cluster from *Moorea producens* JHB, CamABC are encoded in carmabin A biosynthetic gene cluster from *Moorea producens* 3L, and TtuABC are encoded in the genome of *Teredinibacter turnerae* T7901. Genes encoding JamABC and their homologs are highlighted in *gray*, genes encoding polyketide synthases or nonribosomal peptide synthetases are highlighted in *black*, other genes are in *white*. (B) Terminal alkyne biosynthetic pathways. JamABC and TtuABC modify C6 and C10 fatty acids, respectively.

2. ENZYME EXPRESSION, PURIFICATION, AND IN VITRO ASSAYS

2.1 Heterologous Expression and Purification of Proteins

2.1.1 Protein Expression

Since marine organisms are underdeveloped for protein expression and purification, a heterologous host is often employed. *Escherichia coli* is one of the most widely used heterologous expression hosts for recombinant proteins due to its fast growth rate, ample genetic tools, and well-characterized primary metabolism (Rosano & Ceccarelli, 2014). The commercially available strain *E. coli* BL21 (DE3) (Studier & Moffatt, 1986) in combination with pET vectors is typically used for heterologous expression of hexahistidine-tagged recombinant proteins. Commonly used pET vectors include pET24b that carries a C-terminal hexahistidine tag, pET28a and pET30XA/LIC that carry both N- and C-terminal hexahistidine tags. pSV272 vector encoding an N-terminal His_6-maltose-binding protein tag is occasionally used to enhance the solubility of the target recombinant protein in *E. coli* by promoting protein folding (Nallamsetty & Waugh, 2006; Zhu, Su, Manickam, & Zhang, 2015). The gene of interest can be cloned into the vectors under a T7 promoter either through restriction enzyme-dependent or ligation-independent cloning strategies. A T7 RNA polymerase under an isopropyl β-D-1-thiogalactopyranoside (IPTG)-inducible *lac*UV5 promoter is encoded in the genome of *E. coli* BL21 (DE3) and is required for the expression of target recombinant protein cloned on the pET vectors. Codon optimization of the target genes is sometimes required to avoid adverse effects due to differences in tRNA distribution between natural and heterologous hosts (Gustafsson, Govindarajan, & Minshull, 2004).

It is notable that heterologous ACPs expressed in *E. coli* BL21 (DE3) are often in *apo*-forms, and a promiscuous phosphopantetheinyl transferase such as Sfp from *Bacillus subtilis* can be used to ensure ACP posttranslational modification to *holo*-forms (Quadri et al., 1998). Thus, a modified *E. coli* BL21 (DE3) with a chromosomal copy of *sfp* under the control of a T7 promoter (named as *E. coli* BAP1 strain) is usually used to express *holo*-ACPs (Pfeifer, Admiraal, Gramajo, Cane, & Khosla, 2001). A limitation of BAP1 system is that both *apo*- and *holo*-forms of ACPs sometimes can be copurified due to partial conversion of *apo*-ACPs to *holo*-ACPs catalyzed by Sfp. Membrane-bound desaturase homologs can be heterologously expressed in various hosts, such as prokaryotic cells (e.g., *E. coli*) (Garba, Mohamad Ali, Oslan, & Rahman, 2016; Hattab, Warschawski, Moncoq, & Miroux, 2015; Paulucci, Dardanelli, & Garcia de Lema, 2014; Wada, Avelange-Macherel, & Murata, 1993) and eukaryotic cells (e.g., *Saccharomyces cerevisiae*, High Five or Sf9 insect cells, and mammalian cells, etc.) (Andrell & Tate, 2013; Bai et al., 2015; Clark et al., 2010; Sayanova et al., 2001; Wang et al., 2015). *E. coli* BL21 (DE3) is still a popular expression host for recombinant membrane protein expression due to its low cost and ease of manipulation.

A general protein expression procedure for alkyne biosynthetic enzymes is described here. The plasmid containing the gene of interest is transformed into electrocompetent or chemically competent *E. coli* BL21 (DE3) or BAP1 through electroporation or heat-shock transformation. A single colony is picked from LB agar plate supplemented with appropriate antibiotics and grown in 10 mL starter culture of LB plus appropriate antibiotics at 37°C overnight. 1 mL of the overnight culture is mixed with 1 mL of 40% glycerol in a cryovial and saved at −80°C for long-term storage. 7.5 mL of the remaining overnight seed culture is used to inoculate 750 mL of LB supplemented with appropriate antibiotics (~1:100 inoculation ratio). The culture is shaken at 250 rpm at 37°C until OD_{600} reaches 0.4–0.6, which may take ~2–3 h. The culture is then placed on ice for 10 min, followed by the induction with 0.1 m*M* IPTG. The resulting culture is shaken at 220 rpm at 16°C for 16 h. Researchers can adjust induction time, IPTG concentration, culture temperature, and duration for optimal expressions of different proteins.

2.1.2 Protein Purification

2.1.2.1 Soluble Protein Purification

The cells are harvested by centrifugation at ~4500 × *g* at 4°C for 15 min, resuspended in 30 mL of lysis buffer (50 m*M* 4-(2-hydroxyethyl)-1-piperazineethanesulfonic acid (HEPES), pH 8.0, 300 m*M* NaCl, 10 m*M* imidazole), and lysed by homogenization or sonication on ice. Cell debris is pelleted using centrifugation at 15,000 × *g* at 4°C for 1 h, and the supernatant is filtered by syringe filters (0.2 μm). Ni-NTA agarose resin is then added to the filtrate (1.5 mL/L of culture), and the mixture is nutated at 4°C for 1 h and loaded onto the gravity flow column. Proteins are eluted with increasing concentrations of imidazole (ranging from 10 to 250 m*M*) in buffer A (50 m*M* HEPES, pH 8.0, 100 m*M* NaCl). The cell debris and all fractions are collected for SDS-PAGE analysis. An Amicon Ultra filter with an appropriate molecular weight cutoff is used to concentrate and buffer exchange the purified protein into buffer B (50 m*M* HEPES, pH 8.0, 100 m*M* NaCl). Reducing agents (e.g., dithiothreitol (DTT)) and stabilizing elements (e.g., glycerol) can also be added in protein purification buffers, and buffering system, pH, and salt concentration can be optimized for different proteins. Protein concentration is determined by NanoDrop or Bradford assay. The final protein is flash-frozen in liquid nitrogen and stored at −80°C until further usage. Depending on the expression levels of target proteins, the purity of the resulting protein ranges from ~50% to 90%.

Additional chromatographic methods (e.g., ion exchange or size-exclusion chromatography) may be used to further purify the proteins when necessary.

2.1.2.2 Membrane-Bound Desaturase Purification

Although the purification of the alkyne-forming bifunctional desaturase/acetylenase has not been reported, homologous nonheme diiron-dependent membrane-bound desaturases or hydroxylases have been purified for biochemical and structural studies (Alonso & Roujeinikova, 2012; Bai et al., 2015; Naing, Parvez, Pender-Cudlip, Groves, & Austin, 2013; Shanklin, Achim, Schmidt, Fox, & Munck, 1997; Wang et al., 2015). The given protocol here is focused on membrane protein purification from the widely used host *E. coli* but is also applicable to membrane proteins expressed in other heterologous hosts.

2.1.2.2.1 Detergent-Dependent Purification

2.1.2.2.1.1 Membrane Preparations The cells are harvested by centrifugation at ~4500 × *g* at 4°C for 15 min, resuspended in 30 mL of HEPES buffer (50 m*M* HEPES, pH 8.0, 100 m*M* NaCl), and lysed by homogenization on ice. Cell debris and unlysed cells are removed by centrifugation at 10,000 × *g* at 4°C for 10 min. The supernatant was further ultracentrifugated (100,000 × *g*, 1 h, 4°C) to pellet the cell membrane fraction (Harder & Fotiadis, 2012; Zhu, Liu, et al., 2015).

2.1.2.2.1.2 Membrane Protein Solubilization and Purification The membrane pellet is resuspended in membrane suspension buffer (20 m*M* tris(hydroxymethyl)aminomethane (Tris) buffer, pH 8.0, 100 m*M* NaCl, 20% (v/v) glycerol, 1 mL/g of pellet), followed by the dilution in the appropriate solubilization buffer (20 m*M* Tris, pH 8.0, 100 m*M* NaCl, 10% (v/v) glycerol, and appropriate detergent). Different dilution ratios and detergents need to be screened for optimal protein solubilization, and commonly used detergents include 3-[(3-cholamidopropyl)-dimethylammonio]-1-propane sulfonate (CHAPS), *n*-dodecyl-β-D-maltoside (DDM), fos-choline 12 (FC-12), lauryldimethylamine N-oxide (LDAO), and *n*-octyl-β-D-glucoside (OG). The solubilization mixture is then continuously stirred for 12–18 h at 4°C, followed by the centrifugation at 200,000 × *g* and 4°C for 2 h to remove the insoluble pellet. Ni-NTA agarose resin is mixed with the supernatant according to the manufacturer's instruction, and the mixture is loaded onto the gravity flow column. The proteins are eluted with increasing concentrations of imidazole (ranging from 10 to 300 m*M*) in imidazole wash buffer (20 m*M* Tris, pH 8.0, 300 m*M* NaCl, 10% (v/v) glycerol, and appropriate detergent). All fractions are collected for SDS-PAGE analysis. 3 mL of purified protein is loaded onto a desalting column (e.g., Econo-Pac 10DG column) and eluted with 4 mL of size-exclusion buffer (20 m*M* HEPES, pH 7.4, 150 m*M* NaCl, 10% (v/v) glycerol, and appropriate detergent) to remove imidazole. An Amicon spin filter is used to concentrate the membrane protein. Protease inhibitors (e.g., PMSF) and reducing agents (e.g., DTT, 2-mercaptoethanol) can be optionally added in all buffers (Newby et al., 2009; Wang et al., 2015).

2.1.2.2.2 Detergent-Independent Purification A detergent-independent protocol has been used to successfully purify a membrane-bound alkane ω-hydroxylase AlkB, which shares common structural organization with membrane-bound desaturases (Alonso & Roujeinikova, 2012; Shanklin et al., 1997). The reported procedure may be adapted for the purification of membrane-bound desaturases. Specifically, the cells are resuspended in 50 m*M* HEPES buffer (pH 7.5) supplemented with 4 m*M* $MgCl_2$, 2 m*M* PMSF, 10 mg of DNase I, and 50 mg of catalase, and then lysed by homogenization on ice. Unbroken cells and cell debris are removed by centrifugation (43,000 × *g*, 4°C, 45 min). The supernatant is transferred to an ultracentrifuge tube and underlayed with 50 m*M* HEPES buffer (pH 7.5) containing 40% (v/v) glycerol, followed by ultracentrifugation at 285,000 × *g* and 4°C for 90 min. The glycerol cushions are harvested and diluted to 50 m*M* HEPES (pH 7.5) and 1 *M* urea, after which the membrane fraction is pelleted by ultracentrifugation (240,000 × *g*, 4°C, 1 h) and resuspended by sonication in 25 m*M* HEPES (pH 7.5) containing 10% (v/v) ethylene glycol. The suspension is then subjected to ion exchange or affinity chromatographic methods for protein purification. The purified protein can be buffer exchanged and concentrated into appropriate buffer that is compatible with downstream applications.

2.2 In Vitro Activity Assays

2.2.1 ATP-[^{32}P]Pyrophosphate (PP_i) Exchange Assays for Fatty Acyl-ACP Ligases

ATP-[^{32}P]PP_i exchange assays have been conventionally used to determine the substrate specificities of ATP-dependent adenylation enzymes (Fig. 3A), such as acyl-CoA ligase homologs, adenylation (A) domains in nonribosomal synthetases (NRPSs), and aminoacyl-tRNA synthetases (Edwards et al., 2004; Francklyn, First, Perona, & Hou, 2008; Zhang, Ostash, & Walsh, 2010). This assay measures reversible formation of (amino)acyl-AMP and

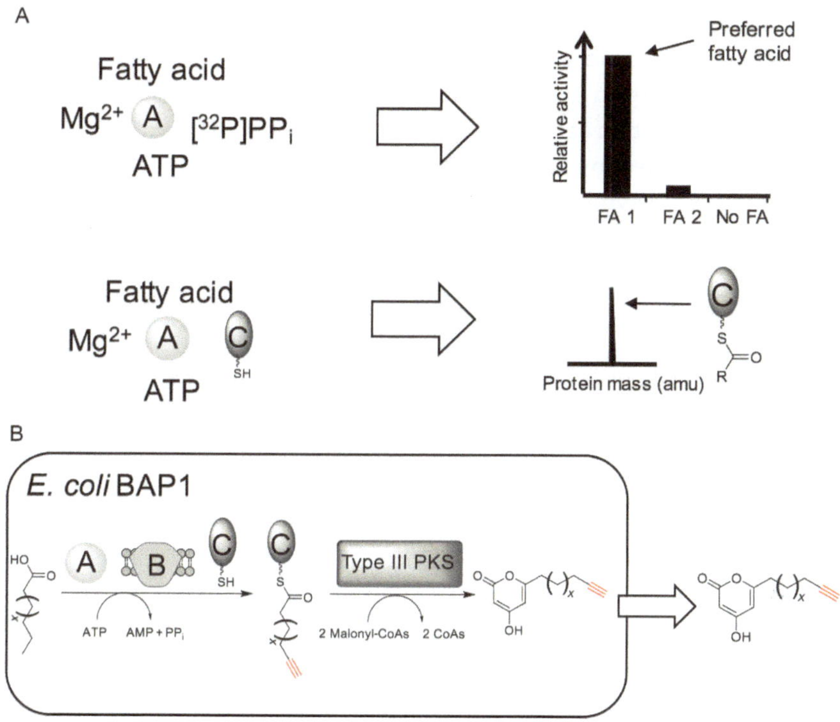

Fig. 3 Schematic of the strategy used for the study of JamABC and their homologs. (A) In vitro reactions can be used to determine the substrate specificities of JamA and its homologs. ATP-[^{32}P]PP$_i$ exchange assay is shown at the *top*, and ACP-loading assay is shown at the *bottom*. (B) Substrates determined from (A) are fed into the *E. coli*-PKS reporting system to reconstitute the in vivo activities of JamABC or their homologs.

allows the determination of substrate selectivity of the adenylation enzymes. A typical reaction mixture contains, in a total volume of 100 μL, 50 m*M* Tris buffer (pH 8.0), 2 m*M* $MgCl_2$, 5 m*M* ATP, 1 m*M* tris(2-carboxyethyl) phosphine (TCEP), 1 m*M* Na_4PP_i (containing 250–400 kcpm of $Na_4[^{32}P]PP_i$), 5 m*M* substrate, and 5 μ*M* of the enzyme. After the incubation at room temperature for 2 h, the reaction is quenched by the addition of a charcoal suspension (100 m*M* Na_4PP_i, 350 m*M* $HClO_4$, and 1.6% (w/v) activated charcoal). The charcoal is pelleted using centrifugation, washed twice with washing solution (100 m*M* Na_4PP_i and 350 m*M* $HClO_4$), resuspended in H_2O, and added into 5 mL of scintillation fluid for the measurement of charcoal-bound [^{32}P]-ATP radioactivity (R_{sub}). A no-substrate control is also carried out in parallel to measure the radioactivity baseline R_{blk}, which usually ranges from 100 to 1000 cpm. In addition, the total radioactivity (R_{tot}) of the no-substrate mixture before quenching is also recorded. Percent of substrate conversion is calculated as $(R_{sub}-R_{blk})/R_{tot}$. Reaction temperature and buffer system can be modified for individual enzyme. Similar reaction conditions can also be used to study enzyme kinetics, in which substrate concentrations, enzyme concentrations, and incubation time need to be modified for different enzymes. It is notable that in some cases, a high radioactivity baseline (R_{blk}) is observed due to the intrinsic ATPase activity of the purified enzyme. An alternative method is to perform product formation assay. In the case of acyl-ACP ligases (e.g., JamA homologs) involved in terminal alkyne formation, an ACP-loading assay (described in Section 2.2.2) is suggested.

An ATP-[^{32}P]PP$_i$ exchange assay in 96-well format has been developed and is useful when a high-throughput substrate profiling of ATP-dependent adenylation enzymes is required (Kries et al., 2014; Otten, Schaffer, Villiers, Stachelhaus, & Hollfelder, 2007; Villiers & Hollfelder, 2011). The reactions are carried out in each well of a conical plate that contains, in a total volume of 60 μL, 50 m*M* HEPES buffer (pH 8.0), 10 m*M* $MgCl_2$, 2 m*M* ATP, ~1 nCi/μL Na_4[^{32}P]PP$_i$, 100 m*M* NaCl, 1 m*M* EDTA, 5 m*M* substrate, and 5 μ*M* enzyme. 60 μL of a charcoal suspension (150 m*M* Na_4PP_i, 840 m*M* $HClO_4$, and 3.6% (w/v) activated charcoal) is added to each well after a 2-h incubation, followed by the addition of 130 μL water and centrifugation at room temperature for 3 min. The plate is inverted to remove the supernatant. The charcoal is washed with 250 μL water three times and resuspended in 150 μL scintillation fluid. The radioactivity is measured by reading in top mode at 1–2 min/well in a MicroBeta 96-well plate liquid scintillation counter.

2.2.2 ACP-Loading Assays for Fatty Acyl-ACP Ligases

ACP-loading assays can also be used to study the substrate specificities of acyl-ACP ligases or NRPS-A domains (Fig. 3A; Dorrestein et al., 2006; Zhu, Su, et al., 2015). A reaction is typically carried out at room temperature for 2 h or overnight in 50 m*M* HEPES (pH 8.0) with 2 m*M* $MgCl_2$, 2 m*M* ATP, 1 m*M* TCEP, 5 m*M* substrate, 100 μ*M* *holo*-ACP, and 5–20 μ*M* enzyme. The mixture is filtered by syringe filters (0.2 μm) to remove any precipitant, and the supernatant is injected into high-performance liquid chromatography (HPLC) or liquid chromatography-high-resolution mass spectrometry (LC-HRMS) for protein analysis using a reversed phase C18 peptide column with mobile phases containing water and acetonitrile supplemented with 0.1% formic acid. UV detection is occurred at 220 nm

(Chan et al., 2006). MassHunter Qualitative Analysis software (or equivalent) is used for mass data analysis, and the intact protein masses are obtained using a maximum entropy deconvolution algorithm.

Radioactive assays can also be used to analyze acyl-ACP ligases (Trivedi et al., 2004). A typical reaction, in a total volume of 15 μL, contains 50 m*M* Tris buffer (pH 8.0), 2 m*M* ATP, 8 m*M* $MgCl_2$, ^{14}C substrate (0.1 μCi), 10 μ*M* enzyme, and 10–100 μ*M* ACP, and is performed at room temperature for 2 h. The reaction is quenched and proteins are separated using SDS-PAGE. The gel is stained with Coomassie blue, dried, and exposed to phosphor screen at room temperature for 3–7 days. Radiolabeled protein is imaged on a phosphorimager. In a control reaction ATP is excluded from the mixture.

2.2.3 Desaturase Assays

An in vitro desaturase assay can be used to study the activity of membrane-bound desaturase/acetylenase (Wada, Schmidt, Heinz, & Murata, 1993; Zhu, Liu, et al., 2015; Zhu, Su, et al., 2015). Followed by the generation of substrates (described in Section 2.2.2), 0.1 m*M* $Fe(NH_4)_2(SO_4)_2$, catalase, and the desaturase-containing membrane fraction or the membrane fraction purified from the *E. coli* strain transformed with an empty vector (prepared using protocol described in Section 2.1.2.2.1.1) is added to the reaction mixtures. To some extent, the electron transport system could be provided by the semipurified *E. coli* membrane to enable the reaction (Wada, Avelange-Macherel, et al., 1993; Zhu, Liu, et al., 2015; Zhu, Su, et al., 2015). The membrane fraction is removed by centrifugation (21,000 × *g*, 4°C) after 20 min of incubation at room temperature. The supernatant is subjected to LC-HRMS for the product of alkynoic ACP detection.

A modified desaturase assay protocol may be used if a pure membrane desaturase is available (Goren & Fox, 2008). The enzyme diiron active site can be reconstituted by the incubation with $Fe(NH_4)_2(SO_4)_2$ and a 10-fold molar excess of ascorbate in degassed 60 m*M* HEPES (pH 7.4). The reaction mixture contains 25 m*M* HEPES (pH 7.2), 180 m*M* NaCl, 100 μ*M* substrate, 20 μ*M* enzyme, 30 μ*M* $Fe(NH_4)_2(SO_4)_2$, 300 μ*M* ascorbate, and a necessary electron transport system. Since the native electron transport system for the alkyne-forming bifunctional desaturase/acetylenase has not been determined, testing known electron transport systems for membrane-bound desaturases such as the ferredoxin system (ferredoxin/NADPH/ferredoxin-$NADP^+$ oxidoreductase) and the cytochrome *b*5 system (cytochrome *b*5/NADH/cytochrome *b*5 reductase) is recommended (Shanklin & Cahoon, 1998; Wada, Schmidt, et al., 1993). Catalase can also be added into the reaction to protect the desaturase system from hydrogen peroxide generated from aerobic oxidation of electron transport partners (Browse & Slack, 1981).

3. AN IN VIVO REPORTING SYSTEM FOR THE FUNCTIONAL RECONSTITUTION OF MEMBRANE-BOUND DESATURASE/ACETYLENASE ACTIVITIES

Due to the difficulty in working with membrane-bound enzyme in vitro, an in vivo reporting system has been developed to study membrane-bound desaturase/acetylenase activities (Fig. 3B; Zhu, Liu, et al., 2015; Zhu, Su, et al., 2015). In this system, a promiscuous simple type III polyketide synthase (PKS) is used to condense an acyl-ACP starter unit with malonyl-CoAs to yield a terminal alkyne-tagged small-molecule polyketide, thus easing the detection and the confirmation of an alkyne product (a small molecule vs a protein). In addition, an in vivo system could circumvent problems associated with in vitro assays such as inactivation of membrane-bound enzymes during purification and inaccessibility of required cosubstrates. An in vivo reporting system is therefore particularly useful for functional reconstitution of putative membrane-bound desaturase/acetylenases when little is known about the activity of these enzymes.

3.1 Type III PKS Selection

It is essential to select a promiscuous PKS that recognizes the desired product of an alkynoic ACP for elongation. An in vitro product assay can be used to screen for a suitable type III PKS compatible with the chain length of the donor acyl-ACP (Zhu, Su, et al., 2015). The reaction is typically carried out at room temperature in 50 mM HEPES (pH 8.0) with 2 mM $MgCl_2$, 2 mM ATP, 1 mM TCEP, 1 mM fatty acid, 100 μM *holo*-ACP (e.g., JamC), 10 μM fatty acyl-ACP ligase (e.g., JamA), 1 mM malonyl-CoA, and 20 μM type III PKS. A final concentration of 5% (v/v) trichloroacetic acid is added to the mixture to quench the reaction. Precipitated proteins are pelleted by centrifugation, and the supernatant is subjected to LC-HRMS analysis. A PKS that leads to an efficient polyketide formation is used in the PKS in vivo reporting system. It is notable that PKSs do not typically distinguish acyl starter units with or without a terminal alkyne functionality. Promiscuous HsPKS1, which accepts medium-chain fatty acyl-ACPs (C6–C11), is often used (Wanibuchi et al., 2007; Zhu, Liu, et al., 2015; Zhu, Su, et al., 2015).

3.2 Host Selection

JamABC homologs are widespread across Gram-negative bacteria, and thus a representative model organism *E. coli* may be used to mimic the native membrane environment of JamB homologs for functional expression of the enzymes (Zhu, Su, et al., 2015). In addition, the well-studied metabolism, genetic tractability, and fast growth rate of *E. coli* render it attractive for engineering the biosynthesis of alkyne-tagged metabolites. Specifically, *E. coli* BAP1 strain has been used to coexpress JamA homologs, JamB homologs, *holo*-JamC homologs, and type III PKSs, and alkyne-tagged polyketides have been made. Attempts have been made to reconstitute the activity of JamABC in *S. cerevisiae* BJ5464-NpgA, a popular eukaryotic host for metabolic engineering and natural product biosynthesis (Yu et al., 2013). Unfortunately, the coexpression of *jamABC* and *hspks1* only yielded polyketides primed with a supplemented fatty acyl chain, suggesting the inactivity of JamABC in *S. cerevisiae*.

3.3 Plasmid Vector Selection

Commercially available Duet vectors (Novagen) are typically used for the coexpression of multiple genes in *E. coli*. The commonly used vectors include pRSFDuet-1, pETDuet-1, pCDFDuet-1, pACYCDuet-1, and pCOLADuet-1, with plasmid copy numbers sorted in descending order. Each Duet vector carries two T7 promoters, and thus is suitable to coexpress at least two proteins. The vectors carry compatible replication origins and antibiotic resistance genes for effective maintenance of multiple plasmids in a single *E. coli* cell. It is notable that target protein expression levels can be affected by plasmid copy numbers, and therefore, it is necessary to optimize combination of Duet vectors when multiple plasmids are required for in vivo pathway reconstitution. In our laboratory, pETDuet-1 and pCDFDuet-1 are generally used in a two-plasmid system, pCOLADuet-1 is often added when a three-plasmid system is required, and pACYCDuet-1 may be used as a fourth plasmid (Liu, Zhu, Seipke, & Zhang, 2015; Zhu, Liu, et al., 2015; Zhu, Su, et al., 2015). A high plasmid copy number may result in extra metabolic burden and reduce bacterial growth rate (Rosano & Ceccarelli, 2014), and therefore, pRSFDuet-1 should be chosen with care in a coexpression system. In an *E. coli*-PKS system, JamABC homologs and a type III PKS are coexpressed on pETDuet-1 and pCDFDuet-1 vectors, with each gene controlled by a T7 promoter (Zhu, Liu, et al., 2015; Zhu, Su, et al., 2015).

3.4 Culture Conditions

3.4.1 Seed Culture Conditions

The plasmids are cotransformed into *E. coli* BAP1 strain, and a single colony is picked and grown in 3 mL starter culture of LB supplemented with appropriate antibiotics at 37°C overnight. 0.5 mL of the overnight culture is mixed with equal volume of 40% glycerol in a cryovial and saved at −80°C for long-term storage. The rest of the seed culture is used to inoculate culture medium for polyketide production. Terrific broth (TB) and F1 minimal medium are used for high-level polyketide production (Pfeifer, Hu, Licari, & Khosla, 2002; Zhang, Li, & Tang, 2008; Zhou, Zhan, Watanabe, Xie, & Tang, 2008; Zhu, Liu, et al., 2015; Zhu, Su, et al., 2015). Culture conditions for the *E. coli*-PKS system in these media (Zhu, Liu, et al., 2015; Zhu, Su, et al., 2015) are described later. Note, LB medium is not chosen for polyketide production in the *E. coli*-PKS reporting system due to insufficient carbon source for malonyl-CoA production.

3.4.2 TB-Based Culture Conditions and Extraction of Small-Molecule Reporter

1 L of TB medium is composed of 24 g yeast extract, 12 g tryptone, 10 g NaCl, 6 mL glycerol, and 9 mL of 1 *M* Tris buffer (pH 7.4). 0.5 mL of overnight seed culture is used to inoculate 50 mL of TB supplemented with appropriate antibiotics. The culture is shaken at 250 rpm at 37°C to an OD_{600} of 0.8–1.0. The culture is then placed on ice for 10 min, followed by the induction with 0.5 m*M* IPTG and supplementation of 1 m*M* fatty acid. The culture is shaken at 220 rpm at 20°C for ~2 days. The culture broth and the cell pellets are then separated by centrifugation, and polyketides are extracted from the supernatant (where most of compounds reside) using ethyl acetate. Note, organic solvent can be optimized for extracting different compounds, and it is recommended to analyze the targeted compound titer in organic extracts of both cell pellets and supernatant during an initial test. The solvent is removed by rotary evaporation, and the residue is redissolved in methanol. The final sample is subjected to LC-HRMS analysis using a reversed phase C18 column with mobile phases containing water and acetonitrile supplemented with 0.1% formic acid.

3.4.3 F1 Minimal Medium-Based Culture Condition

2 mL of overnight seed culture is used to inoculate 200 mL of LB supplemented with appropriate antibiotics. The culture is shaken at 250 rpm at 37°C to an OD_{600} of ~0.4. The cells are pelleted at 4°C and resuspended

in 40 mL fresh F1 medium supplemented with appropriate antibiotics, 0.5 m*M* IPTG, and 1 m*M* fatty acid. 1 L of F1 medium contains 3 g KH_2PO_4, 6.62 g K_2HPO_4, 4 g $(NH_4)_2SO_4$, 150.5 mg $MgSO_4$, 5 g glucose, 1.25 mL trace metal solution, 100 μ*M* $Fe(NH_4)_2(SO_4)_2$, and 10 mL 100 × vitamin solution. Trace metal solution is composed of 40 mg $ZnCl_2$, 200 mg $FeCl_3 \cdot 6H_2O$, 10 mg $CuCl_2 \cdot 2H_2O$, 10 mg $MnCl_2 \cdot 4H_2O$, 10 mg $Na_2B_4O_7 \cdot 10H_2O$, 10 mg $(NH_4)_6Mo_7O_{24} \cdot 4H_2O$ in 1 L water. 100 × vitamin solution contains 100 mg/L biotin, 100 mg/L choline chloride, 100 mg/L D-calcium pantothenate, 100 mg/L folic acid, 100 mg/L nicotinamide, 100 mg/L pyridoxal hydrochloride, 10 mg/L riboflavin, 100 mg/L thiamine hydrochloride, 200 mg/L i-inositol, and 8.5 g/L NaCl. The culture is shaken at 220 rpm at 20°C for ~2 days. The culture broth and the cell pellets are then separated by centrifugation, and polyketides are extracted from supernatant using ethyl acetate. The solvent is removed by rotary evaporation, and the residue is redissolved in methanol. The final sample is subjected to HPLC or LC-HRMS analysis.

4. STRATEGIES FOR DISCOVERING ADDITIONAL TERMINAL ALKYNE BIOSYNTHETIC PROTEINS

4.1 Bioinformatics Analysis

Since secondary metabolite biosynthetic genes are often clustered in microbial organisms, identification and elucidation of functions of gene clusters encoding JamABC homologs might reveal additional terminal alkyne biosynthetic machinery. With the rapid improvement of genome sequencing technology and the wealth of accessible genetic data, bioinformatics analysis is becoming increasingly crucial in new enzyme discovery and characterization, including gene clusters encoding JamABC homologs. Many bioinformatics tools have been developed to facilitate in silico analysis of genetic data and prediction of protein functions (Ziemert, Alanjary, & Weber, 2016). Integrated microbial genomes (IMG) (http://img.jgi.doe.gov) is a commonly used web-based tool for comparative analysis of publicly available genomes (Markowitz et al., 2012; Mavromatis et al., 2009). Draft and complete genomes from archaea, bacteria, and eukarya are all included in the database. To search for the homologs of a target gene, researchers need to locate the gene of interest in the database through "Find Genes," and a "Gene Detail" web page will then be displayed. The web page includes important information of the target gene, such as gene sequence, genome, protein function, pathway information, and gene neighborhood.

The homologous genes can be identified via "homolog display," "show neighborhood regions with this gene's bidirectional best hits," or "show neighborhood regions with the same top COG hit (via top homolog)," and the resulting output is analyzed manually. MultiGeneBlast is a standalone software that allows homology searches of operons or gene clusters, and has been used to successfully identify homologous biosynthetic gene clusters (Leao et al., 2017; Medema, Takano, & Breitling, 2013). The genome database can be customized by researchers by combining both published and unpublished data with the MakeDB program of MultiGeneBlast. The gene cluster of interest is used as a MultiGeneBlast query with default settings, and the output is manually inspected. MultiGeneBlast is available at http://multigeneblast.sourceforge.net/.

4.2 Experimental Validation

In addition to bioinformatics analysis, experimental validation is also essential for discovering new alkyne biosynthetic machinery and understanding their catalytic capabilities. In this section, we briefly describe general experimental procedures for identifying additional homologous gene cassettes that are involved in alkyne biosynthesis revealed from in silico analysis (Zhu, Su, et al., 2015). JamA homologs and *holo*-JamC homologs are first expressed and purified from *E. coli* using the protocol described in Section 2.1. In vitro ACP-loading assays are then used to determine the substrate preference of JamA homologs toward fatty acids of different chain lengths using the protocol described in Section 2.2.2 (Fig. 3A). Note, an ATP-[^{32}P]PP$_i$ exchange assay described in Section 2.2.1 may also be employed to quickly determine the substrate preferences of JamA homologs with no intrinsic ATPase activities. *E. coli*-PKS reporting systems are then used for in vivo analysis of JamB homologs (Fig. 3B). A type III PKS that recognizes an acyl-ACP of the corresponding fatty acyl chain length is selected and is further included in *E. coli*-PKS system to study the activities of JamABC homologs in vivo using the protocols described in Sections 3.3 and 3.4. Using this workflow, eleven gene clusters homologous to *jamABC* have been screened which led to the discovery of TtuABC as new terminal alkyne biosynthetic machinery (Zhu, Su, et al., 2015). TtuABC are encoded in the genome of *Teredinibacter turnerae* T7901, a marine gamma proteobacterium that occurs as an intracellular endosymbiont in the gills of wood-boring marine bivalves of the family Teredinidae (shipworms) (Yang et al., 2009). Analogous to the functions of JamABC, the fatty acid is activated and loaded onto TtuC by TtuA, and it is then modified by TtuB to form a terminal alkyne moiety. Differently, this machinery prefers decanoic acid instead of hexanoic acid for terminal alkyne formation (Fig. 2). Further bioinformatic analysis revealed that *ttuABC* are clustered with genes encoding an additional membrane-bound desaturase, a dehydrogenase, and hypothetic proteins, suggesting that this gene cluster is likely responsible for synthesizing a polyyne metabolite that is yet to be determined. It is notable that this experimental validation method still suffers from possible lack of native fatty acid substrate and inactivity of JamB homologs in *E. coli*, which impedes successful validation of the activities of JamABC homologs.

5. CLICK CHEMISTRY TO IDENTIFY AND QUANTIFY TERMINAL ALKYNES

A facile method to identify and quantify terminal alkynes, in particular directly in microbial cultures, could facilitate understanding and engineering the biosynthesis of terminal alkyne. Terminal alkyne has the ability to react with azide by the catalysis of Cu(I) in a mild condition (known as click chemistry), and this unique feature has allowed facile identification of alkynes in cell cultures (Fig. 4). For example, it has been recently demonstrated that click chemistry can be employed to trap terminal alkyne-bearing metabolites in microbial culture extracts for compound discovery and purification (Ross, Scherlach, Kloss, & Hertweck, 2014). Specifically, 1 L of bacterial culture is extracted with ethyl acetate, and the organic phase is

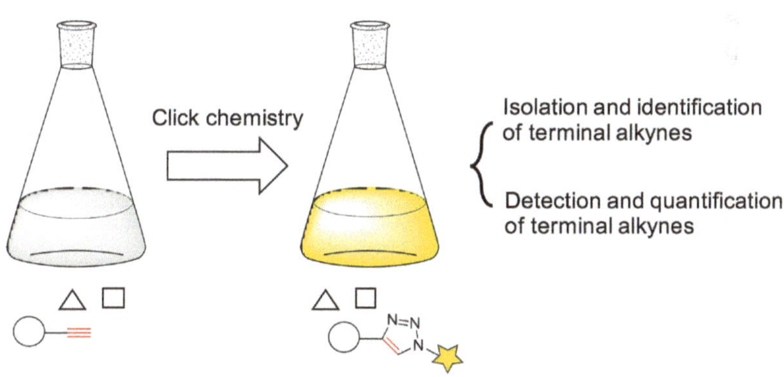

Fig. 4 Applications of click chemistry to identify and quantify terminal alkynes from microbial cultures. The *triangle* and the *square* represent the untagged extracellular metabolites, while the *circle* represents the extracellular metabolite that can be tagged with the alkyne functionality.

concentrated to 30 mL, followed by the addition of 12 mL DMF, and further concentration to 10 mL final volume. The click reaction contains 10 mL culture extracts, 0.5 mL of 1.3 *M* benzyl azide (dissolved in DMF), and 1 mL aqueous solution composed of 0.5 mL of 0.8 *M* $CuSO_4$ and 0.5 mL of 1.1 *M* ascorbic acid. After overnight reaction at room temperature, the mixture is then diluted with ethyl acetate, washed twice with water, and the resulting organic phase is dried and subjected to HPLC analysis and purification. Using this method, caryoynencin and an unstable terminal alkyne-bearing biosynthetic intermediate have been successfully isolated from the carnation pathogen *Burkholderia caryophylli* (Fig. 1). It is notable that the caryoynencin biosynthetic gene cluster also encodes proteins that are homologous to JamABC.

In addition, using an azido fluorogenic probe, terminal alkyne-bearing extracellular metabolites can also be directly quantified in the producing microbial culture media (Zhu et al., 2016). For example, terminal alkynes produced in an *E. coli* culture coexpressing *jamABC* and *hspks1* have been detected and quantified through a fluorogenic click reaction, which further enabled screening of a JamB variant with an improved desaturase/acetylenase activity in *E. coli*. A typical quantification reaction mixture contains, in a total volume of 200 μL, 0.75 μ*M* fluorogenic probe (e.g., di-pegOF), 5 m*M* ascorbic acid, 500 μ*M* $CuSO_4$, 100 μ*M* 2-[4-((bis[(1-*tert*-butyl-1H-1,2,3-triazol-4-yl)methyl]amino)methyl)-1H-1,2,3-triazol-1-yl]acetic acid (BTTAA), and 190 μL of cell-free F1 culture. The reaction is carried out in a black-walled 96-well plate in dark for 10 min at room temperature. The resulting fluorescence is measured using a plate reader with appropriate excitation/emission wavelengths (e.g., 500/520 nm for di-pegOF). Alkynes in quantities as low as 50 n*M* can be detected, and a broad linear range of responses can be achieved by varying the amounts of the fluorogenic probe in the assays. The reaction conditions can be adjusted when alternative culture media and fluorogenic probes are used (Shieh et al., 2015; Shieh, Hangauer, & Bertozzi, 2012; Shieh, Siegrist, Cullen, & Bertozzi, 2014).

6. PERSPECTIVES

While progress has been achieved toward understanding the terminal alkyne biosynthetic machinery from marine microbes, continued effort will facilitate further characterization and engineering of the alkyne biosynthetic pathway. There are still a few intriguing mechanistic questions remained to be tackled, such as how the acyl-ACP ligase and desaturase/acetylenase selectively interact with their cognate ACP, how the bifunctional desaturase/acetylenase differs from typical desaturases that stop at the double-bond formation, and how the acetylenase abstracts two hydrogens from C=C without forming an epoxide (Shanklin, Guy, Mishra, & Lindqvist, 2009). Recently, the first crystal structure of a CoA-dependent membrane-bound desaturase was reported, which could facilitate the mechanistic understanding of membrane-bound bifunctional desaturase/acetylenase (Bai et al., 2015; Wang et al., 2015). The reported membrane enzyme purification and crystallization protocols may also be useful to obtain the structural information of the ACP-dependent membrane-bound desaturase/acetylenase.

The ACP-dependent terminal alkyne biosynthetic pathway may also be employed to tag a broad spectrum of biomolecules through biosynthetic pathway engineering to circumvent the need of alkyne precursor feeding. This strategy, so called de novo biosynthesis, may open the door to biological applications in which in situ enzymatic generation of a terminal alkyne is required or preferred. Tagging of natural products with a terminal alkyne functionality through de novo biosynthesis has been explored recently (Zhu, Liu, et al., 2015). Promiscuous natural product biosynthetic enzymes could possibly incorporate an alkyne precursor into major classes natural products such as polyketides, hybrids of polyketide and nonribosomal peptide, lipopeptides, and lipoglycopeptides (Zhu & Zhang, 2015). However, only limited success has been achieved and it remains a challenge to achieve efficient biosynthesis of alkyne-tagged natural products, whose success relies on the cross talk of terminal alkyne biosynthetic machinery with biosynthetic pathways of natural products of interest. A key mediator in the hybrid biosynthetic pathway is the ACP (JamC homologs), and further understanding and engineering of the protein–protein interactions of ACP with enzyme partners in the terminal alkyne and natural product biosynthetic machinery may be fruitful in improving natural product tagging. The docking domains for ACP and KS interactions in cyanobacterial curacin pathway have been successfully characterized and may direct the use of docking domains in these engineering efforts (Whicher et al., 2013). In addition, further characterization of how the PKS JamE recognizes 5-hexynoyl-JamC, which was proposed to serve as the natural starter unit for the hybrid PKS/NRPS machinery in the marine natural product jamaicamide B biosynthesis, may also suggest strategies for optimizing protein–protein interactions (Edwards et al., 2004).

ACKNOWLEDGMENTS

Research in our laboratory on this topic has been supported by grants from the National Institutes of Health (DP2AT009148), Alfred P. Sloan Foundation, and the Chan Zuckerberg Biohub Investigator Program.

REFERENCES

Alonso, H., & Roujeinikova, A. (2012). Characterization and two-dimensional crystallization of membrane component AlkB of the medium-chain alkane hydroxylase system from Pseudomonas putida GPo1. *Applied and Environmental Microbiology*, *78*(22), 7946–7953. https://doi.org/10.1128/AEM.02053-12.

Andrell, J., & Tate, C. G. (2013). Overexpression of membrane proteins in mammalian cells for structural studies. *Molecular Membrane Biology*, *30*(1), 52–63. https://doi.org/10.3109/09687688.2012.703703.

Bai, Y., McCoy, J. G., Levin, E. J., Sobrado, P., Rajashankar, K. R., Fox, B. G., et al. (2015). X-ray structure of a mammalian stearoyl-CoA desaturase. *Nature*, *524*(7564), 252–256. https://doi.org/10.1038/nature14549.

Browse, J. A., & Slack, C. R. (1981). Catalase stimulates linoleate desaturase activity in microsomes from developing linseed cotyledons. *FEBS Letters*, *131*(1), 111–114. https://doi.org/10.1016/0014-5793(81)80899-X.

Chai, Q. Y., Yang, Z., Lin, H. W., & Han, B. N. (2016). Alkynyl-containing peptides of marine origin: A review. *Marine Drugs*, *14*(11), 216. https://doi.org/10.3390/md14110216.

Chan, Y. A., Boyne, M. T., 2nd, Podevels, A. M., Klimowicz, A. K., Handelsman, J., Kelleher, N. L., et al. (2006). Hydroxymalonyl-acyl carrier protein (ACP) and aminomalonyl-ACP are two additional type I polyketide synthase extender units. *Proceedings of the National Academy of Sciences of the United States of America*, *103*(39), 14349–14354. https://doi.org/10.1073/pnas.0603748103.

Clark, K. M., Fedoriw, N., Robinson, K., Connelly, S. M., Randles, J., Malkowski, M. G., et al. (2010). Purification of transmembrane proteins from Saccharomyces cerevisiae for X-ray crystallography. *Protein Expression and Purification*, *71*(2), 207–223. https://doi.org/10.1016/j.pep.2009.12.012.

Dorrestein, P. C., Blackhall, J., Straight, P. D., Fischbach, M. A., Garneau-Tsodikova, S., Edwards, D. J., et al. (2006). Activity screening of carrier domains within nonribosomal peptide synthetases using complex substrate mixtures and large molecule mass spectrometry. *Biochemistry*, *45*(6), 1537–1546. https://doi.org/10.1021/bi052333k.

Edwards, D. J., Marquez, B. L., Nogle, L. M., McPhail, K., Goeger, D. E., Roberts, M. A., et al. (2004). Structure and biosynthesis of the jamaicamides, new mixed polyketide-peptide neurotoxins from the marine cyanobacterium Lyngbya majuscula. *Chemistry & Biology*, *11*(6), 817–833. https://doi.org/10.1016/j.chembiol.2004.03.030.

Francklyn, C. S., First, E. A., Perona, J. J., & Hou, Y. M. (2008). Methods for kinetic and thermodynamic analysis of aminoacyl-tRNA synthetases. *Methods*, *44*(2), 100–118. https://doi.org/10.1016/j.ymeth.2007.09.007.

Garba, L., Mohamad Ali, M. S., Oslan, S. N., & Rahman, R. N. (2016). Molecular cloning and functional expression of a delta9-fatty acid desaturase from an Antarctic Pseudomonas sp. A3. *PLoS One*, *11*(8), e0160681. https://doi.org/10.1371/journal.pone.0160681.

Goren, M. A., & Fox, B. G. (2008). Wheat germ cell-free translation, purification, and assembly of a functional human stearoyl-CoA desaturase complex. *Protein Expression and Purification*, *62*(2), 171–178. https://doi.org/10.1016/j.pep.2008.08.002.

Gustafsson, C., Govindarajan, S., & Minshull, J. (2004). Codon bias and heterologous protein expression. *Trends in Biotechnology*, *22*(7), 346–353. https://doi.org/10.1016/j.tibtech.2004.04.006.

Harder, D., & Fotiadis, D. (2012). Preparation of detergent-solubilized membranes from Escherichia coli. *Protocol Exchange*. https://doi.org/10.1038/protex.2012.033.

Hattab, G., Warschawski, D. E., Moncoq, K., & Miroux, B. (2015). Escherichia coli as host for membrane protein structure determination: A global analysis. *Scientific Reports*, *5*, 12097. https://doi.org/10.1038/srep12097.

Jiang, Y., Chen, J., Deng, C., Suuronen, E. J., & Zhong, Z. (2014). Click hydrogels, microgels and nanogels: Emerging platforms for drug delivery and tissue engineering. *Biomaterials*, *35*(18), 4969–4985. https://doi.org/10.1016/j.biomaterials.2014.03.001.

Jones, A. C., Monroe, E. A., Podell, S., Hess, W. R., Klages, S., Esquenazi, E., et al. (2011). Genomic insights into the physiology and ecology of the marine filamentous cyanobacterium Lyngbya majuscula. *Proceedings of the National Academy of Sciences of the United States of America*, *108*(21), 8815–8820. https://doi.org/10.1073/pnas.1101137108.

Kolb, H. C., & Sharpless, K. B. (2003). The growing impact of click chemistry on drug discovery. *Drug Discovery Today*, *8*(24), 1128–1137.

Kries, H., Wachtel, R., Pabst, A., Wanner, B., Niquille, D., & Hilvert, D. (2014). Reprogramming nonribosomal peptide synthetases for "clickable" amino acids. *Angewandte Chemie (International Ed in English)*, *53*(38), 10105–10108. https://doi.org/10.1002/anie.201405281.

Leao, T., Castelao, G., Korobeynikov, A., Monroe, E. A., Podell, S., Glukhov, E., et al. (2017). Comparative genomics uncovers the prolific and distinctive metabolic potential of the cyanobacterial genus Moorea. *Proceedings of the National Academy of Sciences of the United States of America*, *114*(12), 3198–3203. https://doi.org/10.1073/pnas.1618556114.

Liu, J., Zhu, X., Seipke, R. F., & Zhang, W. (2015). Biosynthesis of antimycins with a reconstituted 3-formamidosalicylate pharmacophore in Escherichia coli. *ACS Synthetic Biology*, *4*(5), 559–565. https://doi.org/10.1021/sb5003136.

Lutz, J. F. (2007). 1,3-Dipolar cycloadditions of azides and alkynes: A universal ligation tool in polymer and materials science. *Angewandte Chemie (International Ed in English)*, *46*(7), 1018–1025. https://doi.org/10.1002/anie.200604050.

Markowitz, V. M., Chen, I. M., Palaniappan, K., Chu, K., Szeto, E., Grechkin, Y., et al. (2012). IMG: The integrated microbial genomes database and comparative analysis system. *Nucleic Acids Research*, *40*(Database issue), D115–122. https://doi.org/10.1093/nar/gkr1044.

Mavromatis, K., Chu, K., Ivanova, N., Hooper, S. D., Markowitz, V. M., & Kyrpides, N. C. (2009). Gene context analysis in the integrated microbial genomes (IMG) data management system. *PLoS One*, *4*(11), e7979. https://doi.org/10.1371/journal.pone.0007979.

Medema, M. H., Takano, E., & Breitling, R. (2013). Detecting sequence homology at the gene cluster level with MultiGeneBlast. *Molecular Biology and Evolution*, *30*(5), 1218–1223. https://doi.org/10.1093/molbev/mst025.

Moses, J. E., & Moorhouse, A. D. (2007). The growing applications of click chemistry. *Chemical Society Reviews*, *36*(8), 1249–1262. https://doi.org/10.1039/b613014n.

Naing, S. H., Parvez, S., Pender-Cudlip, M., Groves, J. T., & Austin, R. N. (2013). Substrate specificity and reaction mechanism of purified alkane hydroxylase from the hydrocarbonoclastic bacterium Alcanivorax borkumensis (AbAlkB). *Journal of Inorganic Biochemistry*, *121*, 46–52. https://doi.org/10.1016/j.jinorgbio.2012.12.012.

Nallamsetty, S., & Waugh, D. S. (2006). Solubility-enhancing proteins MBP and NusA play a passive role in the folding of their fusion partners. *Protein Expression and Purification*, *45*(1), 175–182. https://doi.org/10.1016/j.pep.2005.06.012.

Newby, Z. E., O'Connell, J. D., 3rd, Gruswitz, F., Hays, F. A., Harries, W. E., Harwood, I. M., et al. (2009). A general protocol for the crystallization of membrane proteins for X-ray structural investigation. *Nature Protocols*, *4*(5), 619–637. https://doi.org/10.1038/nprot.2009.27.

Otten, L. G., Schaffer, M. L., Villiers, B. R., Stachelhaus, T., & Hollfelder, F. (2007). An optimized ATP/PP(i)-exchange assay in 96-well format for screening of adenylation domains for applications in combinatorial biosynthesis. *Biotechnology Journal*, *2*(2), 232–240. https://doi.org/10.1002/biot.200600220.

Paulucci, N. S., Dardanelli, M. S., & Garcia de Lema, M. (2014). Biochemical and molecular evidence of a delta9 fatty acid desaturase from Ensifer meliloti 1021. *Microbiological Research*, *169*(5–6), 463–468. https://doi.org/10.1016/j.micres.2013.08.003.

Pfeifer, B. A., Admiraal, S. J., Gramajo, H., Cane, D. E., & Khosla, C. (2001). Biosynthesis of complex polyketides in a metabolically engineered strain of E. coli. *Science*, *291*(5509), 1790–1792. https://doi.org/10.1126/science.1058092.

Pfeifer, B., Hu, Z., Licari, P., & Khosla, C. (2002). Process and metabolic strategies for improved production of Escherichia coli-derived 6-deoxyerythronolide B. *Applied and Environmental Microbiology*, *68*(7), 3287–3292.

Quadri, L. E. N., Weinreb, P. H., Lei, M., Nakano, M. M., Zuber, P., & Walsh, C. T. (1998). Characterization of Sfp, a Bacillus subtilis phosphopantetheinyl transferase for peptidyl carrier protein domains in peptide synthetases. *Biochemistry*, *37*(6), 1585–1595. https://doi.org/10.1021/Bi9719861.

Rosano, G. L., & Ceccarelli, E. A. (2014). Recombinant protein expression in Escherichia coli: Advances and challenges. *Frontiers in Microbiology*, *5*, 172. https://doi.org/10.3389/fmicb.2014.00172.

Ross, C., Scherlach, K., Kloss, F., & Hertweck, C. (2014). The molecular basis of conjugated polyyne biosynthesis in phytopathogenic bacteria. *Angewandte Chemie (International Ed. in English)*, *53*(30), 7794–7798. https://doi.org/10.1002/anie.201403344.

Sayanova, O., Beaudoin, F., Libisch, B., Castel, A., Shewry, P. R., & Napier, J. A. (2001). Mutagenesis and heterologous expression in yeast of a plant delta6-fatty acid desaturase. *Journal of Experimental Botany*, *52*(360), 1581–1585.

Shanklin, J., Achim, C., Schmidt, H., Fox, B. G., & Munck, E. (1997). Mossbauer studies of alkane omega-hydroxylase: Evidence for a diiron cluster in an integral-membrane enzyme. *Proceedings of the National Academy of Sciences of the United States of America*, *94*(7), 2981–2986.

Shanklin, J., & Cahoon, E. B. (1998). Desaturation and related modifications of fatty acids1. *Annual Review of Plant Physiology and Plant Molecular Biology*, *49*, 611–641. https://doi.org/10.1146/annurev.arplant.49.1.611.

Shanklin, J., Guy, J. E., Mishra, G., & Lindqvist, Y. (2009). Desaturases: Emerging models for understanding functional diversification of diiron-containing enzymes. *The Journal of Biological Chemistry*, *284*(28), 18559–18563. https://doi.org/10.1074/jbc.R900009200.

Shieh, P., Dien, V. T., Beahm, B. J., Castellano, J. M., Wyss-Coray, T., & Bertozzi, C. R. (2015). CalFluors: A universal motif for fluorogenic azide probes across the visible spectrum. *Journal of the American Chemical Society*, *137*(22), 7145–7151. https://doi.org/10.1021/jacs.5b02383.

Shieh, P., Hangauer, M. J., & Bertozzi, C. R. (2012). Fluorogenic azidofluoresceins for biological imaging. *Journal of the American Chemical Society*, *134*(42), 17428–17431. https://doi.org/10.1021/ja308203h.

Shieh, P., Siegrist, M. S., Cullen, A. J., & Bertozzi, C. R. (2014). Imaging bacterial peptidoglycan with near-infrared fluorogenic azide probes. *Proceedings of the National Academy of Sciences of the United States of America*, *111*(15), 5456–5461. https://doi.org/10.1073/pnas.1322727111.

Sletten, E. M., & Bertozzi, C. R. (2009). Bioorthogonal chemistry: Fishing for selectivity in a sea of functionality. *Angewandte Chemie (International Ed. in English)*, *48*(38), 6974–6998. https://doi.org/10.1002/anie.200900942.

Studier, F. W., & Moffatt, B. A. (1986). Use of bacteriophage T7 RNA polymerase to direct selective high-level expression of cloned genes. *Journal of Molecular Biology*, *189*(1), 113–130.

Thirumurugan, P., Matosiuk, D., & Jozwiak, K. (2013). Click chemistry for drug development and diverse chemical-biology applications. *Chemical Reviews*, *113*(7), 4905–4979. https://doi.org/10.1021/cr200409f.

Trivedi, O. A., Arora, P., Sridharan, V., Tickoo, R., Mohanty, D., & Gokhale, R. S. (2004). Enzymic activation and transfer of fatty acids as acyl-adenylates in mycobacteria. *Nature*, *428*(6981), 441–445. https://doi.org/10.1038/nature02384.

Villiers, B., & Hollfelder, F. (2011). Directed evolution of a gatekeeper domain in nonribosomal peptide synthesis. *Chemistry & Biology*, *18*(10), 1290–1299. https://doi.org/10.1016/j.chembiol.2011.06.014.

Wada, H., Avelange-Macherel, M. H., & Murata, N. (1993). The desA gene of the cyanobacterium Synechocystis sp. strain PCC6803 is the structural gene for delta 12 desaturase. *Journal of Bacteriology*, *175*(18), 6056–6058.

Wada, H., Schmidt, H., Heinz, E., & Murata, N. (1993). In vitro ferredoxin-dependent desaturation of fatty acids in cyanobacterial thylakoid membranes. *Journal of Bacteriology*, *175*(2), 544–547.

Wang, H., Klein, M. G., Zou, H., Lane, W., Snell, G., Levin, I., et al. (2015). Crystal structure of human stearoyl-coenzyme A desaturase in complex with substrate. *Nature Structural & Molecular Biology*, *22*(7), 581–585. https://doi.org/10.1038/nsmb.3049.

Wanibuchi, K., Zhang, P., Abe, T., Morita, H., Kohno, T., Chen, G. S., et al. (2007). An acridone-producing novel multifunctional type III polyketide synthase from Huperzia serrata. *FEBS Journal*, *274*(4), 1073–1082. https://doi.org/10.1111/j.1742-4658.2007.05656.x.

Wei, L., Hu, F., Shen, Y., Chen, Z., Yu, Y., Lin, C. C., et al. (2014). Live-cell imaging of alkyne-tagged small biomolecules by stimulated Raman scattering. *Nature Methods*, *11*(4), 410–412. https://doi.org/10.1038/nmeth.2878.

Whicher, J. R., Smaga, S. S., Hansen, D. A., Brown, W. C., Gerwick, W. H., Sherman, D. H., et al. (2013). Cyanobacterial polyketide synthase docking domains: A tool for engineering natural product biosynthesis. *Chemistry & Biology*, *20*(11), 1340–1351. https://doi.org/10.1016/j.chembiol.2013.09.015.

Yang, J. C., Madupu, R., Durkin, A. S., Ekborg, N. A., Pedamallu, C. S., Hostetler, J. B., et al. (2009). The complete genome of Teredinibacter turnerae T7901: An intracellular endosymbiont of marine wood-boring bivalves (shipworms). *PLoS One*, *4*(7) ARTN e6085https://doi.org/10.1371/journal.pone.0006085.

Yu, D., Xu, F., Zi, J., Wang, S., Gage, D., Zeng, J., et al. (2013). Engineered production of fungal anticancer cyclooligomer depsipeptides in Saccharomyces cerevisiae. *Metabolic Engineering*, *18*, 60–68. https://doi.org/10.1016/j.ymben.2013.04.001.

Zhang, W., Li, Y., & Tang, Y. (2008). Engineered biosynthesis of bacterial aromatic polyketides in Escherichia coli. *Proceedings of the National Academy of Sciences of the United States of America*, *105*(52), 20683–20688. https://doi.org/10.1073/pnas.0809084105.

Zhang, W., Ostash, B., & Walsh, C. T. (2010). Identification of the biosynthetic gene cluster for the pacidamycin group of peptidyl nucleoside antibiotics. *Proceedings of the National Academy of Sciences of the United States of America*, *107*(39), 16828–16833. https://doi.org/10.1073/pnas.1011557107.

Zhou, H., Zhan, J., Watanabe, K., Xie, X., & Tang, Y. (2008). A polyketide macrolactone synthase from the filamentous fungus Gibberella zeae. *Proceedings of the National Academy of Sciences of the United States of America*, *105*(17), 6249–6254. https://doi.org/10.1073/pnas.0800657105.

Zhu, X., Liu, J., & Zhang, W. (2015). De novo biosynthesis of terminal alkyne-labeled natural products. *Nature Chemical Biology*, *11*(2), 115–120. https://doi.org/10.1038/nchembio.1718.

Zhu, X., Shieh, P., Su, M., Bertozzi, C. R., & Zhang, W. (2016). A fluorogenic screening platform enables directed evolution of an alkyne biosynthetic tool. *Chemical Communications (Cambridge)*, *52*(75), 11239–11242. https://doi.org/10.1039/c6cc05990b.

Zhu, X., Su, M., Manickam, K., & Zhang, W. (2015). Bacterial genome mining of enzymatic tools for alkyne biosynthesis. *ACS Chemical Biology*, *10*(12), 2785–2793. https://doi.org/10.1021/acschembio.5b00641.

Zhu, X., & Zhang, W. (2015). Tagging polyketides/non-ribosomal peptides with a clickable functionality and applications. *Frontiers in Chemistry*, *3*, 11. https://doi.org/10.3389/fchem.2015.00011.

Ziemert, N., Alanjary, M., & Weber, T. (2016). The evolution of genome mining in microbes—A review. *Natural Product Reports*, *33*(8), 988–1005. https://doi.org/10.1039/c6np00025h.

CHAPTER FOUR

The Biochemistry and Structural Biology of Cyanobactin Pathways: Enabling Combinatorial Biosynthesis

Wenjia Gu*, Shi-Hui Dong†, Snigdha Sarkar*, Satish K. Nair†,1, Eric W. Schmidt*,1

*Department of Medicinal Chemistry, University of Utah, Salt Lake City, UT, United States
†Department of Biochemistry, University of Illinois at Urbana-Champaign, Urbana, IL, United States
[1]Corresponding authors: e-mail address: s-nair@life.illinois.edu; ews1@utah.edu

Contents

Methods in Enzymology, Volume 604
ISSN 0076-6879
https://doi.org/10.1016/bs.mie.2018.03.002

Abstract

Cyanobactin biosynthetic enzymes have exceptional versatility in the synthesis of natural and unnatural products. Cyanobactins are ribosomally synthesized and posttranslationally modified peptides synthesized by multistep pathways involving a broad suite of enzymes, including heterocyclases/cyclodehydratases, macrocyclases, proteases, prenyltransferases, methyltransferases, and others. Here, we describe the enzymology and structural biology of cyanobactin biosynthetic enzymes, aiming at the twin goals of understanding biochemical mechanisms and biosynthetic plasticity. We highlight how this common suite of enzymes may be utilized to generate a large array or structurally and chemically diverse compounds.

1. INTRODUCTION

Cyanobactins are common peptide natural products found in diverse cyanobacterial taxa (Fig. 1) (Donia, Ravel, & Schmidt, 2008; Leikoski, Fewer, & Sivonen, 2009). They are distributed through many different genera in the cyanobacterial lineage and have been estimated to occur in perhaps one in five strains (Donia & Schmidt, 2011; Leikoski et al., 2009). They are found in cyanobacteria that form massive oceanic blooms (Sudek, Haygood, Youssef, & Schmidt, 2006), in terrestrial and freshwater cyanobacteria (Ziemert et al., 2008), in symbionts of marine animals (Schmidt et al., 2005), in food products (Donia & Schmidt, 2011), and elsewhere. Overall, they represent a common but as yet to be understood component of the Earth's biochemistry.

1.1 What Are Cyanobactins?

The cyanobactins are defined not by their structural features, but rather by similarities in their biosynthesis (Arnison et al., 2013). All cyanobactins are ribosomally synthesized and posttranslationally modified (RiPP) natural products and, as such, are derived from a ribosomally synthesized precursor peptide, which is then posttranslationally modified by enzymes to produce elaborate products. The precursor peptide contains a core sequence that is

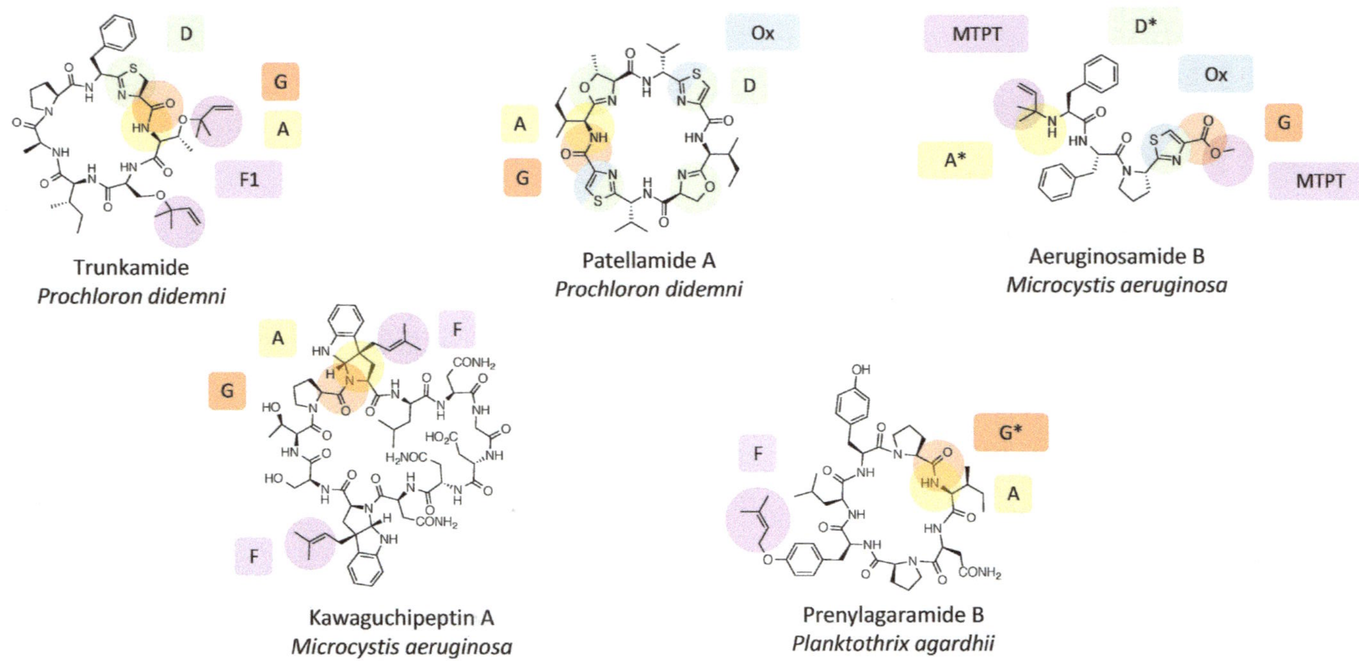

Fig. 1 Representative cyanobactins. Posttranslational modifications are indicated by *colors*, tied to enzymes that catalyze each transformation. * indicates hypothetical reaction; all others are experimentally defined.

found in the final natural product, flanked by leader and/or follower sequences that are not retained. Usually, the larger precursor peptide undergoes proteolysis to release the mature core, the bioactive natural product. Cyanobactins are defined as derived from a peptide similar to PatE, the precursor from the patellamide (*pat*) biosynthetic pathway. In addition, cyanobactins are synthesized by at least one subtilisin-like protease from the PatG family. Initially, cyanobactins were defined as being N–C circular, via the action of PatG-like macrocyclizing proteases, but several linear members have also been identified.

1.2 How Are Cyanobactins Made?

Despite this broad biochemical definition, cyanobactins share several structural features that make them recognizable. All known cyanobactins contain a heterocycle at the C-terminus of the core peptide (either proline or azol(in)e derived from cysteine, serine, or threonine) (Donia & Schmidt, 2010, 2011; Leikoski et al., 2013). Most members of the family are circular via ligation of their N- and C-termini. Commonly, they contain either heterocycles or isoprene units, and sometimes both. The azoline heterocycles may be further oxidized to thiazole or oxazole. A small subset of cyanobactins include linear compounds that are terminated with isoprene on the N-terminus and methyl ester on the C-terminus (Leikoski et al., 2013). Other posttranslational modifications are known beyond this core set, and there are still several proteins and domains with no defined function. Together this, common set of enzymes produces a vast array of structural diversity.

1.3 What Are the Applications of Cyanobactins?

Because they are synthesized from genetic (i.e., ribosomal) precursors, RiPPs in general are thought to provide good platforms for biosynthetic engineering (Burkhart, Kakkar, Hudson, van der Donk, & Mitchell, 2017; Ozaki et al., 2017; Sardar & Schmidt, 2016; Zhang, Li, & Kelly, 2016). Indeed, there is a broad range of mutability evident in nature, and the tolerance of cyanobactin biosynthetic pathways for a range of substrates makes the system useful in biotechnology. Because of this broad substrate tolerance, cyanobactins have seen many applications in the synthesis of designed, engineered compounds. Such biotechnological applications, and the underlying ideas about biosynthetic plasticity, have been reviewed recently (Gu & Schmidt, 2017).

1.4 What Do Cyanobactins Do in Nature?

Beyond the application to synthetic biology, cyanobactins are also of interest for their intrinsic biological activities (Donia & Schmidt, 2010).

Several cyanobactins chelate metals (Bertram & Pattenden, 2007; Comba et al., 2017), while others block P-glycoprotein (Aller et al., 2009; Williams & Jacobs, 1993). Others are cytotoxic (Degnan et al., 1989), and some may even block potassium channels (Tianero et al., 2016). How, and indeed whether, these activities relate to the natural role(s) of cyanobactins remains a mystery.

2. BIOSYNTHETIC DIVERSITY OF THE CYANOBACTIN FAMILY

2.1 General Biosynthetic Pathway

The first cyanobactin biosynthetic pathway to be characterized was that for patellamides A and C (*pat*), so the biosynthetic schemes for cyanobactins follow *pat* nomenclature (see Box 1) (Schmidt et al., 2005). The canonical biosynthetic pathway is thought to proceed in the following order (Fig. 2). The precursor peptide, e.g., PatE, is first synthesized on the ribosome. Commonly, PatE and relatives contain more than one core peptide, flanked by enzyme recognition sequences (RSs), where the core peptides are known as "cassettes," highlighting their repetitive organizational structure. The enzymatic modifications occur at the cassettes. For example, a PatE homologue (i.e., TruE3) containing three cassettes can encode three different natural products of different amino acid sequences. Subsequently, if the natural products contain heterocycles, the heterocyclase, e.g., PatD, acts on PatE to produce a linear peptide product that contains azolines. A subtilisin-like serine protease, e.g., PatA, then liberates free N-termini in each cassette of PatE. A second subtilisin-like serine protease, PatG, cleaves the C-termini of the core peptide in each cassette, usually in tandem with N–C circularization to yield cyclic peptides. Other enzymatic tailoring reactions usually take place after this core set of modifications is installed. In *pat* and related pathways, oxidation by the oxidase domain of PatG is one such potentially late-stage modification. Sometimes, the pathways may diverge from this order or exhibit greater complexity.

2.2 RSs in Precursor Peptides

RSs are short sequence elements in PatE-like precursor peptides that are paired with each enzyme (Fig. 3). As recruitment of the precursor depends only on the conserved RS, rather than the hypervariable core sequence, the modification enzymes are tolerant of diverse core sequences. In patellamide biosynthesis, recognition sequence I (RSI) binds to and directs the PatD

BOX 1 Cyanobactin Terminology

Substrates in cyanobactin biosynthesis

E (e.g., PatE): A precursor peptide encoding the substrates for cyanobactin biosynthetic enzymes.

Precursor peptide: A peptide substrate in ribosomal peptide biosynthesis that encodes the following elements:

Leader peptide: The portion of a precursor peptide that precedes the core peptide, and that is cleaved by proteolysis to afford the mature natural product.

Core peptide: The portion of the precursor peptide that directly encodes the natural product.

Recognition sequence: An element found in some precursor peptides that binds to an enzyme and thus directs the biosynthetic process, but that is not part of the core peptide or the mature natural product.

Cassette: A core peptide and its flanking RSs, used when multiple core peptides are present in a single precursor. For example, in cyanobactin biosynthesis a cassette consists of RSII-core peptide-RSIII.

Follower peptide: An element found in some precursor peptides that follows a core peptide. This term is ambiguous in cyanobactin biosynthesis because of the presence of multiple cassettes in most precursor peptides, and thus "recognition sequence" is preferred.

Proteins in cyanobactin biosynthesis

A (e.g., PatA): Protease proteins responsible for cleaving the leader peptide and liberating free N-termini for cyclization. They recognize RSII and contain two domains: an S8A protease domain and a DUF.

B and C (e.g., PatB and PatC): Short proteins with no known function and no similarity to proteins of known function.

D (e.g., PatD): Proteins responsible for heterocyclizing cysteine (and in some cases serine/threonine) to produce azolines. They recognize RSI and contain three domains: a substrate recognition domain, an E1-like domain, and a YcaO domain.

F (e.g., PatF): Some variants of these proteins have no known function, while others are single domain ABBA prenyltransferases that react with serine/threonine, tyrosine, tryptophan, or the N-terminus.

G (e.g., PatG): Macrocyclases that recognize and cleave RSIII in tandem with circularization. The minimum domain set consists of N- and C-terminal DUFs flanking an S8A protease responsible for circularization or hydrolysis of RSIII. Some variants also contain an oxidase domain that convert azolines to azoles, while others contain domains such as methyltransferases with no known function.

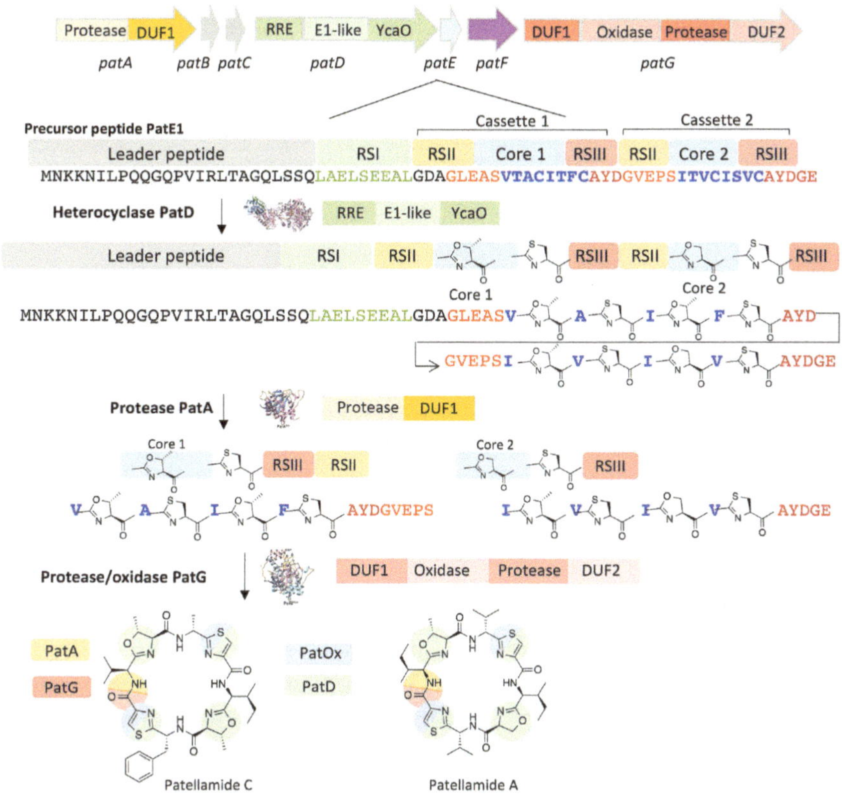

Fig. 2 Canonical cyanobactin biosynthetic pathway. *Top*: The *pat* pathway to patellamides, with an expansion showing the amino acid sequence of PatE1, the substrate for biosynthetic enzymes. The precursor peptide substrate may be demarcated as described in Box 1. The order of the modification occurs such that heterocyclization occurs first, followed by proteolysis and macrocyclization. Late-stage tailoring may further elaborate the modified peptides.

heterocyclase (Donia & Schmidt, 2011; Koehnke et al., 2015), while RSII and RSIII direct the activities of PatA and PatG proteases, respectively (Lee, McIntosh, Hathaway, & Schmidt, 2009). During the course of biosynthesis, the conserved RS elements are excised leaving a natural product peptide that consists only of the hypervariable sequence. For example, of the 29 variants of PatE that were initially described, all had identical sequences except in the cores, resulting in products of different chemical structures (Donia et al., 2006). The model of RS binding is accurate but oversimplified due to the intrinsic binding energy of the core peptide.

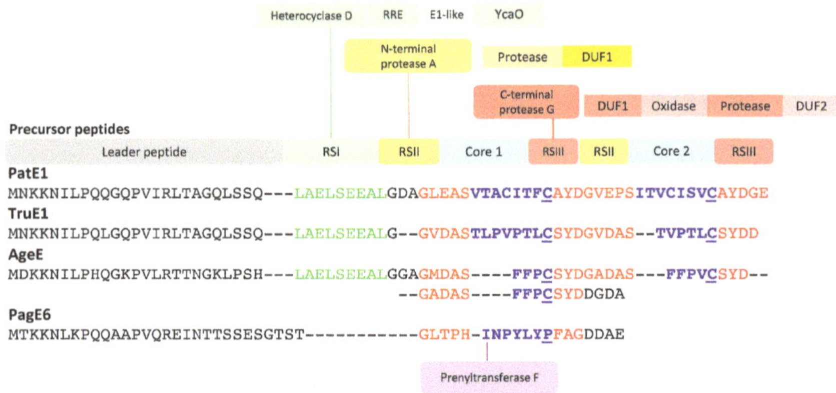

Fig. 3 Recognition sequences and precursor peptides. Shown are example sequences of cyanobactin precursors, with color-coded RSs tied to modifying enzymes. The domain architecture of each enzyme is laid out adjacent to the enzyme name. Prenyltransferase does not require a recognition sequence. Many cyanobactin precursor peptides contain multiple core sequences that are processed into different products.

2.3 Proteins of Unknown Function

Even within the well-characterized *pat* pathway, there are several proteins or domains with no known function (Agarwal, Pierce, McIntosh, Schmidt, & Nair, 2012; Lee et al., 2009; Schmidt et al., 2005). PatA and PatG proteases contain homologous C-terminal domains of unknown function (DUFs). Additionally, PatG and relatives contain an uncharacterized N-terminal domain that flanks the oxidase domain. *patF* encodes an ABBA prenyltransferase-like protein that is not required for biosynthesis in vitro but is required for heterologous expression in *Escherichia coli* (*E. coli*) (Bent et al., 2013; Donia et al., 2006). PatB and PatC are two further proteins that do not prevent synthesis in vivo or in vitro but negatively affect synthesis in *E. coli* if removed (by reducing yield, for example) (Donia et al., 2006; Houssen et al., 2014). B-, C-, and F-like proteins are common in cyanobactin pathways.

2.4 Two Basic Biosynthetic Subdivisions

Following *pat*, many other types of cyanobactin gene clusters have been described (Donia et al., 2008; Donia & Schmidt, 2011). The major division between cyanobactin subtypes is those with heterocyclases and those that lack heterocyclases. When the heterocyclase is present, precursor peptides are cleaved on the C-terminal side at azol(in)e heterocycles derived from cysteine, serine, or threonine. This heterocycle is required for cleavage or macrocyclization by PatG-like enzymes. The structure of the precursor peptide reflects this division: in pathways containing the heterocyclase, RSI is found in the precursor peptide, where it directs heterocyclization of the core (Fig. 3). However, in pathways where the heterocyclase is absent, the precursor peptide lacks RSI. The site of macrocyclization is still a heterocycle, but instead of azol(in)e, it is the amino acid proline.

2.5 Cassette Duplication in Precursor Peptides

PatE relatives vary in sequence, RS content, and in the variability of the core peptide (Donia et al., 2006; Donia & Schmidt, 2011). In addition, other PatE-like precursor peptides may vary in number of cassettes, from as few as one to as many as 10 (Leikoski et al., 2013), and in copy number of PatE-like sequences in the genome. When multiple cassettes are present, they can each encode a different natural product, or potentially multiple copies of the same natural product, or mixtures thereof. When multiple precursors are present, they usually contain different cores and several may be nonfunctional pseudogenes (Donia & Schmidt, 2011; Leikoski et al., 2009). This potentially reflects a duplication–divergence–recombination pathway to functional variation in cyanobactin natural products. For example, multiple precursor peptides may recombine to generate highly conserved RS sequences paired with new cores.

2.6 Other Enzymes

Cyanobactin pathways also often include prenyltransferases, methyltransferases, and/or other enzymes. PatF-like prenyltransferases add dimethylallyl pyrophosphate (DMAPP) or, occasionally, geranyl pyrophosphate (GPP) to the side chains of the core peptide (Leikoski et al., 2012). The prenyltransferases are known to prenylate in either the forward or reverse orientation in respect to DMAPP. TruF is specific to serine and threonine O-prenylation (Donia et al., 2008; Sardar, Lin, & Schmidt, 2015; Tianero, Donia, Young, Schultz, & Schmidt, 2012; Tianero et al., 2016), while other variants prenylate tyrosine (Hao et al., 2016; McIntosh, Donia, Nair, & Schmidt, 2011; McIntosh, Lin, Tianero, & Schmidt, 2013) or tryptophan (Parajuli et al., 2016). While each prenyltransferase appears to be relatively selective for the amino acid, orientation, and isoprene donor, there is much less selectivity as far as peptide structure, and no known associated RS (Hao et al., 2016; McIntosh et al., 2011). One variant homologue carries out prenylation of the N-terminus, rather than the side chain, on peptide substrates (Leikoski et al., 2013; Sardar et al., 2017).

Some pathways contain oxidase enzymes that convert thiazoline, and sometimes also oxazoline, into thiazole and oxazole, respectively (Donia & Schmidt, 2010). Methyltransferases are also very common within the genetic organization of cyanobactin pathways, but thus far only one has been characterized (Donia & Schmidt, 2011; Leikoski et al., 2009, 2013; Sardar et al., 2017).

2.7 Unusual Cyanobactin Pathways

The *tru* biosynthetic pathway to trunkamide has several peculiarities in comparison to other cyanobactin pathways (Donia, Fricke, Ravel, & Schmidt, 2011; Donia et al., 2008). Like *pat*, *tru* is found in symbiotic *Prochloron* living in animals. The products of *pat* pathway are heterocyclic at cysteine, serine, and threonine, whereas the *tru* products are heterocyclic only at cysteine and reverse prenylated on serine and threonine. *tru* is nearly identical to *pat* on the 5′- and 3′-ends of the cluster, and often the clusters are 100% identical in those regions. These nearly identical genes and intergenic sequences include *pat/truA*, *pat/truB*, *pat/truC*, the N-terminal domain of *pat/truD*, and the protease and DUFs of *pat/truG*. Where the pathways are not identical, the resulting enzyme activities explain the differences in chemistry. Notably, in *tru*, the PatF-like protein TruF1 from the *tru* pathway carries out O-prenylation on serine and threonine residues, whereas PatF itself shows no activity (Sardar, Lin, et al., 2015; Tianero et al., 2016).

Another divergent pathway is that involved in the biosynthesis of linear cyanobactins, such as aeruginosamides (Leikoski et al., 2013; Sardar et al., 2017). In the *age* pathway, the PatG homologue AgeG produced a linear peptide product, rather than performing macrocyclization. A bifunctional methyl/prenyl transferase protects the N- and C-termini with prenyl and methyl groups, respectively.

2.8 Stereochemistry

Cyanobactins characterized to date contain all L-amino acids, except in or adjacent to the C=N bond of azoline heterocycles, where epimerization can occur. This process is known to be spontaneous (e.g., not enzyme catalyzed) and occurs after azoline formation and macrocyclization (Donia et al., 2008; Donia, Ruffner, Cao, & Schmidt, 2011; McKeever & Pattenden, 2003; Tianero et al., 2012, 2016). It is likely caused by two factors: (1) the intrinsic chemistry of azoline heterocycles and (2) the thermodynamics of the macrocycle. The C=N bond makes adjacent protons more acidic due to the potential for transient double bond migration. Macrocyclic cyanobactins are known to adopt specific shapes (Donia & Schmidt, 2010). It is likely that epimerization allows the macrocycles to adopt the thermodynamically preferred conformation.

It was long presumed that epimerization is spontaneous, rather than enzymatic (Wipf & Uto, 2000; Zabriskie, Foster, Stout, Clardy, & Ireland, 1990). Because of the facile inversion of configuration that occurs adjacent to thiazoline, the epimerization process has caused some vexation in the field, notably in the assignment of the phenylalanine configuration in trunkamide (Wipf & Uto, 2000). Evidence in favor of nonenzymatic epimerization was achieved during heterologous expression of trunkamide and certain patellins (Donia et al., 2008; Donia, Ruffner, et al., 2011; Tianero et al., 2012, 2016). Notably, in trunkamide, the L-Phe isomer is observed first, whereupon it gradually transforms into the D-isomer over hours to days either in vivo or after organic extraction (Donia et al., 2008; Donia, Ruffner, et al., 2011). Similarly, subsequent in vitro studies of patellamide biosynthesis revealed an identical process, where heterocyclization occurs following thiazoline formation and macrocyclization but before oxidation to thiazole (Houssen et al., 2014). During the expression of hundreds of analyzed trunkamide and patellin derivatives, a doubling of HPLC-MS-derived extracted ion chromatographic peaks is commonly observed that later resolves into single peaks, presumably because spontaneous epimerization converts L to a D–L mixture, followed by all D adjacent to thiazoline (Ruffner, Schmidt, & Heemstra, 2015; Tianero et al., 2012). It is also possible that, in some substrates, epimerization may be fast enough to occur on the macrocyclase.

3. HETEROCYCLASE

3.1 Discovery and Initial Characterization

The prevalence of peptide-based secondary metabolites that contain oxazole and thiazole heterocycles was appreciated from the initial discovery and structural characterization of several classes of natural products that bear such modification. The thiopeptide antibiotics nosiheptide (Pascard, Ducruix, Lunel, & Prange, 1977) and thiostrepton (Dutcher & Vandeputte, 1955) were among the earliest azolic compounds discovered, and metabolic labeling studies were consistent with a ribosomally synthesized peptide as the biosynthetic precursor (Mocek et al., 1989). Studies on the DNA gyrase inhibitor microcin B17, guided by identification of the corresponding biosynthetic gene cluster, identified an unmodified, genetically encoded peptide as the precursor for this molecule (Davagnino, Herrero, Furlong, Moreno, & Kolter, 1986). Subsequent characterization of the posttranslational modifications (Genilloud, Moreno, & Kolter, 1989; San Millan, Kolter, & Moreno, 1985) and partial in vitro reconstitution studies (Li, Milne, Madison, Kolter, & Walsh, 1996) identified a heterotrimeric complex of three polypeptides (McbB/C/D) that carried out posttranslational modifications on Cys, Ser, or Thr residues on the precursor peptide to install the thiazole/oxazole heterocycles found in the final product. Briefly, McbB (a member of the E1-ubiquitin-activating-like gene family) and McbD (an ortholog of the *E. coli* YcaO gene product) were proposed to promote the heterocyclization of nucleophile-containing residues in the precursor peptide, in an ATP-dependent manner, to yield the corresponding azolines. More robust reconstitution experiments firmly established that the YcaO protein catalyzed the ATP-dependent backbone amide activation to facilitate heterocyclization on the peptide substrate engaged by the E1-like protein (Dunbar, Melby, & Mitchell, 2012).

Patellamide cyanobactins are heterocyclic at cysteine, serine, and threonine. The patellamide (*pat*) pathway was identified by sequencing and expressed in *E. coli* (Schmidt et al., 2005). Although the *patD* gene lacked similarity to previously characterized proteins, it was predicted to encode an adenylating enzyme–hydrolase hybrid based upon a short series of ~10 residues similar to the ATP-grasping region of an otherwise unrelated protein from microcin C7 biosynthesis (Gonzalez-Pastor, San Millan, Castilla, & Moreno, 1995). It is now known that PatD represents a class of three domain heterocyclases found in many different types of organisms (Lee et al., 2008). Despite the absence of sequence similarity, PatD and relatives are fusions of the homologous microcin B17 heterocyclase subunits, McbB and McbD (Li et al., 1996). PatG also contains an oxidase domain that bears sequence similarity to microcin B17 oxidase McbC (Schmidt et al., 2005). A second cyanobactin variation of PatD, represented by TruD in trunkamide and patellin biosynthesis, heterocyclizes primarily cysteine, and not serine and threonine (Donia et al., 2008). Both PatD and TruD were expressed as part of cyanobactin pathways in *E. coli*, but as removal of PatD or TruD from the biochemical pathway led to a loss of detectable products, their biochemical roles could not be elucidated in vivo (Donia et al., 2006, 2008). However, the totality of evidence strongly supported their proposed function as heterocyclases. Crucially, work with the *Streptomyces* goadsporin pathway involved a homologous enzyme (Onaka, Nakaho, Hayashi, Igarashi, & Furumai, 2005).

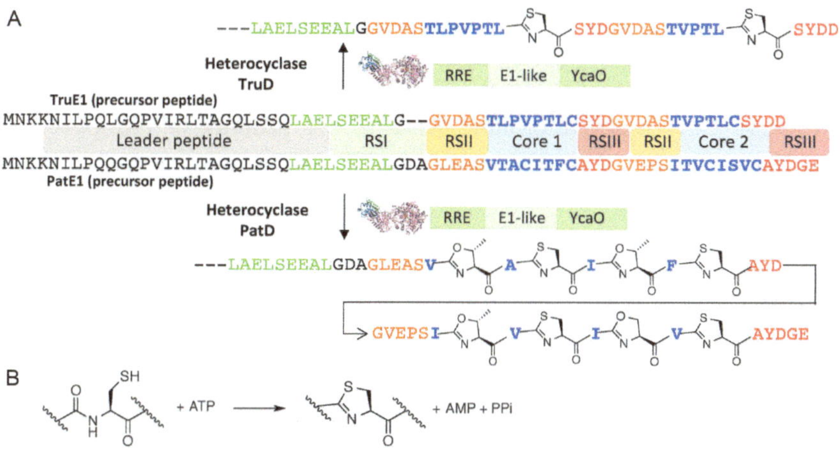

Fig. 4 Heterocyclase reactions. (A) TruD and PatD bind to RSI and catalyze heterocycle formation. TruD, ~100% identical to PatD in the RiPP precursor peptide recognition element (RRE) and E1-like domain, is 77% identical to PatD in YcaO domain. PatD synthesizes thiazoline and oxazoline, while TruD is a thiazoline specialist. As defined by Walsh, thiazoline is energetically more favorable (Belshaw, Roy, Kelleher, & Walsh, 1998). (B) TruD converts ATP to AMP and PPi, whereas some other heterocyclases such as those involved in linear azol(in)e-containing peptides (LAPs) or thiopeptide biosynthesis yield ADP and Pi as the products.

In 2010, PatD and TruD were purified and biochemically analyzed, confirming their roles as heterocyclases (Fig. 4) (McIntosh, Donia, & Schmidt, 2010). This work demonstrated that the heterocyclases dictated which residues were cyclic in the final product, and that the chemical differences resulting from *tru* were not due to other factors such as blocking by prenyltransfer. TruD in particular has since been extensively characterized, as have several other cyanobactin heterocyclases that are similar to either PatD or TruD (Koehnke et al., 2015; McIntosh & Schmidt, 2010; Sardar, Lin, et al., 2015; Sardar, Pierce, McIntosh, & Schmidt, 2015). A recent review on YcaO enzymes provides excellent background and comparison between cyanobactin heterocyclases and those from other types of biochemical pathways (Burkhart, Schwalen, Mann, Naismith, & Mitchell, 2017).

3.2 Substrates and Products

Heterocyclases use the PatE-like precursor peptide and ATP (or to a minor degree, GTP) as substrates (McIntosh, Donia, et al., 2010; McIntosh & Schmidt, 2010). ATP hydrolysis accompanies heterocyclization in a 1:1 manner, in analogy to other YcaO enzymes (McIntosh & Schmidt, 2010).

Unlike earlier characterized heterocyclases, which generate ADP and phosphate, cyanobactin enzymes hydrolyze ATP to AMP and pyrophosphate in tandem with heterocyclization (Koehnke, Bent, et al., 2013; Koehnke et al., 2015). An early report indicating ADP as the product of the D reaction was likely due to a minor amount of contaminating adenylate kinase (unpublished data).

Heterocyclase is thought to act at the first step of the pathway, on the unmodified precursor peptide (McIntosh, Donia, et al., 2010). This was based upon the initial observation that all PatE-like precursors contain the RSI when a heterocyclase is present in the gene cluster (Donia & Schmidt, 2011). Since the subsequent step in the pathway would involve proteolytic removal of RSI, this biochemical step likely proceeds first in vivo. Additionally, initial in vitro data suggested that heterocyclization requires the intact precursor peptide, although in some cases in vitro substrates absent in RSI are still heterocyclized (Goto, Ito, Kato, Tsunoda, & Suga, 2014; Houssen et al., 2014; Koehnke, Bent, et al., 2013; Koehnke et al., 2015; McIntosh, Donia, et al., 2010; McIntosh & Schmidt, 2010; Sardar, Lin, et al., 2015; Sardar, Pierce, et al., 2015). Substantial data described later delineate the interaction between RSI and the RiPP precursor peptide recognition element (RRE) present in the heterocyclases (Koehnke et al., 2015; Sardar, Pierce, et al., 2015).

Upon binding RSI, heterocyclase catalyzes the conversions of one or more cysteine, serine, and/or threonine into the corresponding azoline (McIntosh & Schmidt, 2010). Different heterocyclases are chemoselective, with a subset capable of heterocyclizing both sulfur and oxygen nucleophiles (e.g., PatD and relatives) and another subset specializing in sulfur nucleophiles (e.g., TruD and relatives). It should be emphasized that in artificial substrates, some oxygen heterocyclization was detected in vitro with TruD when serine was in the ideal position for heterocyclization, indicating that oxygen-based heterocyclization is likely much slower, rather than forbidden. The theoretical underpinning of this difference was described in early work by Walsh et al. on microcin biosynthesis (Belshaw et al., 1998; Kelleher, Hendrickson, & Walsh, 1999; Li et al., 1996).

PatE-like precursors contain multiple heterocyclizable residues, but only those within core peptides are heterocyclized (Goto et al., 2014; McIntosh, Donia, et al., 2010; McIntosh & Schmidt, 2010). This indicates a positional dependence for heterocyclase, both in relation to RSI and to other sequence elements. Interestingly, PatD and homologues catalyze heterocyclization of appropriate residues present in multiple cassettes on a single precursor peptide.

The native substrates for PatD are at least 29 different natural precursor peptides, which contain identical RSs and hypervariable core sequences (Donia et al., 2006). At least six natural TruD substrates exhibit identical properties (Donia et al., 2008). Several types of heterocyclases have been identified that process multiple different precursor peptides with single cores (e.g., ThcD), where the RS elements are not identical (Donia & Schmidt, 2011; Koehnke et al., 2015; Sardar, Pierce, et al., 2015). Thus, the characterized PatD-like enzymes are broadly accepting of diverse substrates.

Extensively modified variants of the precursor peptide itself have been examined experimentally. In analogy with work on lanthipeptides (Oman, Knerr, Bindman, Velasquez, & van der Donk, 2012), RSI has been fused with TruD, accelerating the enzymatic reaction rate on short cassette peptides (Sardar et al., 2017). Portions of the precursor pared down to a minimum containing just RSI have been designed and fused to various unnatural derivatives (Goto et al., 2014; Koehnke et al., 2015; Sardar, Lin, et al., 2015; Sardar, Pierce, et al., 2015). In some cases, even cassettes alone without RSI have been successfully cyclized (Koehnke et al., 2015; Sardar, Pierce, et al., 2015). For example, PatD has been used to synthesize up to 18 heterocycles on a single "cassette" (Goto et al., 2014). Mutants of the RSII also resulted in fully heterocyclized products (Goto et al., 2014). Although PatD tolerates noncanonical C-terminus RSIII, it still influenced the catalytic efficiency to some extent (Goto & Suga, 2016).

The order of biosynthetic steps catalyzed by heterocyclases has been measured in various ways. In initial characterization, the enzymes were determined to be distributive when using multicassette substrates (McIntosh, Donia, et al., 2010; McIntosh & Schmidt, 2010). As found in microcin B17 synthetase (Kelleher et al., 1999), TruD may sometimes release the substrate between heterocyclization reactions. As a result, both the partially heterocyclized products and the fully heterocyclized products can be detected by MS spectrometry, SDS-PAGE electrophoresis, or NMR methods (Koehnke, Bent, et al., 2013; McIntosh, Donia, et al., 2010; McIntosh & Schmidt, 2010). When a single-cassette substrate was employed, TruD exhibited a strict directionality (Koehnke, Bent, et al., 2013). This indicates a potential complex mechanism and reaction order underlying heterocyclase activity, which is perhaps similar to that rigorously characterized in prochlorosin lanthipeptide synthesis (Thibodeaux, Ha, & van der Donk, 2014).

The diversity of naturally occurring core sequences that can be modified has been exploited for the synthesis of many derivatives (Houssen et al., 2014; Koehnke, Morawitz, et al., 2013; McIntosh, Donia, et al., 2010;

Ruffner et al., 2015; Sardar, Lin, et al., 2015; Tianero et al., 2012). Artificial substrates can be categorized as those that utilize the native precursor peptide elements, and those with modified precursor peptide arrangements. Using native (or nearly native) precursors, sequence-diverse derivatives have been produced in vivo and in vitro by modifying core peptide sequences. *E. coli* expression libraries encoding millions of derivatives have been prepared and partially characterized (Ruffner et al., 2015). Nonproteinogenic amino acids have been incorporated into backbone peptides in vivo and in vitro (Koehnke, Morawitz, et al., 2013; Tianero et al., 2012). Intriguingly, unnatural azolines have been prepared, including selenazoline (Koehnke, Morawitz, et al., 2013).

3.3 Enzyme Structure and Function

A sequence similarity diagram of fused heterocyclases shows the expected distribution of thiopeptide, linear azol(in)e-containing peptide (LAP), and trifolitoxin like biosynthetic clusters across various genera (Fig. 5). These fusions contain only the E1-like and YcaO proteins and lack the N-terminal extension that harbors the RRE. Likewise, the cyanobactin pathways also cluster together (cluster 4), presumably due to the presence of the RRE in these proteins. Two noteworthy features are evident from

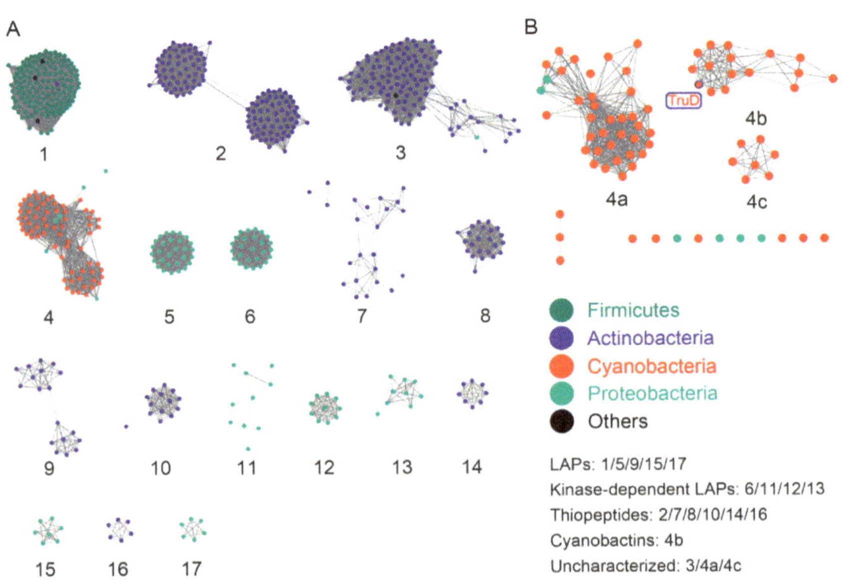

Fig. 5 Sequence similarity analysis of heterocyclases. (A) All heterocyclase domain-containing proteins. (B) Expansion of cyanobacterial heterocyclase group and close relatives. The cyanobactin group is found in 4b of this branch.

this analysis. First, actinobacteria contain a large clade of sequences (cluster 3) that contain a polypeptide with the RRE/E1-like/YcaO architecture, but their corresponding pathways are missing all other biosynthetic genes. The products of these pathways have yet to be determined. Second, the cyanobactin cluster also contains polypeptides with the similar RRE/E1-like/YcaO architecture, but the corresponding pathways contain a large protein (~1900 amino acids) in place of all of the other cyanobacterial biosynthetic genes. Such biosynthetic clusters are all found in cyanobacteria, but the products of these pathways are likewise enigmatic for now.

Experiments using isotopic labeling and substrate analogs firmly establish that the E1-like protein serves as a docking scaffold, while the YcaO protein catalyzes the ATP-dependent amide backbone activation to facilitate heterocyclization (Dunbar et al., 2012). Structural studies of TruD, in the absence of any bound ligands, reveal a tripartite domain organization, in which domains 1 and 2 share the modest structural similarity with the MccB adenylase involved in microcin C7 biosynthesis (Fig. 6) (Regni et al., 2009). However, topological and sequence differences result in the elimination of residues that are necessary for nucleotide binding by MccB, and the binding site for ATP could not be reconciled. Subsequently, the crystal structure of *E. coli* YcaO, an orphan family member that lacks a corresponding docking domain, in complex with a bound ATP analog, identified a new ATP-binding

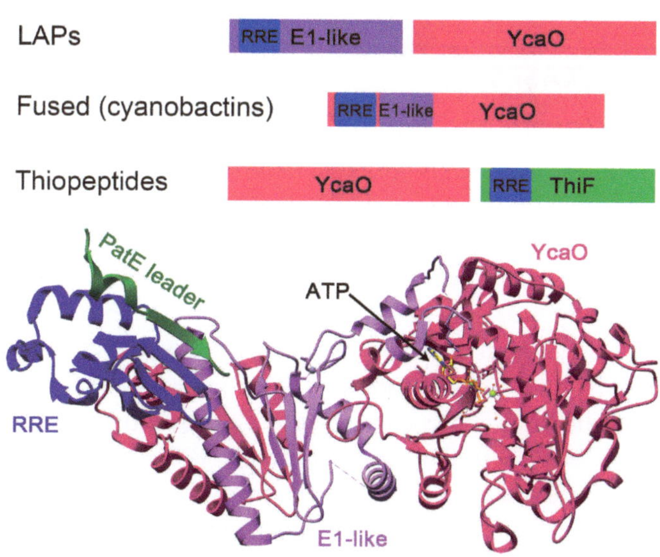

Fig. 6 Structures of the cyanobactin heterocyclase. Crystal structures of LynD (PDB code 4V1T) showing RSI from the PatE precursor peptide (in *green*) bound to the RRE (in *blue*) and ATP in the YcaO active site (*right*, in *purple*). The E1-like domain is in *violet*.

motif (Dunbar et al., 2014) and provided conclusive evidence for the YcaO as the catalytic component of the heterocyclase. In 2015, the structure of LynD, the heterocyclase from the *Lyngbya* sp. aestuaramide pathway, was determined in complex with nucleotide analogs and a synthetic precursor derived from a patellamide cassette (Koehnke et al., 2015). This confirmed the YcaO domain as the site for nucleotide binding, and engagement of the leader sequence by the RRE, as first observed in the cocrystal structure of the dehydratase involved in lanthipeptide biosynthesis (Ortega et al., 2015). Moreover, in this work, a fusion of RSI with the RRE domain led to in *trans* synthesis of derivatives, in analogy to previous work with fusion of lanthipeptide leader peptides to biosynthetic enzymes (Oman et al., 2012).

More recent studies have resulted in the identification of single-chain heterocyclase fusion enzymes present in the biosynthetic pathways for LAPs (Cox, Doroghazi, & Mitchell, 2015) and thiopeptides (Morris et al., 2009; Wieland Brown, Acker, Clardy, Walsh, & Fischbach, 2009; Yu et al., 2009). A distinguishing feature of the fused heterocyclase from cyanobactin pathways is that the RiPP precursor peptide recognition element (RRE), which recruits the substrate peptide (Burkhart, Hudson, Dunbar, & Mitchell, 2015), is universally at the N-terminus of the heterocyclases from cyanobactin clusters, while the enzymes from thiopeptide and LAP pathways require an auxiliary peptide-binding protein (Dunbar, Tietz, Cox, Burkhart, & Mitchell, 2015).

4. PROTEASES

4.1 Discovery and Initial Characterization

Proteases PatA and PatG were first characterized in the discovery of the *pat* gene cluster (Fig. 7) (Schmidt et al., 2005). Heterologous expression of the gene cluster composed of genes *patA–patG* in an *E. coli* strain demonstrated that this cluster carried out the biosynthesis of patellamides. Embedded within the cluster was a gene (*patE*) whose product bore sequences for patellamide C and A, flanked by leader and follower sequences that are not found in the final product. Analyses of the gene cluster failed to identify traditional candidates (such as condensation domains, thioesterase domains, or ATP-grasp enzymes) that could facilitate the necessary α-amino to α-carboxylate transamidation reaction necessary to produce the macrocyclic product. Rather, the *patA* and *patG* genes encoded large polypeptides, each of which contained domains that resembled subtilisin-related (family S8A) serine proteases (Fig. 8). Based on this sequence homology, both PatA

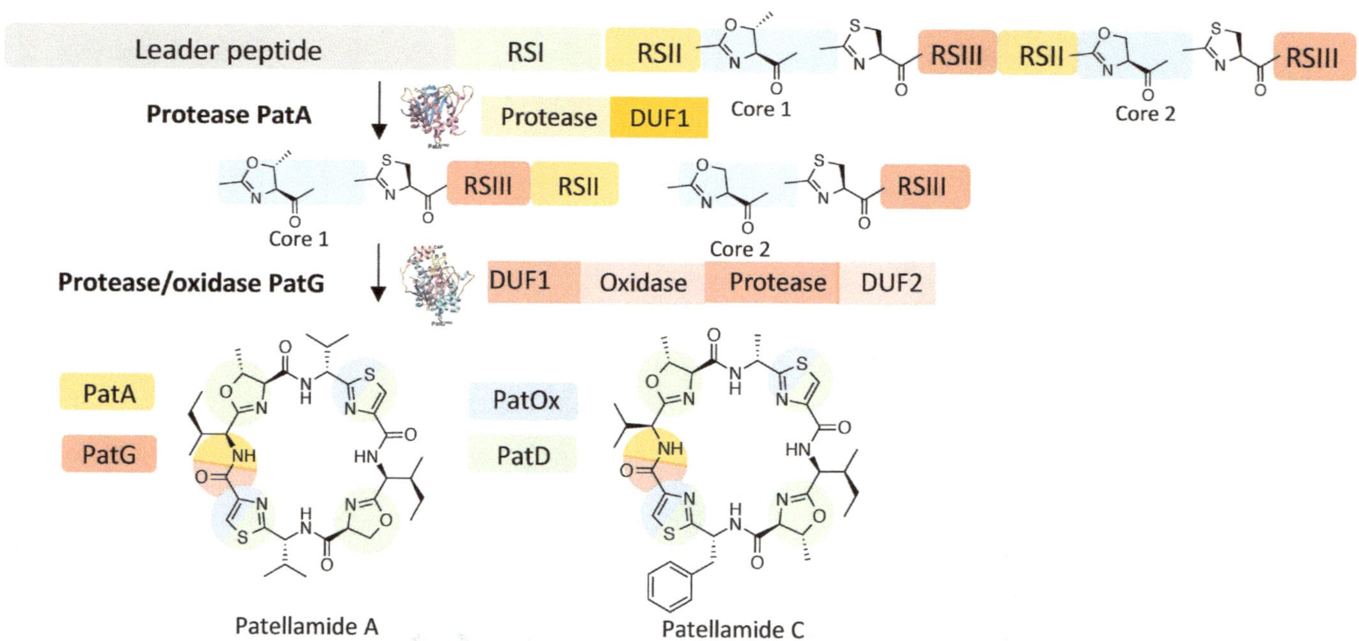

Fig. 7 Macrocyclization by PatA and PatG. PatA frees an N-terminus, which is then available for macrocyclization by PatG.

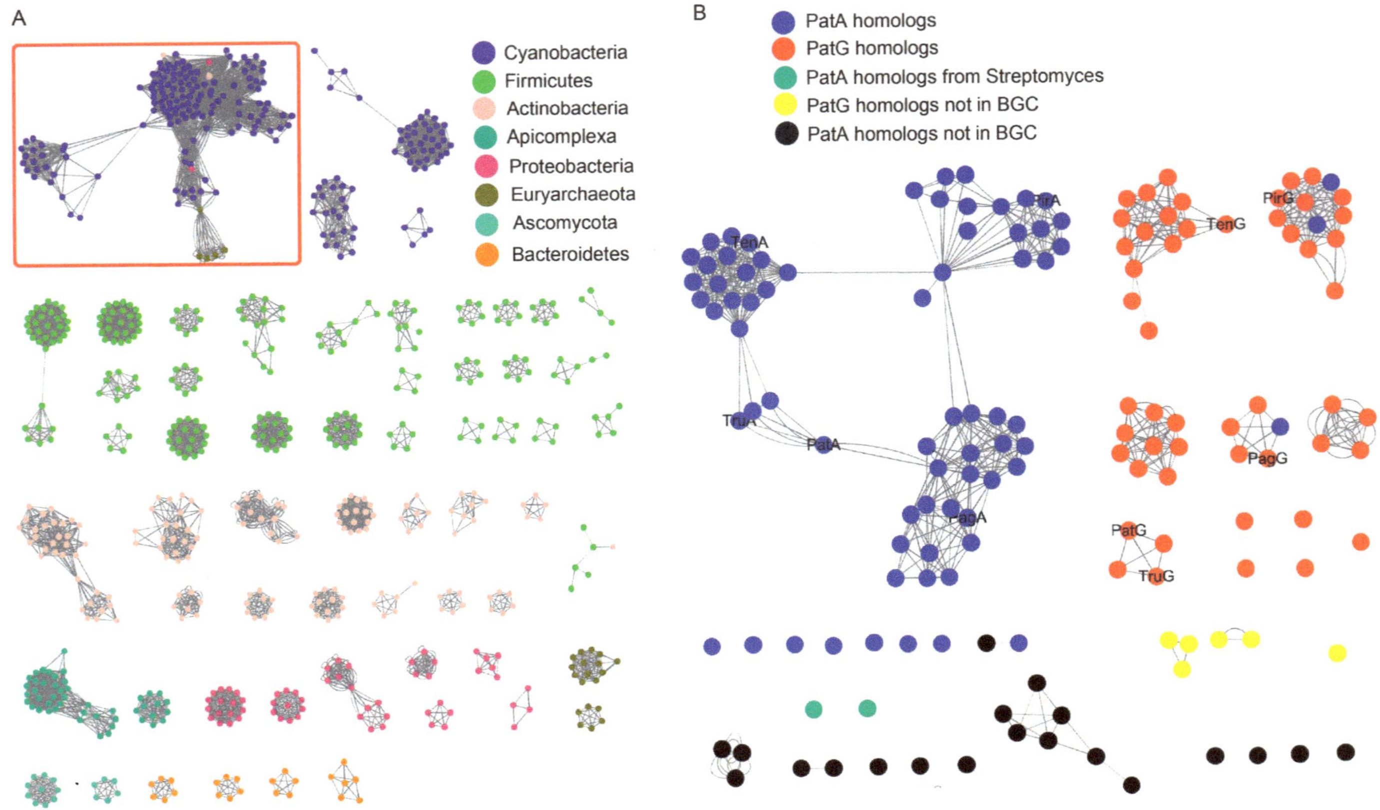

Fig. 8 Sequence similarity diagram of bacterial S8A proteases. (A) Proteases from a broad swath of bacteria. (B) Expansion of *red box* from (A). Showing the close relationship between PatA- and PatG-like proteases. *BGC*, biosynthetic gene cluster.

and PatG were proposed to function in the cleavage and/or maturation of the PatE precursor peptide (Schmidt et al., 2005). Identification of the trunkamide pathway (*tru*) and characterization of the pathway products isolated from a heterologous system further fortified the hypothesis that homologues of the PatA and PatG proteases are involved in the maturation of all cyclic cyanobactins (Donia et al., 2008).

Direct evidence for the specific roles of the two proteases was established through in vitro biochemical studies of various constructs of recombinantly expressed and purified PatA and PatG (Lee et al., 2009). Specifically, the isolated protease domain of PatA processed PatEdm, an artificial precursor peptide substrate with two cassettes flanked by leader and follower sequences (later named RSII and RSIII), to liberate linear peptides corresponding to the embedded cassettes and a follower sequence (RSIII). Mutation of the active site Ser residue abolished proteolytic activity, confirming that PatA processed precursor peptide to remove the N-terminal leader sequence (Lee et al., 2009). Thus, macrocyclization was proposed to proceed via a transpeptidation mechanism with precedent in artificial subtilisin-like enzymes, such as subtiligase (Braisted, Judice, & Wells, 1997; Chang, Jackson, Burnier, & Wells, 1994).

Several other PatA and PatG homologues have been characterized, and all function similarly, although the RSs differ. The major exception to this rule is the aeruginosamide (*age*) pathway, where AgeG performs C-terminal hydrolysis to liberate a linear product, in place of macrocyclization (Leikoski et al., 2013; Sardar et al., 2017).

4.2 Substrates and Products

A directed effort aimed at identifying the diversity of macrocyclic peptides produced by cyanobacterial symbionts revealed a collection of related, but chemically distinct patellamide-like products (Donia et al., 2006, 2008). Notably, sequence analysis of genes encoding PatE-like precursor identified an obvious pattern wherein the amino acid residues found in the final product were highly divergent, but the flanking sequences (encoding the leader RSII and follower RSIII) were essentially identical.

PatA and related proteases cleave after RSII, releasing the free N-terminus of peptides. PatA recognizes several related RSII derivatives, including AVLAS-X, GVDAS-X, GLEAS-X, and GVEPS-X, embedded within longer peptides (-X indicates the position of cleavage) (Lee et al., 2009; McIntosh, Donia, et al., 2010; Sardar, Pierce, et al., 2015). So far, out of hundreds of peptides attempted of various sizes and sequences, only one containing a correct RSII sequence has failed to react (unpublished data). Otherwise, the enzyme has exhibited highly relaxed specificity, as might be expected for a subtilisin-like protease. Because of the simple reaction catalyzed by PatA, few homologues have been investigated, including ThcA, which recognized the sequence AVLAS-X (Sardar, Pierce, et al., 2015). By sequence gazing there are other quite different enzyme–RSII pairs, such as GLTPH-X in the prenylagaramide (*pag*) pathway, but experimental data are largely lacking (Donia & Schmidt, 2011).

Macrocyclase rule 1: Substrates must contain C-terminal RSs. PatG and related proteases cleave before RSIII (Fig. 9). Both PatG and TruG are capable of recognizing the C-terminal sequences X-AYD-COOH, X-SYD-COOH, X-AYDG-COOH, X-AYDGE-COOH, X-SYDD-COOH, and relatives

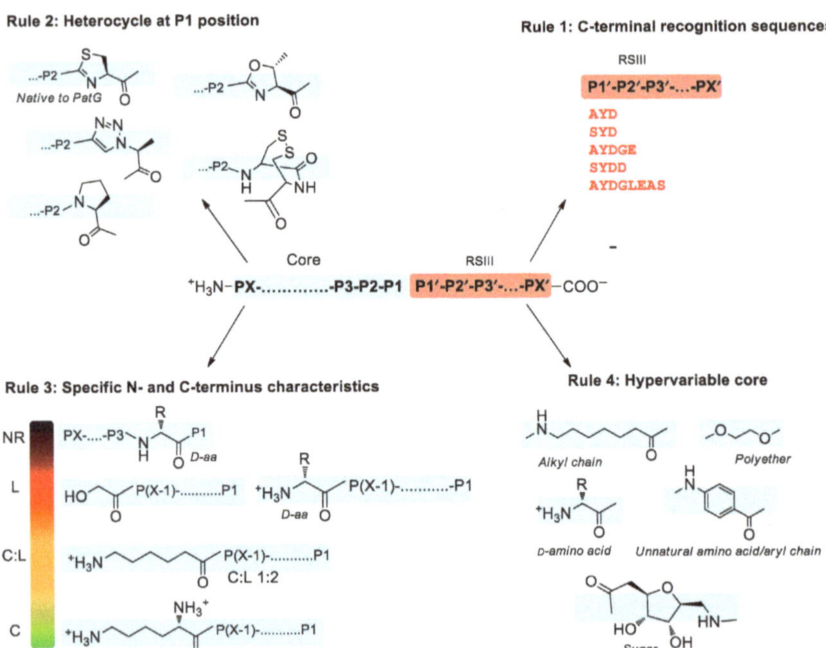

Fig. 9 Macrocyclase rules. Macrocyclase PatG is broadly tolerant of substrates and is capable of generating circular, or sometimes linear products. The above rules are summarized from hundreds of substrates that have been analyzed. Recognition "rules" are useful in synthetic biology and in understanding enzyme mechanism, but there are always exceptions. For example, glycine is disfavored in position P2, but there are examples in which it is successfully used. In rule 3, *NR* = no reaction, *L* = linear product, *C:L* = both linear and cyclic product, and *C* = cyclic product.

(X- indicates the position of cleavage) (Lee et al., 2009; McIntosh, Robertson, et al., 2010; Oueis et al., 2015; Oueis, Jaspars, Westwood, & Naismith, 2016; Oueis, Stevenson, Jaspars, Westwood, & Naismith, 2017; Sardar, Lin, et al., 2015). In addition, they recognize slightly longer sequences resulting from cleavage of multiple cassettes, such as X-AYDGLEAS-COOH (Lee et al., 2009). Thus far, they have not cleaved sequences with C-terminal longer peptides, although interestingly 8-amino-3,6-dioxaoctanoic acid has been appended to the end of the X-AYD sequence leading to successful macrocyclization (Oueis, Nardone, Jaspars, Westwood, & Naismith, 2017). This may prove useful in improving solubility of otherwise challenging substrates.

Macrocyclase rule 2: Substrates must contain heterocycle in the P1 position. PatG catalyzes a more complex reaction than PatA and has more complex substrate requirements. In addition to RSIII, a heterocycle is required in the P1 position (P1 site as per the nomenclature of Schecter and Berger) (Schechter & Berger, 1967), e.g., heterocycle-SYD, etc. In the case of PatG and relatives the natural heterocycle is thiazoline or oxazoline, while in proteins such as PagG it is apparently proline (Donia & Schmidt, 2011). A peptide that lacked the necessary heterocycle at the P1 position failed to react entirely with PatG (McIntosh, Robertson, et al., 2010). Even so, PatG accepts substrates with proline substituting for thiazoline, as well as triazoles and, interestingly, heterocycles comprised of disulfide-linked cysteine–cysteine (e.g., CC-SYD, etc.) (Oueis et al., 2016; Oueis, Stevenson, et al., 2017). These are synthetically useful alternatives in the creation of large molecules. Genome-mining approaches identified a cryptic cyanobactin gene cluster in *Oscillatoria* sp. PCC 6506, and substrate tolerance studies using the protease domain of the corresponding OscG enzyme revealed that this enzyme could still process substrates with the (Cys)- and (Ser)-pseudoprolines in the P1 position, while PatG cannot, further expanding the potential macrocyclase substrates (Alexandru-Crivac et al., 2017). The minimal substrate requirement for macrocyclization by PatG (i.e., minimal three residue follower sequence preceded by a heterocycle) has prompted several efforts focused on utilizing the isolated protease domain as a biotechnology tool (Alexandru-Crivac et al., 2017; McIntosh, Robertson, et al., 2010; Oueis et al., 2016; Oueis, Nardone, et al., 2017; Oueis, Stevenson, et al., 2017; Sardar, Lin, et al., 2015). For example, the macrocyclase was predicted to be useful in producing discrete compounds such as peptides that mimic hormones or neuropeptides, or libraries of natural–unnatural products.

Macrocyclase rule 3: For macrocyclization to occur, the N- and C-terminal residues have specific characteristics. McIntosh and coworkers tested the tolerance of the PatG protease domain against 23 different synthetic peptides containing 5-residue follower sequence but with variable cassette residues (McIntosh, Robertson, et al., 2010). At the N-terminus, peptides terminating in a D-amino acid or in glycolate (OH nucleophile) in place of an amine nucleophile were hydrolyzed and linearized, instead of circularized. In addition, substrates were partially linearized and partially cyclized through the side chain when the terminal residue contained side-chain amines (such as diaminopropionate or lysine). These results showed that the position and nucleophilicity of the N-terminus are crucial in the circularization process.

At the C-terminus, a peptide that lacked the necessary heterocycle at the P1 position failed to react entirely, as described earlier. The P2 position is also somewhat fastidious, in that substrates containing D-amino acids or glycine in that position were not substrates for PatG (McIntosh, Robertson, et al., 2010). A later study suggested that the L-amino acid is critical for correct positioning of the peptide in the active site (Oueis, Stevenson, et al., 2017).

Macrocyclase rule 4: The central residues of PatG interact with a large pocket; specific directionality or hydrogen bonds are not required, explaining the broad substrate tolerance of the enzyme. In the McIntosh work (McIntosh, Robertson, et al., 2010), the 23 substrates included D-amino acids in the central residues. While the C- and N-termini could not tolerate such modifications, in the central residues D-amino acids were readily accommodated. Aminohexanoic acid could be placed at the N-terminus or in the middle of the peptide, revealing that even hydrogen bonds to the peptide backbone are not required. PatG tolerated D-amino acids in the interior of the peptide, but not at the N-terminal or two C-terminal amino acids of the core peptide. The semisynthetic approach was further refined to carry out the macrocyclic peptide hybrids, which incorporated synthetic chemical building blocks such as aryl rings, alkyl chains, sugars, and polyethers (Oueis, Nardone, et al., 2017). The ring size synthesized by G enzymes is very flexible, ranging from 5 to 22 amino acids for PatG and 6 to 30 for OscG (Alexandru-Crivac et al., 2017; Sardar, Lin, et al., 2015). Because nonproteinaceous compounds can also be included, another way to describe the size variability is between 15 and 90 atoms. Taken together, it is clear that PatG and related enzymes exhibit broad tolerance for changes in length and composition of interior residues. Other wild-type cyanobactins such as anacyclamide have been discovered to encompass 7–20 amino acids and have potentially different macrocyclization rules (Leikoski et al., 2010), indicating that the natural cyclases should be capable of affording a wide range of products.

Much has been learned about TruG (macrocyclase domain nearly identical to PatG) via expression in *E. coli*. A heterologous production system from the *tru* pathway was modified with genes encoding for orthogonal tRNA/aaRS pairs, in order to incorporate nonproteinogenic amino acids within the macrocyclic core (Tianero et al., 2012). As proof of concept, the nonnatural amino acid *p*-acetylPhe was incorporated into the TruE precursor peptide. Modified, macrocyclic patellins with the nonproteinogenic amino acids were subsequently manufactured through differential induction for the aaRSs, resulting in the production of patellin derivatives with substituted Phe derivatives and trunkamide with *p*-bromoPhe (Tianero et al., 2012).

The genetic nature of both the precursor peptide and the modification enzymes renders the cyanobactin pathways as ideal toolkits for combinatorial production of small molecule libraries. A combinatorial approach was utilized to generate a series of random double and quadruple variants of the TruE precursor peptide in the heterologous *tru* system (Ruffner et al., 2015). This approach resulted in the production of more than 300 new macrocyclic hexapeptides. More importantly, analysis of variants that failed to produce the desired products provided guidelines for selectivity, which will guide rational design of new libraries. These studies demonstrate the potential of the system for the production of libraries of cyclized small molecules. Owning to the tolerance of the macrocyclase for residues within the cassette, PatG could also be employed to generate cyclic peptides that mimic hormones or neuropeptides.

Macrocyclase rules: Summary. With this large amount of selectivity data now collected for PatG (and TruG/OscG), we propose an overarching scheme explaining substrate tolerance in the G series. More than a hundred substrates have been used in these studies, more than can be described in this review. Taken together, these data indicate that the appropriate heterocycle and adjacent L-amino acid at the C-terminal side, in addition to an appropriately located nitrogen nucleophile at the N-terminus or side chain, are the basic required elements. These C-terminal elements enable binding to the enzyme active site and reaction with the serine oxyanion, while the N-terminal elements are required to position the nucleophile appropriately for reaction. Between these two elements, a vast array of different amino acids and synthetic linkers have been used successfully, indicating a lack of a requirement for preorganization or for precise hydrogen bonding resulting from these intermediate elements. While there are clearly preferences for some amino acids above others, especially immediately adjacent to the heterocycle, substrate tolerance is very broad.

4.3 Enzyme Structure and Function

Both the PatA and PatG were first identified as large polypeptides that housed the respective protease domains (Schmidt et al., 2005). Full-length PatA consists of a C-terminal DUF, which encompass roughly 300 amino acids (Lee et al., 2009). A linker region of approximately 120 residues, for which no function has yet been assigned, separates the protease and DUF domains. The domain architecture of full-length PatG consists of two functionally annotated domains: a flavin-dependent oxidase domain (265 residues) followed by the protease domain (~350 residues), flanked by 210 residue N-terminal and 280 residue C-terminal DUFs. The oxidase domain serves to carry out dehydrogenation of thiazoline residues installed by the heterocyclase to yield the corresponding thiazoles and is functionally similar to the oxidases that function in LAP biosynthesis (Li et al., 1996).

Incubation of the PatA reaction product (namely, the linear peptide cassette followed by the follower sequence) with the PatG protease domain produced two fragments: the linear follower peptide and a macrocycle containing the cassette sequence (Lee et al., 2009). Incubation of PatG with intact PatEdm, containing both the leader and follower sequences flanking the cassette, did not produce any proteolytic product. Hence, the order by which the linear precursor peptide is processed into the macrocyclic product occurs through the removal of the N-terminal leader sequence by PatA to produce a linear peptide product bearing the cassette and follower sequence. Modification of the C-terminal linear product by PatG excises the five-residue linear follower peptide concomitant with the macrocyclization of the cassette sequence to yield the final cyanobactin product. Metagenomic analysis of patellamide pathways shows that PatG can process 29 different substrates to form macrocycles that are seven or eight residues in size (Donia et al., 2006). The cyclic peptides produced in nature are diverse in sequence but all contain a heterocyclic residue (either a Pro or an azol(in)e derived from Ser/Cys/Thr) immediately preceding the follower peptide at the scission site. Consequently, substrates that conform to these requirements can be processed by PatG to produce engineered macrocyclic natural products (Donia et al., 2006).

Based on bioinformatics analyses that identified sequence similarities to subtilisin-like S8A proteases, macrocyclization of the cassette sequences was proposed to occur via the intermediacy of an acyl-enzyme adduct (McIntosh, Robertson, et al., 2010). Formation of the macrocycle is afforded via the attack of the α-amine onto the intermediate, rather than through hydrolysis as is typical for canonical proteases. The importance of the PatG protease domain in the N–C transamidation is further supported through mutational analysis, in which alteration of residues in the requisite catalytic triad resulted in a loss of activity. The mechanism is conceptually analogous to off-loading thioesterase domains that catalyze macrolactam formation of linear substrates that are bound as activated thio- or oxo-esters (Agarwal et al., 2012).

The protease domains of PatA and PatG share greater than 40% sequence identity, which is intriguing given that the two enzymes catalyze very different reactions. PatA recognizes and cleaves *after* the leader sequence preceding the cassette to yield two linear products, while PatG cleaves *before* the follower sequence to produce a linear product (the follower peptide) and a macrocyclic cassette. Details into this discrepancy were afforded through structural elucidation of the protease domains of PatA and PatG (Agarwal et al., 2012), as a complex of the PatG macrocyclase bound to a substrate mimic (Koehnke et al., 2012). While the structure of the PatA protease domain recapitulates the canonical α/β-hydrolase fold common among S8A proteases (Agarwal et al., 2012; Houssen et al., 2012), the structure of the PatG protease domain reveals a similar fold decorated with a helix-turn-helix (capping helices) positioned directly above the active site (Fig. 10). The cocrystal structure with the substrate mimic reveals that the heterocyclic residue at the P1 position adopts a *cis* conformation, which helps to direct the cassette peptide away from the enzyme (Koehnke et al., 2012). Based upon

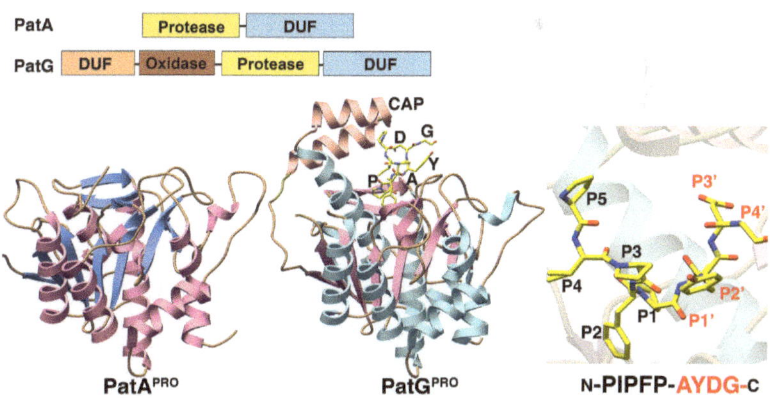

Fig. 10 Crystal structures of PatA and PatG protease domains. The proteins have very similar architectures except for a cap in PatG, shown here interacting with an artificial substrate (shown as *yellow sticks*). The panel on the *right* shows a close-up view of the PatG active site, in the vicinity of the scissile bond.

structural and mutational analysis, the capping helix binds with RSIII and orients the α-amine back toward the active site Ser. This binding cap is essential, since otherwise subtilisin proteases are known to recognize the N-terminal side of the substrate, rather than the C-terminal side as in PatG (Fig. 11) (Chang et al., 1994).

The mechanism of macrocyclization vs hydrolysis has been discussed (Agarwal et al., 2012; Booth et al., 2017; Koehnke et al., 2012; Sardar et al., 2017). Certain substrates (e.g., those with terminal D-amino acids or hydroxyl rather than amine) are linearized rather than cyclized (McIntosh, Robertson, et al., 2010). PatG also naturally catalyzes transamidation of linear peptides in the presence of glycylglycine buffer (Agarwal et al., 2012). From these and other data, it is likely that PatG functions by preferentially activating amines, and not hydroxyls, to cleave the intermediate enzyme-bound ester. Such a mechanism has precedent in subtiligase, where two mutations in subtilisin switch the protease from being OH to amine preferring (Braisted et al., 1997). A second proposed mechanism involves exclusion of water by the unusual capping helix. This latter idea is controversial among workers in the field.

One issue with PatG and its analog so far is their exceptionally slow rate, both in vitro (Lee et al., 2009; Sardar et al., 2017; Sardar, Lin, et al., 2015) and in vivo (Tianero et al., 2016), although this may be an artifact of experimental conditions. In the first characterization of PatG, the best conditions still led to a rate of ~1 turnover per enzyme molecule per day, and the yield was optimized to 75% with a 3:1 substrate to enzyme ratio (Lee et al., 2009). The reaction could be completed in 1 day with a 50% catalyst load (McIntosh, Robertson, et al., 2010). This work used an artificial substrate with proline at the macrocyclization C-terminus, which is far from the ideal PatG substrate. The rate was improved by optimizing the enzymatic reaction solution (Koehnke et al., 2012). Ideal substrates terminating in thiazoline react more readily and do not show such a striking buffer dependence (Sardar et al., 2017; Sardar, Lin, et al., 2015). Thiazoline-containing substrates were quantitatively converted to the cyclic product in 1 day with a 20% catalyst load (Sardar, Lin, et al., 2015). Similarly, PatA was challenging to use, especially in the in vitro synthesis of patellamides (Houssen et al., 2014). This problem was demonstrated to be caused by an interesting redox problem (Sardar, Lin, et al., 2015). While PatE-like substrates need to be reduced, and enzymes such as TruD are accelerated in the presence of dithiothreitol (DTT) reductant, PatA is greatly inhibited by DTT. By manipulating DTT concentration, it was possible to use PatA efficiently

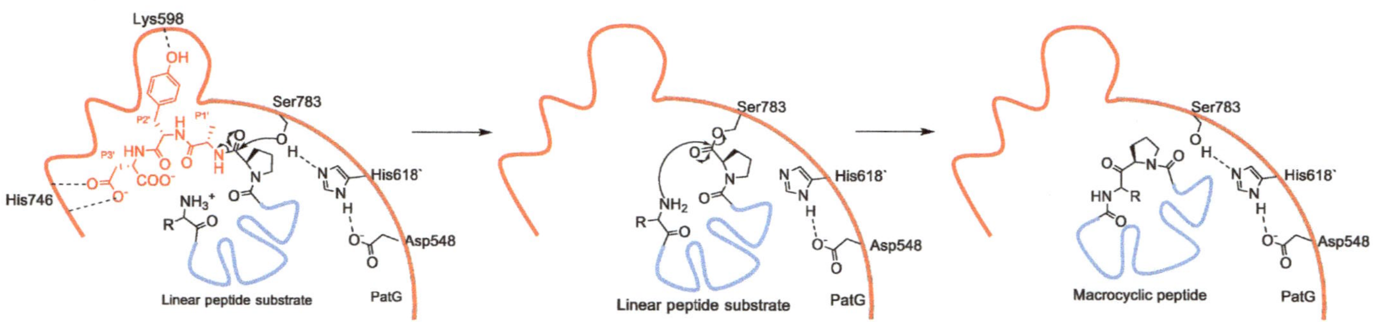

Fig. 11 Mechanism of macrocyclase. The cap (*wavy red*) interacts with RSIII (*red*) to hold the substrate in place. A catalytic triad typical of S8A proteases forms a covalent adduct with P1 at Ser783; this is displaced by an internal nucleophilic nitrogen to afford a macrocycle, while in most proteases hydrolysis would occur instead. The participation of individual residues in catalysis (other than Ser783) remains speculative.

as a synthetic tool. Perhaps a similarly simple problem hinders PatG; alternatively, perhaps an unknown role of the DUFs, or enzyme complex formation, is required.

5. PRENYLTRANSFERASES

5.1 Discovery and Initial Characterization

The PatF family of prenyltransferases, often called "F enzymes," comprises a group of small ABBA-type proteins that prenylate or geranylate serine, threonine, tyrosine, or tryptophan residues in the forward or reverse orientations (McIntosh et al., 2011). Recent characterization also identifies a novel F enzyme that can also prenylate the α-amine of peptide substrates (Sardar et al., 2017). Other F enzymes cluster in groups related to the amino acid that is prenylated, with the network analysis indicating potentially novel prenylation chemistry that has yet to be characterized (Fig. 12).

As far as has been described, all such prenyltransferases act at the final step of biosynthesis, after the precursor peptide has been cyclized (McIntosh et al., 2011). Potential exceptions to this rule include the aforementioned AgeMTPT, which prenylates the N-terminus of the linear peptide substrate after cleavage by AgeA (Leikoski et al., 2013; Sardar et al., 2017). While PatF was the first in the cyanobactin F-group to be identified, it was not characterized as a prenyltransferase because of its lack of sequence similarity to other characterized proteins and the absence of any isoprene moieties in patellamides (Schmidt et al., 2005). When the *tru* pathway was identified, one of the major differences with *pat* was the presence of two PatF-like proteins, TruF1 and TruF2 (Donia et al., 2008). As expression of the *tru* pathway in *E. coli* leads to products that were prenylated in the reverse orientation on serine and threonine, it was proposed that these proteins exercised prenyltransferase activity. Subsequently, the LynF protein was characterized as a tyrosine prenyltransferase, with the substrates being DMAPP and the tyrosine residue on a circular peptide from the *lyn* pathway (McIntosh et al., 2011). The crystallization of inactive PatF revealed that the family adopts a modified ABBA prenyltransferase fold (Bent et al., 2013), while the structures of active prenyltransferase PagF with and without substrates revealed the mechanistic basis of broad substrate tolerance (Hao et al., 2016).

5.2 Substrates and Products

Cyanobactin prenyltransferases appear to be highly promiscuous regarding peptide sequence and structure, but highly specific with regard to the donor

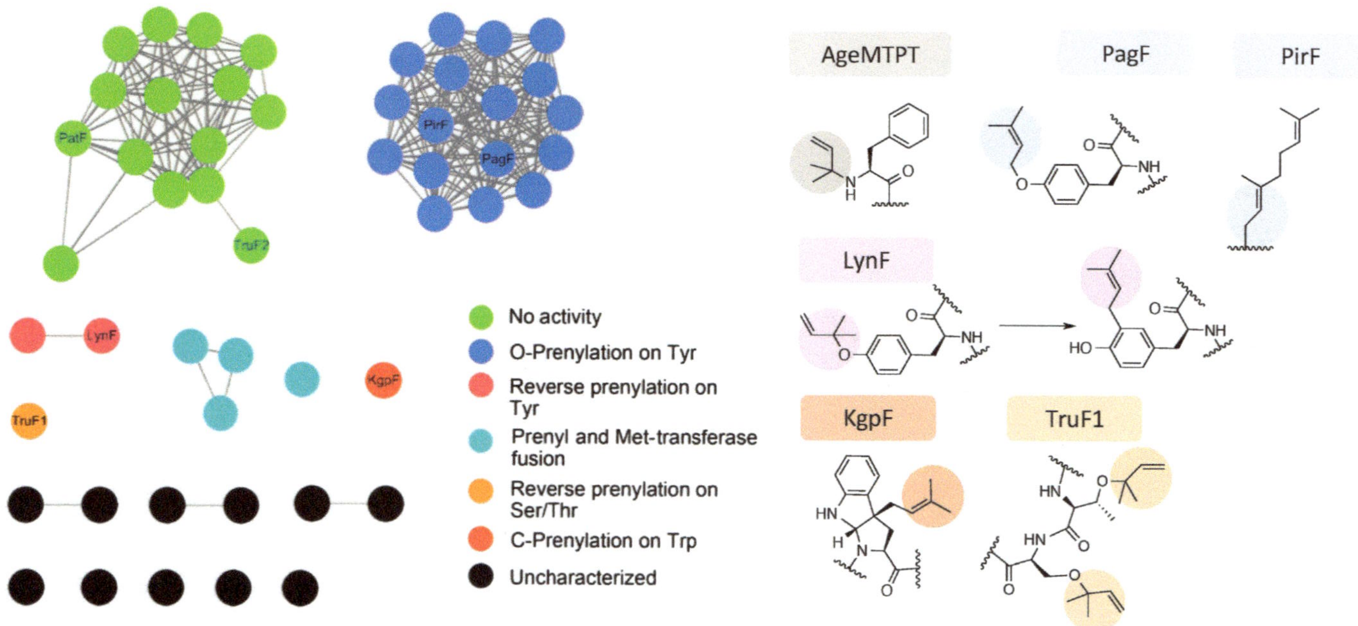

Fig. 12 Sequence similarity diagram of PatF-like prenyltransferases. Note that sequence clusters shown in network diagram (*left*) so far fairly accurately predict the chemistry catalyzed by the enzymes within each cluster (*right*).

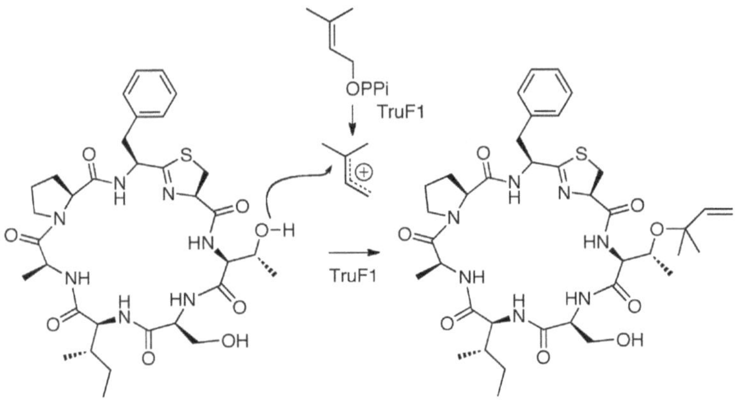

Fig. 13 Prenylation by cyanobactin enzymes. Shown is a single prenylation of the trunkamide precursor peptide by TruF1.

isoprene and the amino acid site of prenylation on the peptide substrate (Fig. 13). As such, they are good tools for the prenylation of specific residues on diverse peptide substrates. All characterized prenyltransferases use DMAPP (or potentially IPP in some cases) as the substrate, and so these will be the subjects of discussion here. An additional family of geranylating prenyltransferases exists, but the product of the pathway has been characterized only via genetic means and mass spectrometry, and therefore the chemistry is not yet certain (Leikoski et al., 2012).

The prenyltransferases known so far can be broadly classified as those that naturally utilize cyclic substrates (most known F proteins), and those that use linear substrates (aeruginosamide pathway) (Leikoski et al., 2013). Among those enzymes that use circular substrates, in vitro studies demonstrate a capability of prenylation on both individual amino acids and close variants of those amino acids (e.g., various phenols in the case of tyrosine-prenylating enzymes) (McIntosh et al., 2011). They also can prenylate many different variants of linear peptides, but show no activity on amino acids.

5.2.1 Threonine–Serine Prenylation

Several natural 6–8 residue, macrocylic peptides from marine *Prochloron* symbiotic cyanobacteria were known to be prenylated on serine and threonine (Carroll et al., 1996). All of these products were shown to originate in the *tru* pathway, which uses hypervariable substrates and identical enzymes (Donia et al., 2008). The *tru* pathway was expressed in *E. coli*, leading to production of the natural products, with the isoprene group shown by feeding studies to originate from DMAPP (Tianero et al., 2016). Since the *tru* pathway contained two potential prenyltransferases, TruF1 and TruF2, genetic studies in *E. coli* were required to demonstrate activities of the enzyme (Tianero et al., 2016). TruF2 could be deleted without untoward effect, while deletion of TruF1 led only to macrocyclic peptide products lacking prenylation on serine or threonine. Thus, TruF1 was active as a peptide prenyltransferase, while TruF2 showed no such activity.

TruF1 was functional in vitro, but its low activity and solubility have precluded detailed studies. Using heterologous expression in *E. coli*, prenylation is shown to be the last step, with accumulation of unprenylated macrocycle in the cytoplasm, all of which could undergo prenylation upon subsequent addition of DMAPP (Donia, Ruffner, et al., 2011; Tianero et al., 2016). Although the substrate selectivity of TruF1 has not been addressed in vitro, the *tru* pathway has been used to create thousands of derivatives in *E. coli* (Ruffner et al., 2015; Tianero et al., 2012). A limitation of these studies is that they address the selectivity of all pathway enzymes, as well as stability and compatibility in *E. coli*, and thus are not specific to prenyltransferase activity. Nonetheless, a few potential rules were gleaned: (1) TruF1 is absolutely selective for serine and threonine above other amino acids; (2) TruF1 is highly promiscuous with regard to the choice of peptide substrate; (3) TruF1 has been demonstrated to prenylate up to three times on a single substrate; and (4) TruF1 seems to prefer certain flanking sequences adjacent to the modified serine or threonine, although this was only rigorously tested in one set of substrates. These sequences contained the amino acids arginine, leucine, isoleucine, methionine, serine, or threonine.

5.2.2 Tyrosine Prenylation

Many cyanobactins are prenylated on tyrosine, and such molecules contain a diversity of amino acid sequences and ring sizes (Donia & Schmidt, 2011; Leikoski et al., 2009). The characterized natural representatives are all forward prenylated on tyrosine. Identification of these biosynthetic pathways strongly implicated a PatF-like enzyme in prenylation. Characterization of products from a reconstituted prenylagaramide (*pag*) gene cluster led to identification of the F enzymes from this pathway (PagF) as bona fide tyrosine prenyltransferases (Donia & Schmidt, 2011).

Notably, biochemical characterization of a homologous enzyme from *Lyngbya aestuarii* showed that LynF prenylates tyrosine on oxygen in the reverse orientation (McIntosh et al., 2011, 2013). This leads to a product with the isoprene group poised to undergo a Claisen rearrangement to yield forward C-prenylated tyrosine. The rearrangement occurs spontaneously, and the rate of the rearrangement varies depending upon substrate and takes place within a few minutes using boc-tyrosine but requires potentially up to weeks in some of the natural products, as both O- and C-prenylated variants can be isolated from the cyanobacteria. By contrast, PagF forward O-prenylates the tyrosine phenolic oxygen. In the forward orientation, the products are poor substrates for the Claisen rearrangement and do not under any further rearrangements. Most members of F-family of prenyltransferases are likely to be forward O-prenylating as found in PagF.

The substrate scopes of LynF and PagF have been investigated in vitro (Hao et al., 2016; McIntosh et al., 2011). LynF used boc-D- and boc-L-tyrosine, as well as a variety of cyclic peptides containing tyrosine. Linear biosynthetic intermediates were not modified, but other linear peptides proved to be reasonable substrates. PagF exhibited a similar substrate scope. In contrast to LynF, PagF accepted linear peptides with vastly different amino acid sequences than found in the parent compound, making it potentially useful for biotechnology.

5.2.3 Tryptophan Prenylation

Kawaguchipeptins A and B are 10-membered ring macrocyclic peptides originally isolated from *Microcystis aeruginosa* NIES-88 (Ishida, Matsuda, Murakami, & Yamaguchi, 1997), which contain prenylation on two tryptophan residues. Sequencing of the producing strain suggested that both of the products are RiPPs that are produced by a single-cyanobactin pathway, which was confirmed by heterologous expression of the constituent genes in *E. coli* (Parajuli et al., 2016). A single enzyme, KgpF, functions on a range of cyclic and linear peptide substrates and prenylates C3 on the amino acid tryptophan, which is a unique modification in RiPP products. Notably, upon prenylation, a subsequent reaction forms a second reaction five-membered ring with the backbone amide nitrogen. Consequently, the KpgF enzyme does not accept the isolated amino acid L-tryptophan as a substrate. The stereospecificity of the modification was confirmed by comparison with synthetic standards (Okada et al., 2016). Both DMAPP and, to a lesser extent, IPP were reported to be used by KpgF, but GPP could not be used as an isoprene donor.

5.2.4 N-Terminal Prenylation

AgeMTPT, from the aeruginosamide pathway, is the only known prenyltransferase that modified the α-amino moiety of peptides (Leikoski et al., 2013). As described earlier, aeruginosamides stand out from other cyanobactins because they are linear, rather than cyclic products produced by the G-group protease AgeG. Thus, proteolysis liberates a free N-terminus that can be forward prenylated. AgeMTPT is not named with "F" nomenclature because it is a didomain protein consisting of a C-terminal, F-type prenyltransferase (PT) fused with an N-terminal methyltransferase (MT). AgeMTPT uses DMAPP to prenylate the N-terminus with a relatively restricted substrate scope, possibly requiring an N-terminal phenylalanine residue to react (Sardar et al., 2017). Additionally, the protein prenylates *S*-adenosyl-L-methionine (SAM) and other adenosine analogs, presumably due to the close placement of the active site in the MT in comparison to that of PT.

5.2.5 Potentially Nonprenylating PatF Homologues

A large group of PatF homologues, including PatF itself, have no known activity (Donia et al., 2006; Houssen et al., 2014; Tianero et al., 2016). They have not been shown to bind to DMAPP or other isoprene donors. A comparison of the structure of PatF (Bent et al., 2013) with that of the fully functional PagF (Hao et al., 2016) shows that the former enzyme lacks many critical residues. Nonetheless, PatF is essential for the production of patellamide in *E. coli* heterologous system (Donia et al., 2006). The only other nonfunctional F variant that has been characterized in any way is TruF2, which also lacks key residues necessary for functionality (Tianero et al., 2016). Unlike PatF, TruF2 can be deleted from the heterologous production system without loss of patellin production. Notably, there are cyanobactin biosynthetic clusters (such as that for trichamide) (Sudek et al., 2006) that lack any F homologue, suggesting that these enzymes are not necessary for the production of heterocycles and/or macrocycles. Hence, while PatF is necessary for patellamide biosynthesis, the general role for other nonfunctional homologues has yet to be elucidated. Hypotheses about their potential roles include a chaperone/carrier function for shuttling intermediates or products, a role in multiprotein complex formation, as well as a number of other possibilities.

5.3 Enzyme Structure and Function

The crystal structure of PatF revealed an overall architectural similarity to the ABBA-prenyltransferase family, but the basis for substrate selectivity could not be reconciled as PatF is not a functional enzyme (Bent et al., 2013). Subsequent structure determination of PagF, in complex with both linear and cyclic peptide substrates, and structure-guided mutational analysis provided the context for functional analysis (Fig. 14) (Hao et al., 2016). The enzyme shares the same ABBA fold (Kuzuyama, Noel, & Richard, 2005;

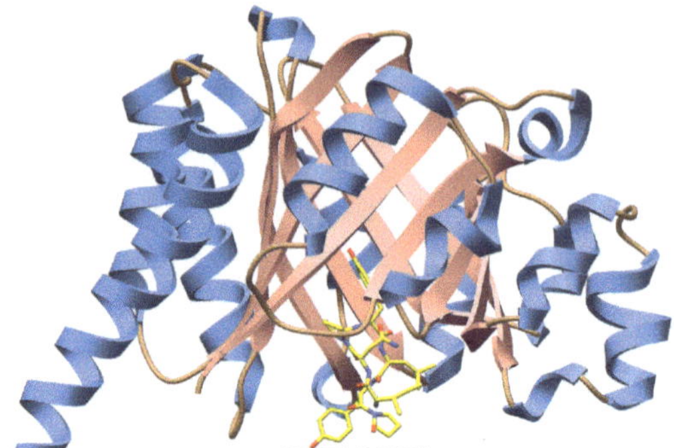

Fig. 14 Structure of PagF bound to substrate. While most ABBA prenyltransferases contain a hydrophobic active site that precludes solvent, the structure of PagF reveals a solvent-exposed cavity. Binding of the peptide substrate is necessary to form a solvent-excluded cavity where productive prenyl transfer can occur.

Metzger et al., 2009), but unlike other ABBA-type enzymes that accept small molecule substrates, PagF lacks a hydrophobic cavity for productive isoprene transfer. Notably, binding of the peptide substrate itself provided the suitable hydrophobic pocket to form a catalytically competent active site. The degree of surface area buried by the substrate corresponded directly with the catalytic efficiency, explaining why isolated amino acids are not turned over, but cyclic peptides that can suitably occlude the solvent exposed active site are kinetically competent. The structure also provides a rationale for the lack of function for PatF as this homologue has mutations at several active site residues that are shown to be critical in PagF for binding to the isoprene donor.

6. OTHER ENZYMES

6.1 Discovery and Initial Characterization

Beyond the core enzymes described earlier, two classes of enzymes are relatively common in cyanobactin biosynthesis: oxidases and methyltransferases (Donia & Schmidt, 2011). In addition, a number of proteins and DUFs are commonly found among several cyanobactin clusters but are not typically prevalent across all pathways.

The oxidase domain was first identified as the N-terminal domain in the polypeptide that also harbored the PatG protease domain (Schmidt et al., 2005). It was the only domain or protein in the pathway with obvious homology to characterized RiPP proteins, showing similarity to the microcin B17 oxidase involved in the oxidative conversion of azoline moieties to the aromatic azoles (Li et al., 1996). While most cyanobactin oxidases are found as fusions with the corresponding G protease, there are also instances of standalone oxidases that carry out the same function, as inferred from the structures of natural products associated with those pathways. Expression of the *pat* pathway in *E. coli* led to the natural products that contained both oxidized thiazoles and unmodified oxazolines, validating the sufficiency of the gene cluster for the oxidation reaction (Donia et al., 2006; Schmidt et al., 2005). Moreover, a pathway lacking the oxidase domain, *tru*, does not contain oxidized heterocycles and instead contain thiazoline, both in the producing organisms and when the pathway is heterologously expressed in *E. coli* (Donia et al., 2008). Oxidase Thc_{Oxi} from *Cyanothece* cyanobactin pathway only oxidizes thiazoline-containing cyclic peptides in vitro, but a homologous enzyme Ap_{oxi} from a microcin biosynthetic pathway, *Arthrospira* pathway, was able to oxidize the thiazoline on both cyclic and linear cyanobactin substrates (Houssen et al., 2014).

Another class of ubiquitous modification enzymes that are commonly found in cyanobactin biosynthetic clusters is methyltransferases (Donia & Schmidt, 2011). In sequenced pathways they are often fused with other functional proteins, such as homologues of PatG. However, the function is known for only one such fusion, AgeMTPT, which is consistent with the known natural product associated with its pathway. The *age* pathway for aeruginosamides was reconstituted in vitro and shown to yield linear cyanobactin natural products in which the terminal α-carboxylate is methyl esterified (Fig. 15) (Sardar et al., 2017). Biochemical analysis with the purified protein showed that AgeMTPT carried methylation of the C-terminus using SAM as the cofactor and could also prenylate the α-amine of the same peptide using DMAPP. Other similar enzymes that contain SAM-binding

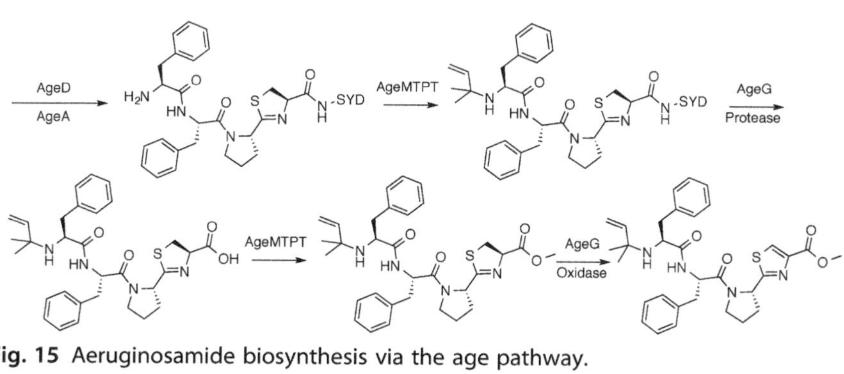

Fig. 15 Aeruginosamide biosynthesis via the age pathway.

motifs have not yet been characterized, and much remains to be learned about the role of methyltransferases in cyanobactin biosynthesis.

Several proteins and domains, including PatB and PatC and the C-terminal DUFs on PatA and PatG, are virtually universally found in cyanobactin pathways but have no known function (Donia et al., 2006). PatB and PatC are not required for synthesis in vitro, and they can be deleted from the pathway in vivo when proteins are under control of individual promoters (albeit with low yields). However, they are apparently essential in expressing the *pat* and *tru* operons in *E. coli*. Similarly, the DUFs can be deleted from PatA and PatG with retention of protein function in vitro (Lee et al., 2009). No published data exist concerning their essentiality in vivo. Crystal structures are available for the C-terminal DUFs, but functional studies failed to elucidate any binding between this domain and the PatE precursor peptide (Mann et al., 2014). Finally, an N-terminal DUF is present in PatG and homologues; its function and structure are unknown, but it is not essential for macrocyclase activity in vitro (Lee et al., 2009). Because these proteins and domains are universally found in cyanobactin biosynthesis, it is certain that they are essential or at least play an important biological role in the natural cyanobacterial producers.

6.2 Substrates and Products

A wide range of thiazole-, and occasionally oxazole-containing cyanobactins have been isolated, but relatively little biosynthetic data exist. Probably the best studied in terms of substrate scope is the patellamide oxidase embedded in PatG (Schmidt et al., 2005). Since about 50 natural hepta- and octapeptide substrates are known to be modified by identical PatG oxidase, this implies that the oxidase is catalytically competent on a variety of substrates. Although many cyanobactins contain oxazole, PatG oxidase is apparently incapable of oxidizing oxazolines. Thc_{oxi} only modified the thiazolines on cyclic substrates. The specificity for thiazoline vs oxazoline was previously explained by Walsh and coworkers in their study of the homologous microcin B17 oxidase, where oxidation to thiazole is intrinsically faster (Belshaw et al., 1998).

The AgeMTPT methyltransferase natively modifies a series of short linear peptides with closely related structures (Sardar et al., 2017). AgeMTPT was tested with 17 linear peptide substrates in vitro, but only modified a close relative of the native substrate: isoprene-FFP-thiazoline, a presumed intermediate in the natural pathway. Interestingly, the methyltransferase only modified the thiazoline carboxylate, and not the thiazole carboxylate, implying that methylation precedes oxidation.

6.3 Enzyme Structure and Function

Relatively little has been reported concerning the structures of these enzymes. Key exceptions are that structures of the PatG DUF and ThcG oxidase were reported (Bent et al., 2016; Mann et al., 2014). No biological insight was gained from the DUF structure, but the crystal structure of the *Cyanothece* oxidase structure revealed the site of binding of the flavin mononucleotide and showed a probable biologically relevant dimer. The structure also revealed an N-terminal extension consisting of twin RRE domains, but the functional relevance of these RRE domains has not yet been experimentally verified.

7. IN VITRO CONSTRUCTION OF ARTIFICIAL CYANOBACTINS

Several principles are essential for pathway engineering: (1) *Modularity*. Enzymes from different pathways can be mixed and assembled to make different cyanobactins; (2) *Hypervariability*. In a precursor peptide, the core sequences can encode many different natural products, as long as certain conserved sequence elements are maintained outside of the cores; and (3) *Inverse kinetic order*. In the canonical cyanobactin pathway, the enzymes post-translationally modify the substrate in a defined order, and each step is slower than the previous one.

Based upon these principles and the extensive knowledge gained through biosynthetic and biochemical investigations described earlier, rules have been gleaned to select enzymes and to design suitable precursor peptides. Here, we will define the steps needed to synthesize designed products (Fig. 16). Note that, although the enzymes are quite broad in substrate utilization, not every peptide-like compound will be a substrate. Current knowledge of substrate rules is defined in the text earlier.

Step 1. Choose an enzyme

Step 1a. Enzyme selection

Enzymes are chosen based upon the desired structural features of the products. A pool of well-characterized cyanobactin enzymes are available to program products with desired chemical features (Table 1). For example, PatD installs both thiazoline and oxazoline, whereas LynD, TruD, and ThcD install only thiazoline. The substrate selectivities of these enzymes, while quite broad and overlapping, are also subtly different.

In addition to wild-type enzymes, a series of artificial constructs have been designed. For example, PatG protease domain is fully functional without

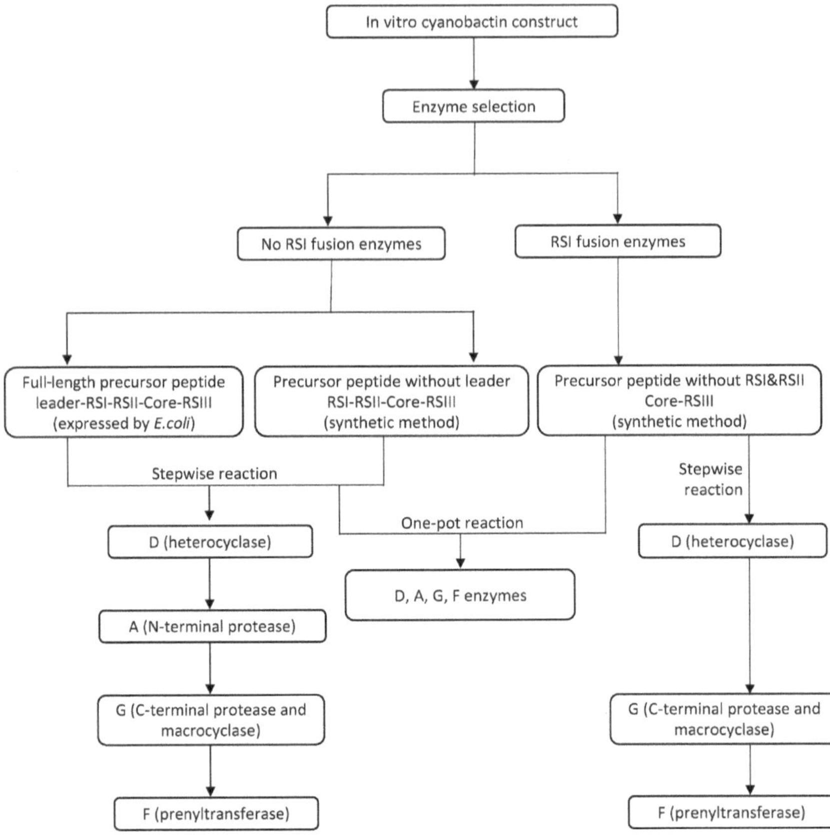

Fig. 16 General steps for in vitro construction of artificial cyanobactins.

addition of other elements (Lee et al., 2009). PatG subtiligase has been designed to increase the efficiency of transamination reactions and some types of circularizations, particularly where there is a flexible linker present. Another subtlety is that RSI can be fused to TruD or LynD, enabling N-terminal cleavage of RSI to be skipped during the process.

Step 1b. Enzyme expression and purification

Methods have been previously described in detail (Sardar, Pierce, et al., 2015). Briefly, genes are codon optimized for the *E. coli* host. All protein sequences are available in GenBank (Table 1).

Step 2. Design the substrate peptide

Step 2a. Synthetic vs E. coli-expressed substrates

Both synthetic and recombinant methodologies are suitable for generating substrate peptides. The substrate should at least contain the RSs for the enzymes and a core sequence encoding the final product. The RSs should

Table 1 Characterized Cyanobactin Enzymes, Their GenBank or PDB Accession Number, Posttranslational Modifications (PTMs), and the Paired Recognition Sequences (RSs)

Enzyme	GenBank Accession	PTM	RS
TruD	ACA04490.1	Cys → thioazoline	RSI: LAELSEEAL
LynD	4V1T (PDB)		
ThcD	WP_012626011.1		
PatD	AAY21153.1	Cys → thioazoline Thr/Ser → oxazoline	
TruA	ACA04487.1	N-Terminal proteolysis	RSII: GVDAS
PatA	AAY21150.1		RSII: GLEAS/GVEPS
PagA	AED99426.1		RSII: GLTPH
ThcA	WP_041236525.1		RSII: AVLAS
PatG (protease domain)	AAY21156.1 (amino acids 513–866)	C-Terminal proteolysis macrocyclization	RSIII: AYD/AYDGE
TruG	ACA04494.1		RSIII: SYD/SYDD
AgeG (protease domain)	CCH92964 (amino acids 609–960)	C-Terminal proteolysis	
TruF1	ACA04492.1	Prenlyation (isoprene)	Thr/Ser
PagF	AED99429.1	O-Prenylation	Tyr
LynF	WP_083798475.1	Claisen rearrangement → C-prenylation	
KgpF	KXS89935.1	Trp prenylation	Trp
AgeMTPT	CCH92966	N-Terminal prenylation C-Terminal methylation	N- and C-terminus
Thc_{Oxi}	5lq4 (PDB)	Thioazoline → thioazole	Thioazoline (cyclic substrate)

PTM, posttranslational modifications processed by individual cyanobactin enzymes; *RS*, recognition sequences known to be processed by the enzyme.

be in certain positions to make the substrate fully functions: (1) RSI should be placed in the front; (2) RSII should be right in front of the core sequence; and (3) RSIII should be right after the core sequence.

When synthesized in *E. coli*, full-length leader sequences should be used, with an N-terminal His_6-tag. An example of such a construct is given as sequence 1 (core peptide in bold):

Sequence 1: MGSSHHHHHHSSGLVPRGSHMNKKNILPQLGQP VIRLTAGQLSSQ LAELSEEALGGVDAS**TSIAPFC**AYDGE

In this sequence, the first residues through QLSSQ (underlined) are not essential for enzymatic reaction but are very useful in stable expression in *E. coli*. Methods to express the constructs have been previously described (Sardar, Pierce, et al., 2015).

By contrast, synthetic peptides can lack such a sequence. An example is given in sequence 2, wherein RS elements are present, but much of the precursor peptide has been excised:

Sequence 2: LAELSEEALGGVDAS**TSIAPFC**AYDGE

The advantage of such a sequence is clearly in its brevity. Moreover, if heterocyclization is not desired (core sequence ending with Pro), or if RSI has been fused in *cis* with TruD/PatD, then an even shorter sequence lacking RSI and RSII is possible, as shown in sequence 3:

Sequence 3: **TSIAPFC**AYDGE

Finally, only a few residues of RSIII are truly required, so that a reasonable substrate is provided in sequence 4.

Sequence 4: **TSIAPFC**AYD

Moreover, synthetic substrates are more readily amenable to modifications with nonprotein elements. A good example of such a synthetic substrate is given by sequence 5, with more details described in Section 4 on PatG substrates.

Sequence 5: **Aminohexanoic acid-TPFC**AYD

This substrate, although short and highly artificial, is fully suitable for reaction by heterocyclase, macrocyclase, and prenyltransferase.

Some enzymes can accept nonnative recognition sequences from other enzymes in the same class. For example, sequence 6 contains the native TruG recognition sequence, but is still a good substrate for reaction with PatG.

Sequence 6: **TSIAPFC**SYD

There are several possible variations of these methods. Most importantly, multiple core peptides can be incorporated into a single precursor. In the case of in vitro synthesis, this will have the effect of increasing the yield per mole of precursor, or of generating multiple different compounds in a single reaction. In these cases, additional core peptides are each flanked by additional RSII and RSIII sequences. An example is given by sequence 7, encoding three different core peptides shown in bold:

Sequence 7: MGSSHHHHHHSSGLVPRGSHMNKKNILPQLGQP VIRLTAGQLSSQ
LAELSEEALGGVDAS**TSIAPFC**SYDGVDAS**TSLAPFC**SYDGV-DAS **SSIAPFC**SYDD

Step 2b. Pairing recognition sequences with enzymes

The recognition sequences required for each enzymatic reaction are given in Table 1. For example, if cleavage by protease PatA is required, the sequence GVDAS or closely related sequences (e.g., GLEAS) shown in the table should be appended prior to the core, as shown in sequence 7.

Step 2c. Synthetic methods

In vivo expression methods have been described in detail (Sardar, Pierce, et al., 2015; Tianero et al., 2016). Substrate synthesis is generally performed by standard solid-phase synthetic methods, although the approach is compatible with many other types of synthetic techniques.

Step 3. Perform reactions

The detailed requirements of each enzyme for reactivity have been previously described (Sardar, Tianero, & Schmidt, 2016). For in vitro cyanobactin construction, the enzyme reaction can undergo either step-by-step reactions or one-pot reaction.

Step 3a. Stepwise method

The order of enzyme addition should follow the canonical order of enzymatic reactions: D → A → G → F/MTPT → Oxi. If one of these enzymatic reactions is not desired, the step can be skipped. For example, if prenylation is not required, the F enzyme step can be left out. Most of the reaction conditions have been described previously (Sardar, Pierce, et al., 2015).

Reactions using AgeMTPT require a slightly different order. When used as a prenyltransferase, DMAPP (5 m*M*) is the isoprene donor. This step should be performed before the C-terminal proteolysis step. DMAPP is chemically synthesized with the established method (Davisson et al., 1986). When used as a methyltransferase, SAM (Sigma-Aldrich, 5–10 m*M*) is used as the methyl donor. This step should follow the C-terminal proteolysis and before the oxidation of the thiazoline. Enzymatic reaction mixtures contain Tris pH 7.5 (50 m*M*), $MgCl_2$ (5 m*M*), substrate (50 μ*M*), and enzyme (2–5 μ*M*). The mixture is incubated at 37°C for up to 2 h for prenyl transfer and up to 18 h for methyl transfer.

The AgeG reaction condition is very similar to that of PatG: enzyme (5–10 μ*M*), substrate (50 μ*M*), Tris pH 7.5 (50 m*M*), and $MgCl_2$ (5 m*M*) at 37°C for 24 h (Sardar et al., 2017).

For Trp prenyltransferase KgpF reaction, enzyme (5 μ*M*) is added to the reaction mixture containing substrate (100 μ*M*), 1% dimethyl sulfoxide (DMSO), DMAPP or isopentenyl pyrophosphate (IPP, 1 m*M*), $MgCl_2$ (12 m*M*), NaCl (150 m*M*), HEPES pH 7.5 (10 m*M*), and tris(2-carboxyethyl)phosphine (TCEP) (3 m*M*) (Parajuli et al., 2016). Since the prenyltransferase is a slower enzyme, the reactions should be incubated at 37°C for 40 h. DMAPP is a better cofactor than IPP for the native substrate kawaguchipeptin B.

Oxidase Thc_{oxi} has been used to oxidize thiazoline to thiazole in cyclic peptide substrates (Houssen et al., 2014). Cyclic peptide substrate dissolved in DMSO is diluted to 20 μ*M* final concentration with 0.5% DMSO, in a reaction mixture containing FMN cofactor (50 μ*M*), Thc_{oxi} (10 μ*M*), NaCl (500 m*M*), and Tris pH 8.0 (20 m*M*). The reaction mixture is incubated overnight at 37°C with shaking. Oxidase Ap_{Oxi} from a microcin pathway has been used to oxidize thiazoline into thiazole on either linear or cyclic cyanobactin precursors (Houssen et al., 2014). Enzyme (20 μ*M*) is incubated with heterocyclic peptide (200 μ*M*) and FMN cofactor (1 m*M*) overnight at 37°C.

Alternatively, thiazoline rings can be oxidized by a chemical method using MnO_2: dissolve product (500 μg) in anhydrous DMSO (0.25 mL) and add activated MnO_2 (3 mg) (Houssen et al., 2014; Sardar et al., 2017). Stir the mixture at 150 rpm at 30°C for 72–96 h. Remove MnO_2 by centrifugation and wash the pellet three times with methanol (1 mL each). Combine DMSO and methanol fractions and dry under vacuum. Purify the mixture using HPLC. Other chemical oxidation methods such as DDQ/DCM method and K_2CO_3/DMF sometimes result in unstable products or side products and are not recommended.

It is not necessary to purify the product after each step except in the case of N-terminal protease A, where yield is improved if the product of PatA cleavage is purified. In the heterocyclization step, reducing agent DTT is required to prevent intramolecular disulfide bridge formation in the substrate peptide. However, DTT will inhibit the activity of the following modifying enzyme protease A. To maintain the activity of protease A, several different methods can be used to remove DTT:

I. HPLC method: The mobile phase generally consists of acetonitrile (1%–99% in water over 20 min). The solid phase is C4 for longer peptides (i.e., over 60 amino acids) and C18 for shorter peptides. In the case of azoline-containing peptides, a diagnostic 255-nm shoulder in the UV spectrum aids in purification.

II. Dialysis method: Dialysis tubing such as SnakeSkin, 3.5K MWCO (Thermo Fisher Scientific) is soaked in a buffer (1 L) consisting of the same composition as the reaction buffer: Tris pH 7.5 (50 m*M*) and $MgCl_2$ (5 m*M*). The dialysis is performed at 4°C with stirring for 6 h, followed by placement in fresh buffer for a total of 18 h of dialysis. For small-scale reactions, a 96-well microdialysis plate (Thermo Fisher Scientific, Pierce) can be used with similar conditions.

III. Desalting method: The reaction mixture is passed through a size-exclusion desalting column such as Bio-Gel P-6 (Bio-rad) following the manufacturer's protocol. After removing the storage buffer, the reaction mixture is passed through the column, and the flow-through is collected.

Although the HPLC method is useful and provides a pure compound, recovery suffers due to the intrinsic limitations of chromatography. Dialysis is useful for larger substrates (3.5 kDa) but is time consuming. Desalting columns using size-exclusion resins are often ideal, as they are faster and provide higher recovery, although not of pure compounds.

Step 3b. One-pot synthesis method

One-pot reaction provides a good alternative approach to quickly produce the cyanobactins. Enzymes (D + A + G) are mixed with the substrate in the same concentrations as described in the stepwise methods earlier. When the substrate does not tend to form the disulfide bridge, DTT can be left out from the reaction mixture, simplifying the process, but it is essential if disulfide bridges might be formed in the substrate. With DTT, N-terminal protease A is partially inhibited and results in a side product. This side product is formed by modifications only with heterocyclase D and N-terminal protease/macrocyclase G. In a typical one-pot synthesis reaction, the following conditions are used: Tris pH 7.5 (50 m*M*), $MgCl_2$ (5 m*M*), $CaCl_2$ (10 m*M*), ATP (1 m*M*), heterocyclase D (2 μ*M*), protease A (2 μ*M*), and protease and macrocyclase G (10 μ*M*). If prenyltransferase is used, then 10 μ*M* of the enzyme and DMAPP (10 m*M*) would be added. The reaction mixture is commonly incubated at 37°C for 18 h (Sardar, Lin, et al., 2015). Up till now only monoprenylated product can be observed.

8. CONCLUSION

Cyanobactin biosynthetic pathways are highly promiscuous and useful in biotechnology, where they have been used to synthesize many natural products, designed compounds, and libraries in vivo and in vitro. Ultimately, such tools will be useful in the directed genetic synthesis of desired compounds without constraints. In order for this goal to be realized, many different enzymes and biosynthetic pathways are needed, and the rules underlying their combination will need to be elucidated. These goals, and progress toward them with cyanobactin enzymes, have recently been reviewed (Gu & Schmidt, 2017; Sardar & Schmidt, 2016).

Since the discovery of *pat* in 2005, much has been learned about the genetics, chemistry, and biochemistry of this intriguing group of compounds. Here, we hope that we have provided an unbiased catalog of what is known about cyanobactin enzyme structure and function. In turn, we hope this facilitates the broad accessibility of cyanobactin enzymes to nonspecialists for use in synthetic chemistry and synthetic biology applications.

REFERENCES

Agarwal, V., Pierce, E., McIntosh, J., Schmidt, E. W., & Nair, S. K. (2012). Structures of cyanobactin maturation enzymes define a family of transamidating proteases. *Chemistry & Biology*, *19*(11), 1411–1422.

Alexandru-Crivac, C. N., Umeobika, C., Leikoski, N., Jokela, J., Rickaby, K. A., Grilo, A. M., et al. (2017). Cyclic peptide production using a macrocyclase with enhanced substrate promiscuity and relaxed recognition determinants. *Chemical Communications (Cambridge)*, *53*(77), 10656–10659.

Aller, S. G., Yu, J., Ward, A., Weng, Y., Chittaboina, S., Zhuo, R., et al. (2009). Structure of P-glycoprotein reveals a molecular basis for poly-specific drug binding. *Science*, *323*(5922), 1718–1722.

Arnison, P. G., Bibb, M. J., Bierbaum, G., Bowers, A. A., Bugni, T. S., Bulaj, G., et al. (2013). Ribosomally synthesized and post-translationally modified peptide natural products: Overview and recommendations for a universal nomenclature. *Natural Product Reports*, *30*(1), 108–160.

Belshaw, P. J., Roy, R. S., Kelleher, N. L., & Walsh, C. T. (1998). Kinetics and regioselectivity of peptide-to-heterocycle conversions by microcin B17 synthetase. *Chemistry & Biology*, *5*(7), 373–384.

Bent, A. F., Koehnke, J., Houssen, W. E., Smith, M. C., Jaspars, M., & Naismith, J. H. (2013). Structure of PatF from *Prochloron didemni*. *Acta Crystallographica. Section F, Structural Biology and Crystallization Communications*, *69*(Pt 6), 618–623.

Bent, A. F., Mann, G., Houssen, W. E., Mykhaylyk, V., Duman, R., Thomas, L., et al. (2016). Structure of the cyanobactin oxidase ThcOx from *Cyanothece* sp. PCC 7425, the first structure to be solved at Diamond Light Source beamline I23 by means of S-SAD. *Acta Crystallographica. Section D, Structural Biology*, *72*(Pt 11), 1174–1180. https://doi.org/10.1107/S2059798316015850.

Bertram, A., & Pattenden, G. (2007). Marine metabolites: Metal binding and metal complexes of azole-based cyclic peptides of marine origin. *Natural Product Reports*, *24*(1), 18–30. https://doi.org/10.1039/b612600f.

Booth, J., Alexandru-Crivac, C. N., Rickaby, K. A., Nneoyiegbe, A. F., Umeobika, U., McEwan, A. R., et al. (2017). A blind test of computational technique for predicting the likelihood of peptide sequences to cyclize. *Journal of Physical Chemistry Letters*, *8*(10), 2310–2315. https://doi.org/10.1021/acs.jpclett.7b00848.

Braisted, A. C., Judice, J. K., & Wells, J. A. (1997). Synthesis of proteins by subtiligase. *Methods in Enzymology*, *289*, 298–313.

Burkhart, B. J., Hudson, G. A., Dunbar, K. L., & Mitchell, D. A. (2015). A prevalent peptide-binding domain guides ribosomal natural product biosynthesis. *Nature Chemical Biology*, *11*(8), 564–570. https://doi.org/10.1038/nchembio.1856.

Burkhart, B. J., Kakkar, N., Hudson, G. A., van der Donk, W. A., & Mitchell, D. A. (2017). Chimeric leader peptides for the generation of non-natural hybrid RiPP products. *ACS Central Science*, *3*(6), 629–638.

Burkhart, B. J., Schwalen, C. J., Mann, G., Naismith, J. H., & Mitchell, D. A. (2017). YcaO-dependent posttranslational amide activation: Biosynthesis, structure, and function. *Chemical Reviews*, *117*(8), 5389–5456.

Carroll, A., Coll, J., Bourne, D., Macleod, J., Zabriskie, T., Ireland, C., et al. (1996). Patellins 1-6 and trunkamide A: Novel cyclic hexa-, hepta- and octa-peptides from colonial ascidians, *Lissoclinum* sp. *Australian Journal of Chemistry*, *49*(6), 659–667.

Chang, T. K., Jackson, D. Y., Burnier, J. P., & Wells, J. A. (1994). Subtiligase: A tool for semisynthesis of proteins. *Proceedings of the National Academy of Sciences of the United States of America*, *91*(26), 12544–12548.

Comba, P., Eisenschmidt, A., Gahan, L. R., Herten, D. P., Nette, G., Schenk, G., et al. (2017). Is Cu(II) coordinated to patellamides inside *Prochloron* cells? *Chemistry*, *23*(50), 12264–12274.

Cox, C. L., Doroghazi, J. R., & Mitchell, D. A. (2015). The genomic landscape of ribosomal peptides containing thiazole and oxazole heterocycles. *BMC Genomics*, *16*, 778.

Davagnino, J., Herrero, M., Furlong, D., Moreno, F., & Kolter, R. (1986). The DNA replication inhibitor microcin B17 is a forty-three-amino-acid protein containing sixty percent glycine. *Proteins*, *1*(3), 230–238.

Davisson, V. J., Woodside, A. B., Neal, T. R., Stremler, K. E., Muehlbacher, M., & Poulter, C. D. (1986). Phosphorylation of isoprenoid alcohols. *The Journal of Organic Chemistry*, *51*(25), 4768–4779.

Degnan, B. M., Hawkins, C. J., Lavin, M. F., McCaffrey, E. J., Parry, D. L., van den Brenk, A. L., et al. (1989). New cyclic peptides with cytotoxic activity from the ascidian *Lissoclinum patella*. *Journal of Medicinal Chemistry*, *32*(6), 1349–1354.

Donia, M. S., Fricke, W. F., Ravel, J., & Schmidt, E. W. (2011). Variation in tropical reef symbiont metagenomes defined by secondary metabolism. *PLoS One*, *6*(3), e17897.

Donia, M. S., Hathaway, B. J., Sudek, S., Haygood, M. G., Rosovitz, M. J., Ravel, J., et al. (2006). Natural combinatorial peptide libraries in cyanobacterial symbionts of marine ascidians. *Nature Chemical Biology*, *2*(12), 729–735.

Donia, M. S., Ravel, J., & Schmidt, E. W. (2008). A global assembly line for cyanobactins. *Nature Chemical Biology*, *4*(6), 341–343.

Donia, M. S., Ruffner, D. E., Cao, S., & Schmidt, E. W. (2011). Accessing the hidden majority of marine natural products through metagenomics. *ChemBioChem*, *12*(8), 1230–1236.

Donia, M. S., & Schmidt, E. W. (2010). Cyanobactins—Ubiquitous cyanobacterial ribosomal peptide metabolites. In: L. Mander & H.-W. (Ben) Liu (Eds.), *Comprehensive natural products II: Chemistry and biology: Vol. 2.* (pp. 539–558). Elsevier.

Donia, M. S., & Schmidt, E. W. (2011). Linking chemistry and genetics in the growing cyanobactin natural products family. *Chemistry & Biology*, *18*(4), 508–519.

Dunbar, K. L., Chekan, J. R., Cox, C. L., Burkhart, B. J., Nair, S. K., & Mitchell, D. A. (2014). Discovery of a new ATP-binding motif involved in peptidic azoline biosynthesis. *Nature Chemical Biology*, *10*(10), 823–829.

Dunbar, K. L., Melby, J. O., & Mitchell, D. A. (2012). YcaO domains use ATP to activate amide backbones during peptide cyclodehydrations. *Nature Chemical Biology*, *8*(6), 569–575.

Dunbar, K. L., Tietz, J. I., Cox, C. L., Burkhart, B. J., & Mitchell, D. A. (2015). Identification of an auxiliary leader peptide-binding protein required for azoline formation in ribosomal natural products. *Journal of the American Chemical Society*, *137*(24), 7672–7677.

Dutcher, J. D., & Vandeputte, J. (1955). Thiostrepton, a new antibiotic. II. Isolation and chemical characterization. *Antibiotics Annual*, *3*, 560–561.
Genilloud, O., Moreno, F., & Kolter, R. (1989). DNA sequence, products, and transcriptional pattern of the genes involved in production of the DNA replication inhibitor microcin B17. *Journal of Bacteriology*, *171*(2), 1126–1135.
Gonzalez-Pastor, J. E., San Millan, J. L., Castilla, M. A., & Moreno, F. (1995). Structure and organization of plasmid genes required to produce the translation inhibitor microcin C7. *Journal of Bacteriology*, *177*(24), 7131–7140.
Goto, Y., Ito, Y., Kato, Y., Tsunoda, S., & Suga, H. (2014). One-pot synthesis of azoline-containing peptides in a cell-free translation system integrated with a posttranslational cyclodehydratase. *Chemistry & Biology*, *21*(6), 766–774.
Goto, Y., & Suga, H. (2016). A post-translational cyclodehydratase, PatD, tolerates sequence variation in the C-terminal region of substrate peptides. *Chemistry Letters*, *45*(11), 1247–1249.
Gu, W., & Schmidt, E. W. (2017). Three principles of diversity-generating biosynthesis. *Accounts of Chemical Research*, *50*(10), 2569–2576.
Hao, Y., Pierce, E., Roe, D., Morita, M., McIntosh, J. A., Agarwal, V., et al. (2016). Molecular basis for the broad substrate selectivity of a peptide prenyltransferase. *Proceedings of the National Academy of Sciences of the United States of America*, *113*(49), 14037–14042.
Houssen, W. E., Bent, A. F., McEwan, A. R., Pieiller, N., Tabudravu, J., Koehnke, J., et al. (2014). An efficient method for the in vitro production of azol(in)e-based cyclic peptides. *Angewandte Chemie (International Ed. in English)*, *53*(51), 14171–14174.
Houssen, W. E., Koehnke, J., Zollman, D., Vendome, J., Raab, A., Smith, M. C., et al. (2012). The discovery of new cyanobactins from *Cyanothece* PCC 7425 defines a new signature for processing of patellamides. *ChemBioChem*, *13*(18), 2683–2689.
Ishida, K., Matsuda, H., Murakami, M., & Yamaguchi, K. (1997). Kawaguchipeptin B, an antibacterial cyclic undecapeptide from the cyanobacterium *Microcystis aeruginosa*. *Journal of Natural Products*, *60*(7), 724–726.
Kelleher, N. L., Hendrickson, C. L., & Walsh, C. T. (1999). Posttranslational heterocyclization of cysteine and serine residues in the antibiotic microcin B17: Distributivity and directionality. *Biochemistry*, *38*(47), 15623–15630.
Koehnke, J., Bent, A., Houssen, W. E., Zollman, D., Morawitz, F., Shirran, S., et al. (2012). The mechanism of patellamide macrocyclization revealed by the characterization of the PatG macrocyclase domain. *Nature Structural & Molecular Biology*, *19*(8), 767–772.
Koehnke, J., Bent, A. F., Zollman, D., Smith, K., Houssen, W. E., Zhu, X., et al. (2013). The cyanobactin heterocyclase enzyme: A processive adenylase that operates with a defined order of reaction. *Angewandte Chemie (International Ed in English)*, *52*(52), 13991–13996.
Koehnke, J., Mann, G., Bent, A. F., Ludewig, H., Shirran, S., Botting, C., et al. (2015). Structural analysis of leader peptide binding enables leader-free cyanobactin processing. *Nature Chemical Biology*, *11*(8), 558–563.
Koehnke, J., Morawitz, F., Bent, A. F., Houssen, W. E., Shirran, S. L., Fuszard, M. A., et al. (2013). An enzymatic route to selenazolines. *ChemBioChem*, *14*(5), 564–567.
Kuzuyama, T., Noel, J. P., & Richard, S. B. (2005). Structural basis for the promiscuous biosynthetic prenylation of aromatic natural products. *Nature*, *435*(7044), 983–987.
Lee, J., McIntosh, J., Hathaway, B. J., & Schmidt, E. W. (2009). Using marine natural products to discover a protease that catalyzes peptide macrocyclization of diverse substrates. *Journal of the American Chemical Society*, *131*(6), 2122–2124.
Lee, S. W., Mitchell, D. A., Markley, A. L., Hensler, M. E., Gonzalez, D., Wohlrab, A., et al. (2008). Discovery of a widely distributed toxin biosynthetic gene cluster. *Proceedings of the National Academy of Sciences of the United States of America*, *105*(15), 5879–5884.

Leikoski, N., Fewer, D. P., Jokela, J., Alakoski, P., Wahlsten, M., & Sivonen, K. (2012). Analysis of an inactive cyanobactin biosynthetic gene cluster leads to discovery of new natural products from strains of the genus *Microcystis*. *PLoS One*, 7(8), e43002.

Leikoski, N., Fewer, D. P., Jokela, J., Wahlsten, M., Rouhiainen, L., & Sivonen, K. (2010). Highly diverse cyanobactins in strains of the genus *Anabaena*. *Applied and Environmental Microbiology*, *76*(3), 701–709.

Leikoski, N., Fewer, D. P., & Sivonen, K. (2009). Widespread occurrence and lateral transfer of the cyanobactin biosynthesis gene cluster in cyanobacteria. *Applied and Environmental Microbiology*, *75*(3), 853–857.

Leikoski, N., Liu, L., Jokela, J., Wahlsten, M., Gugger, M., Calteau, A., et al. (2013). Genome mining expands the chemical diversity of the cyanobactin family to include highly modified linear peptides. *Chemistry & Biology*, *20*(8), 1033–1043.

Li, Y. M., Milne, J. C., Madison, L. L., Kolter, R., & Walsh, C. T. (1996). From peptide precursors to oxazole and thiazole-containing peptide antibiotics: Microcin B17 synthase. *Science*, *274*(5290), 1188–1193.

Mann, G., Koehnke, J., Bent, A. F., Graham, R., Houssen, W., Jaspars, M., et al. (2014). The structure of the cyanobactin domain of unknown function from PatG in the patellamide gene cluster. *Acta Crystallographica Section F Structural Biology and Crystallization Communications*, *70*(Pt 12), 1597–1603.

McIntosh, J. A., Donia, M. S., Nair, S. K., & Schmidt, E. W. (2011). Enzymatic basis of ribosomal peptide prenylation in cyanobacteria. *Journal of the American Chemical Society*, *133*(34), 13698–13705.

McIntosh, J. A., Donia, M. S., & Schmidt, E. W. (2010). Insights into heterocyclization from two highly similar enzymes. *Journal of the American Chemical Society*, *132*(12), 4089–4091.

McIntosh, J. A., Lin, Z., Tianero, M. D., & Schmidt, E. W. (2013). Aestuaramides, a natural library of cyanobactin cyclic peptides resulting from isoprene-derived Claisen rearrangements. *ACS Chemical Biology*, *8*(5), 877–883.

McIntosh, J. A., Robertson, C. R., Agarwal, V., Nair, S. K., Bulaj, G. W., & Schmidt, E. W. (2010). Circular logic: Nonribosomal peptide-like macrocyclization with a ribosomal peptide catalyst. *Journal of the American Chemical Society*, *132*(44), 15499–15501.

McIntosh, J. A., & Schmidt, E. W. (2010). Marine molecular machines: Heterocyclization in cyanobactin biosynthesis. *ChemBioChem*, *11*(10), 1413–1421.

McKeever, B., & Pattenden, G. (2003). Total synthesis of trunkamide A, a novel thiazoline-based prenylated cyclopeptide metabolite from *Lissoclinum* sp. *Tetrahedron*, *59*(15), 2713–2727.

Metzger, U., Schall, C., Zocher, G., Unsold, I., Stec, E., Li, S. M., et al. (2009). The structure of dimethylallyl tryptophan synthase reveals a common architecture of aromatic prenyltransferases in fungi and bacteria. *Proceedings of the National Academy of Sciences of the United States of America*, *106*(34), 14309–14314.

Mocek, U., Chen, L. C., Keller, P. J., Houck, D. R., Beale, J. M., & Floss, H. G. (1989). 1H and 13C NMR assignments of the thiopeptide antibiotic nosiheptide. *Journal of Antibiotics (Tokyo)*, *42*(11), 1643–1648.

Morris, R. P., Leeds, J. A., Naegeli, H. U., Oberer, L., Memmert, K., Weber, E., et al. (2009). Ribosomally synthesized thiopeptide antibiotics targeting elongation factor Tu. *Journal of the American Chemical Society*, *131*(16), 5946–5955.

Okada, M., Sugita, T., Akita, K., Nakashima, Y., Tian, T., Li, C., et al. (2016). Stereospecific prenylation of tryptophan by a cyanobacterial post-translational modification enzyme. *Organic & Biomolecular Chemistry*, *14*(40), 9639–9644.

Oman, T. J., Knerr, P. J., Bindman, N. A., Velasquez, J. E., & van der Donk, W. A. (2012). An engineered lantibiotic synthetase that does not require a leader peptide on its substrate. *Journal of the American Chemical Society*, *134*(16), 6952–6955.

Onaka, H., Nakaho, M., Hayashi, K., Igarashi, Y., & Furumai, T. (2005). Cloning and characterization of the goadsporin biosynthetic gene cluster from Streptomyces sp. TP-A0584. *Microbiology*, *151*(Pt 12), 3923–3933. https://doi.org/10.1099/mic.0.28420-0.

Ortega, M. A., Hao, Y., Zhang, Q., Walker, M. C., van der Donk, W. A., & Nair, S. K. (2015). Structure and mechanism of the tRNA-dependent lantibiotic dehydratase NisB. *Nature*, *517*(7535), 509–512.

Oueis, E., Adamson, C., Mann, G., Ludewig, H., Redpath, P., Migaud, M., et al. (2015). Derivatisable cyanobactin analogues: A semisynthetic approach. *ChemBioChem*, *16*(18), 2646–2650.

Oueis, E., Jaspars, M., Westwood, N. J., & Naismith, J. H. (2016). Enzymatic macrocyclization of 1,2,3-triazole peptide mimetics. *Angewandte Chemie (International Ed. in English)*, *55*(19), 5842–5845.

Oueis, E., Nardone, B., Jaspars, M., Westwood, N. J., & Naismith, J. H. (2017). Synthesis of hybrid cyclopeptides through enzymatic macrocyclization. *ChemistryOpen*, *6*(1), 11–14.

Oueis, E., Stevenson, H., Jaspars, M., Westwood, N. J., & Naismith, J. H. (2017). Bypassing the proline/thiazoline requirement of the macrocyclase PatG. *Chemical Communications (Cambridge)*, *53*(91), 12274–12277.

Ozaki, T., Yamashita, K., Goto, Y., Shimomura, M., Hayashi, S., Asamizu, S., et al. (2017). Dissection of goadsporin biosynthesis by in vitro reconstitution leading to designer analogues expressed in vivo. *Nature Communications*, *8*, 14207.

Parajuli, A., Kwak, D. H., Dalponte, L., Leikoski, N., Galica, T., Umeobika, U., et al. (2016). A unique tryptophan C-prenyltransferase from the Kawaguchipeptin biosynthetic pathway. *Angewandte Chemie (International Ed in English)*, *55*(11), 3596–3599.

Pascard, C., Ducruix, A., Lunel, J., & Prange, T. (1977). Highly modified cysteine-containing antibiotics. Chemical structure and configuration of nosiheptide. *Journal of the American Chemical Society*, *99*(19), 6418–6423.

Regni, C. A., Roush, R. F., Miller, D. J., Nourse, A., Walsh, C. T., & Schulman, B. A. (2009). How the MccB bacterial ancestor of ubiquitin E1 initiates biosynthesis of the microcin C7 antibiotic. *The EMBO Journal*, *28*(13), 1953–1964.

Ruffner, D. E., Schmidt, E. W., & Heemstra, J. R. (2015). Assessing the combinatorial potential of the RiPP cyanobactin tru pathway. *ACS Synthetic Biology*, *4*(4), 482–492.

San Millan, J. L., Kolter, R., & Moreno, F. (1985). Plasmid genes required for microcin B17 production. *Journal of Bacteriology*, *163*(3), 1016–1020.

Sardar, D., Hao, Y., Lin, Z., Morita, M., Nair, S. K., & Schmidt, E. W. (2017). Enzymatic N- and C-protection in cyanobactin RiPP natural products. *Journal of the American Chemical Society*, *139*(8), 2884–2887.

Sardar, D., Lin, Z., & Schmidt, E. W. (2015). Modularity of RiPP enzymes enables designed synthesis of decorated peptides. *Chemistry & Biology*, *22*(7), 907–916.

Sardar, D., Pierce, E., McIntosh, J. A., & Schmidt, E. W. (2015). Recognition sequences and substrate evolution in cyanobactin biosynthesis. *ACS Synthetic Biology*, *4*(2), 167–176.

Sardar, D., & Schmidt, E. W. (2016). Combinatorial biosynthesis of RiPPs: Docking with marine life. *Current Opinion in Chemical Biology*, *31*, 15–21. https://doi.org/10.1016/j.cbpa.2015.11.016.

Sardar, D., Tianero, M. D., & Schmidt, E. W. (2016). Directing biosynthesis: Practical supply of natural and unnatural cyanobactins. *Methods in Enzymology*, *575*, 1–20.

Schechter, I., & Berger, A. (1967). On the size of the active site in proteases. I. Papain. *Biochemical and Biophysical Research Communications*, *27*(2), 157–162.

Schmidt, E. W., Nelson, J. T., Rasko, D. A., Sudek, S., Eisen, J. A., Haygood, M. G., et al. (2005). Patellamide A and C biosynthesis by a microcin-like pathway in *Prochloron didemni*, the cyanobacterial symbiont of *Lissoclinum patella*. *Proceedings of the National Academy of Sciences of the United States of America*, *102*(20), 7315–7320.

Sudek, S., Haygood, M. G., Youssef, D. T., & Schmidt, E. W. (2006). Structure of trichamide, a cyclic peptide from the bloom-forming cyanobacterium *Trichodesmium erythraeum*, predicted from the genome sequence. *Applied and Environmental Microbiology*, *72*(6), 4382–4387.

Thibodeaux, C. J., Ha, T., & van der Donk, W. A. (2014). A price to pay for relaxed substrate specificity: A comparative kinetic analysis of the class II lanthipeptide synthetases ProcM and HalM2. *Journal of the American Chemical Society*, *136*(50), 17513–17529.

Tianero, M. D., Donia, M. S., Young, T. S., Schultz, P. G., & Schmidt, E. W. (2012). Ribosomal route to small-molecule diversity. *Journal of the American Chemical Society*, *134*(1), 418–425.

Tianero, M. D., Pierce, E., Raghuraman, S., Sardar, D., McIntosh, J. A., Heemstra, J. R., et al. (2016). Metabolic model for diversity-generating biosynthesis. *Proceedings of the National Academy of Sciences of the United States of America*, *113*(7), 1772–1777.

Wieland Brown, L. C., Acker, M. G., Clardy, J., Walsh, C. T., & Fischbach, M. A. (2009). Thirteen posttranslational modifications convert a 14-residue peptide into the antibiotic thiocillin. *Proceedings of the National Academy of Sciences of the United States of America*, *106*(8), 2549–2553.

Williams, A. B., & Jacobs, R. S. (1993). A marine natural product, patellamide D, reverses multidrug resistance in a human leukemic cell line. *Cancer Letters*, *71*(1–3), 97–102.

Wipf, P., & Uto, Y. (2000). Total synthesis and revision of stereochemistry of the marine metabolite trunkamide A. *The Journal of Organic Chemistry*, *65*(4), 1037–1049.

Yu, Y., Duan, L., Zhang, Q., Liao, R., Ding, Y., Pan, H., et al. (2009). Nosiheptide biosynthesis featuring a unique indole side ring formation on the characteristic thiopeptide framework. *ACS Chemical Biology*, *4*(10), 855–864.

Zabriskie, T. M., Foster, M. P., Stout, T. J., Clardy, J., & Ireland, C. M. (1990). Studies on the solution- and solid-state structure of patellin 2. *Journal of the American Chemical Society*, *112*(22), 8080–8084.

Zhang, F., Li, C., & Kelly, W. L. (2016). Thiostrepton variants containing a contracted quinaldic acid macrocycle result from mutagenesis of the second residue. *ACS Chemical Biology*, *11*(2), 415–424.

Ziemert, N., Ishida, K., Quillardet, P., Bouchier, C., Hertweck, C., de Marsac, N. T., et al. (2008). Microcyclamide biosynthesis in two strains of *Microcystis aeruginosa*: From structure to genes and vice versa. *Applied and Environmental Microbiology*, *74*(6), 1791–1797.

CHAPTER FIVE

The Enzymology of Prochlorosin Biosynthesis

Silvia C. Bobeica, Wilfred A. van der Donk[1]
Howard Hughes Medical Institute, University of Illinois at Urbana–Champaign, Urbana, IL, United States
[1]Corresponding author: e-mail address: vddonk@illinois.edu

Contents

Abstract

Lanthipeptides are ribosomally synthesized and posttranslationally modified peptides containing thioether cross-links formed through addition of a cysteine to a dehydroalanine (to form lanthionine) or to a dehydrobutyrine (to form 3-methyllanthionine). Genome sequencing of marine cyanobacteria lead to the discovery of 1.6 million open reading frames encoding lanthipeptides. In many cases, a genome encodes a single lanthipeptide synthetase, but a large number of substrates. The enzymatic modification process in *Prochlorococcus* MIT9313 has been reconstituted in vitro, and a variety of experimental approaches have been used to try and understand how one enzyme is capable of

Methods in Enzymology, Volume 604
ISSN 0076-6879
https://doi.org/10.1016/bs.mie.2018.01.038

modifying 30 different substrates. The methods used to characterize this system will be described along with a brief genomic description of the lanthipeptide landscape found in *Prochlorococcus* and *Synechococcus*.

1. INTRODUCTION: LANTHIPEPTIDES IN MARINE CYANOBACTERIA POPULATIONS

Lanthipeptides are ribosomally synthesized and posttranslationally modified natural peptides (RiPPs) containing thioether cross-links. These structures stem from the dehydration of serine or threonine residues to produce dehydroalanine (Dha) or dehydrobutyrine (Dhb) residues, followed by the intramolecular Michael-type addition of a cysteine to the β carbon of the dehydrated amino acid (Fig. 1). The ring topologies found in lanthipeptides are the structural basis for their bioactivities (Chatterjee, Paul, Xie, & van der Donk, 2005). For example, the most studied lanthipeptide to date, nisin, uses its two N-terminal thioether rings to sequester the undecaprenyl pyrophosphate of lipid II, the precursor in cell wall biosynthesis (Hsu et al., 2004). In addition, nisin inserts its C-terminal rings into the membrane to form pores (Wiedemann et al., 2001). Other lantibiotics such as mersacidin, nukacin ISK-1, and lacticin 481 also bind lipid II (Brötz, Bierbaum, Leopold, Reynolds, & Sahl, 1998; Islam et al., 2012; Knerr et al., 2012) but without forming pores in the membranes, presumably due to their smaller size that may not span the membrane (Bierbaum, Brötz, Koller, & Sahl, 1995; Brötz, Bierbaum, Markus, Molitor, & Sahl, 1995; Hasper et al., 2006). In two-component lanthipeptides such as the

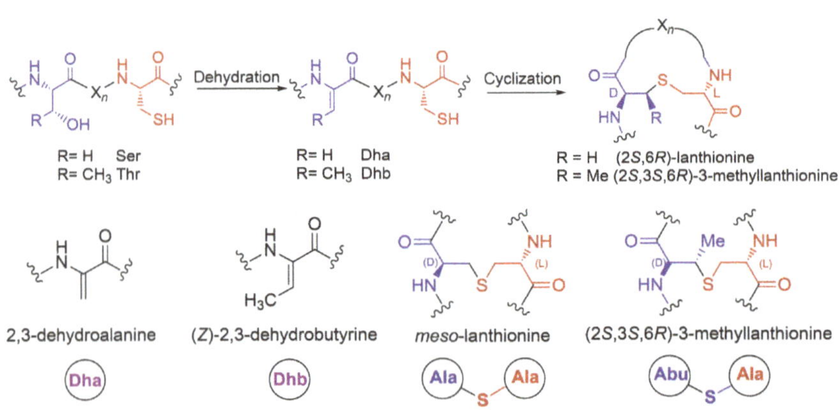

Fig. 1 Structures and shorthand notations for the characteristic posttranslational modifications in lanthipeptides.

lacticin 3147 (Morgan, O'Connor, Cotter, Ross, & Hill, 2005) or haloduracin systems (McClerren et al., 2006), one peptide interacts with lipid II serving as a recognition cue for the second component to bind and form pores (Cooper, McClerren, Chary, & van der Donk, 2008; Garneau, Martin, & Vederas, 2002; Wiedemann et al., 2006).

The biosynthesis of lanthipeptides starts with the ribosomal synthesis of the precursor peptide designated LanA, consisting of a leader peptide of varying length that aids in recognition, modification, and transport of the peptide and a core peptide where the posttranslational modifications take place (Fig. 2) (Oman & van der Donk, 2010). In class I lanthipeptide biosynthesis the dehydration is catalyzed by a dehydratase termed LanB and the cyclization is catalyzed by a cyclase termed LanC, while for class II compounds a bifunctional enzyme, LanM, performs both reactions (Siezen, Kuipers, & de Vos, 1996; Xie et al., 2004). Two-component systems encode two LanAs, each recognized by its own modifying LanM (McAuliffe, Hill, & Ross, 2000; McClerren et al., 2006). After the thioether cross-links are installed, the leader peptide is removed by a bifunctional protease-ATP binding cassette transporter LanT, or a specific protease LanP whose leader-cleavage activity is paired with a separate ATP binding cassette transporter (McAuliffe, Ross, & Hill, 2001).

In contrast to the biosynthetic strategies for lanthipeptide biosynthesis in Firmicutes and Actinobacteria (Zhang, Yu, Velásquez, & van der Donk, 2012) in which one lanthipeptide synthetase makes one product or a series of structurally very closely related products, marine picocyanobacteria of the genera *Prochlorococcus* and *Synechococcus* often harbor a collection of highly diverse genetically encoded precursor peptides that can all be successfully modified by a single lanthipeptide synthetase. The initial discovery of these posttranslationally modified natural products in *Prochlorococcus* led to their naming of prochlorosins; however, similar biosynthetic clusters are also found in strains of *Synechococcus* (Cubillos-Ruiz, Berta-Thompson, Becker, van der Donk, & Chisholm, 2017; Li et al., 2010) and other cyanobacteria (Zhang, Yang, Wang, & van der Donk, 2014). At present, the function of prochlorosins has not yet been elucidated. Whereas many studies had shown that class II bifunctional LanM enzymes have relatively low substrate specificity and allow for substantial changes in the core peptide without impaired dehydration and cyclization as long as certain recognition elements in the leader peptide are preserved (Oman & van der Donk, 2010), Nature has taken this substrate tolerance to a completely new level in these marine cyanobacteria by making combinatorial libraries of products of highly diverse ring topologies with a single enzyme.

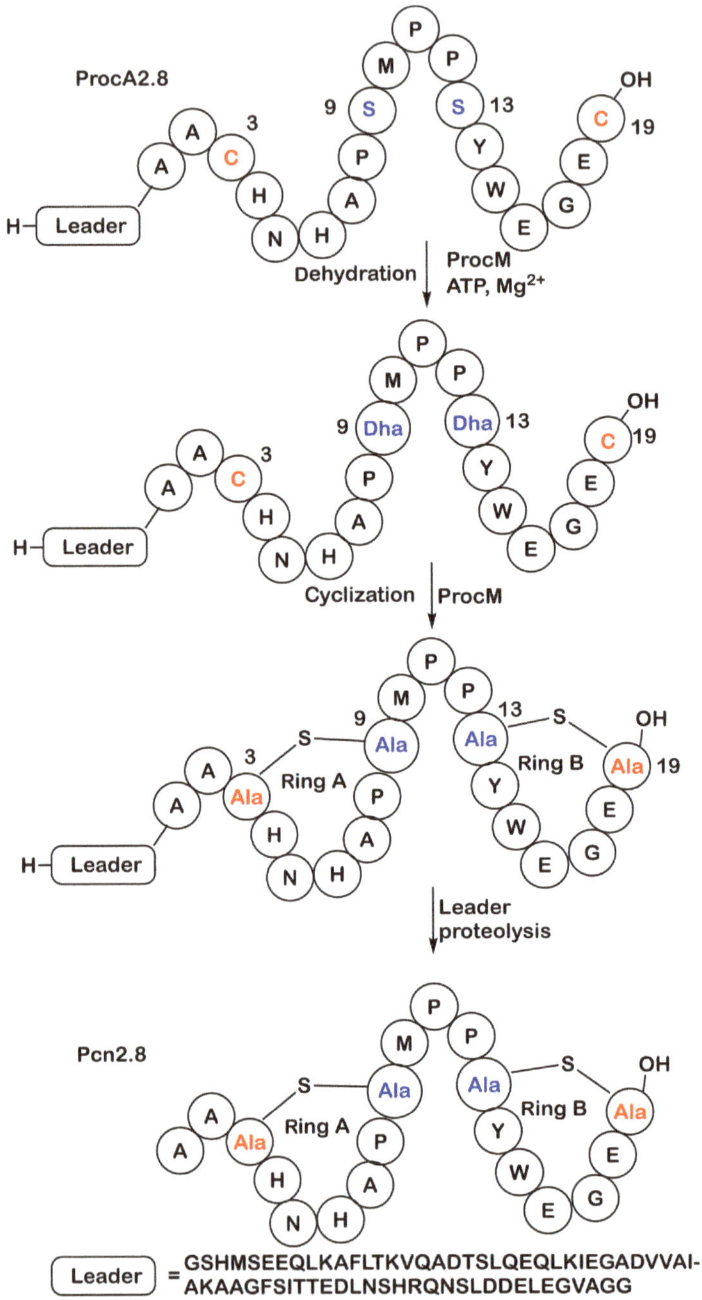

Fig. 2 Biosynthesis of prochlorosin 2.8 (Pcn2.8) as a representative example of post-translational maturation of lanthipeptides. ProcM dehydrates the Ser and Thr residues in the core region of the ribosomally synthesized linear ProcA peptide, and then catalyzes the intramolecular Michael addition of the cysteine thiols onto the dehydro amino acids. Proteolysis of the leader peptide sequence affords the mature lanthipeptide.

The first sequenced genome of a cyanobacterium encoding lanthipeptide biosynthesis contains one gene for a LanM enzyme (called ProcM) and up to 30 genes encoding different LanA peptides termed ProcAs (Fig. 3) (Li et al., 2010; Zhang, Yang, et al., 2014). One set of *procA* genes is clustered with the *procM*, while another >20 putative precursor peptides are encoded in three separate clusters spread across the genome. Initial biochemical studies employed separate expression of a subset of the ProcA peptides and ProcM in *Escherichia coli* to obtain linear precursor peptides that were subsequently modified in vitro by purified ProcM (Li et al., 2010). Subsequent experiments utilized coexpression of an affinity-tagged ProcA peptide with ProcM in *E. coli* (Shi, Yang, Garg, & van der Donk, 2011).

The characterization of the unusual substrate tolerance of ProcM prompted investigation of a wider sample of picocyanobacterial populations to search for similar clusters. Deep sequencing of organisms and metagenomes from 39 environmental samples revealed more than 1.6 million *procA*-like open reading frame sequences stemming from almost 1700 prochlorosin-containing clusters (Cubillos-Ruiz et al., 2017). The abundance of prochlorosin encoding traits varies by geographical origin and shares little overlap between locations, but does not vary significantly over time at each given site. Protein sequence analysis of these clusters discovered a low amino acid conservation of less than 30% of the core peptide sequences, while the leader peptide is more than 60% conserved for half the sequences. Co-occurrence of Cys/Ser and Cys/Thr pairs in core regions that could result in at least one thioether ring was found in more than three quarters of the sequences analyzed (Cubillos-Ruiz et al., 2017). Thus, the products of these precursor peptides collectively represent a tremendously diverse group of cyclic peptides. Analysis of the coevolution of the lanthionine synthetase ProcM and its ProcA substrates revealed that its codons are under strong purifying selection when compared to the coevolution of LanMs from the lacticin 481 group of lanthipeptides with their cognate single-substrate LanAs, which suggests that these promiscuous ProcM enzymes have evolved specifically to accept a wide variety of substrates (Cubillos-Ruiz et al., 2017).

Comparison of a subset of prochlorosin-encoding genes with 100% nucleotide identity in their leader peptide found that the 3′ downstream intergenic region is strongly conserved as well, which may indicate that randomization of the nucleotide sequence of the core peptide is a recent event and results from insertions and deletions downstream of the leader sequence and upstream of the conserved intergenic region. This type of event is consistent with relatively fast diversification of the prochlorosin core peptide

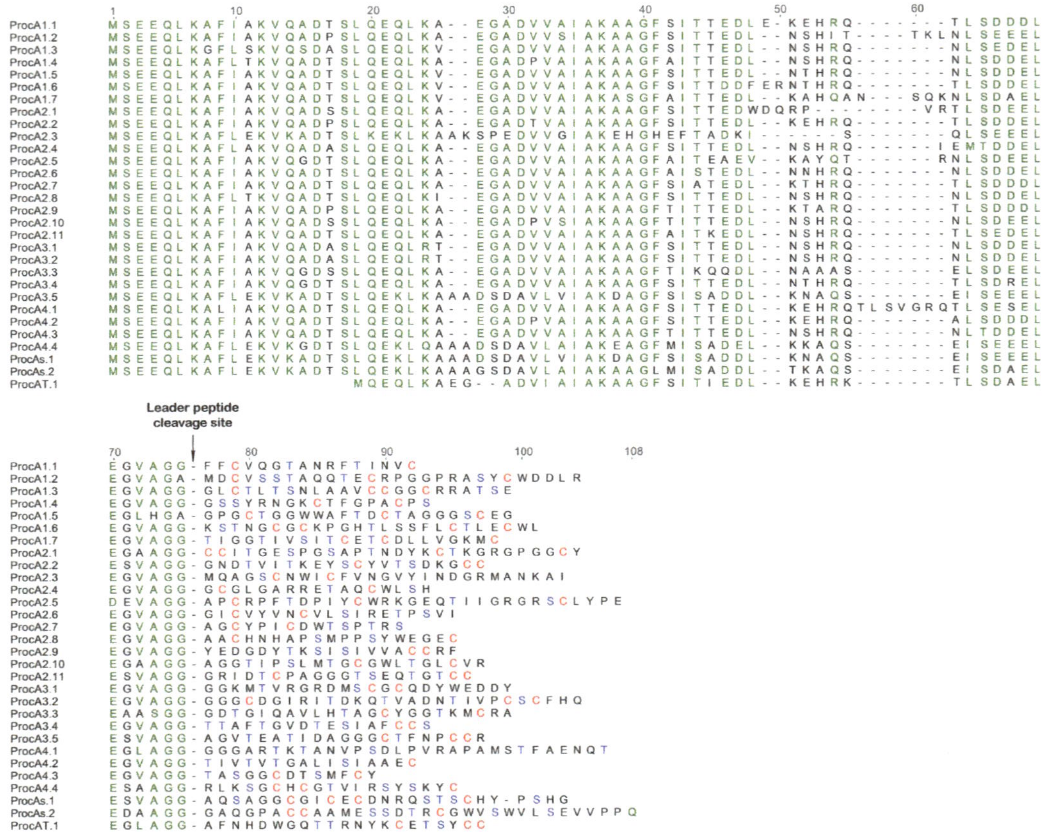

Fig. 3 The prochlorosin precursor peptide sequences encoded in *Prochlorococcus* MIT9313. The leader peptide has regions of more than 80% conserved residues (highlighted in *green*), whereas the core peptides are highly diverse. One substrate, ProcAT.1, has an N-terminally truncated leader peptide. The leader peptide cleavage site is indicated with an *arrow*.

sequence composition while preserving the leader peptide sequence. Similar evolutionary processes were identified for cyanobactins, another RiPP class (Cubillos-Ruiz et al., 2017; Donia et al., 2006; Lin, Torres, Tianero, Kwan, & Schmidt, 2016); however in these organisms the changes in the precursor peptide nucleotide sequence are more gradual than in the *Prochlorococcus* lanthipeptides (Cubillos-Ruiz et al., 2017; Schmidt, Donia, McIntosh, Fricke, & Ravel, 2012).

This chapter will describe approaches to prepare and purify the ProcA substrates and the ProcM enzyme along with the methods used to understand the dehydration activity and cyclization activity of the enzyme, and to characterize the ring topology of the products.

2. EXPRESSION AND PURIFICATION OF ProcA PEPTIDE SUBSTRATES

2.1 Overview

Lanthipeptides are synthesized ribosomally as linear precursor peptides consisting of a leader peptide of variable lengths and a core peptide that is modified by the lanthionine synthetase and eventually secreted to the extracellular environment. The role of the leader peptide can range from serving as a recognition element for the synthetase guiding the correct dehydration and cyclization, to keeping the cyclized product inactive inside the cell in cases in which the proteolyzed mature natural product has strong antibacterial activity (Oman & van der Donk, 2010; Plat, Kuipers, Rink, & Moll, 2013; Yang & van der Donk, 2013). The proteolytic events for class II lanthipeptides take place C-terminal to a double glycine motif (Gly-Gly, Gly-Ser, Gly-Ala) that is often preceded by negatively charged amino acid residues, Asp or Glu (Gebhard, 2012; Håvarstein, Diep, & Nes, 1995).

A variety of tags and engineered cleavage sites have been used to express RiPP precursor peptides such as the maltose binding protein (di Guan, Li, Riggs, & Inouye, 1988; Fox & Waugh, 2003; Sun, Tropea, & Waugh, 2011) and the SUMO tag (Panavas, Sanders, & Butt, 2009) that can help improve the solubility and the stability of the RiPP precursor peptides. The hexahistidine fusion tag (Bell, Engleka, Malik, & Strickler, 2013) has the advantage of a smaller size compared to these protein tags, allowing direct analysis of the fusion peptides by mass spectrometry without the need of removing the affinity tag. Thus far, ProcA peptides have been generated with such His_6 tags.

2.2 Protocol for Heterologous Expression and Purification of ProcA Peptides

2.2.1 Equipment

- Bio-Rad Micropulser Electroporation Apparatus (catalog number 165-2100)
- Sonics & Materials Inc. Vibra Cell VCX 500 or VCX 700 or Avestin C3 Cell Homogenizer
- Rainin Rabbit Peristaltic pump or equivalent
- HiTrap chelating HP nickel affinity column (GE Healthcare) or His_{60} Ni Superflow Resin (Clontech)
- Millipore ziptips containing 0.6 μL of C18 resin (EMD Millipore part number ZTC18S096) or Agilent Bond Elut Omix (Agilent part number A5700310)
- Digital Series SpeedVac Systems (Thermo Fisher catalog number SPD1111V-115) or any equivalent rotary evaporator
- DeltaPak C4 Radial-Pak cartridge (Waters part number WAT037700) assembled in a PrepLC 40 mm assembly (Waters part number WAT022441) or equivalent
- Äkta FPLC system (P-920) or equivalent
- HiLoad 16/60 column with Superdex 200 resin (GE Healthcare)
- Agilent 1260 Infinity HPLC system or equivalent

2.2.2 Materials, Buffers, and Solvents

- Kanamycin (Goldbio)
- LanA Start Buffer (20 m*M* NaH_2PO_4, 500 m*M* NaCl, 0.5 m*M* imidazole, 20% glycerol, pH 7.5)
- LanA B1 Buffer (6.0 *M* guanidine hydrochloride, 0.5 m*M* imidazole, 20 m*M* NaH_2PO_4, 500 m*M* NaCl, pH 7.5)
- LanA B2 Buffer (4.0 *M* guanidine hydrochloride, 30 m*M* imidazole, 20 m*M* NaH_2PO_4, 300 m*M* NaCl, pH 7.5)
- LanA Elute Buffer (4.0 *M* guanidine hydrochloride, 20 m*M* Tris–HCl, 1.0 *M* imidazole, 100 m*M* NaCl, pH 7.5)
- LanA Lysis Buffer (50 m*M* HEPES, 500 m*M* NaCl, pH 7.0, 1 m*M* triscarboxyethylphosphine [TCEP])
- LanA Cleavage Buffer (50 m*M* NaH_2PO_4, 500 m*M* NaCl, 50 m*M* sodium 2-mercaptoethanesulfonate [MESNa, Sigma-Aldrich], pH 7.0, 1 m*M* TCEP)

- Solvent A (0.1% trifluoroacetic acid [TFA] in Millipore H_2O)
- Solvent B (0.1% TFA in LC-grade acetonitrile)
- pRSFDuet-1 (Novagen/Millipore-Sigma)
- *E. coli* BL21 (DE3) (New England Biolabs)

2.2.3 Peptide Expression Procedure

1. *E. coli* BL21 (DE3) cells are transformed with the desired tagged-*lanA*-containing plasmid that may or may not include the gene for the lanthipeptide synthetase. The prochlorosin peptides have yielded the best results by incorporating the *lanA* gene downstream of the hexahistidine tag of multiple cloning site I (MCSI) of a pRSFDuet-1 vector with the synthetase in MCSII without a purification tag.
2. Single colonies of transformants are grown in a 37°C shaker overnight in an appropriate volume of media containing kanamycin at a final concentration of 50 μg/mL for pRSFDuet constructs.
3. After overnight incubation, the starter culture is used to inoculate a 1:50 v/v dilution of a production culture. Luria-Bertani (LB) or Terrific Broth (TB) with 4% w/w glycerol per liter of culture supplemented with 2.0 m*M* final concentration of $MgCl_2$ is used for media. The glycerol in TB can be substituted by 2% glucose as carbon source.
4. The production cultures are grown aerobically at 37°C until the OD_{600} is between 0.6 and 0.8 for LB or 1.2 and 1.5 for TB medium.
5. The cultures are cooled on ice for 10–20 min and IPTG is added to a final concentration of 0.5 m*M*.

6a. For peptides that are coexpressed with the synthetases, the culture is grown aerobically at 18°C for an additional 16–20 h since the synthetase requires lower temperatures for producing soluble protein.

6b. For peptides that are expressed without synthetase, an additional 3 h at 37°C is sufficient, but overnight 18°C works as well, provided the peptides are produced predominantly in inclusion bodies to avoid proteolytic truncation upon longer incubation.

6c. If an ArcticExpress strain is used (see Notes in Section 2.2.5), the cells are grown at 6–10°C for 24 h.

7. The cells are harvested by centrifugation at 5000 × *g* for 10 min at 4°C.
8. At this point the cell paste can be resuspended in the buffer used for peptide purification (see below) using between 30 and 50 mL of buffer for each liter of culture.
9. Cell paste can be stored at −80°C until purification.

2.2.4 Expressed Peptide Purification Procedure

1. For peptides that are not coexpressed with the synthetase it is recommended that 1 m*M* TCEP be added to buffers at the time of use, to prevent oxidative aggregation of the peptide.

2a. If the LanA peptide is expressed in the soluble fraction (which varies for each peptide), the cell paste is suspended in 3 mL of LanA Start Buffer per gram of cell paste.

2b. If the LanA peptide is expressed entirely or partly in the insoluble fraction (inclusion bodies), the cell paste is suspended in 3 mL LanA B1 buffer per gram of cell paste.

3a. For LanA peptide expressed in the soluble fraction, after vigorous resuspension and complete thawing of the frozen cells, 4 μL benzonase nuclease (≥250 U/μL, EMD Millipore) is added per liter of culture along with 20 mg of lysozyme, and the cells are incubated at 4°C for 60 min with occasional mixing. A short sonication (60% amplitude, 2.0 s pulse-on, 8.0 s pause, 3 min total pulse on time) results in lysis. Longer sonication time is recommended if the sample is still viscous, indicating that not all the *E. coli* genomic DNA has been digested by the benzonase.

3b. If the LanA peptide is expressed in the insoluble fraction, one freeze–thaw cycle will usually lyse the cells because of the presence of a strong denaturing agent in the resuspension buffer. Since benzonase cannot be used in denaturing conditions, the lysate after freeze–thaw in LanA B1 Buffer will be more viscous. A longer sonication time of 5–8 min can reduce viscosity.

4. The lysate is centrifuged at 14,000–30,000 × *g* (depending on the tubes and centrifuge used) for 30 min at 4°C.

5a. If the cells were first suspended in LanA Start Buffer, the supernatant should be saved, the pellet should be resuspended in LanA B1 buffer and sonicated or subjected to a freeze–thaw cycle, and centrifuged again to obtain additional peptide from the insoluble fraction.

5b. If the cells were resuspended in LanA Buffer B1, proceed with purification.

6. The supernatants are filtered through 0.45-μm syringe filters (Corning) before using a HiTrap chelating HP nickel affinity column (GE Healthcare). If the lysate is still considerably viscous, a gravity flow column using His_{60} Ni Superflow Resin (Clontech) is preferable. Usually 2–3 mL of resin is sufficient for 6.0 L of culture. The resin can bind

upward of 40 mg/mL His-tagged protein, and therefore, the amount of resin can be adjusted accordingly.

7a. The HiTrap resin is first charged with 2 column volumes (CV) of 0.1 *M* $NiSO_4$, rinsed with at least 5 CV of water, then equilibrated with at least 5 CV of the buffer that the cell lysis was performed in.

7b. If the Clontech Ni Superflow resin is used, then the resin is charged with 2 CV of 0.1 *M* $NiSO_4$ and washed with 10 CV of water before equilibration with 10 CV of the buffer that the cells were lysed in.

8. After applying the lysate to the column, 10 CV of LanA B2 Buffer are used to wash the column, and the flow through is discarded.

9. The peptide can be eluted from the HiTrap HP column with 2–3 CV of LanA Elute Buffer and from the Clontech resin with 10 CV of LanA Elute Buffer.

10. Matrix-assisted laser desorption ionization time-of-flight mass spectrometry (MALDI-ToF MS) can be performed at this time by ziptipping the eluted peptide with either Millipore ziptips or Agilent Bond Elut Omix according to the manufacturer's instructions.

11. Peptide may be stored at −20°C at this point in the purification.

12a. The peptide is desalted (also removing guanidinium chloride and imidazole) using a DeltaPak C4 Radial-Pak cartridge assembled in a PrepLC 40 mm assembly (Waters) as described in the manufacturer's instructions coupled to a high-performance liquid chromatography (HPLC) module using Solvent A and Solvent B. At least 10 min of mostly aqueous wash (2% Solvent B) is recommended as the guanidine-containing salt peak is broad, and hence the general gradient will have a 10–12 min wash with 2% Solvent B, followed by a linear ramp from 2% to 98% Solvent B in 45 min. The His-tagged full-length prochlorosin peptides usually do not elute before 30% Solvent B on these columns.

12b. An alternative to HPLC desalting is use of an Agilent Bond Elut C18 SPE column. Large- or small-scale desalting can be performed with these columns by using the appropriate size. A few mL of eluted peptide solution from the Ni^{2+} affinity chromatography step is loaded onto the Bond Elut column. As per the manufacturer's recommendation, the resin bed is dried by sparging with air before elution. This step is recommended to ensure all bound material elutes in as small a volume as possible. Between 8–10 bed size volumes of 0.1% TFA in water is used for washing and 8–10 bed size volumes of 0.1%

TFA in acetonitrile is used for elution. All solvent is eluted with air pressure until the column bed is completely dry, before application of the next aliquot of sample. The procedure is repeated until all sample eluting from the Ni^{2+} column has been desalted. Agilent lists no binding capacity by bed size, but in our experience a 500 mg bed size can be used for approximately 5 mg of His-tagged peptide. The manufacturer's instructions recommend using methanol as a mobile phase, but acetonitrile is also compatible with no detectable yield decrease, and provides easier processing after desalting. If the mobile phase contains methanol, the methanol should be removed by vacuum concentration before lyophilization. The elution can then be frozen and lyophilized.

13. Yield for modified His_6-ProcAs ranges between 5 and 20 mg/L of culture.

2.2.5 Notes

1. Depending on the peptide, better results are sometimes obtained by using *E. coli* BL21 (DE3) Rosetta 2.
2. If the lanthipeptide synthetase being expressed originates from an organism that grows at colder temperature, the BL21 ArcticExpress strain (Ferrer, Chernikova, Yakimov, Golyshin, & Timmis, 2003) may be considered as it contains chaperones that can help essential *E. coli* protein folding at temperatures between 4°C and 10°C.
3. If the peptide is eluting at low organic content of the solvent, it may be an indicator the DeltaPak C4 cartridge no longer provides good retention and may need to be replaced for optimal desalting.

2.3 Protocol for Preparation of Deuterium-Labeled LanA Analogues

Despite the large number of diverse substrates of ProcM, for all tested substrates, only one specific ring topology and stereochemistry is formed (Yu, Mukherjee, & van der Donk, 2015). Whether the reactions have a specific order and how this process is orchestrated has been the topic of much investigation. Analysis of the enzymatic products by MS methods may not offer any insight into the enzymatic order of dehydration for products with overlapping rings. A valuable tool for determining the order of dehydration features the preparation of substrates with one α-deuterated serine or threonine in the core peptide (Krawczyk, Ensle, Müller, & Süssmuth, 2012). Appropriately protected commercially available [2,3,3-D_3]-Ser or [2,3-

D2]-Thr is used in standard Fmoc-based solid-phase peptide synthesis (SPPS) and manually coupled at the desired position of the substrate peptide (Krawczyk et al., 2012). In vitro incubation with the cognate enzyme will then result in an 18 Da loss for an unlabeled Ser/Thr or a 19 Da loss for the dehydration of a deuterated Ser/Thr (Fig. 4). The difference is unambiguously determined by MS using a properly calibrated method for a mass range between 900 and 4500 Da.

This method developed by Süssmuth and coworkers was applied to two prochlorosins to investigate the activity of ProcM (Mukherjee & van der Donk, 2014). ProcA2.8 leads to two nonoverlapping rings, whereas ProcA3.3 is converted into a product with a ring within a ring (Fig. 5).

2.3.1 Equipment

- Bio-Rad EconoColumn (2.5 × 5 cm)
- Rainin PS3 peptide synthesizer, CEM Liberty peptide synthesizer or equivalent
- Waters 600 system or equivalent with a Phenomenex Luna C18 (catalog number 00G-4253-N0) or C5 (catalog number 00G-4043-N0) semi-preparative column or equivalent
- Agilent 1260 Infinity HPLC system with a Grace Alltech Vydac 214TP C4 column (catalog number AT214TP54), Grace Alltech Vydac C18 column (catalog number AT218TP5410), or Phenomenex Jupiter C18 column (catalog number 00F-4053-A0)

2.3.2 Materials and Solvents

- Fmoc-amino acids, Boc-protected Cys (Chem-Impex)
- 1-[Bis(dimethylamino)methylene]-1*H*-1,2,3-triazolo[4,5-b]pyridinium 3-oxid hexafluorophosphate (HATU) (Chem-Impex)

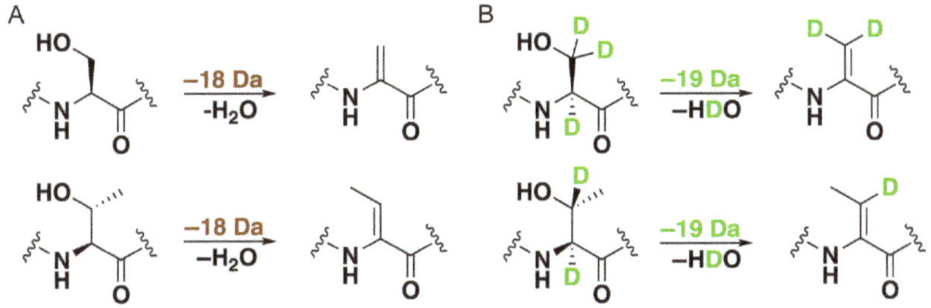

Fig. 4 Incorporation of deuterium-labeled serine and threonines in the core peptide allows monitoring of the directionality of dehydration. (A) An 18 Da loss from an unlabeled Ser/Thr dehydration is differentiated from (B) a 19 Da loss from an α-deuterated residue.

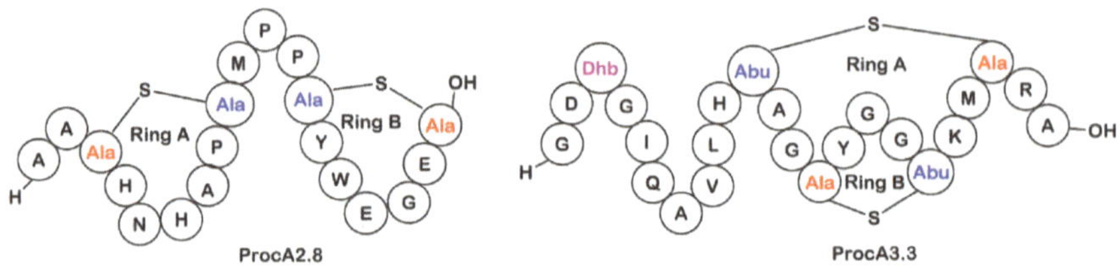

Fig. 5 Ring topologies of prochlorosins 2.8 and 3.3 used to study ProcM activity.

- 2-(6-Chloro-1*H*-benzotriazole-1-yl)-1,1,3,3-tetramethyluronium hexafluorophosphate (HCTU) (Chem-Impex)
- (7-Azabenzotriazol-1-yloxy)tripyrrolidinophosphonium hexafluorophosphate (PyAOP) (Chem-Impex)
- *N*,*N*′-diisopropylcarbodiimide (DIC) (Chem-Impex)
- 1-Ethyl-3-(3-dimethylaminopropyl)carbodiimide hydrochloride (EDC) (Chem-Impex)
- 1-Hydroxy-7-azabenzotriazole (HOAt) and 1-hydroxybenzotriazole (HOBt) (Chem-Impex)
- [2,3,3-D_3]-L-Ser (Cambridge Isotope Laboratories)
- [2,3-D_2]-L-Thr (CDN Isotopes)
- 2-(Ethyldisulfanyl)benzo[d]thiazole (Sigma-Aldrich)
- Chitin resin (New England Biolabs)
- Fmoc-Cys-loaded chlorotrityl resin (Chem-Impex)
- Chlorotrityl resin (Chem-Impex)
- Ampicillin (GoldBio)
- HPLC solvent compositions: Solvent A was water with 0.1% TFA; Solvent B was 0.1% TFA in acetonitrile.
- LanA Cleavage Buffer (50 m*M* NaH_2PO_4, 500 m*M* NaCl, 50 m*M* mercaptoethanesulfonate (MESNa, Sigma-Aldrich), pH 7.0, 1 m*M* TCEP)
- pTXB1 (New England Biolabs)
- pET15b (Novagen/Millipore-Sigma)

2.3.3 Procedure to Prepare Deuterium-Labeled ProcA Core Peptides

1. Because the prochlorosin leader peptide has a length of approximately 65 amino acids, expressed protein ligation (EPL) (Muir, 2003) is used to attach the leader peptide to the deuterium-labeled synthetic core peptides.
2. The core peptides are made by SPPS (see below).
3. The deuterium-labeled Fmoc-protected Ser and Thr are prepared as described in Mukherjee and van der Donk (2014).

4. For SPPS, preloaded H-Cys(Trt)-2-chlorotrityl resin (0.05 mmol) is used for peptides ending in Cys; then the penultimate residue is coupled manually to H-Cys(Trt)-2-chlorotrityl resin using DIC/HOAt chemistry in the absence of base to avoid racemization. We find that chlorotrityl resin provides the best results in terms of stereochemical purity of these peptides. If the C-terminal amino acid is not Cys, we couple the amino acid to activated chlorotrityl resin following the manufacturer's recommendations.
5. All subsequent amino acids (0.25 mmol) are coupled using HCTU (0.25 mmol) as activator and DIPEA (0.5 mmol) as base, except for the deuterium-labeled Ser/Thr and Cys residues (see below). Coupling can be performed on a peptide synthesizer or manually as described below. When using a microwave synthesizer, HCTU is used as activator, 2 *M* *N*,*N*-diisopropylethylamine (DIPEA) in *N*-methyl-2-pyrrolidone (NMP) as the activator base, 20% piperidine in dimethylformamide (DMF) as the deprotection agent, and 90/8/2 DMF/acetic anhydride/DIPEA as the capping agent after coupling.
6. For manual SPPS procedures (steps 6–10) using a simple flask with a glass frit, Fmoc deprotection is performed using 20% piperidine in DMF while sparging the resin with N_2 for agitation (2×5 min).
7. After draining the reaction vessel, the resin is washed with DMF (6×30 s).
8. The appropriate protected amino acid (4 equiv.) is dissolved in DMF (5–10 mL) and preactivated with DIC/HOAt (4 equiv.) for 5 min, then added to the resin, and the reaction is agitated by sparging with N_2 for 1 h.
9. After draining the reaction vessel, the resin is washed with DMF (6×30 s).
10. Kaiser test is performed to test completion of coupling; double (or triple) coupling is performed as needed.
11. Coupling of deuterium-labeled Ser/Thr for both automated and manual synthesis is performed using PyAOP (0.25 mmol) as activator, HOAt (0.25 mmol) as racemization suppressant, and 2,4,6-collidine as base, with a 3 h coupling time. These conditions also prevent partial loss of the α-deuterium during coupling.
12. Cys is coupled manually using diisopropylcarbodiimide (DIC) as activator and HOAt as racemization suppressant.
13. For peptides in which the leader and core peptides are ligated by copper-catalyzed azide–alkyne cycloaddition (Section 2.3.6), in the last

step a synthetic N-terminal azide linker (Mukherjee & van der Donk, 2014) is coupled manually with DIC/HOAt similar to steps 8–10.

14. After the completion of SPPS, the resin is washed with 5 mL of DMF (3 × 30 s), followed by 5 mL of DCM (3 × 30 s). The vessel is drained under vacuum to dry the resin.
15. Cleavage of the peptide from the resin is carried out by a cocktail of 95/2.5/2.5 TFA/triisopropylsilane/H_2O for 1 h. The solution is filtered through a fritted funnel and the filtrate is evaporated under a stream of N_2 to remove most of the TFA.
16. Peptide is precipitated by adding 10–15 mL of cold diethyl ether to the solution. The mixture is centrifuged and the supernatant is discarded.
17. The precipitate is dissolved in 50% acetonitrile, 0.05% TFA in H_2O, flash frozen in liquid N_2, and lyophilized to generate white to pale yellow amorphous powder, which is stored at −20°C.
18. Postcleavage from the resin, free Cys thiols in the crude peptide are protected as ethyldisulfides by reaction with 10 equiv. of (2-(ethyldisulfanyl) benzo[d]thiazole) dissolved in 20–50 μL of EtOH and the reaction is stirred at room temperature for 7 h. This protection helps in both EPL and copper-catalyzed alkyne–azide ligation.
19. Preparatory RP-HPLC is performed on a Waters 600 system or equivalent with a Phenomenex Luna C18 or C5 semi-preparative column using a gradient of 2% Solvent B (98% Solvent A) to 55% Solvent B in 30 min.
20. All HPLC solvents are filtered with a Millipore filtration system equipped with a 0.22-μm nylon membrane filter prior to use.
21. The identity of the peptide is confirmed by MS.

2.3.4 Procedure to Prepare Deuterium-Labeled ProcA Peptide Substrates by EPL

1. For ligation of the synthetic core peptides prepared in Section 2.3.3, the leader peptide of ProcA2.8 including the first two Ala of the core peptide is expressed from a pTXB1 plasmid encoding His_6-ProcA2.8 leader-Ala-Ala-intein-chitin binding domain (CBD) or appropriate leader peptide for another ProcA peptide (Fig. 6).
2. *E. coli* BL21 (DE3) cells are transformed with the plasmid encoding His_6-ProcA-leader-intein-CBD and then the transformed cells are spread on a LB + ampicillin (100 μg/mL) agar plate and incubated overnight at 37°C.

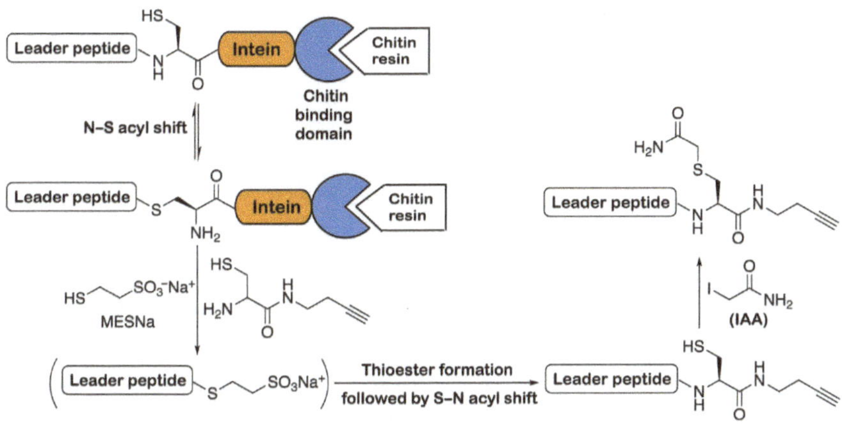

Fig. 6 Generation of the ProcA leader peptide containing a C-terminal thioester for expressed protein ligation. The leader peptide is expressed with a C-terminal intein and chitin binding domain. Purification of the peptide is performed with chitin resin, and cleavage from the resin is performed with MESNa. The thioester can be used for EPL with a core peptide containing an N-terminal Cys, or for ligation with an alkyne-containing fragment for subsequent copper-catalyzed azide–alkyne cycloaddition with an azide-functionalized core peptide (Section 2.3.6). After ligation, the cysteine thiol is capped with iodoacetamide (IAA).

3. Cells from a single colony are used to inoculate two cultures of 25 mL of LB + ampicillin (100 µg/mL), which are shaken overnight. The starter cultures are used to inoculate two 1.5-L cultures in LB + ampicillin (100 µg/mL), and the cells are grown with shaking until the OD at 600 nm reaches 0.6. IPTG is added to a final concentration of 0.5 m*M* and the cultures are incubated in an 18°C shaker for 17 h.
4. The cells are harvested by centrifugation (10,500 × *g*, 30 min).
5. LanA Lysis Buffer (30 mL) is used to resuspend the cells before sonication.
6. The lysate is centrifuged at 14,000–30,000 × *g* (depending on the tubes and centrifuge used), and the supernatant applied to 25 mL of chitin resin placed in a Bio-Rad EconoColumn (2.5 × 5 cm) preequilibrated in LanA Lysis Buffer.
7. The column is placed on a rocker in the cold room, and the resin is incubated with the peptide solution for 2 h.
8. The column is then mounted and washed with 10 CV of the LanA Lysis Buffer, and 30 mL of LanA Cleavage Buffer is added and the column is once again left to rock in the cold room overnight.

9. The column is once more mounted and the elution is drained from the chitin resin.
10. The elution is acidified to 0.1% TFA final concentration, desalted by solid-phase extraction as described in Section 2.2.4, and lyophilized.
11. The peptide is purified by RP-HPLC on a semi-preparative C5-Phenomenex column using a gradient of 2% Solvent B to 55% Solvent B at 7 mL/min over 30 min.
12. The EPL reaction of leader peptide-MESNa thioester with core peptide is performed in a 500 μL volume in an Eppendorf tube containing 0.5 m*M* of the deuterated core peptide from SPPS, 1 m*M* TCEP, 200 m*M* NaH_2PO_4, pH 7.2, 100 m*M* MESNa, 0.5 m*M* EDTA, and 0.05 m*M* leader peptide containing a C-terminal MESNa thioester. For ligation of an alkyne handle (Fig. 6), see Section 2.3.6.
13. The reaction is stirred for 20 h at room temperature and monitored by MALDI-ToF MS for complete consumption of the leader peptide with the C-terminal MESNa thioester.
14. The reaction is acidified with TFA and purified by RP-HPLC using an analytical C4 Grace Alltech Vydac column with a gradient from 2% Solvent B to 65% Solvent B over 45 min at 1 mL/min.

2.3.5 Notes

1. For some leader peptides, better yields are obtained with pET15b. The intein-CBD encoding construct is first obtained by PCR from pTXB1 and then inserted into pET15b containing the gene encoding the ProcA leader peptide (Mukherjee & van der Donk, 2014).

2.3.6 Copper-Catalyzed Alkyne–Azide Ligation Procedure

1. The deuterium-labeled ProcA3.3 core peptide was prepared by SPPS using methods described in Section 2.3.3, but the lack of a convenient cysteine for EPL required the installation of an alkyne and azide such that the leader peptide could be ligated to the core peptide using copper-catalyzed click chemistry (Figs. 6 and 7) (Mukherjee & van der Donk, 2014). The peptide was expressed and purified in a similar manner as described in Section 2.3.4 (steps 1–9) except that the LanA Lysis Buffer included 1 m*M* EDTA, 0.1% Triton-X, 1 m*M* PMSF, and the LanA Cleavage Buffer (40 mL) included 2 m*M* of (*R*)-2-amino-*N*-(but-3-yn-1-yl)-3-mercaptopropionamide (Mukherjee & van der Donk, 2014) (see Fig. 6) and 5 m*M* TCEP.

2. After ligation, the free sulfur atom of the Cys at the C-terminus is capped by incubation with 9 equiv. of iodoacetamide at pH 8.1, 50 m*M* HEPES (Fig. 6) to prevent involvement of this thiol in any ProcM-catalyzed thioether formation.
3. Purification of this product was performed by RP-HPLC using a gradient from 2% to 80% Solvent B in 45 min on a Phenomenex Luna C_{18} or C_5.
4. Equimolar amounts of the azide-containing core peptide and C-terminal alkyne-containing leader peptide (Fig. 7) are dissolved in 30 μL of nitrogen-sparged phosphate buffer, pH 7.8, to a concentration of 1–2 m*M* for each peptide.
5. Copper sulfate (5 μL of 100 m*M* dissolved in degassed water) is mixed with tris(benzyltriazolylmethyl)amine (TBTA, Sigma-Aldrich, 25 μL of 20 m*M* dissolved in methanol) and sodium ascorbate (10 μL of 200 m*M* in degassed water) and the solution is stirred under nitrogen for 5 min.
6. The active catalyst solution from step 5 is added to the peptide solution of step 4.
7. Reaction completion is monitored by MALDI-ToF MS or HPLC. If after 1 h significant starting material remains, a fresh aliquot of active catalyst is prepared as described in step 5 of this section and added to the reaction mixture.
8. The crude reaction is desalted by solid-phase extraction as described in Section 2.2.4, lyophilized, then purified on a semi-preparative Phenomenex Luna C5 column using a gradient of 2%–55% Solvent B over 30 min at 7 mL/min.

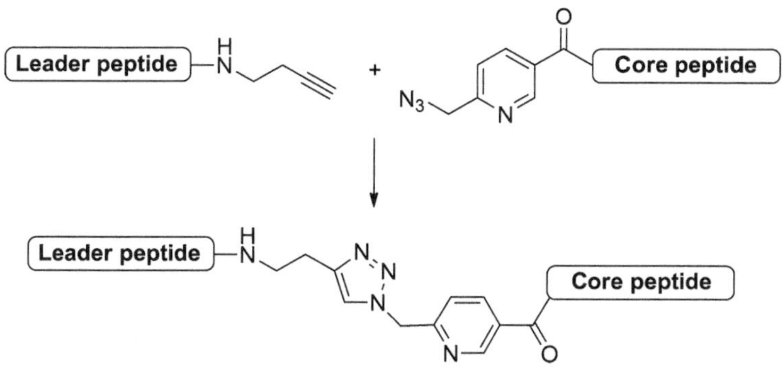

Fig. 7 Preparation of ProcA substrates containing synthetic core peptides by copper-catalyzed azide–alkyne ligation. The preparation of each coupling partner is described in Sections 2.3.3 (azide, step 13) and 2.3.6 (alkyne).

3. EXPRESSION, PURIFICATION, AND ACTIVITY ASSAYS OF ProcM

3.1 Overview

ProcM-like enzymes are ubiquitous in cyanobacteria (Cubillos-Ruiz et al., 2017; Zhang, Yang, et al., 2014), but at present only the ProcM from *Prochlorococcus* MIT 9313 has been characterized in vitro (Li et al., 2010). A few other enzymes from the ProcM clade of LanM enzymes (i.e., containing three Cys ligands to Zn, Fig. 8) (Zhang et al., 2012) from *Cyanothece* sp. PCC 7425 have been coexpressed with their putative substrates in *E. coli* and they have also shown high substrate tolerance. We will here only concentrate on studies that have focused on the original ProcM.

3.2 Protocol for Expression and Purification of ProcM

3.2.1 Materials and Buffers

- ProcM Start Buffer (20 m*M* Tris–HCl, 1.0 *M* NaCl, pH 8.0)
- ProcM Wash Buffer (20 m*M* Tris–HCl, 1.0 *M* NaCl, 30 m*M* imidazole, pH 8.0)
- ProcM Elute Buffer (20 m*M* Tris–HCl, 1.0 *M* NaCl, 200 m*M* imidazole, pH 8.0)
- HiTrap HP Ni column (GE Healthcare)
- Superdex 200 resin (GE Healthcare)
- 0.22-μm nylon membrane filter (Fisherbrand)
- 5,5′-dithio-bis-(2-nitrobenzoic acid) (DTNB, Sigma-Aldrich)
- 4-(2-pyridylazo)resorcinol (PAR, Sigma-Aldrich)

	860		924		970
ProcM MIT9313	GFSHGTAGYA	....	FMA-SWCHGAPGIALGR	....	HLCCGSLGLMVLLE
ProcM KORDI-100	GFSHGTAGYA	....	FMT-FWCHGAPGIALGR	....	HICCGSMGNASVLR
CyanM	GLSHGAAGIA	....	FMV-SWCHGAAGIGLAR	....	HLCCGTMGRIETML
BovM	GFAHGISGIK	....	YTNGSWCNGLLGWITSE	....	CLCHGSFGVLDFLL
CinM	GFSHGSGGIG	....	FSN-AWCNGAAGIGLAR	....	TLCHGTSGNAELLL
CylM	GFGHGIYSYV	....	----SWCKGTVGELLAT	....	CLCHGNAGTLEGLI
GarM	GFSHGAAGIA	....	YPA-LWCHGAAGIGLSR	....	SICHGDFGALALFD
LctM	GIAHGELGYL	....	----GWCNGLSGILMIL	....	SVCHGASGVLQTLL
MrsM	GFAHGASGII	....	FVA--WCNGAAGIGLSR	....	SLCHGDLGNLDILM
NukM	SYAHGESGLM	....	YSV-SWCNGVTGQLISR	....	CLCHGSLGNLLVLN
EpiC	RYDILENYIP	....	----GWCYGDTGIMNTL	....	TFCHGLASHLTIIH
MutC	-------PIR	....	----GWCYGENGIYNTL	....	TFCHGFAGKANFFL
NisC	---------E	....	----AWCYGGPGISLLY	....	MICHGYSGLIEICS
SpaC	EYQYGSPPNA	....	----AWCYGRPGVCLAL	....	TICHGYSGIGQILL

Fig. 8 Conserved residues in the cyclization domains of class II LanM enzymes and class I LanC enzymes. The ProcM-like LanM enzymes that contain a three-cysteine ligand set to the active-site zinc (first three entries, Cys *underlined*) have multiple substrates with diverse sequences.

3.2.2 Procedure

1. *E. coli* BL21 (DE3) cells containing a Rosetta 2 plasmid are transformed with a plasmid (see Notes) encoding the hexahistidine-tagged *procM* gene.
2. Single colonies are picked and grown at 37°C overnight in Luria-Bertani medium supplemented with 25 μg/mL kanamycin and 15 μg/mL chloramphenicol (for the Rosetta 2 plasmid).
3. A 1:100 to 1:50 v/v inoculation of 2–6 L of production culture in LB medium is performed and the culture is grown aerobically with shaking at 37°C until the OD_{600} is 0.8, then the culture is cooled on ice for 20 min.
4. IPTG is added to a final concentration of 0.25 m*M* and the culture is grown aerobically at 18°C overnight with shaking.
5. The cells are harvested by centrifugation at 5000 × *g* for 10 min, the media are discarded, and the cells are resuspended in ProcM Start Buffer, frozen, and kept at −80°C until purification.

6a. Cells are lysed by using 20 mg lysozyme per liter of culture and 4 μL of benzonase (≥250 U/μL, EMD Millipore) per liter of culture for 1 h at 4°C, followed by sonication (3–5 min total pulse on time, 60% amplitude, 2.0 s pulse-on, minimal 8.0 s pause—longer pauses are beneficial as heating of the sample can be avoided).

6b. Alternatively, an Avestin Cell Homogenizer can be used operating between 5000 and 10,000 psi. Usually six passes through the homogenizer are enough for complete cell lysis of up to a 6.0-L culture.

7. The lysates are centrifuged and the supernatant is clarified through 0.45-μm Corning filters and applied to a 5 mL HiTrap HP Ni chelating column (GE Healthcare) that has been preequilibrated in ProcM Start Buffer using a peristaltic pump.
8. The HiTrap Ni column is then rinsed with 10 CV of ProcM Wash Buffer using a peristaltic pump and then the column is transferred to an Äkta FPLC system (P-920).
9. The following gradient is applied: the first 6 CV are 5% ProcM Elute Buffer in ProcM Start Buffer; the next 4 CV are 5% ProcM Elute Buffer linearly increasing to 40% ProcM Elute Buffer; for the next 2 CV the ProcM Elute Buffer content is linearly increased from 40% to 100% ProcM Elute Buffer, followed by 6 CV of 100% ProcM Elute Buffer.
10. Fractions containing protein typically elute between CV 8 and 14 as monitored by absorbance at 280 nm.

11. The contents of the eluates are visualized by SDS-PAGE gel and fractions containing ProcM are concentrated using a 50-kDa cut-off Amicon filter to 2–3 mL.
12. Imidazole is removed by gel filtration using ProcM start buffer on a HiLoad 16/60 column containing Superdex 200 resin (GE Healthcare).
13. The fractions containing the monomer are concentrated to approximately 200 μ*M* without precipitation of the protein.
14. The protein is aliquoted and stored at −80°C for up to 6 months for most batches. In our experience, longer storage can lead to significantly decreased protein activity.

3.2.3 Notes

1. pRSFDuet-1 leads to better yields of His_6-ProcM than pET15b or pET28b (Yu et al., 2015).
2. It is essential to use a Rosetta 2 or a commercially available equivalent plasmid to offset the occurrence of rare codons in the *procM* gene; otherwise, the yield can be as low as 1 mg/L of culture with most of the protein product truncated.
3. Multiple peaks are observed in the gel filtration chromatogram including the ProcM monomer as well as higher molecular weight aggregated protein. Although the dimer and possibly trimer show activity similar to the monomer, higher aggregates do not. We usually isolate the monomer fractions to have homogeneous preparations.

3.3 Characterization of the Enzyme

Several studies have shown that the cyclization and dehydration activities of LanM enzymes can be separated by expression of their individual domains (Dong et al., 2015; Ma, Gao, Zhao, & Zhong, 2015; Shimafuji et al., 2015; Thibodeaux, Wagoner, Yu, & van der Donk, 2016). Unlike most class II lanthipeptide synthetases reported to date that contain two cysteines and one histidine as the zinc-binding residues, ProcM has three cysteines at the homologous positions (Fig. 8) (Yu et al., 2015). This ligand set is also found in other class II LanM synthetases from cyanobacteria, and in many cases but not all, a three-cysteine-zinc-binding LanM has multiple substrate peptides with diverse sequences (Cubillos-Ruiz et al., 2017; Zhang, Yang, et al., 2014; Zhang et al., 2012). Previous reports on increasing the reactivity of thiolates bound to a zinc ion by increasing the number of thiolate ligands (Harris, Derdowski, & Poulter, 2002; Penner-Hahn, 2007) make investigations of these cysteines a prime starting point for understanding enzymes capable of modifying 30 diverse core peptides.

3.3.1 Mutagenesis of the ProcM Zinc-Binding Residues and Determination of Zinc Content

The zinc-binding cysteines of ProcM have been mutated by standard QuikChange PCR to generate C924A, C970A, C971A, and C971H (Yu et al., 2015). The proteins were expressed in *E. coli* Rosetta2 as described in Section 3.2 and the zinc content was compared to that of the wild-type enzyme. Zinc content is determined using the metallochromic indicator 4-(2-pyridylazo)resorcinol (PAR).

3.3.2 Procedure for Zinc Content Determination

1. All flasks and beakers are cleaned with 20% HNO_3 and Milli-Q water before use.
2. Slidalyzers (Pierce) dialysis cassettes are preincubated in 20 m*M* HEPES, pH 8.0 with 10 g/L Chelex-100 (Bio-Rad).
3. ProcM and its mutants are buffer-exchanged four times in these dialysis cassettes in a 4-L beaker for 48 h at 4°C using 20 m*M* HEPES, 1.0 *M* NaCl, 25–50 g/L Chelex-100, pH 8.0.
4. Amicon 50-kDa filters are preincubated with 20 m*M* Tris, pH 7.5 (buffer is treated with 10 g/L Chelex-100) before concentration of the Chelex-treated ProcM samples.
5. Assays to determine metal loading are conducted in 20 m*M* HEPES, 1.0 *M* NaCl, pH 8.0, that has been treated with 25 g/L Chelex-100, and filtered through a 0.22-μm nylon membrane filter.
6. To 40 μL of the protein solutions or standard $ZnCl_2$ solutions of known concentration, 160 μL of 100 μ*M* PAR in 4 *M* guanidine hydrochloride is added.
7. Then 10 μL of 3.5 m*M* DTNB is added to free the enzyme-bound zinc by reaction with the Cys ligands, and incubated for 5–10 min.
8. The absorption of PAR-Zn^{2+} is measured at 500 nm and its concentration is determined from a standard curve derived from the $ZnCl_2$ calibration standards.

This assay determined that wild-type ProcM contained about 1 equiv. of zinc, whereas the C924A and C970A point mutants contained about 0.5 equiv. of Zn^{2+}. Using these ProcM mutants for in vitro studies as described in Section 3.4 demonstrated normal dehydration activity, but impaired cyclization activity on the ProcA2.8 precursor peptide (Yu et al., 2015).

3.3.3 Separation of the Cyclase Domain of ProcM

Investigation of the cyclase domain of ProcM was carried out by cloning residues 655–1068 of WT ProcM and ProcM-C971H, expression of the

His-tagged truncated proteins from a pRSFDuet-1 plasmid, and purification by the standard ProcM purification procedure described in Section 3.2 (Yu et al., 2015). The cyclization activity of these mutant enzymes was tested as described in Section 3.4.2.

3.4 In Vitro Assay Conditions

3.4.1 Standard Procedure for ProcM Activity Assays

1. ProcM Reaction Buffer contains 10 m*M* ATP, 0.1–0.2 m*M* TCEP, 10 m*M* $MgCl_2$, 50 m*M* HEPES, pH 8.0. The concentration of ProcM can range from 0.2 to 20 μ*M* and the concentration of ProcA can range from 10 to 100 μ*M*, depending on the purpose of the assay.
2. Supplemental zinc up to four times the concentration of the enzyme can be added for the zinc-binding cysteine mutants in order to compensate for the inferior zinc loading of the ProcM mutants (Li et al., 2010; Yu et al., 2015).
3. Reactions are quenched with TFA to 1% final concentration before analysis.

3.4.2 Procedure for Testing Cyclase Activity Decoupled From Dehydration

1. ProcA3.3 is oxidized by dissolving in water at a concentration of 6 mg/mL and treating with 5 m*M* oxidized glutathione at room temperature for 3 h. This step forms a disulfide between the two cysteines of ProcA3.3. The 2-Da mass shift can be observed by MS.
2. The peptide is purified by solid-phase extraction on a C4 Grace Vydac column as described in Section 2.2.4.
3. The oxidized ProcA3.3 (50 μ*M*) is reacted with wild-type ProcM (5.0 μ*M*) for 24 h at room temperature to obtain dehydrated ProcA3.3 using conditions described in Section 3.4.1.
4. Wild-type ProcM is removed from the reaction by using a 50-kDa cut-off Amicon filter.
5. The filtrate is buffered with 50 m*M* HEPES pH 6.5 and then treated with 2.5 m*M* TCEP at room temperature for 3 h to reduce the disulfide.
6. The peptide is desalted as described in Section 2.2.4 and lyophilized.
7. The dehydrated peptide is dissolved in HEPES (50 m*M*, pH 8.0) and treated with 1 m*M* TCEP in the ProcM reaction conditions described in Section 3.4.1. The cyclization activity can be determined for WT ProcM and its mutants, as well as the cyclization domain (Section 3.3.3) and its mutants. The peptide can also be used to study nonenzymatic cyclization at pH 8.0.

8. The enzyme is removed after 16h at room temperature by using a 50-kDa cut-off Amicon filter and the filtrates are analyzed by ESI-MS/MS using the methods described in Section 4.

This assay showed that nonenzymatic cyclization is not occurring in detectable amounts, whereas both the full-length ProcM and the cyclase domain of ProcM catalyze thioether ring formation (Yu et al., 2015). The C971H mutants of both full-length and truncated ProcM promote significant formation of a nonnative thioether ring between Cys21 and Dhb18 (see Fig. 10B).

3.4.3 Notes

1. Higher concentrations of TCEP can generate adducts with the dehydro amino acids. Concentrations are adjusted so that minimal adduct formation is observed.

4. PRODUCT AND REACTION CHARACTERIZATION METHODS

4.1 Leader Peptide Removal Strategies

In the native producer of class II lanthipeptides, the leader peptide is removed by a membrane-associated Cys39-peptidase ABC transporter fusion protein (ProcT). To date, no such enzyme has been reconstituted from *Prochlorococcus*. For the initial organism studied, *Prochlorococcus* MIT9313, *procT* is not clustered with any of the *procA* or the *ProcM* genes, but three homologous enzymes are found elsewhere in the genome (Li et al., 2010). In the absence of an easily accessible native tool for leader peptide removal, mutagenesis of the precursor peptides has been used to install protease cleavage sites for commercial endoproteinases such as LysC and GluC at the C-terminus of the leader peptide. ProcM has proven more restrictive than other LanM proteases with respect to the C-terminus of its leader peptides such that larger protease cleavage sites such as Factor Xa have not been well tolerated. The peptide is either overexpressed in *E. coli* with its synthetase or processed in vitro, then purified, and subjected to leader peptide cleavage. The commercial protease can be used in a 1:100 to 1:50 w/w ratio to the peptide to obtain complete leader peptide cleavage. A requirement for this approach is that the core peptide does not contain a recognition motif for the protease such that cleavage does not occur in the modified core peptide, and generally that a thioether ring is not formed next to the introduced artificial cleavage site as it may inhibit cleavage.

4.2 Methods to Determine Cyclization State

Because addition of a Cys to a dehydro amino acid does not result in a change in mass and because cyclization often does not result in sufficiently different chromatographic properties to resolve cyclized and uncyclized peptides, indirect methods must often be used to determine cyclization activity. These usually involve reaction with electrophiles that report on the presence of noncyclized Cys.

4.2.1 Procedure for Cys Alkylation

1. The peptide is dissolved at a concentration between 10 and 100 µ*M* in 100 m*M* HEPES buffer, pH 6.5.
2. A 1.0 m*M* stock solution of TCEP in water is freshly prepared and added to the peptide solution to give a final concentration of 0.1–0.5 m*M*. The reaction is left at room temperature for 15–60 min.
3. A 500 m*M* stock solution of iodoacetamide (IAA) in water or *N*-ethylmaleimide (NEM) in ethanol is diluted 10-fold into the reduced peptide solution. The reaction is left at room temperature for 15–60 min. Iodoacetamide reactions should be kept in the dark until analyzed by MALDI-ToF MS and ESI-tandem MS. These methods can both identify the number of free cysteines and their location in the peptide. Details on the ESI-MS/MS methodology that can be used to quantify species are provided in Section 4.3.
4. Protease can be added simultaneously with IAA or NEM to cleave off the leader peptide as the alkylation reaction progresses.

4.2.2 Notes

1. The TCEP concentration is adjusted experimentally to minimize adduct formation to the dehydro amino acids.
2. Neither alkylating agent (IAA/NEM) inhibits the endoproteinases GluC and LysC.

4.3 Methods for Investigating Dehydration and Cyclization Reactions

4.3.1 Procedure for Determining the Directionality of Dehydration

1. The deuterated peptides described in Section 2.3 are incubated with ProcM in the following conditions: 50 m*M* HEPES, pH 8.0, 10 m*M* ATP, 0.5 m*M* TCEP, 10 m*M* $MgCl_2$, 2 µ*M* ProcM, and 50–100 µ*M* substrate.

2. Aliquots are taken at 5, 10, 30, and 60 min and are quenched by TFA addition to a final concentration of 1%.
3. The aliquots are centrifuged using a 50-kDa cut-off filter (Millipore) to remove the enzyme.
4. The pH of the filtrate containing the peptides is adjusted to 8.0 with 5.0 *M* NaOH and endoproteinase GluC or LysC is added for leader peptide removal.
5. The time aliquots are ziptipped as described in Section 2.2.4 and analyzed by MALDI-ToF MS.
6. When directionality is strict, the mass spectra directly show the order of dehydration using the concepts in Fig. 4. When directionality is not strict, tandem mass spectra are collected and analyzed by hypothetical structure enumeration and evaluation to provide an unbiased interpretation of the data. The R scripts for such analysis can be obtained from Zhang, Ortega, et al. (2014).

Leader peptide removal in the deuterated serine-containing ProcA2.8 variants followed by MALDI-ToF MS showed that Ser13 is dehydrated before Ser9 (Fig. 9). The same C-to-N directionality of the dehydration catalyzed by ProcM was found for the ProcA3.3 variant containing deuterated threonine (Mukherjee & van der Donk, 2014).

4.3.2 Procedure for Determining the Directionality of Cyclization

1. Steps 1–4 are repeated as described in Section 4.3.1, but with unlabeled peptides.
2. After endoproteinase digestion is complete as detected by MALDI-ToF MS, ammonium bicarbonate is added to a concentration of 200 m*M* NH_4HCO_3 to buffer the reaction mixture.
3. Stock solutions of 10 m*M* TCEP and 10 m*M* IAA are diluted 10-fold into the reaction mixtures and samples are reacted for 3 h at room temperature.
4. Samples are frozen and stored at −80°C until tandem MS analysis.

For two prochlorosins, the order of cyclization has been established using this procedure. Time point samples that consisted predominantly of peptide with one iodoacetamide adduct were subjected to tandem ESI-MS. The data suggest that for ProcA2.8 the thioether spanning Cys19-Ser13 is formed before the Cys3-Ser9 ring, which would suggest a C-to-N cyclization directionality. However, the same trend did not hold for ProcA3.3 (Mukherjee & van der Donk, 2014), suggesting that the order of cyclization is not generally directional but instead depends on the sequence of the dehydrated substrates.

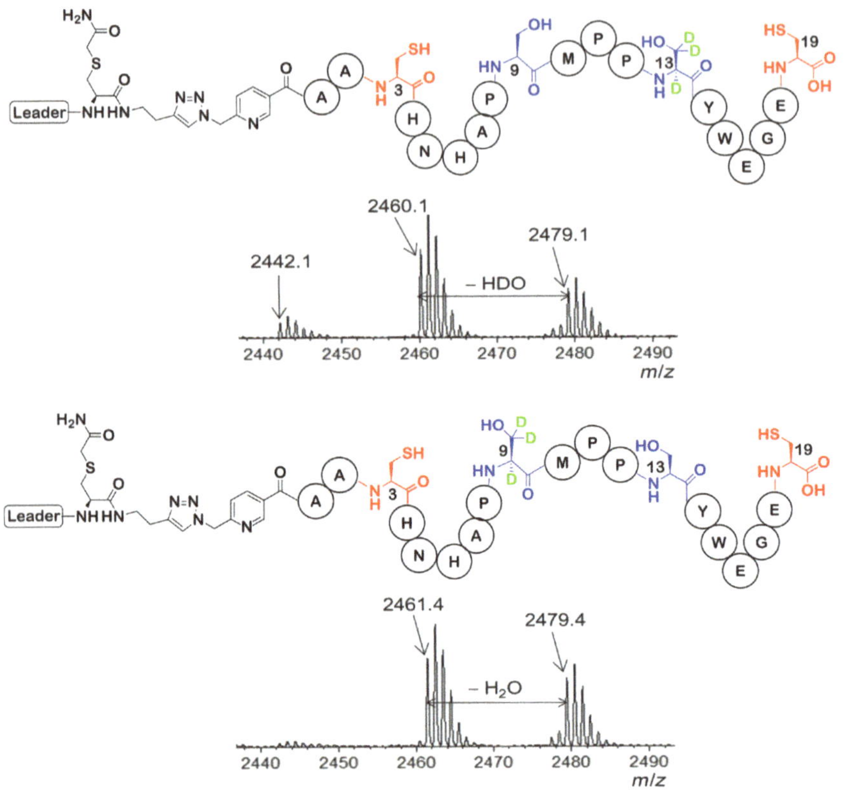

Fig. 9 Determination of dehydration order using incorporation of deuterium-labeled Ser/Thr in the ProcA2.8 core peptide that is attached to the leader peptide via click chemistry (see Fig. 7). Mass spectrometric analysis of the first dehydration clearly shows Ser13 is dehydrated initially, resulting in loss of 19 Da for the top substrate and 18 Da for the bottom substrate.

4.3.3 Procedure for Testing the Reversibility of Cyclization

Previous studies on the class II lanthipeptide synthetase HalM2 have shown that it can catalyze retro-Michael addition (Yang & van der Donk, 2015). This finding opened the possibility that the many different products that are made with high fidelity by ProcM could be the result of a thermodynamically controlled process of reversible Michael-type additions.

1. To test the reversibility of cyclization catalyzed by ProcM, full-length His-tagged ProcA2.8 isolated from *E. coli* overexpression as described in Section 2.2.4 was treated under standard conditions with ProcM (Section 3.4.1) with all stock solutions made in D_2O. During this assay,

enolates formed upon Michael-type addition react with D_2O or a deuterated amino acid in the active site of ProcM to generate an α-deuterated product (Fig. 10A).

2. ProcM is then removed from the assay mixture by ultrafiltration with a 50-kDa cut-off filter (Millipore) and the peptide filtrate is split in half.
3. Half of the sample is digested with GluC for MS analysis, while the other half is lyophilized and resubjected to standard ProcM enzymatic reaction conditions this time performed in buffer made in unlabeled water (Fig. 10A).

When this protocol was applied to ProcA2.8, two deuterium atoms were incorporated after the first step; however, ProcM was not able to catalyze any D–H exchange in the second step (Mukherjee & van der Donk, 2014). The reversibility of cyclization can also be tested by using an incorrectly cyclized substrate formed by ProcM-C971H followed by removal of the mutant enzyme by ultrafiltration in a 50-kDa cut-off filter (Millipore), then incubation with the wild-type enzyme (Fig. 10B). This experiment showed that one incorrectly installed ring was not corrected by the wild-type enzyme (Yu et al., 2015).

4.4 Mass Spectrometry-Based Kinetic Assay for LanM Synthetases

4.4.1 Overview

Until recently, methodology to quantify the concentration changes of substrates and partially modified intermediates during lanthipeptide biosynthesis was not available. Such a method was developed based on electrospray ionization mass spectrometry coupled to liquid chromatography (LC/ESI-MS), which was used to compare ProcM to HalM2, the biosynthetic enzyme responsible for the biosynthesis of haloduracin β (Thibodeaux, Ha, & van der Donk, 2014).

4.4.2 Sample Preparation for Kinetic Studies

1. Reactions containing 1 μ*M* LanM and 40 μ*M* LanA are set up on a 1.5-mL scale based on the standard LanM reaction conditions (5 m*M* ATP, 5 m*M* $MgCl_2$, 100 m*M* HEPES, pH 7.5) with 0.1 m*M* TCEP for the ProcM/ProcA 2.8 assays.
2. The solution is equilibrated at room temperature for 1 h before addition of the enzyme.

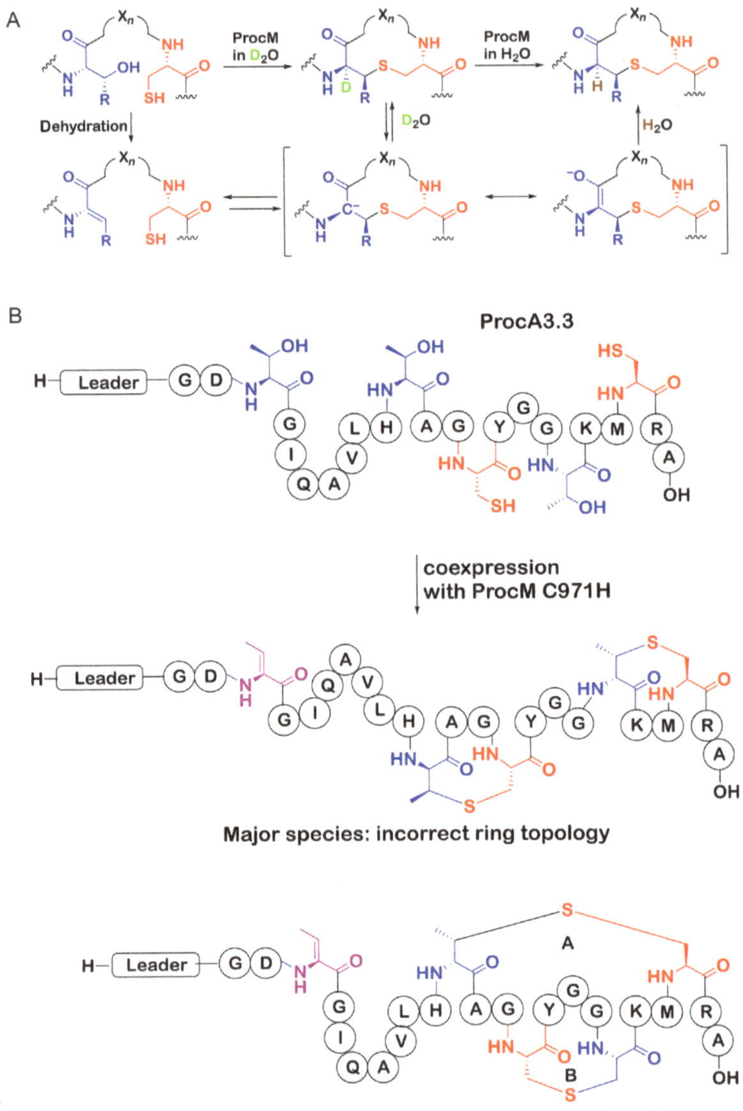

Fig. 10 (A) A potentially reversible Michael-type addition reaction can be tested by use of deuterium labeling at the α-carbons of the Lan/MeLan residues. (B) An alternative approach involves generation of incorrectly cyclized products and subjecting them to the WT ProcM to investigate whether it can correct the ring topology. Coexpression of ProcA3.3 with ProcM C971H yields a prochlorosin 3.3 variant with a nonnative ring topology as the major product and correctly cyclized ProcA3.3 as a minor species. Subjecting the incorrectly cyclized product to wild-type ProcM did not result in correction of the incorrect ring topology, suggesting the products of ProcM catalysis are not the result of thermodynamic control.

3. Aliquots (100 μL) are removed at specific time points and diluted 10-fold into a quenching buffer containing 1.1 m*M* EDTA, 10 m*M* TCEP, 111 m*M* citrate at pH 3.5. The low pH inactivates the LanM while preserving the cysteines in a reduced state.
4. Time point aliquots are kept at −80°C until analysis.
5. To analyze the state of cyclization, the 1 mL time aliquots are thawed at 25°C in a heat block and spiked with 100 μL of 100 m*M* TCEP (final concentration 10 m*M*) and incubated for 10 min at 25°C.
6. First, 35–40 μL of 5 *M* NaOH is used to bring the pH up to 6.2–6.4 before addition of 11 μL of 1.0 *M* *N*-ethylmaleimide (NEM) dissolved in ethanol to test cyclization (Section 4.2.1).
7. After 10 min, 11 μL of pure TFA is added to acidify the reactions to pH 3.0–3.5.
8. The resulting aliquot is desalted by using a Grace Vydac C4 solid-phase extraction column according to the manufacturer's instructions and lyophilized to dryness, then resuspended in 50 μL of water.
9. Aliquot concentrations are estimated based on UV–visible spectroscopy using extinction coefficients calculated for the full-length peptides by using ExPASy ProtParam.
10. Samples are diluted to 10–20 μ*M* with 50% acetonitrile/water, 0.1% formic acid.
11. Samples are analyzed by ESI-MS as described in Section 4.4.4.
12. Extracted ion chromatograms for each mass of interest are used to integrate peak areas.
13. Raw kinetic data are processed with Excel (Microsoft) and OriginPro 9.1 (OriginLab Corporation).

4.4.3 Notes

1. Higher concentrations of TCEP of up to 0.5 m*M* may be used; however, adducts with the dehydro amino acids may start forming, which complicates the analysis.
2. We verified that the quench conditions used for our kinetic assays were sufficient to eliminate LanM activity (Thibodeaux et al., 2014).
3. The autosampler chamber during MS analysis was kept at 8°C to minimize variations caused by solvent evaporation (Thibodeaux et al., 2014).
4. A very early time point should be analyzed to ensure that the peptides of interest show complete NEM alkylation, and an end point must be established, when an aliquot shows no trace of partially alkylated peptide.

4.4.4 Liquid Chromatography and Electrospray Ionization Settings for Kinetic Studies

1. A Synapt-G1 ESI quadrupole TOF Mass Spectrometry system (Waters) with an Acquity Ultra Performance Liquid Chromatography (UPLC) unit is used for sample analysis.
2. As little as 5 μL sample of digested or full-length peptide is injected on a reversed-phase UPLC column (Phenomenex Jupiter C18 or BEH C8 1 μ*M* column 1 × 100 mm).
3. The sample is separated by using a gradient of 2% Solvent B (0.1% formic acid in acetonitrile) to 98% Solvent B over 20 min at a flow rate of 0.090 mL/min. Solvent A is 0.1% formic acid in Milli-Q water.
4. Mass spectra are typically monitored between 50 and 2000 m/z while using a capillary voltage of 2.5–3.5 kV, cone voltage 40 V, 110–120°C source temperature, 150–300°C desolvation temperature, 20–150 L/h cone gas flow, and 500–600 L/h desolvation gas flow. Transfer collision energy for both simple and tandem MS is 4.0 V. Trap collision energy is 6.0 V for simple MS and a 20–50 V ramp for tandem MS (Thibodeaux et al., 2014). These instrument settings will most often provide no fragments inside thioether linkages.
5. Phosphoric acid (0.1%) is used as an external calibrant, and Glu-fibrinopeptide (Sigma) is used as an external calibrant for lock mass.
6. The linear region of the LC/ESI-MS must be established by making serial dilutions of the full-length peptides (in the case of ProcA2.8 from 0.31 to 20 μ*M*) in the same buffer conditions that the kinetic samples are run in.
7. The global kinetic data obtained are then numerically simulated with various possible kinetic models using KinTek Explorer, using FitSpace Explorer to evaluate whether each variable parameter is well constrained by the data for each model (Johnson, Simpson, & Blom, 2009a, 2009b).

4.4.5 Notes

1. Typically 50 μ*M* concentration is required for good fragmentation, but lower concentrations may work depending on the sample.
2. The ramp energy is dependent on peptide and often more sample is required to optimize initial attempts.
3. Using Glu-fibrinopeptide is often not necessary, but recommended if running long lists of samples to check if the initial calibration is drifting.
4. The relatively large masses (~8.5–9.5 kDa) of the full-length LanA peptides preclude the resolution of their individual isotopic peaks by a

Waters Synapt G1 Q/TOF mass spectrometer. To gain more confidence in the assignment of mass spectral signals to specific peptide ions, the 60 (for ProcM) and 8 min (for the faster enzyme HalM2) time points are proteolyzed to remove the leader peptide and thus generate smaller peptides for more accurate mass determination. These time points were chosen because they contain most if not all of the reaction intermediates.

5. To ensure linearity of mass spectrometric signals and peptide concentrations, serial dilutions of peptide samples are made in duplicate in LC solvent (50% acetonitrile in H_2O, 0.1% formic acid) and analyzed by LC-ESI-MS in randomized order. The linear signal ranges of reaction intermediates are similarly determined by serial dilutions of reaction samples. Extracted ion chromatograms (EICs) for each detectable multiply charged ion of each relevant peptide are generated, integrated, and summed together to give the total signal for that particular peptide, which is then plotted vs peptide concentration. A linear correlation was observed for the intermediates and products of ProcM and HalM2 (Thibodeaux et al., 2014).

4.4.6 Data Analysis and Information Gained From Kinetic Studies

For the quantification of LanA/LanM reaction intermediates, extracted ion chromatograms (EICs) are generated by applying a mass window of 1.5–2.5 Da around the center of the most intense isotope peak for each candidate ion. MassLynx is then used to integrate the peaks in the EIC. A corresponding EIC of a blank run is extracted from the EIC of the ion of interest to remove the background (Thibodeaux et al., 2014).

The full-length peptides with the leader peptide attached are necessary in order to observe modification by the lanthipeptide synthetase. These full-length peptides are larger than 8 kDa; therefore, the 50–2000 m/z range will show high charge states ranging from 11^+ to 7^+, providing multiple measurements for each intermediate. The abundance of each peptide species is calculated for each charge state and weighed by the contribution of that charge state to the total ion count (Thibodeaux et al., 2014).

Analysis of time points by this method allowed development of kinetic models for the reactions catalyzed by ProcM on the precursor peptide ProcA 2.8. In contrast to the general N- to C-terminal direction of dehydration and cyclization catalyzed by the HalM2 synthetase on the HalA2 precursor peptide, ProcM displayed the opposite trend on ProcA2.8. The first two dehydrations occur in a C- to the N-terminal direction, followed by the first cyclization near the C-terminus, and then formation of the second ring at the N-terminus of the core peptide. In contrast to HalM2, which displayed faster cyclization rates as HalA2 progresses to the final product, ProcM slows down as catalysis progresses and the final ring formation is the slowest conversion in the process (Thibodeaux et al., 2014).

4.5 Use of Nuclear Magnetic Resonance Spectroscopy for Ring Topology Assignment

Nuclear magnetic resonance (NMR) spectroscopy has the ability to circumvent the difficulty encountered in tandem MS when analyzing intertwined thioether rings, as long as sufficient pure sample can be obtained. Four prochlorosins had their ring topologies determined by NMR spectroscopy after coexpression of their ProcA precursor peptides with ProcM in *E. coli* and in vitro removal of the leader peptide (Tang & van der Donk, 2012).

4.5.1 Procedure for Determination of Ring Topology by NMR Spectroscopy

1. Precursor peptides ProcA1.1-G−1E, ProcA1.7-G−1R, ProcA3.3-G−1 K, and ProcA4.3 wt were purified after heterologous expression with ProcM as described in Sections 2.2.3 and 2.2.4.
2. The majority of their leader peptides were removed by either endoproteinase GluC or LysC digestion. The analogs of prochlorosins 1.7, 3.3, and 4.3 thus obtained contained short N-terminal overhangs from incomplete leader peptide removal.
3. The products are purified by RP-HPLC on a Phenomenex Luna C_{10} HPLC column (Tang & van der Donk, 2012).
4. The peptides are dissolved in concentrations between 2 and 4 m*M* in 90% H_2O and 10% D_2O with 0.01 mg/mL of 4,4-dimethyl-4-silapentane-1-sulfonic acid (DSS), and a series of one- and two-dimensional experiments are performed to fully assign the core peptides (TOCSY and gDQCOSY).
5. The water signal at 4.7 ppm is suppressed either using a Watergate sequence or a double pulsed field gradient spin echo (DPFGSE).
6. All amide protons for these prochlorosins are found from 7 to 10 ppm (using DSS as internal standard) and for the examples discussed here showed good dispersion, which is a prerequisite for full structural assignment.
7. Each vertical line in the TOCSY spectrum represents the protons of one amino acid spin system. Amide protons from dehydrated residues are

found more downfield than 9.0 ppm and have weak correlation with the β,γ protons of Dhb side chains.

8. Connectivity is established by sequential amide-to-amide correlations and sequential amide-to-α proton assignment from a NOESY spectrum with a 200-ms mixing time.
9. For lanthionine and methyllanthionine, correlations are observed between the α protons of the former Dha/Dhb residues and the β protons of the Cys with which they reacted as well as from the α protons of the Cys to the β protons of the former Dha/Dhb. (The β proton of a cyclized Dha or Dhb will have a chemical shift similar to that of the β proton of Cys.) Paired with strong correlations between the β protons of the former Dha/Dhb and the β protons of Cys residues, these correlations are evidence for a thioether linkage. In the case of methyllanthionine the γ protons of the former Dhb also display a correlation with the β protons of the Cys (typically found between 2.7 and 3.6 ppm).

4.5.2 Notes

1. Residues in lanthipeptide precursor peptides like ProcA are numbered with positive numbers starting with the first residue of the core peptide and with negative numbers counting backward from the leader peptide cleavage site.
2. Isotopically labeled samples are not necessary if amide protons are well dispersed and concentrated samples can be prepared (Kwan, Mobli, Gooley, King, & Mackay, 2011).
3. The N-terminal residues are not observed in the TOCSY because of the lack of an amide resonance (the amine is usually not observed).
4. NOESY spectra are acquired in D_2O to simplify analysis of protons obscured by the water peak even when using a water suppression pulse sequence. A mixing time of 400 ms was used.
5. With the MeLan γ protons occurring in a less crowded area of the spectrum (1.0–1.5 ppm), diagnosis of the MeLan topologies is simplified (Tang & van der Donk, 2012).

5. SUMMARY

Recent studies have started to provide an overview of the genomic landscape of combinatorial lanthipeptide biosynthesis in marine picocyanobacteria. These pathways are unique in that a single enzyme acts on a set of highly diverse substrates to make libraries of polycyclic peptides. Given the multitude of reactions that convert a linear precursor peptide into a macrocyclic product, investigation of these systems has required development of new tools to study their mechanism and characterize the products and intermediates. This chapter discusses these tools and the insights they have provided into ProcM catalysis. While the biological role of lanthipeptides identified in *Prochlorococcus* MIT 9313 has yet to be elucidated, the extreme substrate tolerance of ProcM has already been used for the creation of hybrid RiPPs incorporating both thioether rings as well as thiazolines by using chimeric leader peptides that engage the synthetases from two different RiPP classes (Burkhart, Kakkar, Hudson, van der Donk, & Mitchell, 2017). ProcM has also been used to create libraries of lanthipeptides derived from three prochlorosin scaffolds that were modified and selected using a phage display platform to identify urokinase plasminogen activators (Urban et al., 2017). Applications of promiscuous enzymes related to ProcM may be used in future studies to create and select inhibitors of disease-relevant protein–protein interactions.

ACKNOWLEDGMENTS

The authors thank the Howard Hughes Medical Institute and the National Institutes of Health (R37 GM 058822 to W.A.v.d.D) for support of our work on lanthipeptides and prochlorosins. The Bruker UltraFleXtreme MALDI-TOF/TOF MS was purchased with funding from the National Institutes of Health (S10 RR027109).

REFERENCES

Bell, M. R., Engleka, M. J., Malik, A., & Strickler, J. E. (2013). To fuse or not to fuse: What is your purpose? *Protein Science*, *22*(11), 1466–1477.

Bierbaum, G., Brötz, H., Koller, K. P., & Sahl, H. G. (1995). Cloning, sequencing and production of the lantibiotic mersacidin. *FEMS Microbiology Letters*, *127*(1–2), 121–126.

Brötz, H., Bierbaum, G., Leopold, K., Reynolds, P. E., & Sahl, H. G. (1998). The lantibiotic mersacidin inhibits peptidoglycan synthesis by targeting lipid II. *Antimicrobial Agents and Chemotherapy*, *42*(1), 154–160.

Brötz, H., Bierbaum, G., Markus, A., Molitor, E., & Sahl, H.-G. (1995). Mode of action of the lantibiotic mersacidin: Inhibition of peptidoglycan piosynthesis via a novel mechanism? *Antimicrobial Agents and Chemotherapy*, *39*, 714–719.

Burkhart, B. J., Kakkar, N., Hudson, G. A., van der Donk, W. A., & Mitchell, D. A. (2017). Chimeric leader peptides for the generation of non-natural hybrid RiPP products. *ACS Central Science*, *3*(6), 629–638.

Chatterjee, C., Paul, M., Xie, L., & van der Donk, W. A. (2005). Biosynthesis and mode of action of lantibiotics. *Chemical Reviews*, *105*, 633–684.

Cooper, L. E., McClerren, A. L., Chary, A., & van der Donk, W. A. (2008). Structure-activity relationship studies of the two-component lantibiotic haloduracin. *Chemistry & Biology*, *15*, 1035–1045.

Cubillos-Ruiz, A., Berta-Thompson, J. W., Becker, J. W., van der Donk, W. A., & Chisholm, S. W. (2017). Evolutionary radiation of lanthipeptides in marine cyanobacteria. *Proceedings of the National Academy of Sciences of the United States of America, 114*(27), E5424–E5433.

di Guan, C., Li, P., Riggs, P. D., & Inouye, H. (1988). Vectors that facilitate the expression and purification of foreign peptides in *Escherichia coli* by fusion to maltose-binding protein. *Gene, 67*(1), 21–30.

Dong, S. H., Tang, W., Lukk, T., Yu, Y., Nair, S. K., & van der Donk, W. A. (2015). The enterococcal cytolysin synthetase has an unanticipated lipid kinase fold. *eLife, 4*, e07607.

Donia, M. S., Hathaway, B. J., Sudek, S., Haygood, M. G., Rosovitz, M. J., Ravel, J., et al. (2006). Natural combinatorial peptide libraries in cyanobacterial symbionts of marine ascidians. *Nature Chemical Biology, 2*(12), 729–735.

Ferrer, M., Chernikova, T. N., Yakimov, M. M., Golyshin, P. N., & Timmis, K. N. (2003). Chaperonins govern growth of *Escherichia coli* at low temperatures. *Nature Biotechnology, 21*(11), 1266–1267.

Fox, J. D., & Waugh, D. S. (2003). Maltose-binding protein as a solubility enhancer. *Methods in Molecular Biology, 205*, 99–117.

Garneau, S., Martin, N. I., & Vederas, J. C. (2002). Two-peptide bacteriocins produced by lactic acid bacteria. *Biochimie, 84*(5–6), 577–592.

Gebhard, S. (2012). ABC transporters of antimicrobial peptides in firmicutes bacteria—Phylogeny, function and regulation. *Molecular Microbiology, 86*(6), 1295–1317.

Harris, C. M., Derdowski, A. M., & Poulter, C. D. (2002). Modulation of the zinc(II) center in protein farnesyltransferase by mutagenesis of the zinc(II) ligands. *Biochemistry, 41*(33), 10554–10562.

Hasper, H. E., Kramer, N. E., Smith, J. L., Hillman, J. D., Zachariah, C., Kuipers, O. P., et al. (2006). An alternative bactericidal mechanism of action for lantibiotic peptides that target lipid II. *Science, 313*(5793), 1636–1637.

Håvarstein, L. S., Diep, D. B., & Nes, I. F. (1995). A family of bacteriocin ABC transporters carry out proteolytic processing of their substrates concomitant with export. *Molecular Microbiology, 16*(2), 229–240.

Hsu, S. T., Breukink, E., Tischenko, E., Lutters, M. A., De Kruijff, B., Kaptein, R., et al. (2004). The nisin-lipid II complex reveals a pyrophosphate cage that provides a blueprint for novel antibiotics. *Nature Structural & Molecular Biology, 11*(10), 963–967.

Islam, M. R., Nishie, M., Nagao, J., Zendo, T., Keller, S., Nakayama, J., et al. (2012). Ring A of nukacin ISK-1: A lipid II-binding motif for type-A(II) lantibiotic. *Journal of the American Chemical Society, 134*(8), 3687–3690.

Johnson, K. A., Simpson, Z. B., & Blom, T. (2009a). FitSpace Explorer: An algorithm to evaluate multi-dimensional parameter space in fitting kinetic data. *Analytical Biochemistry, 387*, 30–41.

Johnson, K. A., Simpson, Z. B., & Blom, T. (2009b). Global kinetic explorer: A new computer program for dynamic simulation and fitting of kinetic data. *Analytical Biochemistry, 387*, 20–29.

Knerr, P. J., Oman, T. J., Garcia de Gonzalo, C., Lupoli, T. J., Walker, S., & van der Donk, W. A. (2012). Non-proteinogenic amino acids in lacticin 481 analogues result in more potent inhibition of peptidoglycan transglycosylation. *ACS Chemical Biology*, 7(11), 1791–1795.

Krawczyk, B., Ensle, P., Müller, W. M., & Süssmuth, R. D. (2012). Deuterium labeled peptides give insights into the directionality of class III lantibiotic synthetase LabKC. *Journal of the American Chemical Society, 134*, 9922.

Kwan, A. H., Mobli, M., Gooley, P. R., King, G. F., & Mackay, J. P. (2011). Macromolecular NMR spectroscopy for the non-spectroscopist. *The FEBS Journal, 278*(5), 687–703.

Li, B., Sher, D., Kelly, L., Shi, Y., Huang, K., Knerr, P. J., et al. (2010). Catalytic promiscuity in the biosynthesis of cyclic peptide secondary metabolites in planktonic marine cyanobacteria. *Proceedings of the National Academy of Sciences of the United States of America, 107*(23), 10430–10435.

Lin, Z., Torres, J. P., Tianero, M. D., Kwan, J. C., & Schmidt, E. W. (2016). Origin of chemical diversity in prochloron-tunicate symbiosis. *Applied and Environmental Microbiology, 82*(12), 3450–3460.

Ma, H., Gao, Y., Zhao, F., & Zhong, J. (2015). Individual catalytic activity of two functional domains of bovicin HJ50 synthase BovM. *Wei Sheng Wu Xue Bao, 55*(1), 50–58.

McAuliffe, O., Hill, C., & Ross, R. P. (2000). Each peptide of the two-component lantibiotic lacticin 3147 requires a separate modification enzyme for activity. *Microbiology, 146*(Pt. 9), 2147–2154.

McAuliffe, O., Ross, R. P., & Hill, C. (2001). Lantibiotics: Structure, biosynthesis and mode of action. *FEMS Microbiology Reviews, 25*(3), 285–308.

McClerren, A. L., Cooper, L. E., Quan, C., Thomas, P. M., Kelleher, N. L., & van der Donk, W. A. (2006). Discovery and in vitro biosynthesis of haloduracin, a two-component lantibiotic. *Proceedings of the National Academy of Sciences of the United States of America, 103*(46), 17243–17248.

Morgan, S. M., O'Connor, P. M., Cotter, P. D., Ross, R. P., & Hill, C. (2005). Sequential actions of the two component peptides of the lantibiotic lacticin 3147 explain its antimicrobial activity at nanomolar concentrations. *Antimicrobial Agents and Chemotherapy, 49*(7), 2606–2611.

Muir, T. W. (2003). Semisynthesis of proteins by expressed protein ligation. *Annual Review of Biochemistry, 72*(1), 249–289.

Mukherjee, S., & van der Donk, W. A. (2014). Mechanistic studies on the substrate tolerant lanthipeptide synthetase ProcM. *Journal of the American Chemical Society, 136*, 10450–10459.

Oman, T. J., & van der Donk, W. A. (2010). Follow the leader: The use of leader peptides to guide natural product biosynthesis. *Nature Chemical Biology, 6*, 9–18.

Panavas, T., Sanders, C., & Butt, T. R. (2009). SUMO fusion technology for enhanced protein production in prokaryotic and eukaryotic expression systems. *Methods in Molecular Biology, 497*, 303–317.

Penner-Hahn, J. (2007). Zinc-promoted alkyl transfer: A new role for zinc. *Current Opinion in Chemical Biology, 11*, 166–171.

Plat, A., Kuipers, A., Rink, R., & Moll, G. N. (2013). Mechanistic aspects of lanthipeptide leaders. *Current Protein & Peptide Science, 14*, 85–96.

Schmidt, E. W., Donia, M. S., McIntosh, J. A., Fricke, W. F., & Ravel, J. (2012). Origin and variation of tunicate secondary metabolites. *Journal of Natural Products, 75*(2), 295–304.

Shi, Y., Yang, X., Garg, N., & van der Donk, W. A. (2011). Production of lantipeptides in *Escherichia coli*. *Journal of the American Chemical Society, 133*(8), 2338–2341.

Shimafuji, C., Noguchi, M., Nishie, M., Nagao, J. I., Shioya, K., Zendo, T., et al. (2015). In vitro catalytic activity of N-terminal and C-terminal domains in NukM, the post-translational modification enzyme of nukacin ISK-1. *Journal of Bioscience and Bioengineering, 120*(6), 624–629.

Siezen, R. J., Kuipers, O. P., & de Vos, W. M. (1996). Comparison of lantibiotic gene clusters and encoded proteins. *Antonie Van Leeuwenhoek, 69*(2), 171–184.

Sun, P., Tropea, J. E., & Waugh, D. S. (2011). Enhancing the solubility of recombinant proteins in *Escherichia coli* by using hexahistidine-tagged maltose-binding protein as a fusion partner. *Methods in Molecular Biology, 705*, 259–274.

Tang, W., & van der Donk, W. A. (2012). Structural characterization of four prochlorosins: A novel class of lantipeptides produced by planktonic marine cyanobacteria. *Biochemistry, 51*(21), 4271–4279.

Thibodeaux, C. J., Ha, T., & van der Donk, W. A. (2014). A price to pay for relaxed substrate specificity: A comparative kinetic analysis of the class II lanthipeptide synthetases ProcM and HalM2. *Journal of the American Chemical Society*, *136*(50), 17513–17529.

Thibodeaux, C. J., Wagoner, J., Yu, Y., & van der Donk, W. A. (2016). The leader peptide establishes dehydration order, promotes efficiency, and ensures fidelity during lacticin 481 biosynthesis. *Journal of the American Chemical Society*, *138*, 6436–6444.

Urban, J. H., Moosmeier, M. A., Aumuller, T., Thein, M., Bosma, T., Rink, R., et al. (2017). Phage display and selection of lanthipeptides on the carboxy-terminus of the gene-3 minor coat protein. *Nature Communications*, *8*(1), 1500.

Wiedemann, I., Bottiger, T., Bonelli, R. R., Wiese, A., Hagge, S. O., Gutsmann, T., et al. (2006). The mode of action of the lantibiotic lacticin 3147—A complex mechanism involving specific interaction of two peptides and the cell wall precursor lipid II. *Molecular Microbiology*, *61*(2), 285–296.

Wiedemann, I., Breukink, E., van Kraaij, C., Kuipers, O. P., Bierbaum, G., de Kruijff, B., et al. (2001). Specific binding of nisin to the peptidoglycan precursor lipid II combines pore formation and inhibition of cell wall biosynthesis for potent antibiotic activity. *The Journal of Biological Chemistry*, *276*(3), 1772–1779.

Xie, L., Miller, L. M., Chatterjee, C., Averin, O., Kelleher, N. L., & van der Donk, W. A. (2004). Lacticin 481: In vitro reconstitution of lantibiotic synthetase activity. *Science*, *303*(5658), 679–681.

Yang, X., & van der Donk, W. A. (2013). Ribosomally synthesized and post-translationally modified peptide natural products: New insights into the role of leader and core peptides during biosynthesis. *Chemistry—A European Journal*, *19*, 7662–7677.

Yang, X., & van der Donk, W. A. (2015). The Michael-type cyclizations in lantibiotic biosynthesis are reversible. *ACS Chemical Biology*, *10*(5), 1234–1238.

Yu, Y., Mukherjee, S., & van der Donk, W. A. (2015). Product formation by the promiscuous lanthipeptide synthetase ProcM is under kinetic control. *Journal of the American Chemical Society*, *137*(15), 5140–5148.

Zhang, Q., Ortega, M., Shi, Y., Wang, H., Melby, J. O., Tang, W., et al. (2014). Structural investigation of ribosomally synthesized natural products by hypothetical structure enumeration and evaluation using tandem MS. *Proceedings of the National Academy of Sciences of the United States of America*, *111*(33), 12031–12036.

Zhang, Q., Yang, X., Wang, H., & van der Donk, W. A. (2014). High divergence of the precursor peptides in combinatorial lanthipeptide biosynthesis. *ACS Chemical Biology*, *9*(11), 2686–2694.

Zhang, Q., Yu, Y., Velásquez, J. E., & van der Donk, W. A. (2012). Evolution of lanthipeptide synthetases. *Proceedings of the National Academy of Sciences of the United States of America*, *109*(45), 18361–18366.

SECTION II

Biosynthesis by Bacterial Symbionts of Marine Invertebrate Animals

CHAPTER SIX

Chemoenzymatic Dissection of Polyketide β-Branching in the Bryostatin Pathway

Samuel T. Slocum[*,†], Andrew N. Lowell[*], Ashootosh N. Tripathi[*], Vikram V. Shende[*], Janet L. Smith[*,†], David H. Sherman[1,*,‡,§,¶]

[*]Life Sciences Institute, University of Michigan, Ann Arbor, MI, United States
[†]Life Sciences Institute, Department of Biological Chemistry, University of Michigan, Ann Arbor, MI, United States
[‡]Life Sciences Institute, Department of Medicinal Chemistry, University of Michigan, Ann Arbor, MI, United States
[§]Life Sciences Institute, Department of Chemistry, University of Michigan, Ann Arbor, MI, United States
[¶]Life Sciences Institute, Department of Microbiology and Immunology, University of Michigan, Ann Arbor, MI, United States
[1]Corresponding author: e-mail address: davidhs@umich.edu

Contents

Abstract

β-Branching is an expansion upon canonical polyketide synthase extension that allows for the installation of diverse chemical moieties in several natural products. Several of these moieties are unique among natural products, including the two vinyl methylesters found in the core structure of bryostatins. This family of molecules is derived from an obligate bacterial symbiont of a sessile marine bryozoan, *Bugula neritina*. Within this family, bryostatin 1 has been investigated as an anticancer, neuroprotective, and immunomodulatory compound. We have turned to the biosynthetic gene cluster within the bacterial symbiont to investigate the biosynthesis of bryostatins. Recent sequencing

Methods in Enzymology, Volume 604
ISSN 0076-6879
https://doi.org/10.1016/bs.mie.2018.01.034

efforts resulted in the annotation of two missing genes: *bryT* and *bryU*. Using novel chemoenzymatic techniques, we have validated these as the missing enoyl-CoA hydratase and donor acyl carrier protein, essential components of the β-branching cassette of the bryostatin pathway. Together, this cassette installs the vinyl methylester moieties essential to the activity of bryostatins.

1. INTRODUCTION

Bryostatins are a family of molecules that mediate chemical defense during the larval stage in the marine invertebrate *Bugula neritina* life cycle (Lopanik, Lindquist, & Targett, 2004). Bryostatin-1 is a high-affinity modulator of protein kinase C and has clinical efficacy for many pathologies (Blackhall et al., 2001; Gutiérrez et al., 2016; Lucke-Wold et al., 2014; Mehla et al., 2010; Pagliaro et al., 2000; Schrott et al., 2015; Sun, Hongpaisan, & Alkon, 2009). Through a combination of bacterial enrichment and in situ hybridization experiments (Davidson, Allen, Lim, Anderson, & Haygood, 2001; Davidson & Haygood, 1999; Sharp, Davidson, & Haygood, 2007; Woollacott, 1981), and cloning and sequencing of the biosynthetic gene cluster (Sudek et al., 2007), bryostatin biosynthesis has been attributed to a γ-proteobacterial obligate symbiont, *Endobugula sertula* (Davidson & Haygood, 1999). This bacterium has remained refractory to laboratory culture, adding difficulty to sourcing through direct fermentation (Schaufelberger et al., 1991), and has prevented developing a genetic system for investigating the *bry* system. Attempts to resolve these issues have proven to be either ineffective or prohibitively costly (Mendola, 2003). Syntheses of several bryostatins have been reported (Keck, Poudel, Cummins, Rudra, & Covel, 2010; Lu, Woo, & Krische, 2011; Manaviazar & Hale, 2011; Wender & Schrier, 2011), and a recent "scalable" total synthesis of bryostatin-1 has recently been accomplished (Wender et al., 2017). Bryostatin analogs (termed "bryologs") show promise (DeChristopher, Fan, Felsher, & Wender, 2012; Staveness et al., 2016; Wender et al., 2011), however, a simplified and effective replacement for bryostatin-1 has yet to be developed.

To facilitate the study of bryostatin biosynthesis, we have turned to the natural biosynthetic pathway. Previously, we prepared a biosynthetic hypothesis from a cosmid library that enabled sequencing and bioinformatics analysis of the *bry* gene cluster (Sudek et al., 2007). Like many clinically important natural products, bryostatins are produced by an enzymatic assembly line comprised of polyketide synthases (PKS), multidomain complexes used by many branches of the tree of life to synthesize structurally diverse natural products. PKS have yielded dividends as a rich source of drug leads (Newman & Cragg, 2012) and more recently as effective biocatalysts for chemoenzymatic assembly of macrolide antibiotics (Hansen et al., 2013). The activities of PKS and associated tailoring enzymes hold solutions to difficult synthetic problems, and it is critical to understand the ability of PKSs to generate complex metabolites through regio- and stereoselective catalytic processes.

The basic catalytic cycle of type-I PKS can be related to that of mammalian fatty acid synthase (mFAS) (Maier, Leibundgut, & Ban, 2008). While mFAS is a single multidomain module that acts iteratively, type-I PKS pathways are instead made up by a series of synthetic modules that act as a linear assembly line, with each module containing domain deletions, substitutions, and additions that produce diverse chemical functionality. Another strategy for chemical diversification that a subset of PKS pathways has borrowed from primary metabolism involves the use of a β-branching cassette. Reminiscent of isoprene biosynthesis, β-branching employs a 3-hydroxy-3-methylglutaryl-CoA (HMG-CoA) synthase homologue (HMGS) that catalyzes a two-carbon addition to the β carbonyl of the nascent polyketide chain (Campobasso et al., 2004). This HMG-like intermediate is in most cases modified by sequential dehydration and decarboxylation by enoyl-CoA hydratases (termed ECH_1 and ECH_2, respectively). A variety of activities including oxidation (Simunovic et al., 2006), epoxidation (Zhang et al., 2011), halogenation (Chang et al., 2004), and methylation (Simunovic & Müller, 2007) may also be included at different points in this process, introducing functionality that canonical PKS extensions cannot introduce.

The *bry* cluster is made up of five large *trans*-acyltransferase PKS genes with an adjacent group of smaller genes. *trans*-Acyltransferase PKS lack integrated acyltransferase (AT) domains (Helfrich & Piel, 2016), and extender unit selectivity is provided by the activity of a separate gene product, in this case BryP. The biosynthetic hypothesis is described in Fig. 1, highlighting the role of the β-branching cassette. The present work describes the identification and confirmation of the roles of BryT and BryU in β-branching. The corresponding *bryT* and *bryU* genes were predicted to exist, but remained unidentified until recently when Kwan and coworkers completed a draft genome of *E. sertula* (Miller, Vanee, Fong, Lim-Fong, & Kwan, 2016).

β-Branching is relatively rare among PKS pathways, and the vinyl methylester product of the bryostatin β-branching cassette is especially unusual in this group (Fig. 2). Only a handful of natural products have similar

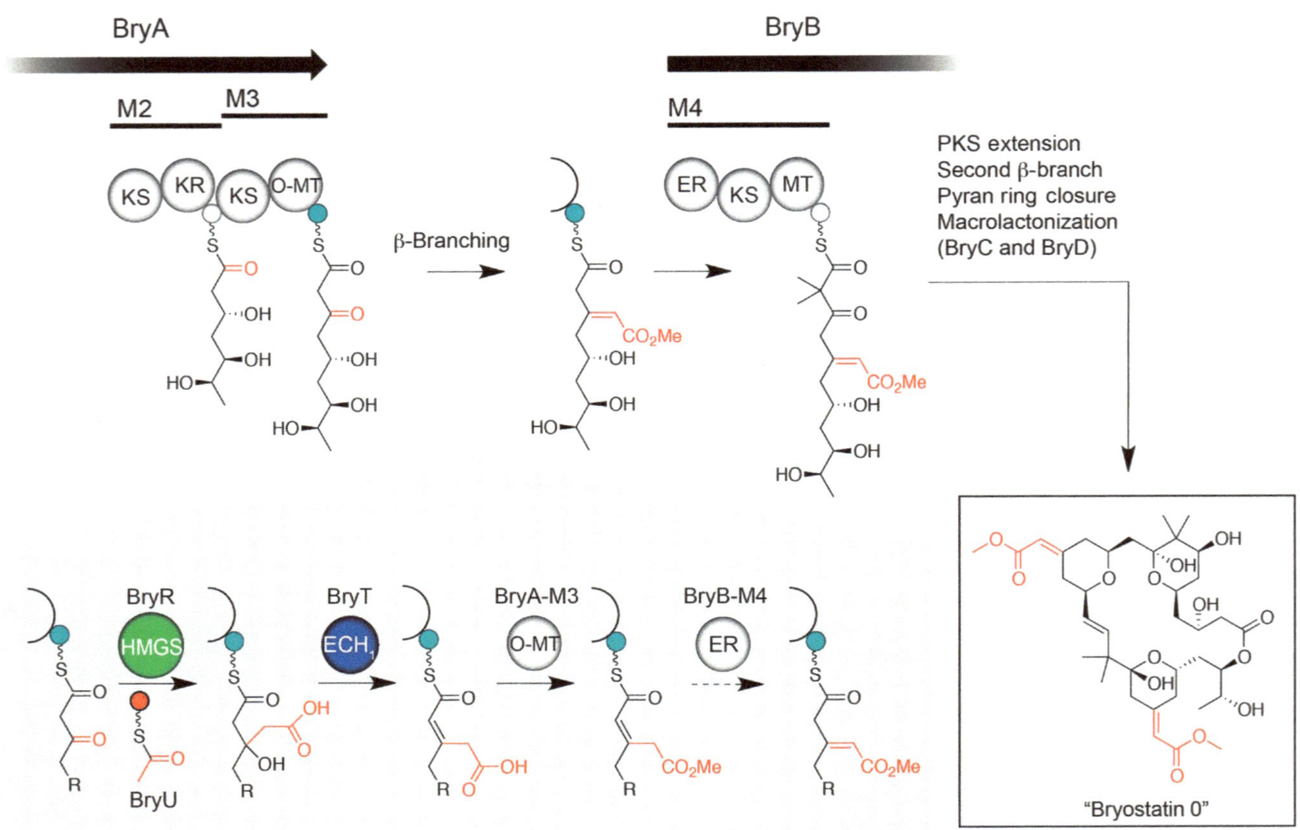

Fig. 1 β-Branching in bryostatin biosynthesis. The β-keto product of BryA module-3 is extended by BryR, an HMGS that uses an acetyl unit carried by BryU, a standalone ACP. BryT dehydrates the HMG-like product, which is then methylated by the *O*-MT domain within BryA module-3. The resulting double bond is isomerized prior to pyran ring closure, most likely by the ER-like domain within BryB module-4.

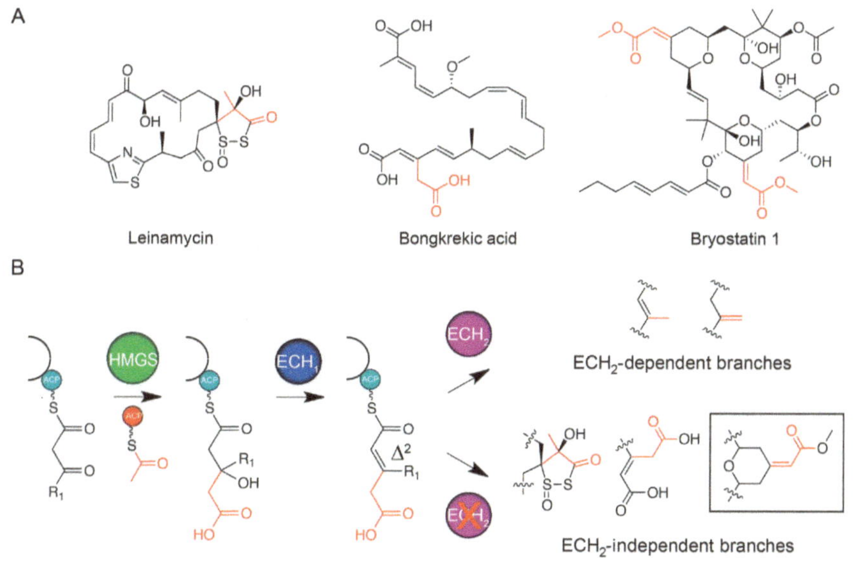

Fig. 2 Unique β-branching cassettes. Only three polyketide natural products contain β-branches that lack an ECH_2 mediated step: leinamycin, bongkrekic acid, and bryostatin 1 (A). These branches are installed using the same HMGS and ECH_1 steps (B), but downstream modifications diverge from other β-branches.

moieties (Higa, Tanaka, Komesu, & Jefford, 1992; Kedei et al., 2015; Ohta et al., 2006), but they currently lack genetically characterized biosynthetic pathways. Additionally, the completed vinyl methylester moieties are situated at two key sites of interaction with its biological target, protein kinase C (Kedei et al., 2015; Kelsey et al., 2016). Understanding the biosynthesis of these molecules and perhaps engineering this ability to other pathways may allow the portability of bryostatin-like biological activity to nonnatural products.

In vitro studies confirmed the role of BryR as the HMGS (Buchholz et al., 2010), however, several key questions remained regarding the conversion of β-keto intermediate to vinyl methylester. The β-branching reaction catalyzed by BryR requires a phylogenetically distinct standalone ACP donor (ACP_D) carrying an acetyl group as a source of branching carbons, but no ACP_D was identified within the cosmid library and a surrogate ACP facilitated previous studies (Buchholz et al., 2010). The source of the final carboxyl O-methylation was unclear, but several candidate enzymes or catalytic domains were identified in the Bry pathway. Additionally, the double bond configuration and lack of decarboxylation of the final product suggest either an isomerization or an unique ECH_1; however, no ECH_1 was identified in the original sequenced and annotated *bry* cluster. A recent draft genome of the *E. sertula* bacterial symbiont revealed candidate genes and corresponding gene products for the missing β-branching components: BryT (ECH_1) and BryU (ACP_D) (Miller, Vanee, et al., 2016). Interestingly, these open reading frames (ORFs) were 15 bp away from one another. However, due to gaps in the genomic assembly and a lack of overlap with any of the earlier draft quality cosmid sequencing data, the distance from the bryostatin PKS genes may only be estimated to be >48 kb.

Both ECH_1 and ECH_2, like most members of the crotonase superfamily, proceed through a vinylogous enolate intermediate of the thioester carbonyl, allowing the selection of Δ^2 or Δ^3 products. ECH_1/ECH_2 pairs can, therefore, generate either a vinyl methyl (Δ^2 isomer) or exomethylene (Δ^3 isomer) moiety. The mechanism of ECH_2 selection can be rationalized structurally by comparing the curacin ECH_2 (Geders et al., 2007) to the primary sequence of the ECH_2 in the closely related jamaicamide pathway (Gu et al., 2009). In principle, ECH_1 may also be able to generate Δ^2 and Δ^3 products, however, this is obfuscated by decarboxylation in most pathways and remains untested biochemically. The discovery of the bryostatin ECH_1, therefore, provides an opportunity to determine if the Δ^3 structure in bryostatins is installed by ECH_1 dehydration or a different biochemical event.

In this chapter, we set out to investigate the unique bryostatin β-branching cassette in order to understand how these unusual vinyl methylester moieties are installed. Specifically, we probed newly identified genes as potential members of a second bryostatin gene locus in *E. sertula.* Using established techniques as well as specific biosynthetic probes, we scrutinized enzymatic products to determine the regiochemistry of the biochemical reaction. The methods described here can be generalized to other biosynthetic pathways to chemoenzymatically determine regiochemistries of complex secondary metabolites.

2. PREVIOUS STUDIES AND IDENTIFICATION OF CRYPTIC COMPONENTS OF BRYOSTATIN BIOSYNTHESIS

The previously described cosmid sequencing assemblies allowed for the preparation of a functional hypothesis of bryostatin biosynthesis (Sudek et al., 2007). The first in vitro biochemical studies focused on confirming the functions of BryP and BryR (Buchholz et al., 2010; Lopanik et al., 2008). Essential to the activity of BryR as HMGS is a standalone donor ACP

(ACP_D) and an acceptor ACP (ACP_A), which is part of a PKS module. A malonyl unit is loaded onto the phosphopantetheine arm of ACP_D, which is then decarboxylated by a specific KS. The resulting acetyl unit is then transferred to a cysteine sidechain of HMGS, where it resides until hydrolysis or interaction with ACP_A. Upon binding of ACP_A, HMGS catalyzes an aldol addition of the α carbon of the acetyl unit to the β ketone carbon, resulting in an HMG-like product. Unlike HMG-CoA synthase that acts on CoA-linked substrates, the activity of HMGS has been found to be very specific to ACP_A and ACP_D, with little tolerance for substitution (Buchholz et al., 2010; Maloney, Gerwick, Gerwick, Sherman, & Smith, 2016). This ACP selectivity presents a challenge when the correct ACP is unknown, but may be overcome by use of surrogate carrier proteins that are closely related phylogenetically (Buchholz et al., 2010).

The functionality of BryR was demonstrated using a set of simplified and commercially available acyl-CoAs that had been used in the study of HMGS from the curacin and jamaicamide pathways: namely, acetyl-, acetoacetyl-, and 3-hydroxy-3-methylglutaryl-CoA (HMG-CoA). The activity and role of BryR in the pathway was confirmed, but several questions about the bryostatin β-branching cassette persisted, as no donor ACP or ECH_1 component were identified in the cosmid library (Buchholz et al., 2010). Somehow, components of the pathway are able to convert the HMG-like product of BryR into a Δ^3 vinyl methylester found in the final product, and several hypotheses were considered. However, in order to more fully understand bryostatin biosynthesis, new insights into the *Ca. E. sertula* genome were required.

Thanks in large part to the development of next-generation sequencing and accompanying analysis software, great advances have been made in metagenomic sequencing (Miller, Weyna, Fong, Lim-Fong, & Kwan, 2016; Schofield & Sherman, 2013). These technologies enable samples containing DNA from the genomes of multiple species to be analyzed directly without the need of cosmid assembly and time-consuming, and significantly costlier Sanger sequencing. The ability to specifically isolate and analyze symbiont sequencing data in silico has led to several interesting results, including the complete genome (~0.63 kb) of the tunicate endosymbiont that produces ET-743, *Candidatus Endoecteinascidia frumentensis* (Schofield, Jain, Porat, Dick, & Sherman, 2015). These techniques are helping to characterize the genetic diversity hidden in symbiont genomes.

Using metagenomic DNA isolated from *B. neritina* larvae and targeted microdissections of adult ovicells, Miller and coworkers were able to achieve sufficient abundance of *Ca. E. sertula* DNA to conduct next-generation sequencing (Miller, Vanee, et al., 2016). Subsequent analysis and careful pooling of reads from the symbiont genome resulted in a draft genome of *Ca. E. sertula* (Miller, Weyna, et al., 2016). Made up of 112 contigs with a total of 3.3 Mbp, this genome is much larger than would be expected for an obligate symbiont (Moran, 2003), and only bears some of the hallmarks of genome reduction, indicating a relatively recent symbiosis. The GC content and pseudogene abundance also suggest that *Ca. E. sertula* may have been able to live independently in the recent past. This is a key question for the isolation of large quantities of bryostatins, as free-living *Ca. E. sertula* may be fermented to facilitate large-scale production of bryostatins. While this is a desirable target, *Ca. E. sertula* is likely dependent upon its host for nutrition, as several deficiencies in amino acid biosynthesis genes were identified. This genome is a very interesting resource for the study of an obligate symbiont that provides an advantage to its host by providing a chemical defense agent for protection (Lopanik et al., 2004).

The Kwan laboratory sequencing effort also delivered two interesting genes: an FAS-like acyl carrier protein adjacent to a PKS enoyl-CoA hydratase. Because of their annotation being unique among ORFs identified in the assembly, these two genes (termed *bryT* and *bryU*) were strong candidates for the missing components from the bryostatin gene cluster. Additionally, the accompanying RNA sequencing provided strong evidence that these genes are actively transcribed while *Ca. E. sertula* is proliferating, as is the entire bryostatin gene cluster.

BryU was identified by automated annotation within the draft genome of *Ca. E. sertula*, and no other hits were found that resemble BryU or any other ACP_D within the draft genome. BryU strongly resembles other FAS-like β-branch donor ACPs (Fig. 3), including mAcpC from the mupirocin pathway (El-Sayed et al., 2003) (40.5% identity) and CurB from the curacin pathway (43.0% identity), both of which were used in previous studies of BryR function. Comparison of BryU and BryR to a cocrystal structure of CurB interacting with the curacin HMGS (CurD, 67.5% identity to BryR) shows that the key residues for HMGS–ACP_D interaction are conserved in both pathways (Maloney et al., 2016). This is perhaps to be expected given the relatively high sequence identity and binding affinity of CurB for BryR as a surrogate ACP_D (Buchholz et al., 2010).

BryT was found immediately adjacent to *bryU* in the draft genome of *Ca. E. sertula*. Automatically annotated as a polyketide synthesis ECH, phylogenetic analysis and sequence comparisons provided strong evidence that this was the missing bryostatin ECH_1 (Fig. 4). Because several ECH_1 and

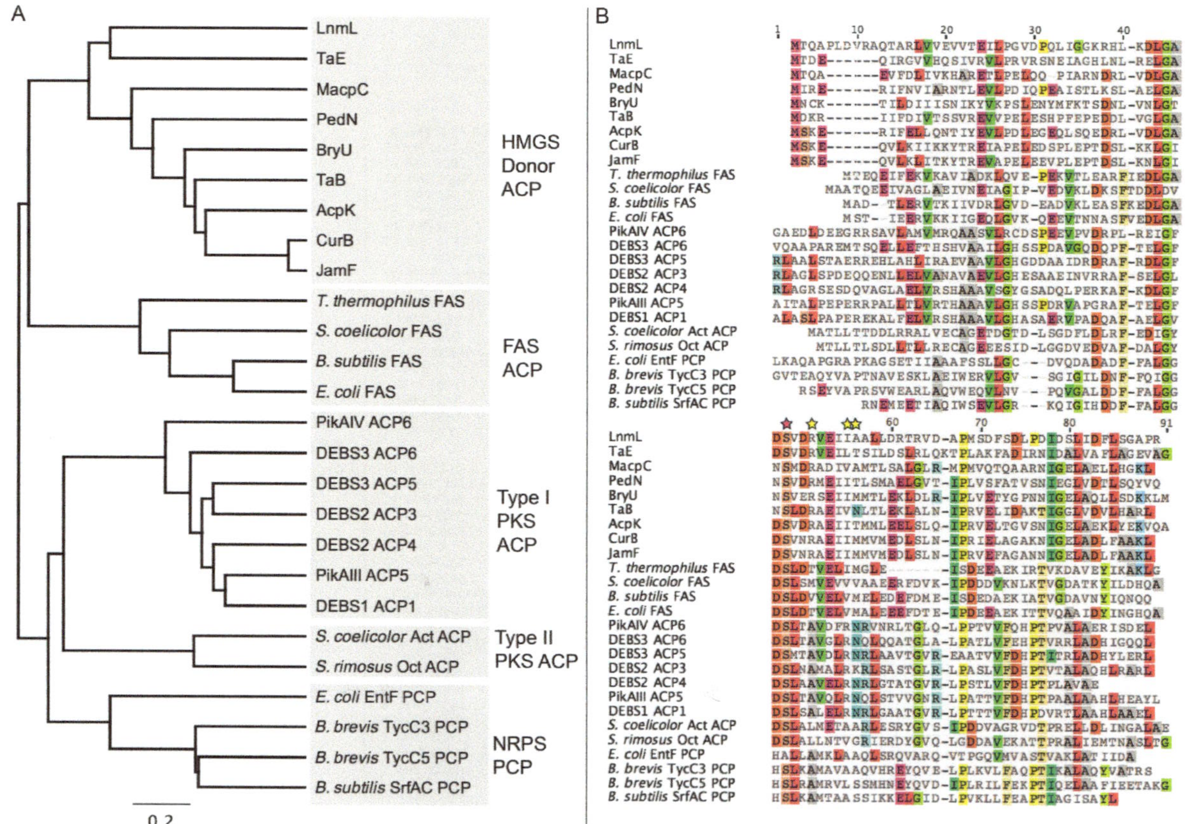

Fig. 3 Phylogenetic analysis of BryU.

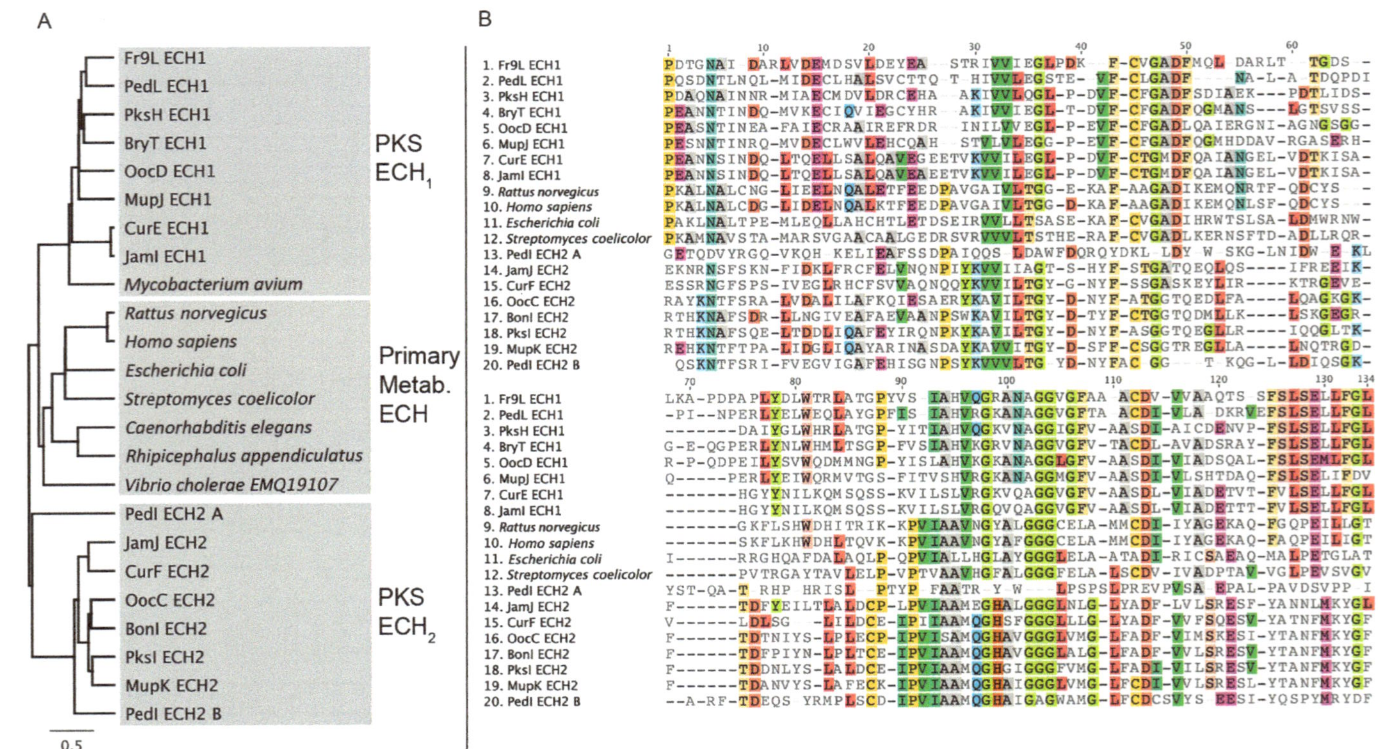

Fig. 4 Phylogenetic analysis of BryT.

ECH_2 proteins from PKS clusters have been biochemically characterized (Calderone, Iwig, Dorrestein, Kelleher, & Walsh, 2007; Gu et al., 2006), primary sequence alone is typically sufficient to classify a protein as ECH_1, ECH_2, or a primary metabolic homologue. In sequence alignment with ECH_1, BryT displays a 38.3% identity with CurE (Chang et al., 2004) and 37.9% identity with JamI (Edwards et al., 2004). Compared to ECH_2, BryT has 15.7% identity with CurF (ECH_2) and 14.5% identity with JamJ (ECH_2), clearly indicating identity with ECH_1 (Gu et al., 2006).

3. INITIAL BIOCHEMICAL ANALYSIS OF NOVEL BRY GENE PRODUCTS

3.1 BryU, β-Branch Donor ACP

BryR, BryU, and BryA module-3 ACP were expressed and purified using methods described previously (Buchholz et al., 2010), and activity was assessed by LC/MS using the phosphopantetheine ejection assay (Meluzzi, Zheng, Hensler, Nizet, & Dorrestein, 2008). Purity of the protein was assessed by SDS-PAGE (Fig. 5). When acetoacetyl-ACP_A (BryA module-3 ACP) was added to BryR and acetyl-BryU, conversion to HMG–ACP_A was observed (Fig. 6C), thus confirming the role of BryU as the donor ACP for the bryostatin β-branching cassette.

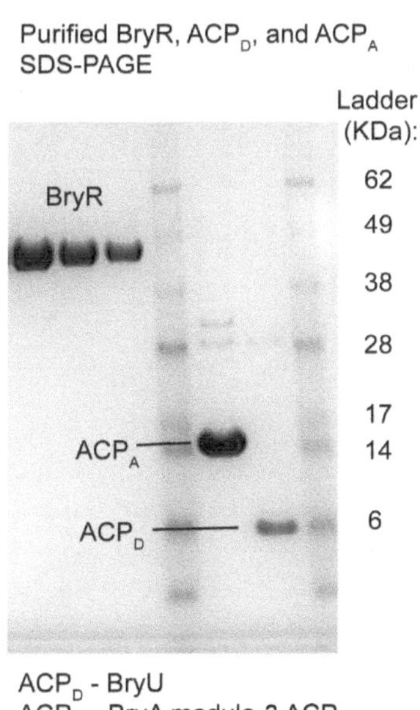

Fig. 5 SDS-PAGE of BryR, BryU, and BryA module-3 ACP.

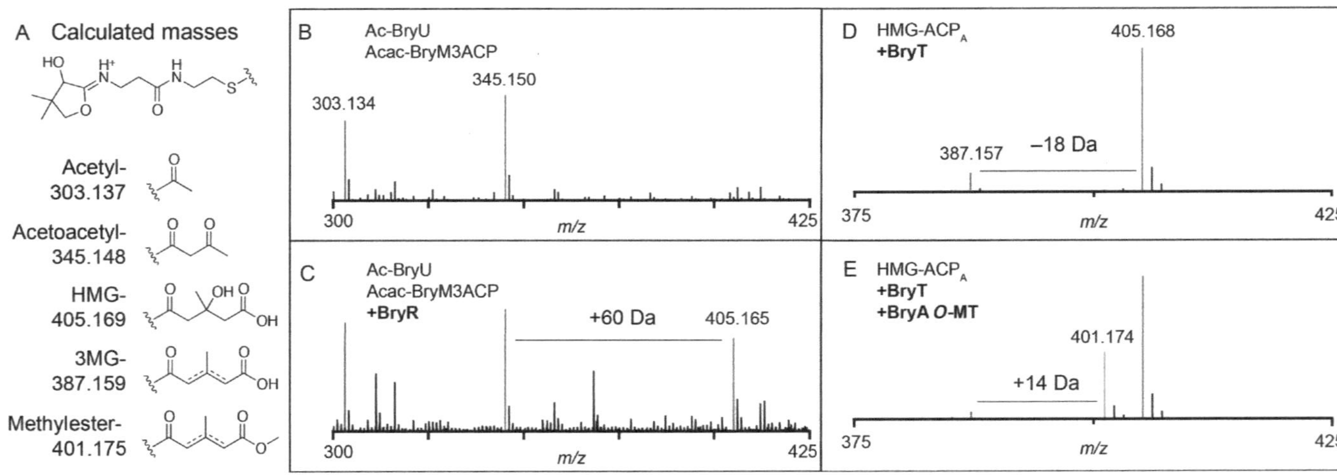

Fig. 6 LC/MS confirmation of the roles of BryU, BryT, and BryA *O*-MT. (A) Calculated phosphopantetheine ejection ion masses for the predicted species. (B) Control reaction with acetyl-BryU, acetoacetyl-BryA module-3 ACP. (C) Incubation with BryR generates HMG-BryA module-3 ACP. (D) Addition of BryT to HMG-BryA module-3 ACP generates 3-methylglutaconyl product. (E) Addition of BryA *O*-MT and SAM to BryT reaction generates 3MG methylester product.

Every β-branching cassette found in type I PKS pathways has been identified with an associated donor ACP that brings an acetyl (or propionyl Simunovic et al., 2006) unit to HMGS for incorporation into the polyketide, and bryostatin is no exception. HMGS has been shown in several pathways, including bryostatin, to have a high degree of selectivity for both donor and acceptor ACPs. The basis of this selectivity has been described structurally in the curacin pathway, and the residues that are key to this binding mode appear to be conserved in both BryR (D214, D222, E225, and R266, CurD numbering) and BryU (R42, I46, and M47, CurB numbering) (Maloney et al., 2016). While the affinity of the curacin donor ACP is far tighter for its native HMGS than for BryR, the measured affinity still suggests that many of these donor ACPs have high degrees of intersystem operability (Buchholz et al., 2010). If future attempts are made to transfer HMGS functionality to other pathways, the matching of donor ACP for native HMGS will likely be relatively straightforward.

3.2 BryT, Enoyl-CoA Hydratase

Rather than attempting to clone from metagenomic DNA, synthetic codon-optimized DNA coding for BryT was designed, and cloned into expression vectors (Sørensen & Mortensen, 2005; Stols et al., 2002). Coexpression with GroEL and GroES at relatively low growth temperatures was determined to be necessary, with low and inconsistent protein yields from other conditions (Fig. 7). Protein purified under optimized conditions was >90% pure, but was sensitive to cooling below ~5°C, and would semireversibly precipitate at low temperatures. Other ECH_1 enzymes from bacterial PKS pathways were also obtained for comparison: CurE and JamI from the curacin and jamaicamide pathways, respectively.

HMG-CoA was used as a substrate as previously described (Gu et al., 2006): both as a reagent used to charge ACPs and as a direct substrate without further modification. Using the phosphopantetheine ejection assay, a mass shift equivalent to a dehydration was observed after incubation with CurE, JamI, or BryT, indicating the generation of 3-methylglutaconyl-CoA (3MG-CoA) (Fig. 6D). Upon detailed review of the data, a small amount of dehydrated mass is still detectable in the absence of ECH_1. This was suspected to be ionization-mediated dehydration occurring during the mass spectrometry analysis.

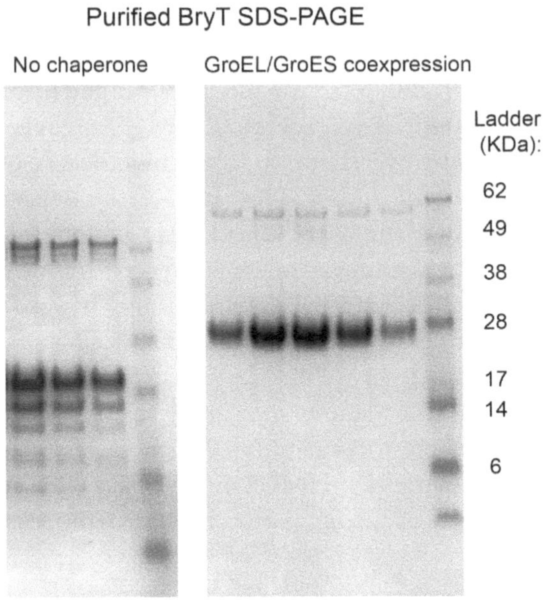

Fig. 7 SDS-PAGE of BryT with and without coexpressed chaperones. Expected mass of BryT with His-tag fusion is 31 kDa

3.3 BryA, Carboxyl *O*-Methyltransferase

Previous attempts to identify the carboxyl *O*-MT that generates the bryostatin vinyl methylesters were unsuccessful. Several candidate *S*-adenosyl methionine (SAM)-dependent methyltransferases were identified within the cluster: BryA module-3 and BryB module-4 both contain MTs, and a standalone MT was identified near the other β-branching cassette genes. None of these domains were able to catalyze methylation of the substrate mimics in initial tests. However, revisiting this search after the discovery of BryT proved fruitful.

After publication of the draft *Ca. E. sertula* genome and accompanying transcriptomic data, BryS could be confidently eliminated as a candidate for *O*-methylation. In a BLAST search, many close homologues (>50% identity) returned as tRNA *C*-methyltransferases. It is notoriously difficult to predict the substrate of methylation for class I MTs by sequence comparison alone, but the large number of BLAST hits for C-MTs raised suspicion. In addition, while transcripts were detected for the rest of the pathway, no mRNA transcripts of BryS were identified in the transcriptomic investigation performed by the Kwan group (Miller, Vanee, et al., 2016). Taken together, these results provide strong evidence that BryS was not the desired carboxyl *O*-MT.

Computational analysis and BLAST searches of other candidate methyltransferases against the MiBIG database (Medema et al., 2015) provided some guidance toward identification of the desired Bry methyltransferase. BryA module-3 MT returned hits for a few *N*- and *O*-standalone methyltransferases with modest sequence identity (~26%). BryB module-7 MT, at 65% identity to BryA module-3 MT, returned similar hits. The other pair of bryostatin MTs provided different results. At 95% identity, BryB module-4 MT and BryC module-9 MT are almost identical and return a large number of module-integrated PKS *C*-methyltransferases with a typical sequence identity of ~45%. Considering that BryA module-3 MT and BryB module-7 MT are located within the PKSs that are predicted to mediate β-branching modifications, these BLAST results strongly suggested their role as the carboxyl *O*-MTs of the Bry β-branching cassette.

With the function of BryT confirmed, we hypothesized that the carboxyl *O*-MT may require dehydration before being able to methylate its substrate. Thus, BryA module-3 was cloned from the bryostatin cosmid library (Sudek et al., 2007), expressed and purified. Additionally, standalone BryA module-3 ACP was purified and charged using Svp (Sánchez, Du, Edwards, Toney, & Shen, 2001) and HMG-CoA. After incubating this HMG–ACP with BryT, BryA module-3, and SAM, a mass shift corresponding to dehydration and methylation was observed in the phosphopantetheine ejection assay (Fig. 6E). Several constructs of the excised *O*-MT domain were subsequently cloned and purified, greatly improving the specific activity and total turnover of the system. Both in the context of the module and as a standalone domain, the *O*-MT acted with a strict dependence upon BryT dehydration, ACP tethering, and supplemental SAM cofactor.

4. CHEMOENZYMATIC REGIOCHEMISTRY DETERMINATION

4.1 Analysis of the BryT Dehydration Product

While the detection of a loss of water is compelling evidence for ECH_1 activity, it does not allow for determination of the regiochemistry of the product. Several approaches were developed that take advantage of the ability of ECH_1 to use CoA-tethered substrates to unambiguously determine whether BryT was catalyzing dehydrase to generate the α/β (Δ^2 product) or β/γ (Δ^3 product) regiochemistry. First, the similarity of PKS HMGS and HMG-CoA synthase in primary metabolism was leveraged to chemoenzymatically synthesize a specifically deuterated HMG-CoA

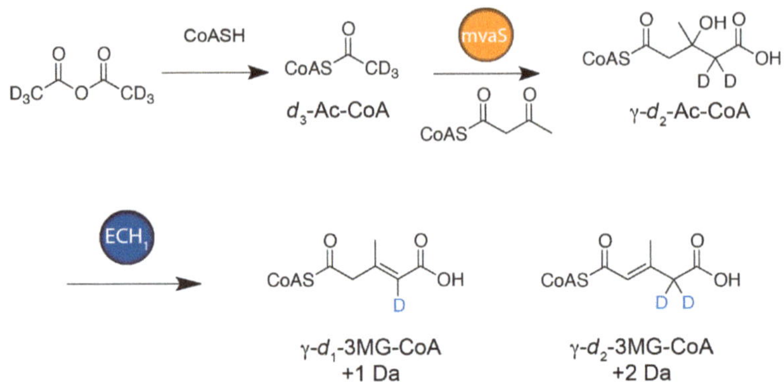

Fig. 8 Chemoenzymatic preparation of dideuterated 3-hydroxy-3-methylglutaryl-CoA. Deuterated acetic anhydride rapidly converts coenzyme A into trideuterated acetyl-CoA, which can be used as a carbon source for HMG-CoA synthase. After reaction with BryT, the double bond configuration can be inferred by mass.

(Fig. 8). Fully deuterated acetic anhydride was used to generate d_3-acetyl-CoA, which was then employed as a carbon donor for MvaS, the HMG-CoA synthase from *Staphylococcus aureus* (Campobasso et al., 2004). This reaction series generated γ-d_2-HMG-CoA, which would result in either a +1 Da or +2 Da difference from the natural isotopic abundance after ECH_1-mediated dehydration. Detection of the product masses was achieved through HPLC separation of the reaction components and MS confirmation of the individual masses.

The addition of BryT to this reaction system resulted in the generation of primarily d_2-3MG-CoA, strongly indicating α/β dehydration and a Δ^2 product (Fig. 9). This is somewhat surprising, as no other enzymatic components are evident within the cluster that may isomerize this double bond to generate the β/γ configuration found in the final product. Because this is a simplified substrate mimic, we cannot rule out that dehydration of the native tetra- and octaketide substrates would result in different regiochemistry. However, this seems unlikely. The shared mechanism and high-sequence identity among BryT and other ECH_1 suggest that nonnative substrate structures would not alter product regiochemistry. Additionally, isolation and NMR analysis discussed in Section 4.1.3 indicate a single product configuration rather than a mixture as might be expected for a nonnative substrate.

Because of proton exchange with solvent, the timing of the reaction has a large impact on the isotope distribution detected in the products. However, proton exchange must be balanced with the generation of sufficient product

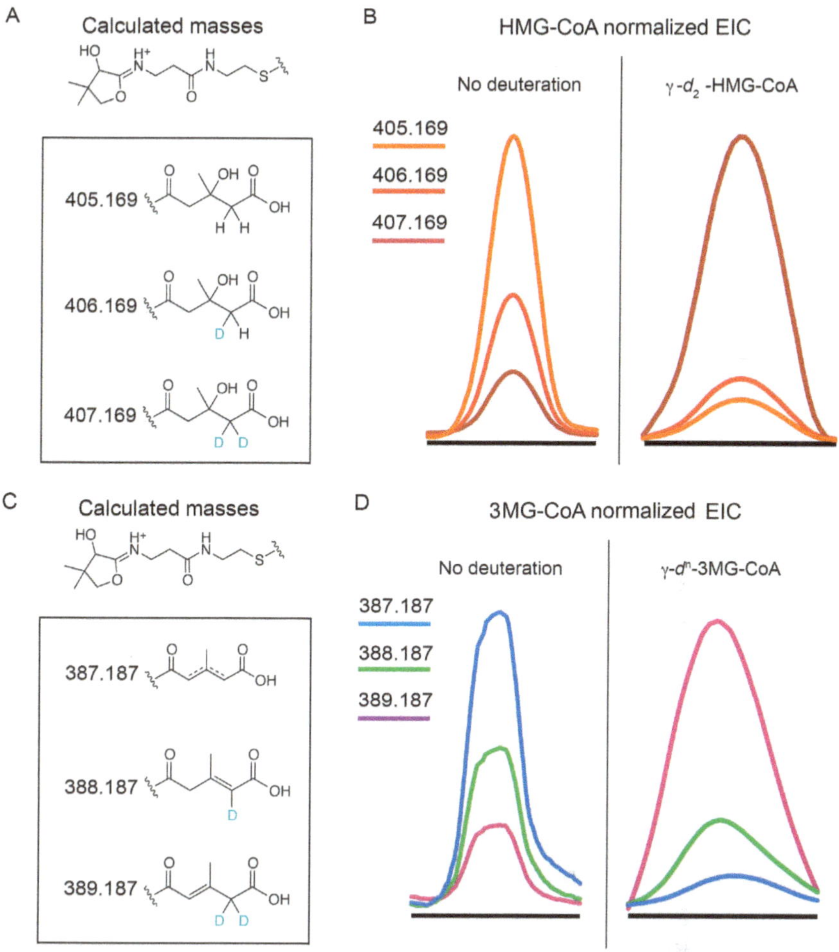

Fig. 9 LC/MS analysis of deuterated HMG-CoA and 3MG-CoA. Calculated masses for 0, 1, or 2 deuterations at the γ-position of HMG-CoA are shown in (A). These masses were used to generate extracted ion chromatograms for each mass from LC/MS data, shown in (B). The same process was employed to prepare the masses for 3MG-CoA in (C) and the extracted ion chromatograms in (D).

to be reliably separated and detected. In the attempt to balance these demands, we found that shorter reaction times led to an even higher ratio of d_2:d_1 3MG-CoA (Fig. 10), further indicating that the dideuterated molecule is the preferred product of BryT.

While analyzing these data, it is important to consider the possibility of a kinetic isotope effect altering the outcome. Because ECH_1 (and most of the crotonase family) proceed through an enolate intermediate stabilized by an

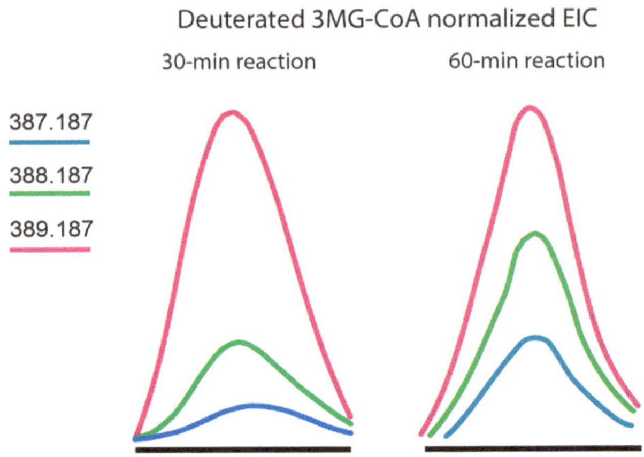

Fig. 10 Varying reaction times changes deuteration of 3MG-CoA. With a 30-min reaction of γ-d_2-HMG-CoA with BryT (*left*), the abundance of Δ^2-3MG-CoA is much higher than that of a 60-min reaction (*right*), likely due to deuterium exchange.

oxyanion hole in the enzyme active site, the rate-limiting step is likely either the formation of the enolate or the first deprotonation of the substrate at the α-carbon. Because the abstraction of a γ-deuteron is highly unlikely to be the rate-limiting step of the reaction, a kinetic isotope effect is unlikely to alter the product distribution sufficiently to explain the abundance of γ-d_2-3MG-CoA detected in these reactions.

4.1.1 Materials and Equipment

1. BryT, CurE, and JamI (ECH_1) were expressed and purified using methods described for CurD (Maloney et al., 2016)
2. *S. aureus* mvaS DNA, codon optimized for *E. coli* (DE3) (IDT, gBlocks)
3. (D/L)-3-Hydroxy-3-methylglutaryl coenzyme A sodium salt hydrate (Sigma #H6132)
4. Acetoacetyl coenzyme A sodium salt hydrate (Sigma #A1625)
5. Acetyl coenzyme A sodium salt hydrate (Sigma #A2056)
6. Coenzyme A sodium salt hydrate (Sigma #C3144)
7. Acetic anhydride-d_6 (Sigma #175641)
8. Magnesium chloride (Fisher #M35-500)
9. Dithiothreitol (DTT) (Sigma #43819)
10. 4-(2-Hydroxyethyl)piperazine-1-ethanesulfonic acid (HEPES) (Fisher #BP310-500)
11. Ammonium formate (Sigma #70221)

12. LC/MS water (Sigma #WX0001)
13. LC/MS methanol (EMD Millipore #MX0475)
14. LC/MS acetonitrile (EMD Millipore #AX0142)
15. LC/MS formic acid (Fisher #A117-50)
16. Reversed-phase C18 HPLC column (Waters, xBridge C18 column 3.6 μ, 2.1 × 250 mm)
17. Q-TOF LC/MS (Agilent G6545A)

4.1.2 In Situ γ-Deuterated HMG-CoA Synthesis and LC/MS Analysis

1. Coenzyme A was dissolved to a concentration of 10 m*M* in 0.5 *M* HEPES pH 7.0.
2. Acetic anhydride-d_6 was added to 1.5 m*M* and incubated for 20 min at 23°C to generate d_3-acetyl-CoA.
3. During the incubation, mvaS reactions were prepared by combining 25 μ*M* purified mvaS with 10 m*M* $MgCl_2$ 2 m*M* DTT, 50 m*M* HEPES pH 7.0, 10 μ*M* ECH_1 (BryT, CurE, or JamI), 2 m*M* acetoacetyl-CoA, and 10% glycerol.
4. Without further purification, crude d_3-acetyl-CoA reaction mixture was added to mvaS reactions to 4% (v/v) and incubated for 20 min at 23°C.
5. Reactions were quenched with three volumes LC/MS methanol with 5% formic acid.
6. Because of proton exchange with the deuterated material, expedience of the protocol is key. Quenched reactions were clarified by high-speed centrifugation and analyzed by LC/MS as quickly as possible.
7. Baseline separation of different CoA thioesters was achieved using a narrow gradient (4%–10% B) where buffer A was water with 50 m*M* ammonium formate pH 5.0 and buffer B was neat acetonitrile.
8. Extracted ion chromatograms were generated using masses calculated for both intact CoA thioesters and phosphopantetheine ejection adducts (Meluzzi et al., 2008). Phosphopantetheine ejection masses were found to be more reliably detected for all species.

4.1.3 Large-Scale Product Isolation and NMR Analysis

In order to corroborate these data, a large-scale reaction and purification effort were undertaken. Fresh BryT was purified and immediately added to a 10-mg scale reaction of HMG-CoA. Because BryT catalyzes both dehydration and the reverse hydration reaction, and also prefers a single stereoisomer of the ~50:50 mixture present in commercial HMG-CoA, the yields of these reactions are typically modest (~10%). Representative HPLC traces

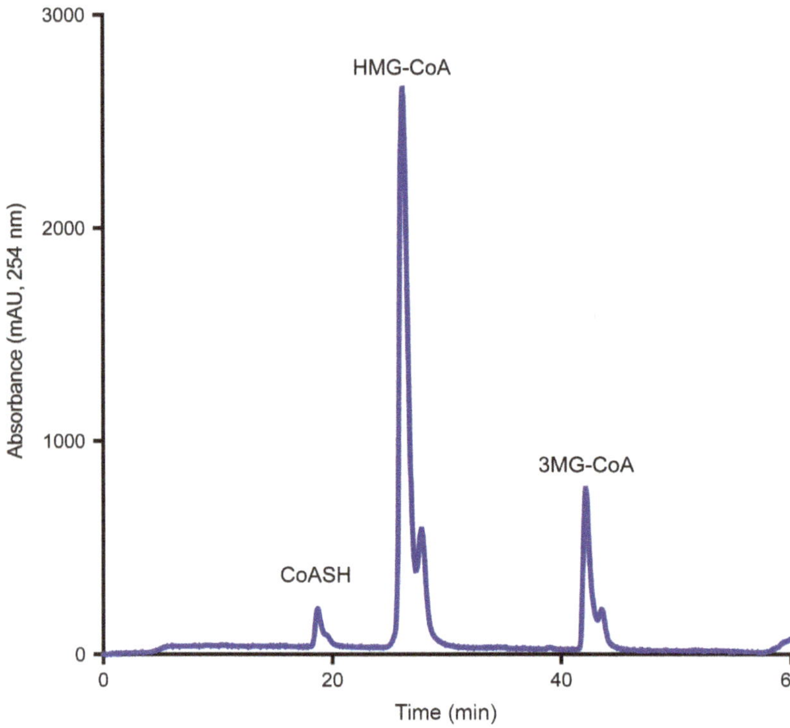

Fig. 11 Representative HPLC of preparative-scale BryT reaction with HMG-CoA. Baseline separation was achieved at the preparative scale using 10 mg of HMG-CoA starting material and 40 μ*M* BryT. 3MG-CoA peak was isolated and used in NMR analysis.

are shown in Fig. 11. Through careful method development and scale-up, a rapid workup and HPLC purification protocol were established that minimized the amount of time 3MG-CoA product was exposed to room temperature. Even so, the low yields and degradation were persistent challenges.

Eventually, NMR spectra were obtained using samples of sufficient concentration and purity. Initial proton NMR data showed conclusively that a methylene signal is lost and a new methane proton appears, indicating a new double bond. To enable analysis of the double bond installed by BryT, a partial assignment of commercial HMG-CoA starting material was carried out. By comparing this assignment to the 3MG-CoA spectra, the site of the lost methylene could be correlated on HMG-CoA (Fig. 12). This analysis led to the conclusion that the methylene signal disappeared from the α-carbon of 3MG-CoA, indicating that the resulting product was derived from an α/β dehydration. This is consistent with the mass shifts in the deuteration experiments and strongly supports that BryT catalyzes α/β dehydration activity.

A

	Number	δ_C	δ_H
Base	1b		
	2b	152.4	8.20 (s, 1H)
	3b		
	4b	149.2	
	5b	118.5	
	6b	155.6	
	7b		
	8b	139.9	8.60 (s, 1H)
	9b		
Sugar	1′	86.8	6.14 (d, J = 6.1 Hz, 1H)
	2′	74.4	4.80 (t, J = 5.6 Hz, 1H)
	3′	74.3	4.93-4.98 (m, 1H)
	4′	86.5	4.47-4.50 (m, 1H)
	5′	65.1	4.23-4.33 (m, 1H)
Substituted pantetheine	1	71.9	4.01 (dd, J = 5.2, 9.8 Hz, 1H)
			3.57 (dd, J = 4.1, 9.8 Hz, 1H)
	2	38.7	
	2Me	18.1	0.83 (s, 3H)
	2Me′	21	1.07 (s, 3H)
	3	73.9	4.10 (s, 1H)
	4	174.6	
	5		
	6	35.2	3.47 (t, J = 6.9 Hz, 2H)
	7	35.3	2.42 (t, J = 6.9 Hz, 2H)
	8	173.1	
	9		
	10	38.6	3.29-3.34 (m, 2H)
	11	28.2	3.00 (t, J = 6.6 Hz, 2H)
	12		
	13	198.3	
	14	54	2.87-2.96 (m, 2H)
	15	70.1	
	16	45.9	2.54 (d, J = 15.4 Hz, 1H)
			2.58 (d, J = 15.4 Hz, 1H)
	17	176.4	
	18	26.4	1.34 (s, 3H)

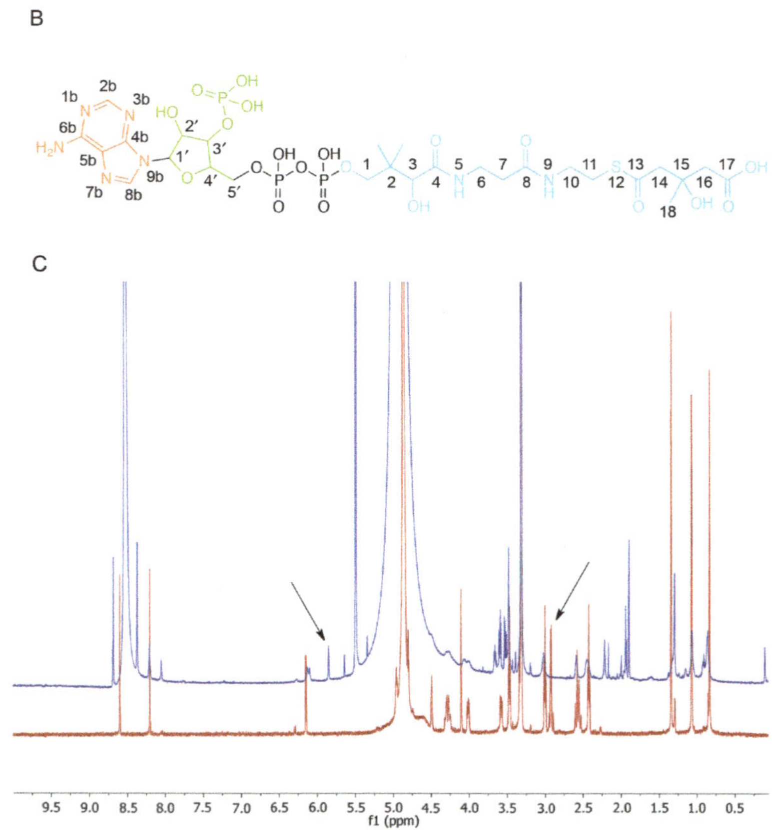

Fig. 12 NMR analysis of HMG- and 3MG-CoA isolated from BryT reactions. Signals listed in (A) were assigned to the chemical structure using commercial HMG-CoA (B). By comparing the 3MG-CoA spectrum (C, *blue*) to HMG-CoA (C, *red*), the lost methylene and gained methine (labeled with *arrows*) signals are apparent. Some impurities are present in the product, most significantly ammonium formate (8.5), water (4.8), and methanol (3.3).

4.2 Analysis of Bry β-Branching *O*-Methylation

Much like the dehydration catalyzed by BryT, mass spectrometry alone cannot confirm the location of a new methyl group by considering mass shift alone. Since BryA *O*-MT was unable to utilize CoA-tethered substrates, scale-up and isolation were not feasible. Thus, we turned again to organic synthesis to address this problem. Diazomethane (DM) is a reagent known to effectively convert carboxylic acids into methyl esters with a high degree of selectivity. Although it is acutely toxic, carcinogenic, and explosive, these risks can be almost completely ameliorated by using a trimethylsilyl-DM (TMS-DM) derivative. The covalently linked TMS is hydrolyzed in situ to generate smaller, safer amounts of DM.

This reaction can be used to probe *O*-methylation by coupling TMS-DM treatment with enzymatic methylation. By treating HMG–ACP with BryT and BryA *O*-MT prior to chemical treatment with TMS-DM and assessing the outcome by phosphopantetheine ejection, the location of methylation may be tested. If BryA *O*-MT is, in fact, a carboxyl *O*-MT, it will preclude methylation by DM, resulting in a mass shift equivalent to one methylation (14 Da). Enzymatic methylation at any other position will not preclude methyl ester formation, resulting in a mass shift indicative of two methylations (28 Da).

TMS-DM is commercially available as solution in hexanes, which presents a challenge for use with aqueous solutions of biomolecules. Initial experiments resulted in a bimodal distribution of methylation of HMG–ACP, with approximately equal populations of ACP bearing ~120 methylations and 0 methylations, with almost no methylation of the phosphopantetheine ejection ion detected. This is perhaps because the only contact made between HMG–ACP and TMS-DM was at the aqueous–organic interface, resulting in the extensive methylation of the fraction of protein in close proximity to the interface. It is interesting that even in the population of extensively methylated HMG–ACP, there was almost no conversion of HMG-phosphopantetheine to methyl ester. This may be due to sequestration of the phosphopantetheine arm away from solvent, a phenomenon that has been studied in detail in type-II FAS and PKS ACPs (Haushalter et al., 2011; Mercer & Burkart, 2007), but remains poorly studied in type-I PKSs.

Serial dilutions of TMS-DM from hexanes into methanol resolved this problem and altered the methylation pattern. HMG–ACP is soluble in methanol at concentrations up to ~70%, and the addition of dilute TMS-DM to

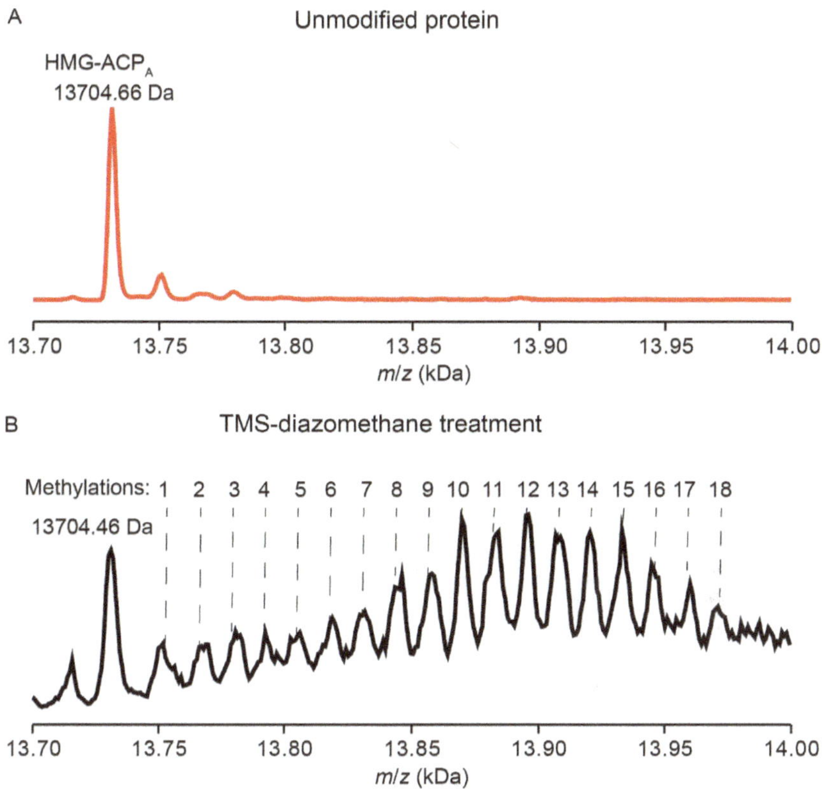

Fig. 13 Deconvoluted intact protein masses of HMG-BryA module-3 ACP with and without TMS-diazomethane treatment. Untreated protein shows a single primary species at 13.7 kDa (A). Diazomethane treatment results in discrete species with masses corresponding to 1 through 18 methylations of HMG-ACP$_A$ (B).

protein that has been prediluted in methanol results in the generation of diverse methylation states of the protein (Fig. 13). Using this method, ~90% of the protein is methylated at least once, with distinct masses corresponding to 1, 2, 3, … and so on methylations up to ~20. The abundance of these species roughly approximates a Gaussian distribution centered at 11 methylations. Interestingly, the frequency of methylation of the HMG moiety on the phosphopantetheine arm is still much lower than protein as a whole, with roughly 60% of the original HMG-phosphopantetheine ion still intact. This, like the previously described experiments, also implies some sort of sequestration of the phosphopantetheine arm away from solvent.

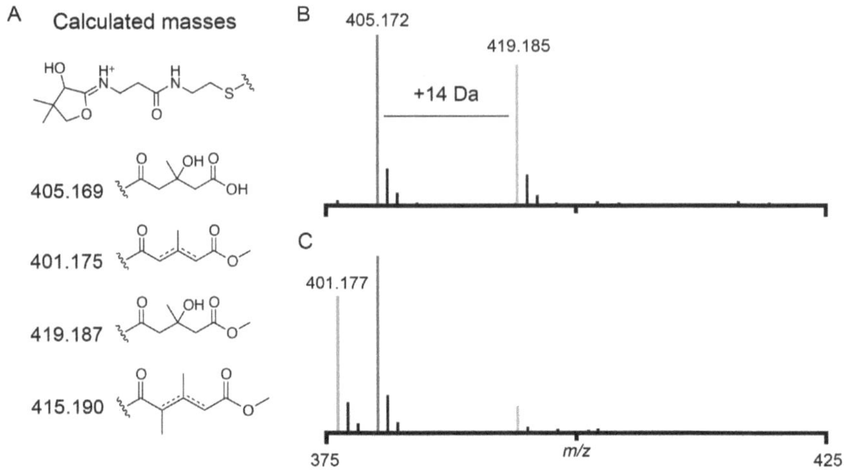

Fig. 14 Phosphopantetheine ejection ions of HMG-ACP$_A$ species treated with TMS-diazomethane. Calculated masses of variously methylated phosphopantetheine ejection ions are shown in (A). Control TMS-diazomethane treatment of HMG-ACP$_A$ (B), and HMG-ACP$_A$ pretreated with BryT and BryA *O*-MT prior to TMS-diazomethane treatment (C).

Most importantly, only one methylation was detected on the phosphopantetheine arm of HMG–ACP that had been treated with BryT and BryA *O*-MT prior to TMS-DM treatment. This indicates that the only site competent for methylation by TMS-DM has been blocked enzymatically by the *O*-MT. Other control samples confirmed that TMS-DM was only able to methylate in one location of the phosphopantetheine arm (Fig. 14). Incubation of HMG–ACP with BryT or no enzyme prior to TMS-DM treatment also resulted in masses corresponding to a single methylation of the phosphopantetheine arm. These results confirm the role of BryA *O*-MT as the carboxyl *O*-MT of the β-branching cassette.

4.2.1 Materials and Equipment

1. ECH_1, *O*-MT, apo-ACP, and Svp were purified as previously described (Gu et al., 2006; Sánchez et al., 2001)
2. *S*-adenosyl-L-methionine *p*-toluenesulfonate (Carbosynth #NA04017)
3. (D/L)-3-Hydroxy-3-methylglutaryl coenzyme A sodium salt hydrate (Sigma #H6132)
4. Magnesium chloride (Fisher #M35-500)
5. (Trimethylsilyl)diazomethane, 2 *M* in hexanes (Sigma #362832)
6. LC/MS methanol (EMD Millipore #MX0475)
7. LC/MS water (Sigma #WX0001)

8. LC/MS formic acid (Fisher #A117-50)
9. LC/MS acetonitrile (EMD Millipore #AX0142)
10. Reversed-phase C4 HPLC column (Waters, Acquity BEH300 C4 column 1.7 μm, 2.1 × 50 mm)
11. Q-TOF LC/MS (Agilent G6545A)

4.2.2 Carrier Protein Methylation and LC/MS Analysis

1. Apo-ACP was loaded with HMG-phosphopantetheine using HMG-CoA, Svp, and $MgCl_2$ using methods previously described (Maloney et al., 2016)
2. (Trimethylsilyl)diazomethane was placed under positive pressure of nitrogen using a syringe needle, and ~200 μL was drawn into a 1-mL syringe using a second needle and transferred to a 1.6-mL microcentrifuge tube
3. 200 μL of LC/MS methanol was added by pipette and vigorously mixed by pipetting to form a crude emulsion.
4. 200 μL of this emulsion was then added to another microcentrifuge tube with 200 μL methanol and mixed again.
5. The dilution procedure was repeated four more times for a final TMS-DM concentration of ~31.25 m*M* in 1.5% hexanes and 98.5% methanol.
6. Enzymatic methylation reactions and control reactions were prepared with final concentrations as follows: 75 μ*M* HMG-BryA module-3 ACP, 30 μ*M* BryA *O*-MT, 40 μ*M* BryT, 1 m*M* SAM, 5 m*M* $MgCl_2$, 50 m*M* HEPES pH 7.0, and 10% glycerol.
7. Enzymatic methylation reactions were incubated for 30 min at 30°C, then mixed with an equal volume of methanol-diluted TMS-DM. Reactions were mixed gently and allowed to incubate for 5 min at 23°C.
8. TMS-DM treatment was carefully quenched with two volumes of 50% methanol with 10% formic acid. Slow addition of this formic acid solution is important, as any unreacted diazomethane will rapidly evolve nitrogen gas upon acid quenching.
9. Reactions were centrifuged for 30 min at 17,000 × *g* to remove debris and precipitation.
10. LC/MS analysis was carried out using a 5 min, 0–100% B gradient where A is 95% water, 5% acetonitrile with 0.1% formic acid, and B is 95% acetonitrile, 5% water with 0.1% formic acid.
11. Masses were monitored during protein elution with m/z from 150 to 3000, enabling facile analysis of phosphopantetheine ejection ions as well as deconvolution of intact protein masses.

5. SUMMARY AND CONCLUSIONS

We have successfully confirmed the roles of BryU, BryT, and BryA *O*-MT in the construction of the unique vinyl methylester moieties found in bryostatin-1 and related natural product analogs. These enzymes act in a strict order during the β-branching of the polyketide intermediate, beginning with donation of an acetyl subunit by BryU to β-branching, then dehydration by BryT, followed by *O*-methylation by BryA *O*-MT. We have also confirmed the regiochemistry of both dehydration and *O*-methylation steps using novel chemoenzymatic tools. The generation and processing of γ-d_2-HMG-CoA revealed α/β (Δ^2) position of the double bond resulting from BryT dehydration. Chemical generation of the methyl ester of HMG- and 3MG-ACP_A combined with enzymatic methylation by BryA *O*-MT conclusively demonstrated carboxyl *O*-methylation with a strict dependence upon α/β dehydrated 3MG-CoA. These results support the proposal of a modified model of bryostatin β-branching that includes post-β-branching double bond isomerization prior to spontaneous hemiketalization to generate the unique vinyl methylester pyran rings found in all natural product bryostatins.

In the majority of β-branching cassettes, the double bond configuration of ECH_1 products is obfuscated by downstream modifications. In most cases, this means ECH_2 decarboxylation, which has been shown to be the determinant of product regiochemistry in the cases where it occurs (Geders et al., 2007). In newly discovered pathways, the knowledge that ECH_1 catalyzes α/β dehydration can inform biosynthetic hypotheses about intermediates and biosynthetic strategies.

These results have implications for the future of PKS pathway functional analysis and engineering. Because β-branch acceptor ACPs can be readily identified by a primary sequence motif, one can envision the modification of a normal modular ACP to transform it into a β-branch acceptor. This method would result in the introduction of a novel chemical structure at a specified β-keto position during polyketide synthesis and ultimately a novel polyketide product. In addition, the chemoenzymatic methods developed for these studies may be applied to a number of challenging questions relating to β-branching schemes in other PKS pathways. These methods provide tools to logically determine regiochemistry of several PKS intermediates that remain difficult to prepare and are only available on microscale levels.

ACKNOWLEDGMENTS

NIH Grants R35 GM118101, GM076477, and the Hans W. Vahlteich Professorship (to D.H.S.) are gratefully acknowledged. S.T.S. was supported in part by a Biological Chemistry Ph.D. Graduate Student Award from the University of Michigan Rackham Graduate School and the Cellular Biotechnology Training Program (T32 GM008353).

REFERENCES

Blackhall, F. H., Ranson, M., Radford, J. A., Hancock, B. W., Soukop, M., McGown, A. T., et al.Cancer Research Campaign Phase I/II Committee, (2001). A phase II trial of bryostatin 1 in patients with non-Hodgkin's lymphoma. *British Journal of Cancer*, *84*, 465–469. https://doi.org/10.1054/bjoc.2000.1624.

Buchholz, T. J., Rath, C. M., Lopanik, N. B., Gardner, N. P., Håkansson, K., & Sherman, D. H. (2010). Polyketide β-branching in bryostatin biosynthesis: Identification of surrogate acetyl-ACP donors for BryR, an HMG-ACP synthase. *Chemistry & Biology*, *17*, 1092–1100. https://doi.org/10.1016/j.chembiol.2010.08.008.

Calderone, C. T., Iwig, D. F., Dorrestein, P. C., Kelleher, N. L., & Walsh, C. T. (2007). Incorporation of nonmethyl branches by isoprenoid-like logic: Multiple β-alkylation events in the biosynthesis of myxovirescin A1. *Chemistry & Biology*, *14*, 835–846. https://doi.org/10.1016/j.chembiol.2007.06.008.

Campobasso, N., Patel, M., Wilding, I. E., Kallender, H., Rosenberg, M., & Gwynn, M. N. (2004). *Staphylococcus aureus* 3-hydroxy-3-methylglutaryl-CoA synthase: Crystal structure and mechanism. *Journal of Biological Chemistry*, *279*, 44883–44888. https://doi.org/10.1074/jbc.M407882200.

Chang, Z., Sitachitta, N., Rossi, J. V., Roberts, M. A., Flatt, P. M., Jia, J., et al. (2004). Biosynthetic pathway and gene cluster analysis of curacin A, an antitubulin natural product from the tropical marine cyanobacterium *Lyngbya majuscula*. *Journal of Natural Products*, *67*, 1356–1367. https://doi.org/10.1021/np0499261.

Davidson, S. K., Allen, S. W., Lim, G. E., Anderson, C. M., & Haygood, M. G. (2001). Evidence for the biosynthesis of bryostatins by the bacterial symbiont "*Candidatus Endobugula sertula*" of the bryozoan *Bugula neritina*. *Applied and Environmental Microbiology*, *67*, 4531–4537.

Davidson, S. K., & Haygood, M. G. (1999). Identification of sibling species of the bryozoan *Bugula neritina* that produce different anticancer bryostatins and harbor distinct strains of the bacterial symbiont "*Candidatus Endobugula sertula*." *The Biological Bulletin*, *196*, 273–280.

DeChristopher, B. A., Fan, A. C., Felsher, D. W., & Wender, P. A. (2012). "Picolog," a synthetically-available bryostatin analog, inhibits growth of MYC-induced lymphoma *in vivo*. *Oncotarget*, *3*, 58–66.

Edwards, D. J., Marquez, B. L., Nogle, L. M., McPhail, K., Goeger, D. E., Roberts, M. A., et al. (2004). Structure and biosynthesis of the jamaicamides, new mixed polyketide-peptide neurotoxins from the marine cyanobacterium *Lyngbya majuscula*. *Chemistry & Biology*, *11*, 817–833. https://doi.org/10.1016/j.chembiol.2004.03.030.

El-Sayed, A. K., Hothersall, J., Cooper, S. M., Stephens, E., Simpson, T. J., & Thomas, C. M. (2003). Characterization of the mupirocin biosynthesis gene cluster from *Pseudomonas fluorescens* NCIMB 10586. *Chemistry & Biology*, *10*, 419.

Geders, T. W., Gu, L., Mowers, J. C., Liu, H., Gerwick, W. H., Hakansson, K., et al. (2007). Crystal structure of the ECH2 catalytic domain of CurF from *Lyngbya majuscula*: Insights into a decarboxylase involved in polyketide chain branching. *Journal of Biological Chemistry*, *282*, 35954–35963. https://doi.org/10.1074/jbc.M703921200.

Gu, L., Jia, J., Liu, H., Håkansson, K., Gerwick, W. H., & Sherman, D. H. (2006). Metabolic coupling of dehydration and decarboxylation in the curacin A pathway: Functional identification of a mechanistically diverse enzyme pair. *Journal of the American Chemical Society*, *128*, 9014–9015. https://doi.org/10.1021/ja0626382.

Gu, L., Wang, B., Kulkarni, A., Geders, T. W., Grindberg, R. V., Gerwick, L., et al. (2009). Metamorphic enzyme assembly in polyketide diversification. *Nature*, *459*, 731–735. https://doi.org/10.1038/nature07870.

Gutiérrez, C., Serrano-Villar, S., Madrid-Elena, N., Pérez-Elías, M. J., Martín, M. E., Barbas, C., et al. (2016). Bryostatin-1 for latent virus reactivation in HIV-infected patients on antiretroviral therapy. *AIDS*, *30*, 1385–1392. https://doi.org/10.1097/QAD.0000000000001064.

Hansen, D. A., Rath, C. M., Eisman, E. B., Narayan, A. R. H., Kittendorf, J. D., Mortison, J. D., et al. (2013). Biocatalytic synthesis of pikromycin, methymycin, neomethymycin, novamethymycin, and ketomethymycin. *Journal of the American Chemical Society*, *135*, 11232–11238. https://doi.org/10.1021/ja404134f.

Haushalter, R. W., Filipp, F. V., Ko, K., Yu, R., Opella, S. J., & Burkart, M. D. (2011). Binding and pKa modulation of a polycyclic substrate analog in a type II polyketide acyl carrier protein. *ACS Chemical Biology*, *6*, 413.

Helfrich, E. J. N., & Piel, J. (2016). Biosynthesis of polyketides by trans-AT polyketide synthases. *Natural Product Reports*, *33*, 231–316. https://doi.org/10.1039/C5NP00125K.

Higa, T., Tanaka, J., Komesu, M., & Jefford, C. W. (1992). Miyakolide: A bryostatin-like macrolide from a sponge, *Polyfibrospongia* sp. *Journal of the American Chemical Society*, *114*, 7587–7588. https://doi.org/10.1021/ja00045a055.

Keck, G. E., Poudel, Y. B., Cummins, T. J., Rudra, A., & Covel, J. A. (2010). Total synthesis of bryostatin 1. *Journal of the American Chemical Society*, *133*, 744–747.

Kedei, N., Kraft, M. B., Keck, G. E., Herald, C. L., Melody, N., Pettit, G. R., et al. (2015). Neristatin 1 provides critical insight into bryostatin 1 structure–function relationships. *Journal of Natural Products*, *78*, 896–900. https://doi.org/10.1021/acs.jnatprod.5b00094.

Kelsey, J. S., Cataisson, C., Chen, J., Herrmann, M. A., Petersen, M. E., Baumann, D. O., et al. (2016). Biological activity of the bryostatin analog merle 23 on mouse epidermal cells and mouse skin. *Molecular Carcinogenesis*, *55*, 2183–2195. https://doi.org/10.1002/mc.22460.

Lopanik, N., Lindquist, N., & Targett, N. (2004). Potent cytotoxins produced by a microbial symbiont protect host larvae from predation. *Oecologia*, *139*, 131–139. https://doi.org/10.1007/s00442-004-1487-5.

Lopanik, N. B., Shields, J. A., Buchholz, T. J., Rath, C. M., Hothersall, J., Haygood, M. G., et al. (2008). In vivo and in vitro trans-acylation by BryP, the putative bryostatin pathway acyltransferase derived from an uncultured marine symbiont. *Chemistry & Biology*, *15*, 1175–1186. https://doi.org/10.1016/j.chembiol.2008.09.013.

Lu, Y., Woo, S. K., & Krische, M. J. (2011). Total synthesis of bryostatin 7 via C-C bond-forming hydrogenation. *Journal of the American Chemical Society*, *133*, 13876–13879. https://doi.org/10.1021/ja205673e.

Lucke-Wold, B. P., Logsdon, A. F., Smith, K. E., Turner, R. C., Alkon, D. L., Tan, Z., et al. (2014). Bryostatin-1 restores blood brain barrier integrity following blast-induced traumatic brain injury. *Molecular Neurobiology*, *52*, 1119–1134. https://doi.org/10.1007/s12035-014-8902-7.

Maier, T., Leibundgut, M., & Ban, N. (2008). The crystal structure of a mammalian fatty acid synthase. *Science*, *321*, 1315–1322. https://doi.org/10.1126/science.1161269.

Maloney, F. P., Gerwick, L., Gerwick, W. H., Sherman, D. H., & Smith, J. L. (2016). Anatomy of the β-branching enzyme of polyketide biosynthesis and its interaction with an acyl-ACP substrate. *Proceedings of the National Academy of Sciences*, *113*(37), 10316–10321. https://doi.org/10.1073/pnas.1607210113.

Manaviazar, S., & Hale, K. J. (2011). Total synthesis of bryostatin 1: A short route. *Angewandte Chemie International Edition*, *50*, 8786–8789. https://doi.org/10.1002/anie.201101562.

Medema, M. H., Kottmann, R., Yilmaz, P., Cummings, M., Biggins, J. B., Blin, K., et al. (2015). Minimum information about a biosynthetic gene cluster. *Nature Chemical Biology*, *11*, 625–631. https://doi.org/10.1038/nchembio.1890.

Mehla, R., Bivalkar-Mehla, S., Zhang, R., Handy, I., Albrecht, H., Giri, S., et al. (2010). Bryostatin modulates latent HIV-1 infection via PKC and AMPK signaling but inhibits acute infection in a receptor independent manner. *PLoS One5.* , https://doi.org/10.1371/journal.pone.0011160.

Meluzzi, D., Zheng, W. H., Hensler, M., Nizet, V., & Dorrestein, P. C. (2008). Top-down mass spectrometry on low-resolution instruments: Characterization of phosphopantetheinylated carrier domains in polyketide and non-ribosomal biosynthetic pathways. *Bioorganic & Medicinal Chemistry Letters*, *18*, 3107–3111. https://doi.org/10.1016/j.bmcl.2007.10.104.

Mendola, D. (2003). Aquaculture of three phyla of marine invertebrates to yield bioactive metabolites: Process developments and economics. *Biomolecular Engineering*, *20*, 441–458.

Mercer, A. C., & Burkart, M. D. (2007). The ubiquitous carrier protein—A window to metabolite biosynthesis. *Natural Product Reports*, *24*, 750–773. https://doi.org/10.1039/B603921A.

Miller, I. J., Vanee, N., Fong, S. S., Lim-Fong, G. E., & Kwan, J. C. (2016). Lack of overt genome reduction in the bryostatin-producing bryozoan symbiont, "*Candidatus Endobugula sertula*" *Applied and Environmental Microbiology*, *82*(22), 6573–6583. https://doi.org/10.1128/AEM.01800-16.

Miller, I. J., Weyna, T. R., Fong, S. S., Lim-Fong, G. E., & Kwan, J. C. (2016). Single sample resolution of rare microbial dark matter in a marine invertebrate metagenome. *Scientific Reports*, *6*, 34362. https://doi.org/10.1038/srep34362.

Moran, N. A. (2003). Tracing the evolution of gene loss in obligate bacterial symbionts. *Current Opinion in Microbiology*, *6*, 512–518.

Newman, D. J., & Cragg, G. M. (2012). Natural products as sources of new drugs over the 30 years from 1981 to 2010. *Journal of Natural Products*, *75*, 311–335. https://doi.org/10.1021/np200906s.

Ohta, S., Uy, M. M., Yanai, M., Ohta, E., Hirata, T., & Ikegami, S. (2006). Exiguolide, a new macrolide from the marine sponge *Geodia exigua*. *Tetrahedron letters*, *47*(12), 1957–1960.

Pagliaro, L., Daliani, D., Amato, R., Tu, S. M., Jones, D., Smith, T., et al. (2000). A phase II trial of bryostatin-1 for patients with metastatic renal cell carcinoma. *Cancer*, *89*, 615–618.

Sánchez, C., Du, L., Edwards, D. J., Toney, M. D., & Shen, B. (2001). Cloning and characterization of a phosphopantetheinyl transferase from *Streptomyces verticillus* ATCC15003, the producer of the hybrid peptide–polyketide antitumor drug bleomycin. *Chemistry & Biology*, *8*, 725–738. https://doi.org/10.1016/S1074-5521(01)00047-3.

Schaufelberger, D. E., Koleck, M. P., Beutler, J. A., Vatakis, A. M., Alvarado, A. B., Andrews, P., et al. (1991). The large-scale isolation of bryostatin 1 from *Bugula neritina* following current good manufacturing practices. *Journal of Natural Products*, *54*, 1265–1270. https://doi.org/10.1021/np50077a004.

Schofield, M. M., Jain, S., Porat, D., Dick, G. J., & Sherman, D. H. (2015). Identification and analysis of the bacterial endosymbiont specialized for production of the chemotherapeutic natural product ET-743. *Environmental Microbiology*, *17*, 3964–3975. https://doi.org/10.1111/1462-2920.12908.

Schofield, M. M., & Sherman, D. H. (2013). Meta-omic characterization of prokaryotic gene clusters for natural product biosynthesis. *Current Opinion in Biotechnology*, *24*, 1151–1158. https://doi.org/10.1016/j.copbio.2013.05.001.

Schrott, L. M., Jackson, K., Yi, P., Dietz, F., Johnson, G. S., Basting, T. F., et al. (2015). Acute oral bryostatin-1 administration improves learning deficits in the APP/PS1 transgenic mouse model of Alzheimer's disease. *Current Alzheimer Research, 12*, 22–31.

Sharp, K. H., Davidson, S. K., & Haygood, M. G. (2007). Localization of "*Candidatus Endobugula sertula*" and the bryostatins throughout the life cycle of the bryozoan bugula neritina. *International Society for Microbial Ecology Journal, 1*, 693–702. https://doi.org/10.1038/ismej.2007.78.

Simunovic, V., & Müller, R. (2007). Mutational analysis of the myxovirescin biosynthetic gene cluster reveals novel insights into the functional elaboration of polyketide backbones. *Chembiochem, 8*, 1273–1280. https://doi.org/10.1002/cbic.200700153.

Simunovic, V., Zapp, J., Rachid, S., Krug, D., Meiser, P., & Müller, R. (2006). Myxovirescin A biosynthesis is directed by hybrid polyketide synthases/nonribosomal peptide synthetase, 3-hydroxy-3-methylglutaryl–coa synthases, and trans-acting acyltransferases. *Chembiochem, 7*, 1206–1220. https://doi.org/10.1002/cbic.200600075.

Sørensen, H. P., & Mortensen, K. K. (2005). Advanced genetic strategies for recombinant protein expression in *Escherichia coli*. *Journal of Biotechnology, 115*, 113–128. https://doi.org/10.1016/j.jbiotec.2004.08.004.

Staveness, D., Abdelnabi, R., Near, K. E., Nakagawa, Y., Neyts, J., Delang, L., et al. (2016). Inhibition of chikungunya virus-induced cell death by salicylate-derived bryostatin analogues provides additional evidence for a PKC-independent pathway. *Journal of Natural Products, 79*, 680–684. https://doi.org/10.1021/acs.jnatprod.5b01017.

Stols, L., Gu, M., Dieckman, L., Raffen, R., Collart, F. R., & Donnelly, M. I. (2002). A new vector for high-throughput, ligation-independent cloning encoding a tobacco etch virus protease cleavage site. *Protein Expression and Purification, 25*, 8–15. https://doi.org/10.1006/prep.2001.1603.

Sudek, S., Lopanik, N. B., Waggoner, L. E., Hildebrand, M., Anderson, C., Liu, H., et al. (2007). Identification of the putative bryostatin polyketide synthase gene cluster from "*Candidatus Endobugula sertula*," the uncultivated microbial symbiont of the marine bryozoan *Bugula neritina*. *Journal of Natural Products, 70*, 67–74. https://doi.org/10.1021/np060361d.

Sun, M.-K., Hongpaisan, J., & Alkon, D. L. (2009). Postischemic PKC activation rescues retrograde and anterograde long-term memory. *Proceedings of the National Academy of Sciences, 106*, 14676–14680. https://doi.org/10.1073/pnas.0907842106.

Wender, P. A., Baryza, J. L., Brenner, S. E., DeChristopher, B. A., Loy, B. A., Schrier, A. J., et al. (2011). Design, synthesis, and evaluation of potent bryostatin analogs that modulate PKC translocation selectivity. *Proceedings of the National Academy of Sciences, 108*, 6721–6726. https://doi.org/10.1073/pnas.1015270108.

Wender, P. A., Hardman, C. T., Ho, S., Jeffreys, M. S., Maclaren, J. K., Quiroz, R. V., et al. (2017). Scalable synthesis of bryostatin 1 and analogs, adjuvant leads against latent HIV. *Science, 358*, 218–223. https://doi.org/10.1126/science.aan7969.

Wender, P. A., & Schrier, A. J. (2011). Total synthesis of bryostatin 9. *Journal of the American Chemical Society, 133*, 9228–9231. https://doi.org/10.1021/ja203034k.

Woollacott, R. M. (1981). Association of bacteria with bryozoan larvae. *Marine Biology, 65*, 155–158. https://doi.org/10.1007/BF00397080.

Zhang, F., He, H.-Y., Tang, M.-C., Tang, Y.-M., Zhou, Q., & Tang, G.-L. (2011). Cloning and elucidation of the FR901464 gene cluster revealing a complex acyltransferase-less polyketide synthase using glycerate as starter units. *Journal of the American Chemical Society, 133*, 2452–2462. https://doi.org/10.1021/ja105649g.

CHAPTER SEVEN

Radical *S*-Adenosylmethionine Peptide Epimerases: Detection of Activity and Characterization of D-Amino Acid Products

Brandon I. Morinaka[*,†], **Anna L. Vagstad**[*], **Jörn Piel**[*,1]
[*]Institute of Microbiology, Eidgenössische Technische Hochschule (ETH) Zurich, Zurich, Switzerland
[†]Department of Pharmacy, National University of Singapore, Singapore, Singapore
[1]Corresponding author: e-mail address: jpiel@ethz.ch

Contents

Abstract

The identification of the polytheonamide (*poy*) gene cluster led to the discovery of the enzyme PoyD, a member of the radical *S*-adenosylmethionine superfamily capable of introducing D-amino acids into a ribosomally synthesized peptide. This enzyme was used as a starting point to identify additional radical *S*-adenosylmethionine peptide epimerases in other cyanobacterial genomes, which show different epimerization patterns compared to PoyD. During the course of studying these enzymes by heterologous expression in *Escherichia coli*, we developed a two-step strategy to (1) detect epimerase activity and (2) localize where epimerization occurs based on an in vivo deuterium labeling strategy. The procedures for these two methods are described in the following chapter and will set the stage for further study of these enzymes.

Methods in Enzymology, Volume 604
ISSN 0076-6879
https://doi.org/10.1016/bs.mie.2018.01.033

1. INTRODUCTION

Nonproteinogenic amino acid residues were once thought only to occur in peptides of nonribosomal origin. They are increasingly also found as posttranslational modifications of biomolecules ribosomally synthesized from standard L-amino acids (Walsh, 2014). Owing to its sensitivity and high-information content, tandem mass spectrometry (MS/MS) is usually the method of choice for mapping such posttranslational modifications using characteristic mass shifts in the peptide fragmentation ions (Larsen, Trelle, Thingholm, & Jensen, 2006). However, this technique is inoperative with isobaric modifications, such as D-configured amino acid residues. These residues are notoriously difficult to detect in proteins and peptides, since this modification contributes no mass difference relative to canonical L-configured residues. In recent years, an appreciation for the natural occurrence of D-amino acids within peptides of diverse and important biological functions has grown, spurring the development of new analytical techniques for their detection (Ollivaux, Soyez, & Toullec, 2014).

Ribosomally synthesized and posttranslationally modified peptide (RiPP) natural products are an emerging resource for bioactive peptides and new enzymology (Arnison et al., 2013; Ortega & van der Donk, 2016). Canonical RiPP natural products stem from a ribosomally produced precursor protein usually comprised of an N-terminal leader region and a C-terminal core region. The core is subject to posttranslational modifications and is proteolytically released to give the mature compound. The proteusin family of RiPPs are characterized by a nitrile hydratase-like leader peptide. A subset of these pathways contains radical *S*-adenosylmethionine peptide epimerases (RSPEs) that unidirectionally and regiospecifically install D-amino acids in the core of precursor protein substrates (Freeman, Vagstad, & Piel, 2016; Freeman et al., 2012; Morinaka et al., 2014). The first characterized RSPE comes from the biosynthetic pathway of polytheonamides A and B (Fig. 1) (Freeman et al., 2012). These potent cytotoxic peptides are highly morphed from their ribosomal origin with 28 of the 49 structural core residues subject to modification, including 18 D-amino acids installed by a single RSPE enzyme, PoyD, at alternating positions excluding Gly residues. Although at first expected to be from a nonribosomal pathway, due to the high number of nonproteinogenic residues, polytheonamides are actually the product of a remarkably streamlined RiPP biosynthetic gene cluster from a bacterial

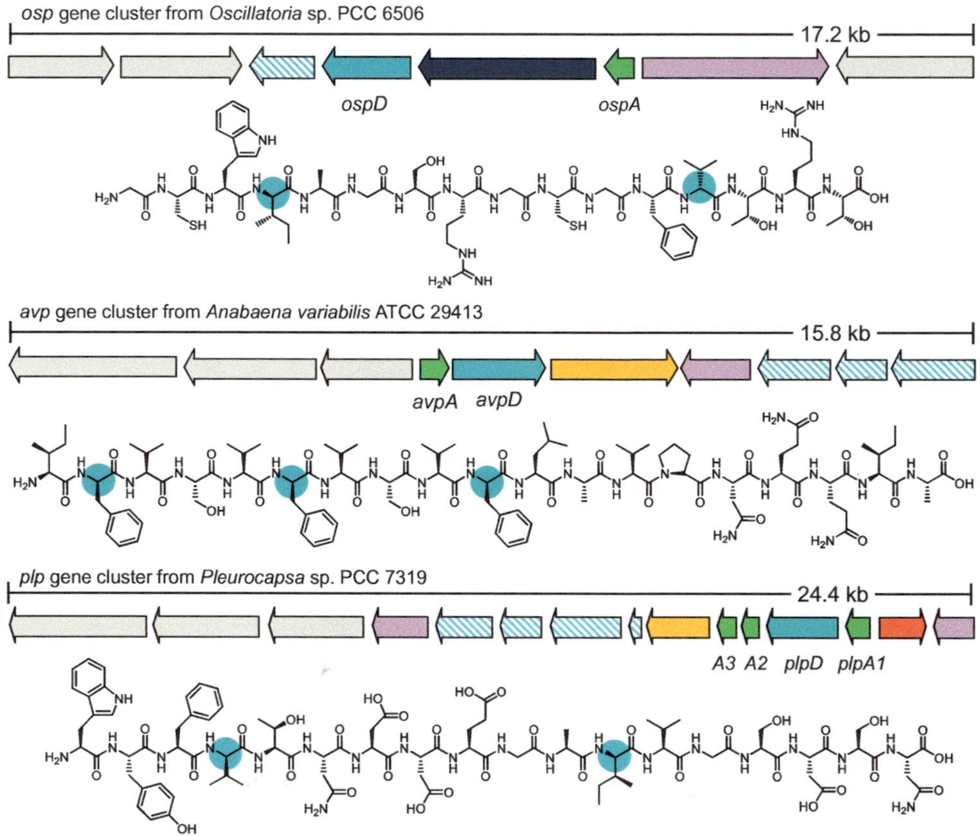

Fig. 1 Proteusin gene clusters containing RSPEs. Precursors and RSPEs are denoted with an *"A"* and *"D,"* respectively. *Blue circles* indicate residues epimerized by their corresponding RSPEs. The structures depicted for *osp*, *avp*, and *plp* gene clusters are derived from coexpression of precursor and RSPE only. Polytheonamides A and B are the product of the *poy* gene cluster and differ in the configuration at the sulfoxide moiety.

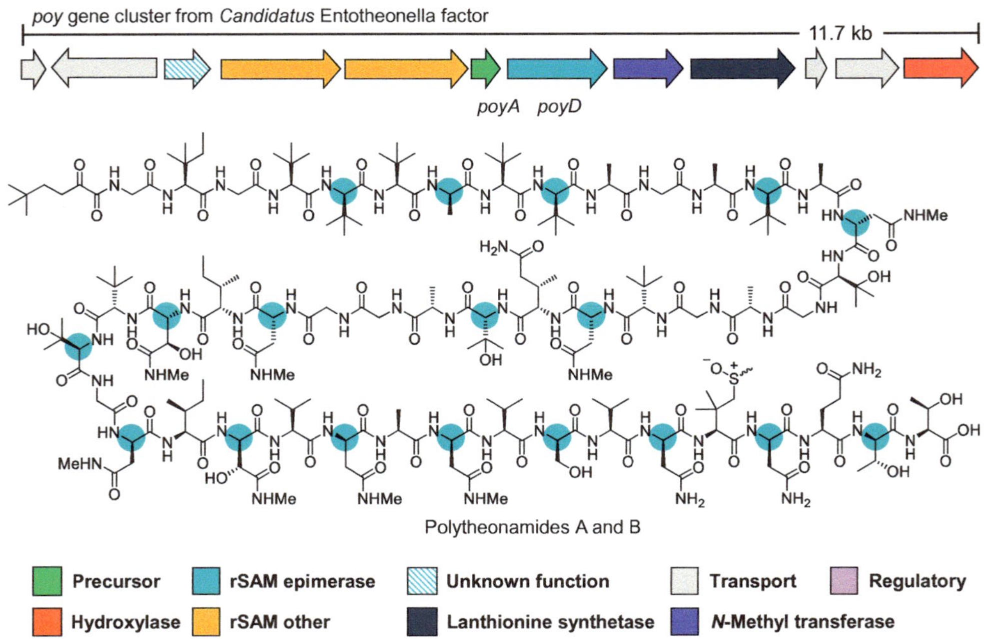
poy gene cluster from Candidatus Entotheonella factor
11.7 kb
poyA
poyD
Polytheonamides A and B
Precursor
rSAM epimerase
Unknown function
Transport
Regulatory
Hydroxylase
rSAM other
Lanthionine synthetase
N-Methyl transferase

Fig. 1—Cont'd

symbiont of the marine sponge *Theonella swinhoei* (Freeman et al., 2012; Wilson et al., 2014). In the polytheonamide pathway, a set of only six post-translational modifying enzymes, coencoded in a biosynthetic gene cluster with the precursor PoyA, transform its core residues prior to proteolytic release of the mature product by a pathway-specific protease (Freeman, Helf, Bhushan, Morinaka, & Piel, 2017).

Original characterization of the biosynthetic gene cluster was conducted by heterologous production of the PoyA precursor with selected modifying enzymes in *Escherichia coli* using a bottom-up approach to determine enzyme function (Freeman et al., 2012). Peptide products of PoyA + PoyD coexpressions were subjected to acid hydrolysis, Marfey-type derivatization, and chromatographic separation to demonstrate the presence of D-amino acids consistent with the identity of residues in the mature polytheonamide natural product (Bhushan & Bruckner, 2011; Marfey, 1984). Chromatographic separation of derivatized acid hydrolysates commonly allows comparison to authentic D-amino acid standards for identification and quantification; however, positional information within the parent peptide is lost. In a follow-up study, three homologs of PoyD from cyanobacterial proteusin pathways (Fig. 1) were similarly characterized using *E. coli* coexpressions of pathway precursors and RSPEs (Morinaka et al., 2014). Bulk D-amino acid content from the isolated core peptides was determined by Marfey-type analysis. However, to localize the position of D-amino acids, a combination of two additional techniques was adopted. Products were either compared to candidate synthetic peptides or analyzed for loss of deuterium at epimerized residues by MS/MS in stable isotope amino acid labeling studies. The isotope studies were conducted in minimal media in a transaminase knockout strain (Kegler et al., 2014), which prevents rapid loss of the α-position label by the action of transaminases in a limited set of amino acids. The three analyzed RSPEs acting on cognate and engineered precursor protein substrates showed high regioselectivity, substrate promiscuity, and irreversibility to install diverse patterns of D-amino acids (Fig. 1) (Morinaka et al., 2014). Despite these successes, peptide synthesis and stable isotope feeding studies require expensive reagents, time-consuming preparations, or are restricted to a subset of amino acids, limiting their general usefulness. Thus, it was highly desirable to develop a more facile and rapid, sequence-independent method for identification of RSPE-catalyzed epimerizations in diverse protein substrates.

Radical *S*-adenosylmethionine (rSAM) enzymes canonically utilize a 5′-deoxyadenosine radical intermediate to abstract a substrate hydrogen to

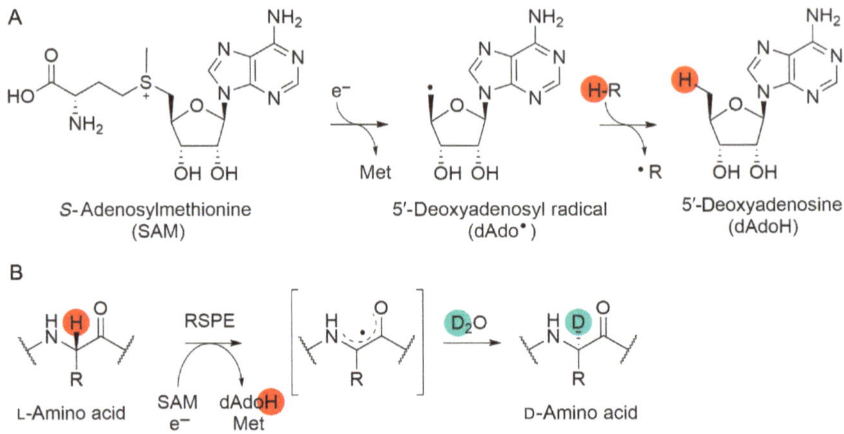

Fig. 2 Putative radical-mediated enzyme mechanism of RSPEs. (A) Canonical initiation step of radical SAM enzymes. A one-electron reduction of the [4Fe–4S] cluster facilitates hemolytic cleavage of SAM to L-methionine (Met) and 5′-deoxyadenosyl radical (dAdo•). It abstracts a substrate hydrogen to generate the radical intermediate (R•) and coproduct 5′-deoxyadenosine (dAdoH). (B) In RSPEs, the substrate is a precursor peptide core residue (*left*) where hydrogen abstraction is expected to generate a transient peptide radical intermediate (*in brackets, center*). An exchangeable hydrogen is incorporated into the D-amino acid product (*right*) enabling deuterium labeling when the reaction is carried out in D_2O.

initiate diverse chemical reactions (Fig. 2) (Broderick, Duffus, Duschene, & Shepard, 2014). The deoxyadenosine radical originates from homolytic cleavage of *S*-adenosylmethionine, instigated by a one-electron reduction of the universally conserved $[4Fe–4S]^{2+}$ cluster. Thus, ^2H-labeled substrates are subject to loss of deuterium by the hydrogen abstraction step, in agreement with the loss of mass observed in our RSPE isotope-labeled amino acid feeding experiments (Morinaka et al., 2014). This loss of deuterium from the labeled L-residue and gain of a different hydrogen in the product D-residue suggested that the backside hydrogen that effects epimerization likely comes from an exchangeable source. With this information in hand, we initiated efforts to design an inverse labeling experiment that could be carried out in vivo and would be applicable to all amino acids. Further support and motivation that these experiments would be successful came from in vitro studies on the aminoglycoside radical SAM epimerase NeoN (Kudo, Hoshi, Kawashima, Kamachi, & Eguchi, 2014), which showed stable incorporation of deuterium from epimerizations in D_2O. We hypothesized that peptidyl epimerizations by RSPEs would mirror this result. We therefore developed an orthogonal system that allows separate induction of the precursor protein substrate (e.g., PoyA) in H_2O medium, followed by induction of the epimerase (e.g., PoyD) in D_2O medium. A heterologous expression system with the precursor under control of the T7 promoter, induced by isopropyl β-D-1-thiogalactopyranoside (IPTG), and the epimerase independently under control of the pBAD promoter, induced by arabinose, was developed and tested.

As suspected, the RSPE-catalyzed epimerizations indeed resulted in gain of deuterium at sites of epimerization in D_2O, thus enabling rapid MS/MS mapping of the modified residues (Morinaka, Verest, Freeman, Gugger, & Piel, 2017). Using this orthogonal D_2O-based induction system (ODIS), we were able to analyze a number of native, nonnative, and engineered substrates to quickly identify the suite of epimerized products from coexpression experiments. The net results showed that the D-amino acid pattern is largely, but not exclusively, dictated by the sequence of the core peptide, while the order of epimerization is more dependent on the enzyme–leader pair (Morinaka et al., 2017). Furthermore, RSPEs were found to be generally highly promiscuous. The RSPEs tolerate well precursor substrates that are N-terminally His-tagged and that are optionally mutated to incorporate a nonnative protease cleavage site (e.g., factor Xa) at the leader–core junction, enabling facile purification and release of the core, respectively, for MS/MS analysis (Fuchs et al., 2016; Morinaka et al., 2017, 2014). The stepwise procedure is described here.

2. EQUIPMENT, CONSUMABLES, AND REAGENTS

2.1 Equipment

- Infors Multitron Pro Incubator Shaker with cooling
- Beckman Allegra 6R centrifuge
- Eppendorf Centrifuge 5415D
- Eppendorf Centrifuge 5417R
- Eppendorf Centrifuge 5810R
- Qsonica Q700 sonicator with 3.2-mm microprobe
- Dionex HPLC coupled to a QExactive mass spectrometer
- BTX ECM 399 Electroporation System
- Gel electrophoresis system for agarose gels
- Gel electrophoresis system for SDS-PAGE gels
- Genevac EZ-2 elite
- Biometra T3000 Thermal cycler

2.2 Consumables

- HPLC column: Phenomenex, Kinetex, C18-XB, 2.6 μm, 150 × 4.6 mm
- HPLC solvents: A = H_2O + 0.1% formic acid; B = CH_3CN + 0.1% formic acid
- Vivaspin 500, MWCO 5000 spin columns
- Strains of *E. coli*: DH5α and BL21(DE3) electrocompetent cells
- Plasmids and corresponding resistance
  - Novagen pET-28b(+) (kanamycin)
  - Novagen pCDFDuet-1 (spectinomycin)
  - Invitrogen pBAD/*Myc*-His A (ampicillin)
- Macherey-Nagel NucleoSpin Plasmid kit
- Macherey-Nagel NucleoSpin Gel and PCR Clean-up kit
- Macherey-Nagel Ni-NTA resin
- Thermo Fisher Scientific Phusion DNA Polymerase
- Restriction enzymes (*Nde*I, *Nco*I, *Xho*I, *Hin*dIII) from New England Biolabs
- Thermo Fisher Scientific T4 DNA Ligase
- Thermo Fisher Scientific FastAP Thermosensitive Alkaline Phosphatase
- Roti®-Nanoquant for protein quantitation
- Bio-Rad Poly-Prep® Chromatography Columns
- GE PD MiniTrap G-25 desalting columns
- Syringe filters, 0.2 μm, polyethersulfone (PES)

2.3 Reagents

- Lysogeny broth (LB) medium with appropriate antibiotics
- LB agar medium with appropriate antibiotics
- Terrific broth (TB) medium with appropriate antibiotics
- Isopropyl β-D-1-thiogalactopyranoside (IPTG, 1 *M* in H_2O, filter sterilized)
- 1000 × antibiotic solutions in H_2O (filter sterilized)
  - Kanamycin (25 mg/mL)
  - Spectinomycin (25 mg/mL)
  - Ampicillin (100 mg/mL)
- NPI-10 (1 L): 50 m*M* NaH_2PO_4, 300 m*M* NaCl, 10 m*M* imidazole, 10% glycerol—adjust pH to 8.0 using NaOH
- NPI-20 (1 L): 50 m*M* NaH_2PO_4, 300 m*M* NaCl, 20 m*M* imidazole, 10% glycerol—adjust pH to 8.0 using NaOH

- NPI-250 (1 L): 50 m*M* NaH_2PO_4, 300 m*M* NaCl, 250 m*M* imidazole, 10% glycerol—adjust pH to 8.0 using NaOH
- TB medium in D_2O: Prepared in 15-mL aliquots. 15 mL of TB medium in a 20-mL glass vial is evaporated in a Genevac EZ-2 Elite centrifugal concentrator at 60°C. The mixture is reconstituted in 2 mL D_2O by sonication, and the D_2O removed in the concentrator again. The mixture is reconstituted in 15 mL D_2O and sterile filtered using a 0.2-μm filter.
- Antibiotics and IPTG in D_2O. Prepared at the concentration above in D_2O and filter sterilized.

3. DETECTION OF EPIMERASE ACTIVITY FROM rSAM EPIMERASES

3.1 Discussion

Our investigations into the *poy* gene cluster revealed the impressive ability of the RSPE PoyD to catalyze the epimerization of 18 residues on the PoyA precursor peptide (Freeman et al., 2012). However, we soon realized that this system was particularly difficult to work with for several reasons. The yield of the epimerized precursor was sufficient for initial analytical studies, but posed challenges with respect to analysis of individual intermediates. This core peptide was significantly hydrophobic, gave somewhat broad peaks on the HPLC, and was difficult to completely remove from reversed-phase HPLC columns. Finally, other modifications from coexpressions with further *poy*-modifying enzymes introduced additional complications when trying to study each enzyme in detail.

At this point, we turned to a genome mining approach to survey a number of epimerases that (1) might offer a more robust and simplified system to study the nature of epimerization and (2) gain initial insights into the patterns and selectivity of epimerization. It was therefore unknown whether radical epimerization was limited to products with regular DL-patterns, as observed for polytheonamides, or could be used to generate a broader range of diastereomers.

Homologs from three cyanobacterial strains were selected, termed AvpD (*Anabaena variabilis* ATCC 29413), OspD (*Oscillatoria* sp. PCC 6506), and PlpD (*Pleurocapsa* sp. PCC 7319) (Morinaka et al., 2014). These originated from gene clusters encoding a proteusin precursor peptide as well as other presumed modifying enzymes (Fig. 1). The core region of the precursors and the sets of maturation genes found in the three gene clusters were highly dissimilar, indicating significant structural diversity among cyanobacterial proteusins. Developing a robust system for detection of activity and determination of the products from these pathways would open the door to further study of these enzymes as well as those in other pathways.

These studies resulted in the development of a two-step procedure relying on routinely used vectors (pET-28b and pCDFDuet-1) and *E. coli* BL21 (DE3) (Morinaka et al., 2017, 2014). The first step detects activity of the epimerase based on comparison of retention times of core peptides resulting from expression without and with the epimerase. The second step is for characterization of the products based on a deuterium-labeling assay described later. Although these assays are carried out under in vivo conditions, assays can be carried out under anaerobic in vitro conditions similar to those of Kudo et al. (2014).

We initiate our investigation of a precursor/epimerase pair by cloning precursor and epimerase genes into pET-28b and pCDFDuet-1, respectively. For example, the precursor gene *ospA* was cloned into pET-28b (at *Nde*I/*Hin*dIII restriction sites), which allows the precursor to be produced with an N-terminal His_6-tag ($NHis_6$). In addition, a Factor Xa site (IDGR, denoted Fx) is engineered at the interface of the leader and core to ease the release of the leader prior to liquid chromatography-mass spectrometry (LC-MS) analysis. The epimerase gene, *ospD*, is cloned as an untagged protein into the second multiple cloning site (MCS2) of pCDFDuet-1 (at *Nde*I/*Xho*I restriction sites). Two separate transformations are carried out with *E. coli* BL21(DE3). One sample introduces only the precursor, while the second transformation includes both the precursor and epimerase gene. The two transformants are subjected to protein expression with induction of proteins with IPTG and shaken overnight at 16°C. The $NHis_6$-precursors are purified by Ni-affinity chromatography, and the leader is cleaved off using Factor Xa. Unlike PoyA, for which all attempts to produce a soluble unmodified protein had failed, $NHis_6$-OspA-Fx expressed well and gave soluble protein irrespective of the presence of OspD.

Reversed-phase LC-MS analysis of core peptides cleaved from $NHis_6$-OspA-Fx and $NHis_6$-OspA-Fx + OspD expression experiments showed remarkably clean conversion into a new product in the coexpression (Figs. 3 and 4) with a different retention time and the same molecular formula, indicating irreversible action of the modifying enzyme. We characterized these products by Marfey's-type analysis (Marfey, 1984) along

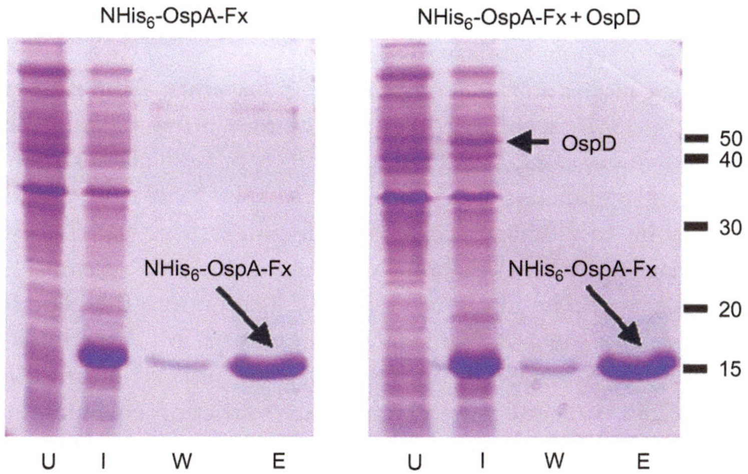

Fig. 3 SDS-PAGE gels for expression and Ni^{2+}-affinity purification of NHis$_6$-OspA-Fx ± OspD. *E*, elution; *I*, induced; *U*, uninduced; *W*, wash.

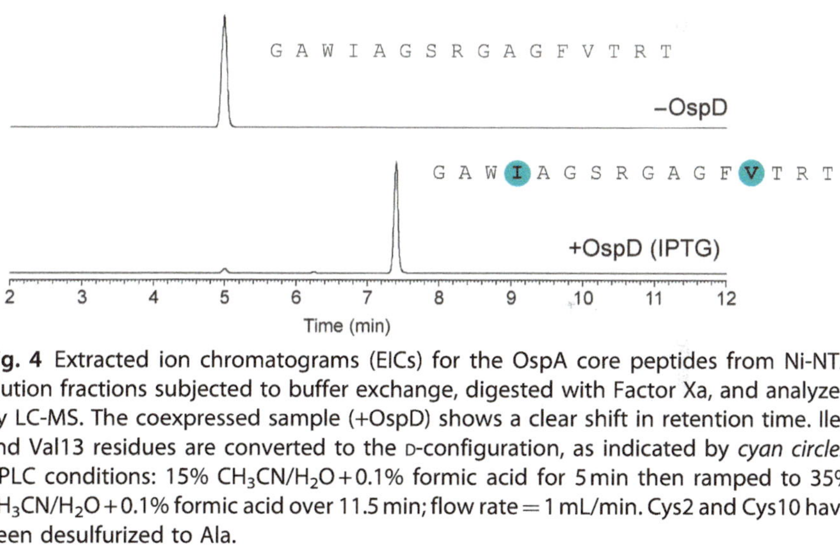

Fig. 4 Extracted ion chromatograms (EICs) for the OspA core peptides from Ni-NTA elution fractions subjected to buffer exchange, digested with Factor Xa, and analyzed by LC-MS. The coexpressed sample (+OspD) shows a clear shift in retention time. Ile4 and Val13 residues are converted to the D-configuration, as indicated by *cyan circles*. HPLC conditions: 15% CH_3CN/H_2O + 0.1% formic acid for 5 min then ramped to 35% CH_3CN/H_2O + 0.1% formic acid over 11.5 min; flow rate = 1 mL/min. Cys2 and Cys10 have been desulfurized to Ala.

with comparison to a peptide standard prepared by chemical synthesis. However, our ODIS method described in the next step avoids the tedious characterization by these methods.

3.2 Protocol

We will describe the protocol used for precursor OspA and epimerase OspD. OspA is cloned with an NHis$_6$-tag and a Factor Xa site (IDGR) at the interface of leader and core. This workflow can be generally applied to any precursor/epimerase pair.

1. *Preparation of precursor and epimerase gene fragments.* Amplify genes of interest by PCR according to manufacturer's protocol using Phusion DNA polymerase and either genomic DNA or fresh cells as a template. Purify gene fragments on a 1% agarose gel, excise bands, use gel and PCR clean-up kit to extract purified gene fragment from gel. Digest with appropriate restriction enzymes, and use gel and PCR clean-up kit to purify reaction. The inserts are ready for ligation into the plasmids.
2. *Preparation of plasmid.* A 5-mL overnight culture in LB medium of the empty vector in DH5α is purified using the plasmid purification kit. The plasmid is digested with appropriate restriction enzymes and gel-purified and extracted using the gel and PCR clean-up kit. The digested vector is dephosphorylated with FastAP thermosensitive alkaline phosphatase and purified with the gel and PCR clean-up kit. The vector is now ready for ligation.
3. *Ligation of gene into cut vector and transformation into cloning strain.* The insert is ligated into the vector prepared above using T4 DNA ligase using the manufacturer's protocol and subsequently transformed via electroporation into *E. coli* DH5α. Colonies are picked and grown overnight in 5-mL LB medium with appropriate antibiotic. The following day, the cultures are purified using the plasmid purification kit and the purified plasmids sent for sequencing to verify insert of gene. The plasmids carrying genes of interest are now ready for transformation into an expression strain.
4. *Transformation of expression strain.* $NHis_6$-precursor/pET-28b alone and with epimerase/pCDFDuet-1 are introduced into *E. coli* BL21(DE3) via electroporation and plated on LB agar containing appropriate antibiotic(s) and incubated at 37°C overnight. The following day, colonies are picked and cultures of LB medium with appropriate antibiotics are inoculated.
5. *Protein expression.* In a typical experiment, 30-mL TB medium with appropriate antibiotic(s) in a 100-mL Erlenmeyer flask is inoculated with 300 μL of overnight culture. This is shaken at 250 rpm until the optical density at 600 nm is approximately 1.5–2.0. This is usually around 2.5–3 h. At this point, an aliquot is taken for the uninduced SDS-PAGE sample. The cultures are then cooled on ice for 30 min and afterward 30 μL IPTG (1 *M*) is added and the cultures are placed in an incubator shaker set to 250 rpm and 16°C. The cultures are shaken for 18–24 h. At this point, an aliquot is taken for an induced SDS-PAGE sample.

6. *Protein purification.* NHis$_6$-precursor proteins are purified in parallel with NHis$_6$-precursor + epimerase coexpressions according to the manufacturer's procedure. The cells are collected by centrifugation (5000 × *g*, 10 min) and the supernatant is removed. At this point, the samples are kept cool using ice or in a cold room. The cell pellet is reconstituted in NPI-10 buffer (4 mL/g cell pellet) and cells are lysed by sonication (amplitude 20, 10 cycles of 10 s on and off). The debris is removed by centrifugation (20,000 × *g*, 30 min), and 200 μL Ni-NTA resin is added. The samples are gently rocked or rotated at 4°C. The resin is collected by centrifugation (850 × *g*, 30 min), and the supernatant is carefully poured off. The resin is transferred to a Poly-Prep® Chromatography Column, allowed to settle, and then the remaining supernatant allowed to flow through. The resin is washed with NPI-20 buffer (2 × 1 mL). The proteins are eluted with NPI-250 buffer (2 × 500 μL) and maintained on ice. Expression and purification fractions were diluted with SDS-sample loading dye (e.g., uninduced, induced, Ni-NTA wash, and Ni-NTA elution) and separated over a 15% acrylamide SDS-PAGE gel. Proteins were visualized with Coomassie stain, showing a clear induction band and good yields of protein in the elution fraction, Fig. 3.
7. *Protease digestion.* The protein elution fraction is buffer exchanged using PD MiniTrap G-25 desalting column into 100 m*M* Tris or sodium phosphate containing 1 m*M* calcium chloride, pH 8. The protein concentration of these solutions is measured by the Roti®-Nanoquant Protein quantitation assay using the manufacturer's protocol. Protease of interest is added, typically at a ratio of 20:1 precursor to enzyme (w/w). The sample is allowed to digest at room temperature or 37°C for at least 4 h and up to overnight.
8. *Optional removal of disulfide bond(s).* In the case of OspA and many other core peptides, a disulfide bond is always present in the starting material and product, which limits the observed fragments for characterization by MS/MS. Therefore, the disulfide can either be reduced and alkylated or desulfurized. Core peptides subjected to radical desulfurization were our favored method (Wan & Danishevsky, 2007). Following in vitro cleavage with Factor Xa, reaction mixtures were desalted using Phenomenex Strata-X Polymeric Reversed-Phase SPE (30 mg) according to the manufacturer's protocol. To the cleaved peptide were added 100 μL of water, 100 μL of 0.5 *M* bond-breaker® TCEP solution (Pierce), 10 μL of ethanethiol, and 5 μL of 2,2′-azobis[2-(2-imidazolin-2-yl)propane] dihydrochloride (0.1 *M* in water). The reaction mixture was shaken or

stirred for 2 h at 50°C. The mixtures were cooled to room temperature and desalted using Phenomenex Strata-X Polymeric Reversed-Phase SPE (30 mg) according to the manufacturer's protocol. The solvent was evaporated using a Genevac concentrator.

9. *Detection of activity*. For detection of activity, we analyze the digested $NHis_6$-precursor from precursor alone and precursor + epimerase coexpressions. In our specific example, the analyzed samples are $NHis_6$-OspA-Fx vs $NHis_6$-OspA + OspD, respectively. The HPLC gradient is 15% CH_3CN/H_2O + 0.1% formic acid for 5 min then ramped to 35% CH_3CN/H_2O + 0.1% formic acid over 11.5 min. To simplify analysis, the mass for the precursor peptide is extracted to generate an extracted ion chromatogram (EIC). When epimerized, there is usually a clear shift in retention time of the core peptide when coexpressed with the epimerase and D-amino acids were installed (Fig. 4), as obvious from the comparison of the two EICs. Now that epimerization has been inferred by a shift in retention time, the materials are prepared for the ODIS experiment.

4. ORTHOGONAL D_2O-BASED INDUCTION SYSTEM

4.1 Discussion

Standard methods for D-amino acid analysis (i.e., Marfey's method, chiral HPLC, etc.) rely on peptide hydrolysis, during which positional information is lost. To obtain this information for products resulting from RSPEs, we previously reported a selective labeling strategy that involves feeding of fully deuterated amino acids to a quintuple transaminase knockout strain of *E. coli* cultured in synthetic medium (Kegler et al., 2014; Morinaka et al., 2014). For a range of deuterated amino acids incorporated into cores, we observed selective single deuterium–hydrogen exchange at the corresponding epimerized position. These results demonstrated that the hydrogen effecting epimerization comes from an exchangeable source and was also consistent with D_2O-based in vitro assays conducted with NeoN, an aminoglycoside RSPE reported by Kudo et al. (2014). Although our method permits the localization of D-residues in a single LC-MS/MS run, amino acids other than Ile, Leu, Val, Tyr, and Lys are susceptible to loss of label at the α-position (Kegler et al., 2014), suggesting that inactivation of additional transaminase genes is required to make this method universal.

To obviate this drawback and the need for selective feeding and cultivation in synthetic medium, we developed the ODIS method that was constructed as follows. Precursor and epimerase genes were cloned into separate expression plasmids that utilize orthogonal induction systems, that is, IPTG-inducible pCDFDuet-1 for precursor genes and L-arabinose-inducible pBAD/*Myc*-His A for epimerase genes. We found that this combination provided a more uniformly labeled product presumably because of tighter control of the epimerase in the pBAD/*Myc*-His A vector.

After cotransforming the expression host *E. coli* BL21(DE3) with the plasmids, cultures were first grown in TB medium and induced with IPTG for precursor production. Subsequently, cells were collected and transferred to TB medium prepared in D_2O, followed by induction of the epimerase gene with L-arabinose. At 16 h postinduction, cells were lysed, and NHis$_6$-tagged proteins were subjected to Ni-affinity purification, in vitro cleavage at the leader–core interface with a protease (Factor Xa, GluC, or the protease PoyH from the *poy* gene cluster), and LC-MS/MS analysis of core peptides (Figs. 5–7).

4.2 Procedure

1. *Cloning of constructs for ODIS experiment using standard restriction enzyme and T4 DNA ligase cloning.* The precursor gene is cloned into the multiple cloning site 1 (MCS1) of pCDFDuet-1. The insert from NHis$_6$-precursor/pET-28b construct derived from Section 3.2, step 3 is obtained by plasmid extraction, digestion with *Nco*I and *Hin*dIII, and gel purification of the insert. This insert is ligated into pCDFDuet-1, which has been prepared as above, digesting with restriction enzymes *Nco*I and *Hin*dIII. The insert of the epimerase gene follows a similar protocol and is inserted into pBAD/*Myc*-His A using the *Nco*I/*Xho*I cloning sites. These constructs are ready for transformation into the expression strain.
2. *Transformation into expression strain.* NHis$_6$-precursor/pCDFDuet-1 is transformed alone and together with epimerase/pBAD/*Myc*-His A into *E. coli* BL21(DE3). Same procedure as Section 3.2, step 4.
3. *Expression of proteins in ODIS experiment.* Two separate 50-mL Falcon tubes containing TB medium (10 mL) containing appropriate antibiotics were inoculated with overnight culture (100 μL), and shaken (250 rpm, 37°C) until an OD of 1.5–2.0. Cultures were cooled on ice for 30 min, induced with 1 m*M* IPTG (10 μL, 1 *M*), and shaken (250 rpm, 16°C) for 16 h. The cultures were centrifuged (10 min, 3220 × *g*) and

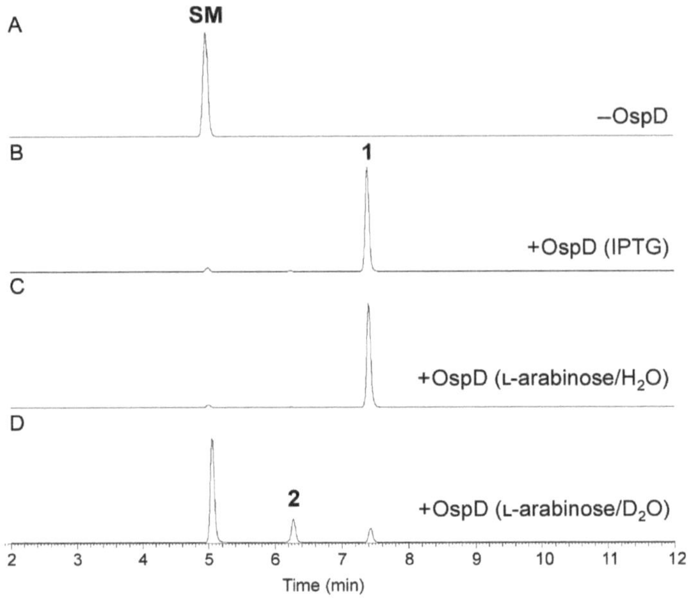

Fig. 5 LC-MS analysis of ODIS experiment. Extracted ion chromatograms for NHis$_6$-OspA-Fx cleaved with Factor Xa derived from: (A) NHis$_6$-OspA-Fx/pET-28b; (B) NHis$_6$-OspA-Fx/pET-28b + OspD/pCDFDuet-1, induced with IPTG; (C) NHis$_6$-OspA-Fx/pCDFDuet-1 + OspD/pBAD/*Myc*-His A induced with IPTG then L-arabinose in H_2O; (D) NHis$_6$-OspA-Fx/pCDFDuet-1 + OspD/pBAD/*Myc*-His A induced with IPTG then L-arabinose in D_2O. Product **1** is the major product observed under standard conditions with two epimerized residues at Ile4 and Val13 (B). Product **2** contains one epimerized residue at Ile4 only.

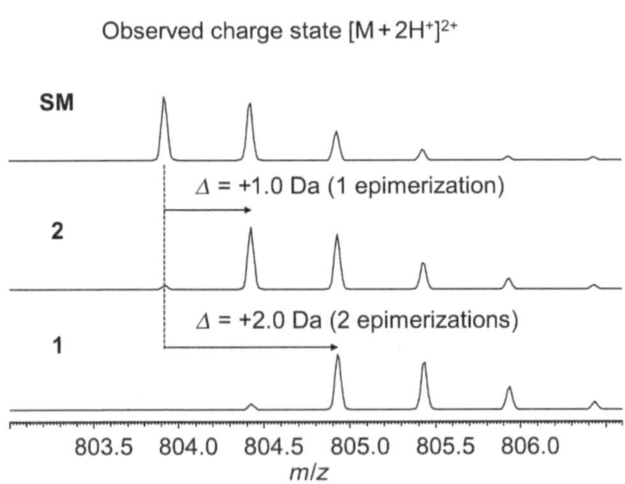

Fig. 6 Mass spectra derived from LC-MS of ODIS experiment (Fig. 5D). Product **2** shows exchange of one hydrogen for one deuterium and contains one epimerized residue. Product **1** shows the exchange of two hydrogens for two deuteriums and contains two epimerized residues.

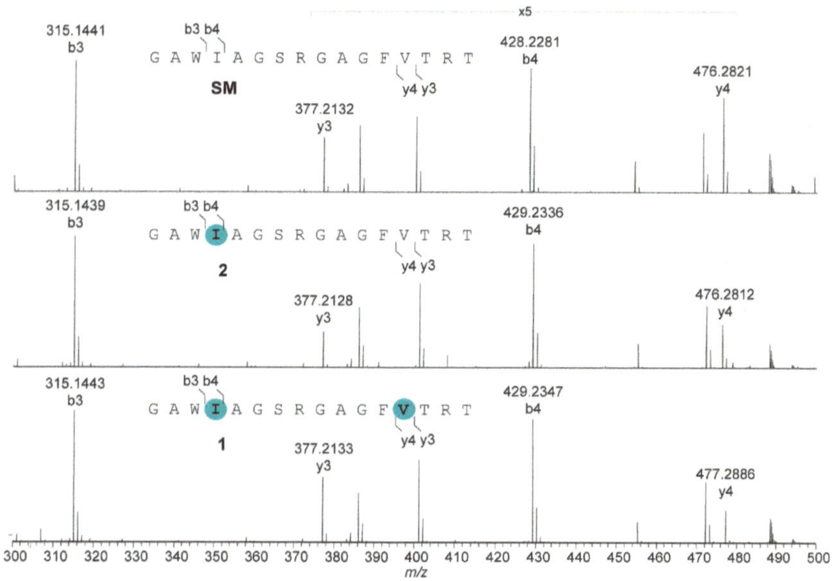

Fig. 7 Localization of epimerized residues. LC-MS/MS spectra derived from ODIS experiments. The corresponding mass shifts observed for core peptides can be localized to Ile4 in product **2** and Ile4 and Val13 in product **1**. Cys2 and Cys10 have been desulfurized to Ala.

the supernatant removed. One sample was kept for induction of the precursor only. In our experience, negligible epimerized products are observed from this expression and were comparable to the protein obtained from NHis$_6$-OspA-Fx/pET-28b. The second sample is rinsed with TB medium (2 × 25 mL) to remove residual IPTG. The rinsed cell pellet was reconstituted in TB medium in D_2O (10 mL) containing spectinomycin (10 μL, 25 mg/mL in D_2O), ampicillin (10 μL, 100 mg/mL in D_2O), and L-arabinose (100 μL, 20% w/v in D_2O) and shaken (250 rpm, 16°C) for 24 h.

4. The proteins are purified, digested, and desulfurized according to Section 3.2, steps 6, 7, and 8, respectively.
5. *Detection of labeled products.* For detection of incorporation of deuterium, the samples are run under the same HPLC conditions as Fig. 4 with a gradient of 15% CH_3CN/H_2O + 0.1% formic acid for 5 min then ramped to 35% CH_3CN/H_2O + 0.1% formic acid over 11.5 min (Fig. 5). This is carried out in full scan mode, and the mass spectra deriving from the core peptides of starting material are compared to those of the products

(Fig. 6). Products **1** and **2** contain one and two epimerized residues, respectively. Product **2** is the major product observed under standard (IPTG, H_2O) conditions.

6. *Localization of deuterium incorporation/epimerized residues.* The masses obtained from step 5 are used in a subsequent LC-MS run in parallel reaction monitoring mode. This allows for targeted isolation of mass peaks of interest to obtain fragmentation data. The highest peak in the isotopic pattern for the starting material (SM), products **1** and **2** are used in an inclusion list with a time window to capture MS/MS data for each peak of interest. In this case, a normalized collision energy of 30 eV is used for the $[M+3H]^{+3}$ m/z. The data are acquired and shown in Fig. 7.

Note: Although not explicitly tested for all precursor + epimerase pairs, our data suggest that the precursor/pCDFDuet-1 and epimerase/pBAD/*Myc*-His A plasmid combination used in Section 4 for ODIS analyses could also be used for the initial detection of epimerase activity from rSAM epimerases (Section 3) where the coexpression culture would be induced with both IPTG and arabinose in parallel. This would prevent the need for additional cloning steps in the two experimental stages outlined in Sections 3 and 4.

5. CONCLUSION

RSPEs represent powerful biochemical tools to install D-amino acids into ribosomally synthesized peptides. A fundamental challenge when studying these enzymes is the characterization of products that occur without a change in molecular formula. Exploiting the nature of epimerization, we have developed an efficient two-step strategy to assign the configuration of products by LC-MS/MS. This method allows characterization of any epimerized amino acid resulting from RSPEs. The materials used in this protocol consist of routinely used medium, vectors, and expression strains. The only drawback is the necessity to use deuterium oxide during the characterization step, which can become costly for larger cultures or high-throughput analyses. We would like to mention it should possible to arrive at the same results by cloning precursor genes directly into pCDFDuet-1 and epimerase into pBAD/*Myc*-His A (Fig. 5B and C), however, we have not undertaken extensive comparative studies. We describe this method in detail in this *Methods in Enzymology* chapter and hope this information can be used to guide further exploration of these enzymes.

We envision that ODIS can be utilized to map additional types of isobaric posttranslational modifications as long as the enzyme meets the requirement that it stably incorporates hydrogen into the product from an exchangeable position. For example, a more recently discovered distinct family of RSPEs from epipeptide RiPPs could be adapted to ODIS since in vitro studies showed deuterium incorporation from D_2O reactions (Benjdia, Guillot, Ruffié, Leprince, & Berteau, 2017).

ACKNOWLEDGMENTS

B.I.M. is a fellowship recipient of the Alexander von Humboldt foundation. A.L.V. is the recipient of an ETH Zurich Postdoctoral Fellowship.

REFERENCES

Arnison, P. G., et al. (2013). Ribosomally synthesized and post-translationally modified peptide natural products: Overview and recommendations for a universal nomenclature. *Natural Product Reports*, *30*(1), 108–160. https://doi.org/10.1039/c2np20085f.

Benjdia, A., Guillot, A., Ruffié, P., Leprince, J., & Berteau, O. (2017). Post-translational modification of ribosomally synthesized peptides by a radical SAM epimerase in *Bacillus subtilis*. *Nature Chemistry*, *9*, 698. https://doi.org/10.1038/nchem.2714.

Bhushan, R., & Bruckner, H. (2011). Use of Marfey's reagent and analogs for chiral amino acid analysis: Assessment and applications to natural products and biological systems. *Journal of Chromatography. B, Analytical Technologies in the Biomedical and Life Sciences*, *879*(29), 3148–3161. https://doi.org/10.1016/j.jchromb.2011.05.058.

Broderick, J. B., Duffus, B. R., Duschene, K. S., & Shepard, E. M. (2014). Radical S-adenosylmethionine enzymes. *Chemical Reviews*, *114*(8), 4229–4317. https://doi.org/10.1021/cr4004709.

Freeman, M. F., Helf, M. J., Bhushan, A., Morinaka, B. I., & Piel, J. (2017). Seven enzymes create extraordinary molecular complexity in an uncultivated bacterium. *Nature Chemistry*, *9*(4), 387–395. https://doi.org/10.1038/nchem.2666.

Freeman, M. F., Vagstad, A. L., & Piel, J. (2016). Polytheonamide biosynthesis showcasing the metabolic potential of sponge-associated uncultivated 'Entotheonella' bacteria. *Current Opinion in Chemical Biology*, *31*(Suppl. C), 8–14. https://doi.org/10.1016/j.cbpa.2015.11.002.

Freeman, M. F., et al. (2012). Metagenome mining reveals polytheonamides as post-translationally modified ribosomal peptides. *Science*, *338*(6105), 387–390. https://doi.org/10.1126/science.1226121.

Fuchs, S. W., et al. (2016). A lanthipeptide-like N-terminal leader region guides peptide epimerization by radical SAM epimerases: Implications for RiPP evolution. *Angewandte Chemie (International Ed. in English)*, *55*(40), 12330–12333. https://doi.org/10.1002/anie.201602863.

Kegler, C., Nollmann, F. I., Ahrendt, T., Fleischhacker, F., Bode, E., & Bode, H. B. (2014). Rapid determination of the amino acid configuration of xenotetrapeptide. *ChemBioChem*, *15*(6), 826–828. https://doi.org/10.1002/cbic.201300602.

Kudo, F., Hoshi, S., Kawashima, T., Kamachi, T., & Eguchi, T. (2014). Characterization of a radical S-adenosyl-L-methionine epimerase, NeoN, in the last step of neomycin B biosynthesis. *Journal of the American Chemical Society*, *136*(39), 13909–13915. https://doi.org/10.1021/ja507759f.

Larsen, M., Trelle, M., Thingholm, T., & Jensen, O. (2006). Analysis of posttranslational modifications of proteins by tandem mass spectrometry. *BioTechniques*, *40*(6), 790–798. https://doi.org/10.2144/000112201.

Marfey, P. (1984). Determination of D-amino acids. II. Use of a bifunctional reagent, 1,5-difluoro-2,4-dinitrobenzene. *Carlsberg Research Communications*, *49*, 591–596. https://doi.org/10.1007/BF02908688.

Morinaka, B. I., Verest, M., Freeman, M. F., Gugger, M., & Piel, J. (2017). An orthogonal D_2O-based induction system that provides insights into D-amino acid pattern formation by radical *S*-adenosylmethionine peptide epimerases. *Angewandte Chemie (International Ed. in English)*, *56*(3), 762–766. https://doi.org/10.1002/anie.201609469.

Morinaka, B. I., et al. (2014). Radical *S*-adenosyl methionine epimerases: Regioselective introduction of diverse D-amino acid patterns into peptide natural products. *Angewandte Chemie (International Ed. in English)*, *53*(32), 8503–8507. https://doi.org/10.1002/anie.201400478.

Ollivaux, C., Soyez, D., & Toullec, J. Y. (2014). Biogenesis of D-amino acid containing peptides/proteins: Where, when and how? *Journal of Peptide Science*, *20*(8), 595–612. https://doi.org/10.1002/psc.2637.

Ortega, M. A., & van der Donk, W. A. (2016). New insights into the biosynthetic logic of ribosomally synthesized and post-translationally modified peptide natural products. *Cell Chemical Biology*, *23*(1), 31–44. https://doi.org/10.1016/j.chembiol.2015.11.012.

Walsh, C. T. (2014). Blurring the lines between ribosomal and nonribosomal peptide scaffolds. *ACS Chemical Biology*, *9*(8), 1653–1661. https://doi.org/10.1021/cb5003587.

Wan, Q., & Danishevsky, S. (2007). Free-radical-based, specific desulfurization of cysteine: A powerful advance in the synthesis of polypeptides and glycopolypeptides. *Angewandte Chemie (International Ed. in English)*, *46*(48), 9248–9252. https://doi.org/10.1002/anie.200704195.

Wilson, M. C., et al. (2014). An environmental bacterial taxon with a large and distinct metabolic repertoire. *Nature*, *506*, 58. https://doi.org/10.1038/nature12959.

FURTHER READING

Bode, H. B., et al. (2012). Determination of the absolute configuration of peptide natural products by using stable isotope labeling and mass spectrometry. *Chemistry*, *18*(8), 2342–2348. https://doi.org/10.1002/chem.201103479.

Ewing, M. A., Wang, J., Sheeley, S. A., & Sweedler, J. V. (2008). Detecting D-amino acid-containing neuropeptides using selective enzymatic digestion. *Analytical Chemistry*, *80*, 2874–2880.

Hamada, T., Matsunaga, S., Yano, G., & Fusetani, N. (2005). Polytheonamides A and B highly cytotoxic linear polypeptides with unprecedented structural features. *Journal of the American Chemical Society*, *127*(12), 110–118. https://doi.org/10.1021/ja045749e.

Heck, S. D., Faraci, W. S., Kelbaugh, P. R., Saccomano, N. A., Thadeio, P. F., & Volkmann, R. A. (1996). Posttranslational amino acid epimerization: Enzyme-catalyzed isomerization of amino acid residues in peptide chains. *Proceedings of the National Academy of Sciences of the United States of America*, *93*(9), 4036–4039.

Jeanne Dit Fouque, K., et al. (2017). Fast and effective ion mobility–mass spectrometry separation of D-amino-acid-containing peptides. *Analytical Chemistry*, *89*(21), 11787–11794. https://doi.org/10.1021/acs.analchem.7b03401.

Jia, C., Lietz, C. B., Yu, Q., & Li, L. (2014). Site-specific characterization of D-amino acid containing peptide epimers by ion mobility spectrometry. *Analytical Chemistry*, *86*(6), 2972–2981. https://doi.org/10.1021/ac4033824.

Livnat, I., et al. (2016). A D-amino acid-containing neuropeptide discovery funnel. *Analytical Chemistry*, *88*(23), 11868–11876. https://doi.org/10.1021/acs.analchem.6b03658.

Miyamoto, T., Sekine, M., Ogawa, T., Hidaka, M., Homma, H., & Masaki, H. (2010). Detection of D-amino acids in purified proteins synthesized in *Escherichia coli*. *Amino Acids*, *38*(5), 1377–1385. https://doi.org/10.1007/s00726-009-0348-2.

Mohimani, H., & Pevzner, P. A. (2016). Dereplication, sequencing and identification of peptidic natural products: From genome mining to peptidogenomics to spectral networks. *Natural Product Reports*, *33*(1), 73–86. https://doi.org/10.1039/c5np00050e.

Sachon, E., Clodic, G., Galanth, C., Amiche, M., Ollivaux, C., Soyez, D., et al. (2009). D-Amino acid detection in peptides by MALDI-TOF-TOF. *Analytical Chemistry*, *81*(11), 4389–4396. https://doi.org/10.1021/ac9002886.

CHAPTER EIGHT

Cobalamin-Dependent *C*-Methyltransferases From Marine Microbes: Accessibility via Rhizobia Expression

Michael F. Freeman[1]
Department of Biochemistry, Molecular Biology, and Biophysics & BioTechnology Institute, University of Minnesota-Twin Cities, St. Paul, MN, United States
[1]Corresponding author: e-mail address: mffreema@umn.edu

Contents

Methods in Enzymology, Volume 604
ISSN 0076-6879
https://doi.org/10.1016/bs.mie.2018.02.013

Abstract

Cobalamin-dependent radical *S*-adenosylmethionine (rSAM) methyltransferases catalyze chemically challenging methylation reactions on diverse natural products at unactivated carbon centers. In vivo reconstitution and biosynthetic studies of natural product gene clusters encoding these enzymes are often severely limited by ineffective heterologous expression hosts, including the otherwise versatile *Escherichia coli*. In this chapter, we describe the use of rhizobia bacteria as effective expression hosts for cobalamin-dependent rSAM *C*-methyltransferases. We chose the natural product pathway encoding the heavily modified cytotoxic peptides, the polytheonamides, as our model pathway due to the presence of two methyltransferases responsible for a total of 17 *C*-methylations. Detailed protocols are given for vector construction, transformation, and heterologous expression in *Rhizobium leguminosarum* bv. *viciae* 3841. Additional methods pertaining to analytical separation and mass spectrometric analysis of modified peptides are also entailed. As genomics continues to uncover new enzymes and pathways from unknown and uncultivated microbes, use of metabolically distinct heterologous expression hosts like rhizobia will be a necessary tool to unravel the catalytic and metabolic diversity of marine microbial life.

ABBREVIATIONS

EIC	extracted ion chromatogram
FA	formic acid
HCD	high-energy collisional dissociation
LB	lysogeny broth
LC–ESI–MS/MS	ultra-high pressure liquid chromatography coupled with electrospray ionization tandem mass spectrometry
MCS	multiple cloning site
Me	methylation
MeCN	acetonitrile
m/z	mass-to-charge ratio
nanoLC	nanoscale liquid chromatography
NCE	normalized collision energy
PCA	polymerase cycling assembly
PCR	polymerase chain reaction
***Rlv*3841**	*Rhizobium leguminosarum* bv. *viciae*
rSAM	radical *S*-adenosylmethionine
TY	tryptone-yeast extract

1. INTRODUCTION

Cobalamin-dependent (vitamin B_{12}) *C*-methyltransferases comprise a recently discovered enzyme class in the rapidly expanding radical *S*-adenosylmethionine (rSAM) superfamily (Zhang, van der Donk, & Liu, 2012). These B_{12}-dependent enzymes catalyze radical-mediated methyl transfer to unactivated sp^2- or sp^3-hybridized carbon centers (or phosphorus) on a wide variety of small-molecule and peptide natural products (Zhou, Alkhalaf, de Los Santos, & Challis, 2016). The enzymes harbor a conserved $CX_{3-7}CX_2C$ motif shown to coordinate a [4Fe–4S] iron–sulfur cluster that likely donates an electron for the reductive cleavage of SAM to produce a 5′-deoxyadenosyl radical. The 5′-deoxyadenosyl radical abstracts a hydrogen from the target carbon atom on the substrate. Methylcobalamin, formed by cobal(I)amin methyl abstraction from another SAM molecule, then donates the methyl group to the activated substrate. A recent set of in vitro characterizations have begun to reveal the diversity of mechanisms and substrates encompassed by this family of enzymes (Benjdia et al., 2015; Marous et al., 2015; Parent et al., 2016; Werner et al., 2011). One B_{12}-dependent rSAM class member was observed to be not involved in methyl transfer, but in nucleoside ring contraction during oxetanocin biosynthesis (Bridwell-Rabb, Zhong, Sun, Drennan, & Liu, 2017).

Marine ecosystems, highlighted by sponge and tunicate symbioses, are a treasure trove of diverse and distinct bioactive metabolites (Blunt et al., 2018). As more metagenomes are being sequenced and putative encoded pathways identified, B_{12}-dependent rSAM *C*-methyltransferases are increasingly being linked to the biosynthesis of marine-derived natural products (Fig. 1). Most notably, the cytotoxic membrane-spanning peptides the polytheonamides are posttranslationally tailored with 17 *C*-methylations at nearly all unactivated sp^3 carbon centers (Freeman et al., 2012; Hamada et al., 2010). Isolated from the Japanese sponge *Theonella swinhoei* (Hamada, Sugawara, Matsunaga, & Fusetani, 1994a, 1994b) and produced by the uncultivated bacterial symbiont "*Candidatus* Entotheonella factor" (Wilson et al., 2014), polytheonamides were found to be synthesized, in part, by two B_{12}-dependent rSAM *C*-methyltransferases encoded in their biosynthetic gene cluster (Freeman, Helf, Bhushan, Morinaka, & Piel, 2017).

Much like polytheonamides, many putative natural product pathways discovered from either uncultivated or genetically intractable microbes ultimately require heterologous expression to probe the biosynthesis of the encoded metabolite in entirety. Hindrances expressing gene clusters encoding B_{12}-dependent *C*-methyltransferases is that there are no cases thus far of in vivo activity achieved with these enzymes in the otherwise robust laboratory strain, *Escherichia coli*. In spite of this, *E. coli* is an adequate host for producing soluble B_{12}-dependent *C*-methyltransferases that can be

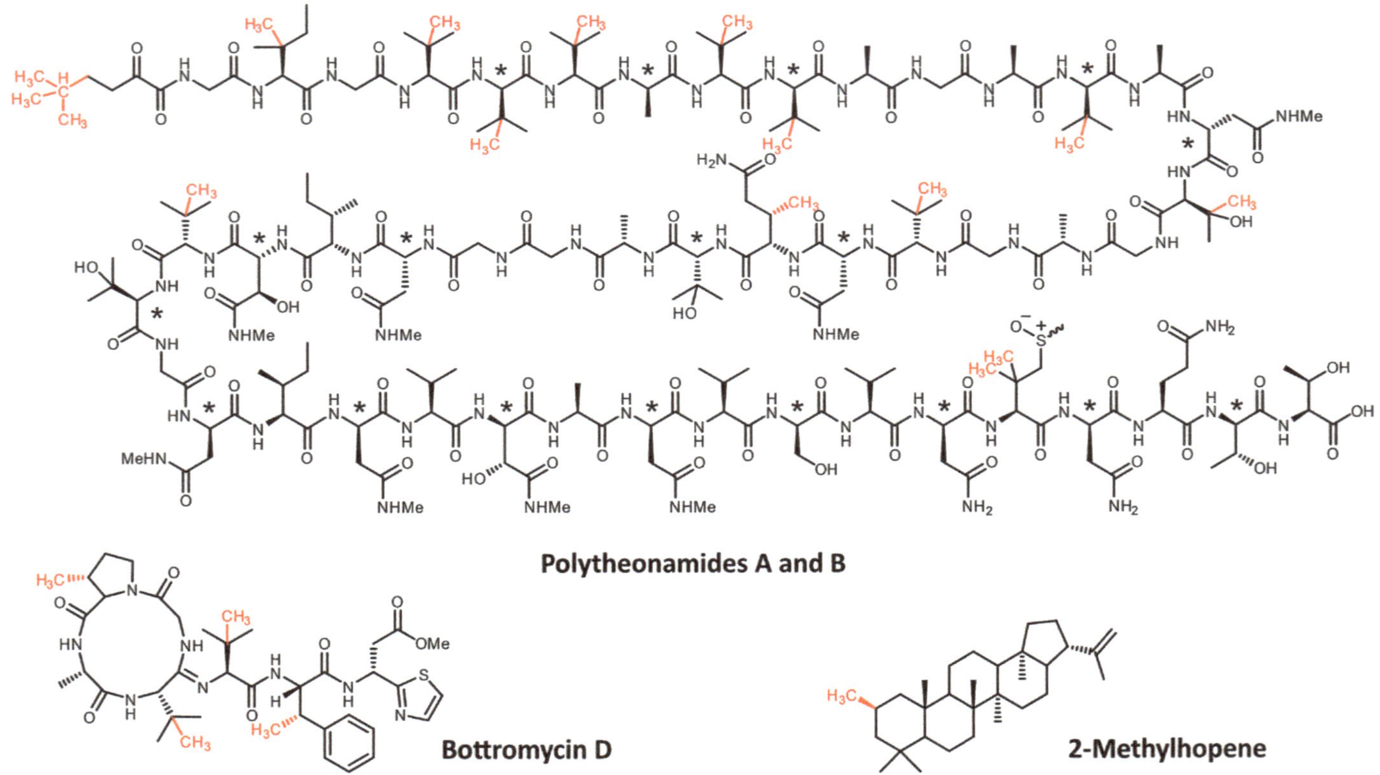

Fig. 1 Structures of marine-derived metabolites methylated (denoted as $CH_{(3)}$) by B_{12}-dependent rSAM methyltransferases. Polytheonamides A and B (differing the configuration of the sulfoxide) are ribosomally synthesized and posttranslationally modified peptides (RiPPs) that harbor 17 *C*-methylations and an additional 29 posttranslational modifications (PTMs) including 18 L-to-D epimerizations (marked with *asterisks*). Bottromycins are RiPPs produced by *Streptomyces* spp. and have been isolated from marine ascidians (Hou et al., 2012) as well as from terrestrial sources (Crone et al., 2016). Methylated hopanoids are produced by *Rhodopseudomonas palustris* isolated from marine coastal sediments (Larimer et al., 2004; Welander, Coleman, Sessions, Summons, & Newman, 2010).

reconstituted in vitro (Parent et al., 2016). While the precise reason for this discrepancy is as-of-yet unknown, possibilities include the strict regulation and low levels of cobalamin within *E. coli* (Di Girolamo, Kadner, & Bradbeer, 1971), as well as possible differences in iron–sulfur cluster maintenance among different organisms (Masepohl et al., 1992). In this chapter, we detail the use of *Rhizobium leguminosarum* bv. *viciae* (*Rlv*3841) as a new and effective heterologous host for natural product gene clusters containing B_{12}-dependent rSAM *C*-methyltransferases. Using the polytheonamide gene cluster as our model system, we detail host and vector selection considerations, vector design, and an effective method for plasmid conjugal transfer into *Rlv*3841. In addition, we describe expression conditions, in-depth sample preparation, and mass spectrometry-based identification strategies used to identify the activity profiles for the polytheonamide *C*-methyltransferases PoyB and PoyC. We hope this work facilitates the discovery and characterization of other cobalamin-dependent natural product pathways through appropriate choice and use of suitable heterologous hosts such as rhizobia.

2. HETEROLOGOUS HOST AND VECTOR CONSIDERATIONS

2.1 Host Choice

Host choice for heterologous expression is perhaps the first and most important consideration for the effective study of natural product biosynthesis (Zhang, Boghigian, Armando, & Pfeifer, 2011). Metabolic precursor pools, cofactor availability, accessibility to molecular biological tools, ease of genetic manipulation, growth rates, and maximal cell densities are some of the most important factors to consider when selecting a heterologous host. Owing to many of the mentioned complications, expressions of natural products in closely related microbes like actinobacteria have often been successful without further genetic manipulation, including clusters encoding B_{12}-dependent rSAM *C*-methyltransferases (Kamigiri, Hidaka, Imai, Murakami, & Seto, 1992).

Although not originally known at its discovery (Freeman et al., 2012), the polytheonamide gene cluster was soon shown to be encoded by the uncultivated *T. swinhoei* bacterial symbiont "*Candidatus* Enthotheonella swinhoei" (Wilson et al., 2014). This large and morphologically distinct Gram-negative bacterium is the first member of a newly proposed phylum termed "Tectomicrobia," which does not have a single cultivated representative (Hug et al., 2016). Therefore, a suitable expression host for the polytheonamide gene cluster posed a distinct challenge. After successfully demonstrating in vivo activity of the rSAM epimerase PoyD and other enzymes modifying the polytheonamide precursor protein PoyA (Freeman et al., 2012), we extensively tested *E. coli* for functional coexpressions with the B_{12}-dependent rSAM *C*-methyltransferases PoyB and PoyC. Our initial criterion was to use a Gram-negative host like Entotheonella, since the mechanism of native host resistance to pore formation of the toxin is currently unknown. Coexpression of the *Azotobacter vinelandii isc* operon for [4Fe–4S] iron–sulfur cluster insertion and assembly, supplementation of hydroxocobalamin, and attempts to increase intracellular concentrations of cobalamin using ethanolamine as the primary carbon source (Marous et al., 2015) did not yield detectable in vivo activity of the *C*-methyltransferases.

Ultimately, our subsequent host selection contained the following criteria:

(1) Gram-negative hosts capable of aerobic, microaerobic, and anaerobic growth to maximize expressions of the oxygen-sensitive *C*-methyltransferases.

(2) Sequenced microorganisms that encode a full biosynthetic pathway for cobalamin. A recent in-depth genomic and proteomic analysis of the polytheonamide producer "*Candidatus* Entotheonella factor" revealed a complete and expressed oxygen-independent cobalamin biosynthetic pathway (Lackner, Peters, Helfrich, & Piel, 2017).

(3) Strains that have been previously shown capable of robust plasmid transformation (either by electroporation or conjugation). Accessibility to vectors containing inducible promoters was also heavily considered.

(4) A heterologous host encoding homologs of B_{12}-dependent rSAM *C*-methyltransferases; this was a conservative criterion attempting to maximize chances for achieving in vivo activity.

When all the above criteria were taken into account, two main bacterial groups were considered for heterologous hosts, cyanobacteria, and rhizobia (see Note 2 in Section 2.3). *Rlv*3841 was finally chosen due to its fast growth rate, ease of transformation, and inducible vector selection (Kaczmarczyk, Vorholt, & Francez-Charlot, 2013; Tett, Rudder, Bourdès, Karunakaran, & Poole, 2012).

2.2 Vector Construction

We chose the taurine-inducible, conjugative low-copy plasmid pLMB51 for all of our rhizobia expressions (Tett et al., 2012). Owing to the uncertain

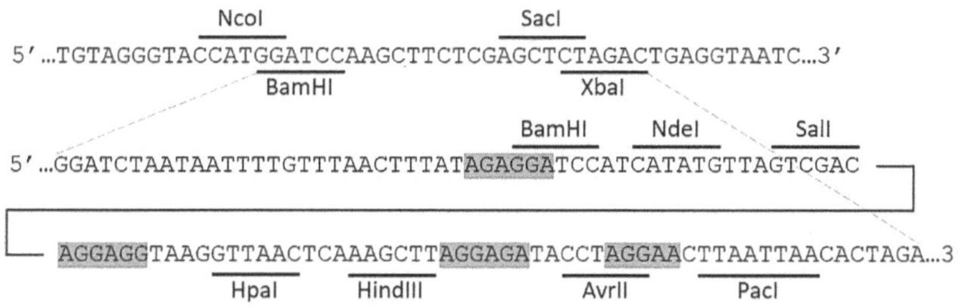

Fig. 2 Diagram of the original MCS found in pLMB51 (*top*) compared to the customized MCS inserted between the *gray-dashed lines* (*bottom*). Restriction enzymes are listed along with their recognition sites (*black solid lines*). Inserted ribosomal binding sites are highlighted with *gray boxes*.

biosynthetic timing of *C*-methylation in polytheonamide biosynthesis, our strategy for producing *C*-methylated intermediates was to construct a combinatorial series of polycistronic expression vectors utilizing both wild-type and mutant genes. Thus, we decided to create a customized multiple cloning site (MCS) that incorporated additional ribosomal binding and restriction enzyme sites (see Note 1 in Section 2.3). A variety of cloning methods can be used to design and assemble polycistronic constructs for these purposes including overlap extension polymerase chain reaction (PCR) (Ho, Hunt, Horton, Pullen, & Pease, 1989), polymerase cycling assembly (PCA) (Kodumal et al., 2004; Prodromou & Pearl, 1992), Gibson Assembly (Gibson, 2011), Golden Gate cloning (Weber, Engler, Gruetzner, Werner, & Marillonnet, 2011), and MethyloBrick technologies (Schada von Borzyskowski, Remus-Emsermann, Weishaupt, Vorholt, & Erb, 2015). The following procedure outlines the construction of a pLMB51-customized MCS using PCA (Fig. 2).

2.2.1 Equipment

- PCR machine
- DNA agarose electrophoresis apparatus
- Nanodrop (or equivalent spectrophotometer)
- Electroporator
- Shaking and static incubators

2.2.2 Buffers, Solutions, and Reagents

- DNA primers
- Plasmid pLMB51 (Addgene #40083)
- Phusion DNA polymerase and supplied buffer
- Deoxyribonucleotides (dNTPs, 10 m*M* each)

- TAE agarose gel with 0.5 μg/mL ethidium bromide or equivalent, TAE buffer (40 m*M* Tris, 20 m*M* acetic acid, 1 m*M* EDTA, pH 8.0)
- DNA gel extraction kit
- Restriction enzymes BamHI-HF, XbaI, BglII, SpeI-HF, and associated buffers
- PCR purification kit
- Antarctic phosphatase
- T4 DNA ligase
- *E. coli* cloning strain such as DH5α
- Lysogeny broth (LB) medium and LB-agar plates with 100 μg/mL of ampicillin (final concentration)
- DNA miniprep kit

2.2.3 Procedure for Creating a New MCS in pLMB51

1. Setup and run PCA using PCR primers 5′-TAAGATCTAATAATTTTGTTTAACTTTATAGAGGATCCAT-CATATGTTAGTCGACAGGAGGTAAGGTTAACTC-3′ and 5′-TTACTAGTGTTAATTAAGTTCCTAGGTATCTCCTAAGCTT-TGAGTTAACCTTACCTCCTGTCGACTA-3′ (see Note 3 in Section 2.3). In a 50-μL reaction, add a final concentration of: 4 μ*M* each primer, 0.2 μ*M* each dNTP, 5% (v/v) DMSO, and 1X Phusion buffer. In a PCR machine, run a program of: (1) 5 min at 98°C (adding 0.6 units of Phusion polymerase at 2 min); (2) 20 s at 98°C; (3) 30 s at 60°C; (4) 10 s at 72°C; (5) repeat (2)–(4) four times; and (6) 2 min at 72°C.
2. Run a 1% (w/v) agarose gel of your PCA reaction, cut out the expected 114 base pair (bp) product, and extract the DNA using a DNA gel extraction kit.
3. Digest the PCA product with restriction enzymes BglII and SpeI-HF and digest pLMB51 with BamHI-HF and XbaI. For New England Biolab-purchased enzymes, BglII and SpeI-HF require tandem digestion in buffer 3.1 and Cutsmart buffer, respectively. Use a PCR purification kit for buffer exchange. Let digests run for 1 h at 37°C.
4. Add 5 units of Antarctic phosphatase and Antarctic phosphatase buffer (1X, final concentration) to pLMB51 digest and incubate an additional 1 h at 37°C.
5. In a 10-μL reaction, ligate DNA in a vector:PCA insert mole ratio of 1:5 with 400 units of T4 DNA ligase and 1X T4 DNA ligase buffer for either 1 h at 37°C or overnight at 4°C.

6. Electroporate a 5-μL aliquot of the ligation reaction in prechilled cuvettes at 1800–2000 V into electrocompetent DH5α cells. Add 1 mL LB medium and shake at 250 rpm at 37°C. Plate cells on LB-agar medium with 100 μg/mL of ampicillin (final concentration). Verify positive clones by plasmid isolation, restriction digest, and DNA sequencing.

2.3 Notes

1. With the advent of affordable gene synthesis, the synthetic biology rationale of gene cluster refactoring is gaining traction (Smanski et al., 2016). Refactoring is a process of deconstructing a pathway to individual genetic components, removing unwanted regulatory elements and non-coding DNA, and recombining the series of parts with defined regulatory elements for a desired output. While most refactoring methodology is currently focused in using *E. coli* as a heterologous host, certain facets such as host-tailored codon-optimization of genes (Chaney & Clark, 2015) and predictive models for the strength of ribosomal binding sites (Espah Borujeni, Channarasappa, & Salis, 2014) can be of great use in designing constructs for heterologous expression. However, for genes encoded in polycistronic bacterial operons, simply changing host expression strains is enough to disrupt the coordinated production of proteins that assemble in higher-order complex formation (Shieh et al., 2015; Wells, Bergendahl, & Marsh, 2016).
2. While certain cyanobacterial strains encode homologs of PoyB/C and PoyD, some cyanobacteria synthesize and use a modified pseudo-cobalamin, where dimethyl-benzimidazole is replaced with adenine as the lower axial ligand (Heal et al., 2017; Helliwell et al., 2016). Although it is possible that the lower axial ligand is in a base-off coordination in PoyC (Parent et al., 2016), the production and use of different cobamides could be a source of potential cofactor incompatibility for *C*-methyltransferases from unknown or distantly related organisms.
3. The MCS sequence described is slightly altered compared to the one created in Freeman et al. (2017) for ease of description.

3. RHIZOBIA TRANSFORMATION

A wide variety of transformation procedures have been developed for various rhizobia. These include conjugation (Freeman et al., 2017), electroporation (Garg, Dogra, & Sharma, 1999), freeze–thaw (Vincze & Bowra, 2006), and chemical transformation (Selvaraj & Iyer, 1981) methods.

We chose a conjugative biparental mating procedure using the *E. coli* strain SM10 (Simon, Priefer, & Pühler, 1983), which harbors a chromosomal copy of the IncP-type RP4 plasmid conjugative machinery (see Note 1 in Section 3.4). The procedure described below was successfully used to transfer pLMB51-derived plasmids from SM10 to both *Rlv*3841 and *Sinorhizobium meliloti* strain 1021.

3.1 Equipment

- Shaking and static incubators
- Centrifuge capable of 5000 × *g* spins for 50-mL tubes
- Microcentrifuge
- Nanodrop (or equivalent spectrophotometer)
- Sterile loop
- Biosafety cabinet (optional)

3.2 Buffers, Solutions, and Reagents

- pLMB51-derived construct
- Strains *E. coli* SM10 and *Rlv*3841 (see Note 2 in Section 3.4)
- LB and LB-agar media
- Tryptone-yeast extract (TY) (5.0 g tryptone, 3.0 g yeast extract, 10 m*M* $CaCl_2$ per 1.0 L) and TY-agar media
- 0.9% (w/v) NaCl (sterilized)
- Streptomycin, tetracycline, and kanamycin

3.3 Procedure

1. Start a 5.0-mL *Rlv*3841 culture in TY media containing 200 μg/mL streptomycin (final concentration) in a culture tube from an *Rlv*3841 cell stock or fresh colony. Grow at 30°C shaking at 200–250 rpm for ~3 days; the culture should be at saturation.
2. On the third day, inoculate 50-mLs of TY medium containing 200 μg/mL streptomycin (final concentration) with all 5.0 mL of *Rlv*3841 saturated culture in a 250 mL Erlenmeyer flask. Incubate at 30°C with shaking at 200–250 rpm overnight.
3. Also on third day of incubation, inoculate 15-mL cultures (in 50-mL conical tubes) of LB containing 10–12 μg/mL tetracycline and 25–50 μg/mL kanamycin (final concentrations) with SM10 harboring your pLMB51-derived expression construct. Incubate at 37°C with shaking at 200–250 rpm overnight.

4. Harvest both SM10 and *Rlv*3841 cultures at 5000 × *g* for 15 min and resuspend in 15 mL sterile 0.9% (w/v) NaCl. Measure the OD_{600} of each sample and dilute them to an OD_{600} of ~4.0.
5. In 1-mL final volumes, mix donor (SM10) and recipient (*Rlv*3841) strains in ratios of 1:9, 3:7, and 5:5 (v/v). Vortex thoroughly and spin down at 10,000 × *g* for 10 min. Resuspend pellets in ~50 μL of sterile 0.9% (w/v) NaCl.
6. Add 50-μL mixed cultures in three individual, ~1.0 cm spots on dry TY-agar medium. (Optional: Do this step in a biosafety cabinet and allow spots to completely dry over ~30 min.) Incubate at 30°C for 2 days.
7. After the second day incubation, use a sterile loop to scrape off each of the spotted mixed-culture growths and transfer all into a single tube containing 1.0 mL of sterile 0.9% (w/v) NaCl.
8. 100-mL aliquots should then be streaked on TY-agar medium containing 10–12 μg/mL tetracycline and 200 μg/mL streptomycin (final concentrations) to yield single colonies. Incubate at 30°C until colonies form ~3–5 days (see Note 3 in Section 3.4).

3.4 Notes

1. Prior to conjugal transfer, the antibiotic sensitivities of the rhizobia and *E. coli* strains must be tested to determine appropriate negative selection conditions. While SM10 is kanamycin resistant and is streptomycin sensitive at a 200 μg/mL final concentration, *Rlv*3841 was resistant to streptomycin up to 2000 μg/mL. Therefore, streptomycin was used to counterselect the *E. coli* strain after conjugation. These antibiotic sensitivity tests should be performed in both liquid and solid media and monitored for 7 days to ensure true inhibitory conditions are determined. Alternatively, one could use a diaminopimelic acid auxotroph, such as *E. coli* strain WM3064, for strong counterselection.
2. *E. coli* strain SM10 was a generous gift from Hans-Martin Fischer (ETH Zurich, Switzerland); SM10 can be purchased at Biomedical Lifescience (catalog # BS-3303). *Rlv*3841 was a generous gift from Philip S. Poole (John Innes Centre, Norwich, UK). Several strains of *R. leguminosarum* exist at repositories such as ATCC (e.g., ATCC 700901). However, we cannot guarantee that these are the exact strains used in this work.
3. Transformations should yield a least 1000 clones using this procedure. *Rlv*3841 appears slightly shiny or waxy compared to *E. coli*, but

transformants should be verified by colony PCR for both *Rlv*3841- and plasmid-specific sequences to guarantee successfully transformed clones.

4. RHIZOBIA EXPRESSION AND PROTEIN PURIFICATION

The taurine-inducible promoter P_{tauA} is one of the strongest and tightest inducible promoters (see Note 1 in Section 4.5) currently available for α-proteobacteria (Kaczmarczyk et al., 2013; Mostafavi et al., 2014; Tett et al., 2012). While most studies focus on in vivo expressions and do not report the amount of heterologously produced protein, the choice of medium, growth chamber, and level of aeration can severely affect the amount of cell mass and protein yields from α-proteobacteria. To promote the most robust cell growth and (hopefully) sufficient quantities of cobalamin for our B_{12}-dependent *C*-methyltransferases, we chose to use complex media (TY medium) for all of our expressions. We also supplemented metabolic precursors for iron–sulfur cluster assembly to provide sufficient substrates for this purpose as well. Individual shake flasks were chosen in lieu of fermenters since it was necessary to perform dozens of individual expressions. In addition, we decided to minimize aeration upon protein induction since B_{12}-dependent *C*-methyltransferases are sensitive to oxygen inactivation. While the compilation of expression parameters detailed below were not exhaustively optimized, inducing up to seven individual proteins on a single operon using this procedure consistently yielded ~3.5–4.5 g cell mass and ~150–300 μg precursor protein per liter rhizobia culture.

4.1 Equipment

- Large shaking incubators
- 500 mL Erlenmeyer flasks
- 2.5 L Ultra Yield flasks (Tomson Instrument Company, catalog #931136-B)
- Identi plug foam stoppers (VWR, catalog #89092-870) or equivalent
- Centrifuge and microcentrifuge
- Liquid nitrogen Dewar flask
- Water bath
- Sonicator with microtip probe or equivalent cell-lysis apparatus
- Rotator or rocker
- Small polypropylene columns (Fisher Scientific, catalog #P129922)
- SDS-PAGE apparatus

- Disposable centrifugation molecular weight cutoff filters (e.g., Amicon Ultra-4, 10K MWCO)
- Nanodrop (or equivalent spectrophotometer)

4.2 Buffers, Solutions, and Reagents

- *Rlv*3841 strain transformed with pLMB51-derived construct
- TY (5.0 g tryptone, 3.0 g yeast extract, 10 m*M* $CaCl_2$ per 1.0 L) media
- Streptomycin, tetracycline
- Taurine
- Ferrous ammonium sulfate, $(NH_4)_2Fe(SO_4)_2$
- Cysteine
- Cobalt chloride, $CoCl_2$
- Lysis buffer (20 m*M* imidazole, 300 m*M* NaCl, 50 m*M* potassium phosphate, 10% (v/v) glycerol, 0.1% (v/v) triton X-100, pH 8.0)
- Lysozyme
- Nickel NTA resin
- Wash buffer (20 m*M* imidazole, 300 m*M* NaCl, 50 m*M* potassium phosphate, 10% (v/v) glycerol, pH 8.0)
- Elution buffer (250 m*M* imidazole, 300 m*M* NaCl, 50 m*M* potassium phosphate, 10% (v/v) glycerol, pH 8.0)

4.3 Protein Expression Procedure

1. From a frozen cell stock or fresh colony, start a 50.0-mL *Rlv*3841 culture harboring your pLMB51-derived expression construct in a 500-mL Erlenmeyer flask containing TY media with 200 μg/mL streptomycin and 10 μg/mL tetracycline (final concentrations). Shake at 200–250 rpm at 30°C and grow until saturation, ~3 days.
2. Spin down the saturated *Rlv*3841 primary culture at 4000 × *g* for 15 min and resuspend in fresh TY medium before adding to 1.0 L TY medium with 10 μg/mL tetracycline (final concentration) in a 2.5-L Ultra Yield flask capped with a foam plug. Shake at 200–250 rpm at 30°C until an OD_{600} of ~0.6–0.8 is reached, ~8–12 h.
3. Submerge the Ultra Yield flask in an ice bath for ~10–20 min, ensuring the temperature of the culture is <16°C.
4. Add the following components to the 1.0-L culture such that the final concentrations of each are: 150 μ*M* cysteine, 100 μ*M* $(NH_4)_2Fe(SO_4)_2$, 100 μ*M* $CoCl_2$, and 10 m*M* taurine. Shake at 125 rpm at 16°C for the desired expression duration; our optimum was 3–7 days.

5. Spin down the *Rlv*3841 expression culture at 4000 × *g* for 15–30 min and either flash freeze in liquid nitrogen and store at −80°C until needed, or proceed directly to the protein purification procedure.

4.4 Protein Purification Procedure (See Note 2 in Section 4.5)

1. Cell pellets should be resuspended in a 50-mL conical tube with a ratio of 1.0 g wet cell weight to 4.0 mL lysis buffer. Manually resuspend with a serological pipette or shake at 30°C for 15–30 min to resuspend the cell mass.
2. Add 1.0 mg/mL lysozyme (final concentration) and incubate on ice for 30 min.
3. Incubate the resuspended solutions in a water bath at 95°C for 30 min. While using thermally protective gloves or holders, vent the caps of the 50-mL conical tubes several times to prevent pressure build-up. Be careful during this process as small amounts of hot liquid can escape during this process.
4. Cool tubes on ice for 15 min or until the contents are roughly room temperature. If the 50-mL conicals are spun down without this cooling step, the plastic will likely deform in the centrifuge, making subsequent steps difficult.
5. If needed, sonicate the cell slurry for five, 10 s pulses at 30% amplitude with a microtip for a more complete lysis (see Note 3 in Section 4.5).
6. Spin down lysate in appropriately rated centrifuge tubes at 25,000 × *g* for 45 min. Quickly transfer the supernatant to a new tube.
7. Add 250 μL of Ni-NTA resin per sample to the supernatant described above; incubate at 4°C with gentle rocking for 1 h (see Note 4 in Section 4.5).
8. Batch purify resin at 800 × *g* for 10–20 min. Quickly decant supernatant to minimize loss of resin. Add 1.0 mL wash buffer and transfer resin by pipetting to a fritted polypropylene column.
9. Slowly add 5–8 mL wash buffer to the column and allow buffer to pass through resin by gravity flow. Repeat twice.
10. Elute protein with 1.0 mL elution buffer three times. Run SDS-PAGE of all samples, combine appropriate fractions, and concentrate to ~1–5 mg/mL prior to flash freezing and storage at −80°C or short-term storage at 4°C.

4.5 Notes

1. Synthetic biological tools are becoming more available that may allow for stronger transcriptional control in rhizobia and other α-proteobacteria (Kushwaha & Salis, 2015).
2. The protein purification procedure described is specific for heterologously produced PoyA from the polytheonamide gene cluster. PoyA was shown to be very heat stable (Freeman et al., 2017); use of triton X-100 and a heat-lysis step may not be appropriate for other target proteins. Triton X-100 should be avoided if possible since it is a problematic contaminant in downstream mass spectrometric analyses.
3. Compared to *E. coli*, *Rlv*3841 produces a thicker, waxier lysate that might require more sonication cycles or additional passes through a French press. Centrifugation times and speeds may also need to be varied or increased due to these same properties.
4. The amount of Ni-NTA resin added to samples should be adjusted to the expected yield of protein in relation to the resin's binding capacity. However, a minimum of 200–250 μL of resin per 50 mL lysate should be used to ensure adequate resin recovery from the sample.

5. DATA ACQUISITION AND ANALYSIS OF PEPTIDE AND PROTEIN TARGETS

The vitamin B_{12}-dependent rSAM *C*-methyltransferases PoyB and PoyC from the polytheonamide cluster are responsible for 17 methylations of various amino acids on its PoyA substrate. These methylations occur at C_β (13), C_γ (1), and C_δ (3) sp^3 centers (Fig. 1). The high numbers of possible methylation states, in combination with the highly hydrophobic nature of the peptide and presence of other variable modifications (C_a epimerizations, e.g.), created unique challenges. As some peptide species were incompletely modified, determining the methylation patterns on PoyA was particularly difficult. An effective method for identifying PTMs is through the enzymatic digestion of the proteins/peptides followed by ultra-high pressure liquid chromatography coupled with electrospray ionization tandem mass spectrometry (LC–ESI–MS/MS) (Kersten et al., 2011). This highly sensitive method can often localize PTMs to single residues through *b*- and *y*-ion fragment production, but other methods utilizing nuclear magnetic resonance are usually needed to precisely define the structure of unknown metabolites. The use of proteases trypsin and the polytheonamide-encoded peptidase PoyH afforded the identifications of some methylations in the polytheonamide sequence (Freeman et al., 2012, 2017). However, the large size of the proteolytically cleaved fragments (4500–7000 Da) and the difficulty of chromatographically separating different peptides with the same number of methylations required a different strategy to identify the majority of *C*-methylations. Incomplete and complete digestions with proteinase K, a protease with broad specificity of aromatic and aliphatic amino acids, created smaller and variably cleaved peptides that could be better chromatographically separated and identified (see also Section 5.2). Importantly, proteinase K does not cleave at sites of D-amino acids (Brömme, Peters, Fink, & Fittkau, 1986), which provided an additional level of separation since a suite of differentially epimerized peptides are produced during coexpressions of PoyA and the epimerase PoyD in *E. coli* and *Rlv*3841 (Freeman et al., 2017).

Owing to the high number of possible methylated species and the unexpected sizes of fragments from proteinase K digestion, the proteomics software MaxQuant (Cox & Mann, 2008; Cox et al., 2011) was used to identify possible methylated targets by data-dependent LC–ESI–MS/MS fragmentation. There are many different strategies that can be used for protein, peptide, and metabolite identification (Chicooree, Unwin, & Griffiths, 2015; Covington, McLean, & Bachmann, 2017; Krug & Müller, 2014; Schwämmle, Verano-Braga, & Roepstorff, 2015), but the robust freeware MaxQuant was ultimately chosen as its graphical interface was user friendly and compatible with our Windows operating systems. The following sections detail sample preparation for proteinase K digestion, LC–ESI–MS/MS methods, and use of MaxQuant to identify methylated peptides.

5.1 Equipment

- Water bath or thermal cycler
- Microcentrifuge
- Vacuum concentrator for 1.5 mL tubes
- Phenomenex Kinetex 2.6 μm C18 100 Å ($150 \times 4.6\,mm^2$) column
- Dionex Ultimate 3000 UHPLC system coupled to a Thermo Scientific Q Exactive mass spectrometer

- Preferred: computer running a Windows operating system. 64 bit, quad core, 8 gb RAM, >50 gb free space (on current requirements for MaxQuant refer to Jürgen Cox's documentation at his current website: http://www.biochem.mpg.de/5111795/maxquant)
- Xcalibur software (Thermo Fisher Scientific) and MaxQuant software (version 1.5.0.0 or higher)

5.2 Buffers, Solutions, and Reagents

- Proteinase K
- 10 × proteinase K digestion buffer (500 m*M* Tris pH 8.0, 20 m*M* $CaCl_2$)
- 1.5 mL LoBind tubes (Eppendorf, catalog #022431081)
- C18 or C4 ZipTips (Millipore Sigma, catalog #ZTC18S096)
- LC–MS grade water, acetonitrile (MeCN), and formic acid (FA)
- 50-μL HPLC vials (Chrom Tech, catalog #CTI-2405)
- Caps for HPLC vials (Agilent Technologies, catalog #5190-7024)

5.3 Procedure for Proteinase K Digestion, Sample Preparation, and LC–ESI–MS/MS Analysis

1. Make a fresh 1.0 mg/mL proteinase K stock in 1 × proteinase K digestion buffer.
2. For complete proteinase K digestion: prepare proteinase K reactions (20 μL) in LoBind tubes containing 8 μg purified PoyA protein with 2 μg proteinase K in 1 × proteinase K digestion buffer. Incubate for at least 4 h at 40°C in a water bath (see Note 1 in Section 5.5). For incomplete digestions add 10 × less enzyme and incubate at room temperature for 1–2 h. Place samples on ice.
3. Spin down sample at 10,000 × *g* for 5 min at 4°C to remove any precipitated material and transfer 10 μL to a new LoBind tube.
4. ZipTip 10-μL sample according to manufacturer's specifications; elute into 10 μL of 90% (v/v) MeCN, 0.1% (v/v) FA.
5. Evaporate samples to dryness in a vacuum concentrator. Resuspend each sample in 10 μL of 5% (v/v) MeCN, 0.1% (v/v) FA, and transfer them to glass HPLC vials. Freeze at −20°C until needed.
6. Inject up to 4 μg PoyA (10 μL) onto a 50°C preheated Phenomenex Kinetex 2.6 μm C18 100 Å (150 × 4.6 mm^2) column (see Note 2 in Section 5.5) with a flow rate of 0.5 mL/min. A solvent gradient (solvent A: acetonitrile with 0.1% (v/v) FA; solvent B: water with 0.1% (v/v) FA)

was used for separation consisting of 5% solvent A from 0 to 2 min, 5%–98% solvent A from 2 to 15 min, 98% solvent A from 15 to 20 min, 98%–5% solvent A from 20 to 22 min, and 5% solvent A from 22 to 24.5 min.

7. Run the Q Exactive mass spectrometer with the following parameters: positive ion mode; spray voltage 3500; capillary temperature 268.75°C; probe heater temperature 350–437.5°C; S-lens level 70, full MS 70,000 resolution (AGC target 3e6, maximum IT 200 ms, range 500–2000 *m/z*).
8. To fragment known or expected masses with a targeted mass inclusion list (see Notes 3 and 4 in Section 5.5), calculate the theoretical masses of your target peptides in all relevant charge states and incorporate the mass list (see Note 5 in Section 5.5).
9. Perform data-dependent high-energy collisional dissociation (HCD) MS/MS at a resolution of 17,500 (minimum AGC target of 2e5, maximum IT between 100 and 500 ms, isolation window of 2.2 *m/z*) using a stepped normalized collision energy (NCE) from 16 to 20 or 20 to 30. Scan ranges, mass inclusion lists, charge exclusions, and dynamic exclusions were adjusted as needed (see Note 6 in Section 5.5).
10. For expected mass fragments, manually identify peptides using the Thermo Fisher Scientific Xcalibur software using both MS and MS/MS spectra (see Note 7 in Section 5.5); for identification of MS/MS spectra for unknown and unexpected fragments, see Section 5.4.

5.4 MaxQuant Identification of Methylated Peptides (See Note 8 in Section 5.5)

1. Create a folder on the computer and hard drive for the MaxQuant search. In this folder, add all of the Thermo Fisher RAW MS files to be parsed, a fasta file with your desired peptide or protein sequence named "YOURTARGETS.fasta," and the parameters file created as described below.
2. Modify or create a parameters file (mqpar.xml); this is loaded and saved in the upper left-hand drop-down menu of MaxQuant.
3. Go to the "Configuration" tab, select "Sequence databases," find "contaminants.fasta" or another fasta file that has the "Identifier parse rule" >([^]*) and click "duplicate." Rename "YOURTARGETS. fasta." You need to select "Modify table" then "Save changes" to save any changes in the "Configuration" Table.
4. Under "Modifications" in the "Configuration" tab, create all of the modified amino acid substitutions that will be searched for in your peptides. For example, a methylated valine residue is the equivalent substitution "Val → Xle," and therefore a new modified amino acid

substitution is not needed. However, to search for a methylated threonine residue, copy the "Thr → Ser" substitution and modify the mass to ultimately add two hydrogens and one carbon to the mass of threonine. Change the name of the substitution as desired and save as in step 2 (see Note 9 in Section 5.5).

5. In the "Raw files" tab, add all MS files to be searched, making sure they are in the appropriate folder.
6. In the "Group-specific parameters" tab under "Digestion Mode," select unspecific (see Note 10 in Section 5.5). In "Modifications," select all desired amino acid substitutions and type in the maximum number of modifications expected (see Note 11 in Section 5.5). In the "Instruments" tab, select the appropriate type of MS used.
7. In the "Global parameters" tab under "Sequences," add the "YOURTARGETS.fasta" file in the appropriate folder. Adjust the "Min." and "Max." values in this subtab as needed, and any fixed modifications in case samples were treated with iodoacetamide, for example. Under "Folder locations," select the "Temporary folder" as the same folder where all of your MaxQuant data for these samples are stored (see Note 12 in Section 5.5).
8. In the bottom left corner of the MaxQuant window, set the "Number of threads" to a maximum of one fewer than the processing cores on your computer (see Note 13 in Section 5.5). All other default parameters in MaxQuant should be sufficient for initial peptide searches.
9. Save the parameters file, close MaxQuant, and restart the program prior to starting the search to ensure the run is properly initiated.
10. After the run is complete, you can view the results under the "Viewer" tab. Click on "Identifications" and "Evidence." A graphical representation for fragmentation of an identified sequence can be visualized by selecting a sequence, and then clicking "Display best spectrum."
11. Using the retention times and MS/MS scan numbers from the "Evidence" tab, view the original RAW file in the Xcalibur software (see Note 14 in Section 5.5). Manually search and annotate the MS/MS data, since many false positives and suboptimal MS/MS spectra can often be initially identified by MaxQuant (see Note 15 in Section 5.5).
12. Once a methylated peptide is verified by MS/MS analysis, carefully look back at individual MS/MS spectra from the beginning, middle, and end of the eluting peak. Despite enhanced separation from proteinase K digestion, mixtures of differentially methylated peptides can still coelute. Clear and well-defined spectra can often only be obtained at the extremes of the eluting peak (Fig. 3).

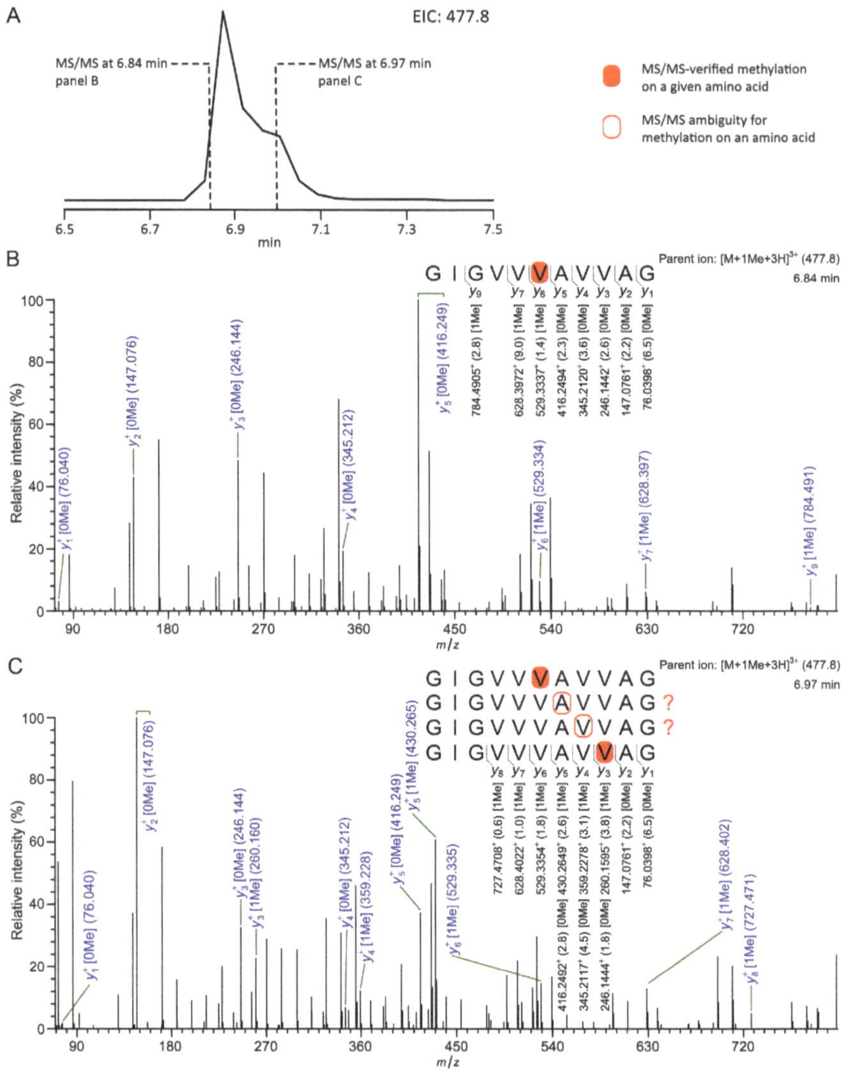

Fig. 3 MS/MS analysis of a mixture of *C*-methylated peptides. Panel A: Magnified image of an extracted ion chromatogram (EIC) (477.80 Da) for a singly methylated PoyA peptide fragment with the sequence GIGVVAVVAG. Individual MS/MS spectra from the beginning (Panel B) and end (Panel C) of the eluting 477.80 Da peak. *Y*-ions are listed in both the MS/MS spectra and their corresponding positions within the peptide sequence depicted above the spectra. The differences (ppm) between observed and theoretical masses are listed in *parentheses*. In panel B, a single methylated species is observed, while in panel C, at least two singly methylated species appear to be coeluting. Positions of verified (*solid circles*) or possible (*open circles*) methylations (Me) are depicted.

5.5 Notes

1. In lieu of a water bath, a thermal cycler can be used for the proteinase K incubation. This has the added benefit of minimizing condensation at the top of the tube during extended incubations. However, depending on the properties of your peptide fragments and if they stick to normal plastics, LoBind tubes incubated in a water bath should be used instead.
2. For large (>30 mer) and hydrophobic peptides, detection may only be achieved when using a larger-pore column (e.g., 300 Å) or shorter-chain resins such as C8 or C4. In addition, nanoscale liquid chromatography (nanoLC) (Ishihama, 2005) is an attractive alternative method to analytical-scale liquid chromatography due to potential increases in sensitivity, chromatographic separation, and decreased use of solvent. However, nanoLC systems can often clog and require specialized handling.
3. For many peptides, the monoisotopic mass is not the most abundant ion in the MS spectra and therefore targeting the highest isotope peak may be a better strategy for maximizing detection and fragmentation. This is also helpful when setting the isolation width for fragmentation so the maximum number of ions can be included in the analysis without including contaminating peaks.
4. All fragmentation data acquired with data-independent mass inclusion lists will not be seen or identified by the program MaxQuant. MaxQuant only detects data-dependent fragmentation. Therefore, the options "dynamic exclusion" and "if idol, pick others" should be used to collect a mixture of MS/MS data in a single run. Alternatively, successive LC–ESI–MS/MS runs with and without a mass inclusion list should be performed. If using a Thermo Fusion mass spectrometer, the "if idol, pick others" option is not available. However, the software for this instrument allows for independent steps to include both data-independent and data-dependent MS/MS fragmentation.
5. Multiple charge states may be detected for a given peptide using this mass spectrophotometer within a mass-to-charge ratio (m/z) of approximately 2000 Da. Salt adducts including sodium, potassium, and ammonium adducts might also be present. Different charge states of a given peptide can give different fragmentation patterns, and therefore lower abundant peaks should be targeted and analyzed. It should also be noted that the fragmentation energies for MS/MS analysis might need to be adjusted for different charge states as well as different salt

adducts. In our hands, the sodium and potassium adducts for PoyA peptides did not fragment well under the described conditions.

6. We found relatively high numbers of ions collected in the Orbitrap (as reflected in the high AGC target and long maximum IT values) were needed to get high quality fragmentation of our peptides. Also, using stepped normalized collision energies (stepped NCE) with three microscans also helped to get fuller coverage of *b*- and *y*-ions in the MS/MS data.

7. To look for masses of expected fragments, first look for EICs of all reasonable charge states. This function is under "ranges" when right clicking an active "chromatograph" window. Select "base peak" under the "Plot type:" drop-down menu and type in the expected mass. This method of identification is dependent on the relative abundance of your desired fragment, and often did not allow for easy identification of PoyA fragments within our data. Another often helpful method was scanning the chromatograph over short timeframes when viewing the "spectrum" window to look for a series of peaks coeluting with a relative mass difference of 14 Da, since polytheonamide fragments often contained multiple methylations in a given peptide. Be aware that differences in amino acids such as glycine and alanine also result in a mass differences of 14 Da. (This complexity could, in theory, be reduced by use of [^{13}C-methyl]-L-methionine (Millipore Sigma, catalog #299146) for in vivo incorporation into SAM to subsequently produce mass shifts of 15 Da, but we have not yet performed these experiments.) Typically, the more methylated peptides will elute slightly later than the nonmethylated sequence on a reverse phase column. Once the elution time is known for your desired fragment, go to "ranges" and select the identified mass in the drop-down menu "Scan filter." After selecting the appropriate mass, begin looking at MS/MS spectra within the identified timeframe.

8. The version of MaxQuant described here is 1.5.3.30, however all versions of MaxQuant from 1.5.0.0 can be used with these guidelines. Please note that the parameter mqpar.xml files created in one version of MaxQuant will likely not work in a different version. Also note that reading the publisher's original instructions is heavily advised, since this description is far from comprehensive.

9. If an amino acid has multiple possible modifications in your peptide, all combinations of individual substitutions need to be created and searched for unless the precise biosynthetic order is known.

10. If a specific protease was used in your digestion, you can select it in this tab. However, we have found that selecting "unspecific" digestion identifies more peptides even when a specific protease is used. It should also be noted that "unspecific" digestion can greatly increase the time necessary for the MaxQuant run and increase the amount of space necessary for the temporary files created for the theoretical mass lists.
11. In our experience, selecting more than seven modifications resulted in prohibitively long MaxQuant runs.
12. In searching for methylated PoyA peptides, we would sometimes have temporary data in the 100s of gigabytes. It is imperative to have plenty of space set aside for this purpose.
13. Be aware that MaxQuant can significantly slow your computer when running near the maximum recommended threads. Also, the searches for methylated PoyA peptides took as long as 2 weeks due to the excessive number of PTMs possible in some of our coexpressions. Use of a computer network with many cores can, in theory, speed up the search, although this depends on the configuration of the network.
14. It can be useful to view MaxQuant data in a Microsoft Excel table. To do so, go the folder created for the MaxQuant run and navigate into the newly created folders "combined" and "txt." Open the file "evidence.txt" in a simple text editor such as Notepad. Select all the text and copy and paste it into Excel. Select row one and set to "filter" in the "Sort & Filter" subtab of "Home" in Excel. This will allow more convenient handling of the data.
15. MaxQuant will not identify every peptide in your spectra; we have consistently found peptides not identified by this program. However, the manually verified peptide fragments initially identified by MaxQuant will dramatically aid in the search for chemically related peptides that elute at similar retention times.

6. SUMMARY AND CONCLUSIONS

Recent advances in DNA sequencing technologies and synthetic biology have reinvigorated the study and discovery of natural products (Smanski et al., 2016). A shift from activity-based, small-molecule discovery to genomics based identification has required a suite of new tools to effectively verify natural product gene clusters from the vast majority of marine microorganisms recalcitrant to laboratory cultivation or genetic manipulation (Medema & Fischbach, 2015). In this chapter, we have described a robust means to verify the activity of genes encoding B_{12}-dependent rSAM

C-methyltransferases through heterologous expression in rhizobia. The protocols detailed in the work should be broadly applicable for the functional validation of pathways encoding rSAM superfamily enzymes, since the rSAM epimerase PoyD was also active in this host along with the B_{12}-dependent *C*-methyltransferases PoyB and PoyC. In addition, we provide specialized analytical strategies for determining precise *C*-methylations on protein and peptide substrates. B_{12}-dependent *C*-methyltransferases comprise an expanding class of rSAM enzymes capable of chemically challenging methylations of unactivated carbon centers on a wide variety of metabolites. The potential catalytic diversity of this rSAM family is enormous, with estimates of less than 1% enzymes characterized thus far (Bauerle, Schwalm, & Booker, 2015). As new catalytic mechanisms gleaned from B_{12}-dependent rSAM enzymes are postulated to be one of the new frontiers in chemistry (Walsh, 2015), expressions in tailored hosts like rhizobia will play a key part to facilitate their discovery and functional validation.

ACKNOWLEDGMENTS

I first would like to thank Joern Piel for all of his advice, guidance, and support throughout the development of this work. I would warmly like to thank Hans-Martin Fischer and Raphael Ledermann for their time and expertise concerning rhizobia. I would also like to thank Maximilian J. Helf, Brandon I. Morinaka, Agneya Bhushan, Alexander O. Brachmann, and Eric J.N. Helfrich for their technical expertise and important contributions. I thank Marissa R. Quijano and Maximilian J. Helf for reviewing this manuscript. I would like to acknowledge the Human Frontiers Science Program and the University of Minnesota and BioTechnology Institute for their generous financial support.

REFERENCES

Bauerle, M. R., Schwalm, E. L., & Booker, S. J. (2015). Mechanistic diversity of radical *S*-adenosylmethionine (SAM)-dependent methylation. *Journal of Biological Chemistry*, *290*(7), 3995–4002. https://doi.org/10.1074/jbc.R114.607044.

Benjdia, A., Pierre, S., Gherasim, C., Guillot, A., Carmona, M., Amara, P., et al. (2015). The thiostrepton A tryptophan methyltransferase TsrM catalyses a cob(II)alamin-dependent methyl transfer reaction. *Nature Communications*, *6*, 8377. https://doi.org/10.1038/ncomms9377.

Blunt, J. W., Carroll, A. R., Copp, B. R., Davis, R. A., Keyzers, R. A., & Prinsep, M. R. (2018). Marine natural products. *Natural Product Reports*, *35*, 8–53. https://doi.org/10.1039/C7NP00052A.

Bridwell-Rabb, J., Zhong, A., Sun, H. G., Drennan, C. L., & Liu, H. (2017). A B12-dependent radical SAM enzyme involved in oxetanocin A biosynthesis. *Nature*, *544*(7650), 322–326. https://doi.org/10.1038/nature21689.

Brömme, D., Peters, K., Fink, S., & Fittkau, S. (1986). Enzyme-substrate interactions in the hydrolysis of peptide substrates by thermitase, subtilisin BPN', and proteinase K. *Archives of Biochemistry and Biophysics*, *244*(2), 439–446. https://doi.org/10.1016/0003-9861(86)90611-9.

Chaney, J. L., & Clark, P. L. (2015). Roles for synonymous codon usage in protein biogenesis. *Annual Review of Biophysics*, *44*(1), 143–166. https://doi.org/10.1146/annurev-biophys-060414-034333.

Chicooree, N., Unwin, R. D., & Griffiths, J. R. (2015). The application of targeted mass spectrometry-based strategies to the detection and localization of post-translational modifications. *Mass Spectrometry Reviews*, *34*(6), 595–626. https://doi.org/10.1002/mas.21421.

Covington, B. C., McLean, J. A., & Bachmann, B. O. (2017). Comparative mass spectrometry-based metabolomics strategies for the investigation of microbial secondary metabolites. *Natural Product Reports*, *34*(1), 6–24. https://doi.org/10.1039/c6np00048g.

Cox, J., & Mann, M. (2008). MaxQuant enables high peptide identification rates, individualized p.p.b.-range mass accuracies and proteome-wide protein quantification. *Nature Biotechnology*, *26*(12), 1367–1372. https://doi.org/10.1038/nbt.1511.

Cox, J., Neuhauser, N., Michalski, A., Scheltema, R. A., Olsen, J. V., & Mann, M. (2011). Andromeda: A peptide search engine integrated into the MaxQuant environment. *Journal of Proteome Research*, *10*(4), 1794–1805. https://doi.org/10.1021/pr101065j.

Crone, W. J. K., Vior, N. M., Santos-Aberturas, J., Schmitz, L. G., Leeper, F. J., & Truman, A. W. (2016). Dissecting bottromycin biosynthesis using comparative untargeted metabolomics. *Angewandte Chemie International Edition*, 1–6. https://doi.org/10.1002/anie.201604304.

Di Girolamo, P. M., Kadner, R. J., & Bradbeer, C. (1971). Isolation of vitamin B12 transport mutants of *Escherichia coli*. *Journal of Bacteriology*, *106*(3), 751–757.

Espah Borujeni, A., Channarasappa, A. S., & Salis, H. M. (2014). Translation rate is controlled by coupled trade-offs between site accessibility, selective RNA unfolding and sliding at upstream standby sites. *Nucleic Acids Research*, *42*(4), 2646–2659. https://doi.org/10.1093/nar/gkt1139.

Freeman, M. F., Gurgui, C., Helf, M. J., Morinaka, B. I., Uria, A. R., Oldham, N. J., et al. (2012). Metagenome mining reveals polytheonamides as posttranslationally modified ribosomal peptides. *Science*, *338*(6105), 387–390. https://doi.org/10.1126/science.1226121.

Freeman, M. F., Helf, M. J., Bhushan, A., Morinaka, B. I., & Piel, J. (2017). Seven enzymes create extraordinary molecular complexity in an uncultivated bacterium. *Nature Chemistry*, *9*(4), 387–395. https://doi.org/10.1038/nchem.2666.

Garg, B., Dogra, R. C., & Sharma, P. K. (1999). High-efficiency transformation of *Rhizobium leguminosarum* by electroporation. *Applied and Environmental Microbiology*, 65(6), 2802–2804. Retrieved from http://www.ncbi.nlm.nih.gov/pubmed/10347085.

Gibson, D. G. (2011). Enzymatic assembly of overlapping DNA fragments. *Methods in Enzymology*, *498*, 349–361. https://doi.org/10.1016/B978-0-12-385120-8.00015-2.

Hamada, T., Matsunaga, S., Fujiwara, M., Fujita, K., Hirota, H., Schmucki, R., et al. (2010). Solution structure of polytheonamide B, a highly cytotoxic nonribosomal polypeptide from marine sponge. *Journal of the American Chemical Society*, *132*(37), 12941–12945. https://doi.org/10.1021/ja104616z.

Hamada, T., Sugawara, T., Matsunaga, S., & Fusetani, N. (1994a). Polytheonamides, unprecedented highly cytotoxic polypeptides, from the marine sponge *Theonella swinhoei* 1. Isolation and component amino acids. *Tetrahedron Letters*, *35*(5), 719–720. https://doi.org/10.1016/S0040-4039(00)75799-6.

Hamada, T., Sugawara, T., Matsunaga, S., & Fusetani, N. (1994b). Polytheonamides, unprecedented highly cytotoxic polypeptides from the marine sponge Theonella swinhoei 2. *Structure elucidation, Tetrahedron Letters*, *35*(4), 609–612. https://doi.org/10.1016/S0040-4039(00)75851-5.

Heal, K. R., Qin, W., Ribalet, F., Bertagnolli, A. D., Coyote-Maestas, W., Hmelo, L. R., et al. (2017). Two distinct pools of B12 analogs reveal community interdependencies in the ocean. *Proceedings of the National Academy of Sciences of the United States of America*, *114*(2), 364–369. https://doi.org/10.1073/pnas.1608462114.

Helliwell, K. E., Lawrence, A. D., Holzer, A., Kudahl, U. J., Sasso, S., Kräutler, B., et al. (2016). Cyanobacteria and eukaryotic algae use different chemical variants of vitamin B12. *Current Biology*, *26*(8), 999–1008. https://doi.org/10.1016/j.cub.2016.02.041.

Ho, S. N., Hunt, H. D., Horton, R. M., Pullen, J. K., & Pease, L. R. (1989). Site-directed mutagenesis by overlap extension using the polymerase chain reaction. *Gene*, 77(1), 51–59. https://doi.org/10.1016/0378-1119(89)90358-2.

Hou, Y., Tianero, M. D. B., Kwan, J. C., Wyche, T. P., Michel, C. R., Ellis, G. A., et al. (2012). Structure and biosynthesis of the antibiotic bottromycin D. *Organic Letters*, *14*(19), 5050–5053. https://doi.org/10.1021/ol3022758.

Hug, L. A., Baker, B. J., Anantharaman, K., Brown, C. T., Probst, A. J., Castelle, C. J., et al. (2016). A new view of the tree of life. *Nature Microbiology*, *1*(5), 16048. https://doi.org/10.1038/nmicrobiol.2016.48.

Ishihama, Y. (2005). Proteomic LC–MS systems using nanoscale liquid chromatography with tandem mass spectrometry. *Journal of Chromatography A*, *1067*(1–2), 73–83. https://doi.org/10.1016/j.chroma.2004.10.107.

Kaczmarczyk, A., Vorholt, J. A., & Francez-Charlot, A. (2013). Cumate-inducible gene expression system for sphingomonads and other alphaproteobacteria. *Applied and Environmental Microbiology*, *79*(21), 6795–6802. https://doi.org/10.1128/AEM.02296-13.

Kamigiri, K., Hidaka, T., Imai, S., Murakami, T., & Seto, H. (1992). Studies on the biosynthesis of bialaphos (SF-1293) 12. C-P bond formation mechanism of bialaphos: Discovery of a P-methylation enzyme. *The Journal of Antibiotics*, *45*(5), 781–787. https://doi.org/10.7164/antibiotics.45.781.

Kersten, R. D., Yang, Y.-L., Xu, Y., Cimermancic, P., Nam, S.-J., Fenical, W., et al. (2011). A mass spectrometry-guided genome mining approach for natural product peptidogenomics. *Nature Chemical Biology*, 7(11), 794–802. https://doi.org/10.1038/nchembio.684.

Kodumal, S. J., Patel, K. G., Reid, R., Menzella, H. G., Welch, M., & Santi, D. V. (2004). Total synthesis of long DNA sequences: Synthesis of a contiguous 32-kb polyketide synthase gene cluster. *Proceedings of the National Academy of Sciences of the United States of America*, *101*(44), 15573–15578. https://doi.org/10.1073/pnas.0406911101.

Krug, D., & Müller, R. (2014). Secondary metabolomics: The impact of mass spectrometry-based approaches on the discovery and characterization of microbial natural products. *Natural Product Reports*, *31*(6), 768. https://doi.org/10.1039/c3np70127a.

Kushwaha, M., & Salis, H. M. (2015). A portable expression resource for engineering cross-species genetic circuits and pathways. *Nature Communications*, *6*(7832) https://doi.org/10.1038/ncomms8832.

Lackner, G., Peters, E. E., Helfrich, E. J. N., & Piel, J. (2017). Insights into the lifestyle of uncultured bacterial natural product factories associated with marine sponges. *Proceedings of the National Academy of Sciences of the United States of America*, *114*(3), E347–E356. https://doi.org/10.1073/pnas.1616234114.

Larimer, F. W., Chain, P., Hauser, L., Lamerdin, J., Malfatti, S., Do, L., et al. (2004). Complete genome sequence of the metabolically versatile photosynthetic bacterium *Rhodopseudomonas palustris*. *Nature Biotechnology*, *22*(1), 55–61. https://doi.org/10.1038/nbt923.

Marous, D. R., Lloyd, E. P., Buller, A. R., Moshos, K. a., Grove, T. L., Blaszczyk, A. J., et al. (2015). Consecutive radical *S*-adenosylmethionine methylations form the ethyl side chain in thienamycin biosynthesis. *Proceedings of the National Academy of Sciences of the United States of America*, *112*(33), 201508615. https://doi.org/10.1073/pnas.1508615112.

Masepohl, B., Kutsche, M., Riedel, K. U., Schmehl, M., Klipp, W., & Pühler, A. (1992). Functional analysis of the cysteine motifs in the ferredoxin-like protein FdxN of *Rhizobium meliloti* involved in symbiotic nitrogen fixation. *Molecular & General Genetics*, *233*(1–2), 33–41. https://doi.org/10.1007/BF00587558.

Medema, M. H., & Fischbach, M. A. (2015). Computational approaches to natural product discovery. *Nature Chemical Biology*, *11*(9), 639–648. https://doi.org/10.1038/nchembio.1884.

Mostafavi, M., Lewis, J. C., Saini, T., Bustamante, J. A., Gao, I. T., Tran, T. T., et al. (2014). Analysis of a taurine-dependent promoter in *Sinorhizobium meliloti* that offers tight modulation of gene expression. *BMC Microbiology*, *14*(1), 295. https://doi.org/10.1186/s12866-014-0295-2.

Parent, A., Guillot, A., Benjdia, A., Chartier, G., Leprince, J., & Berteau, O. (2016). The B12-radical SAM enzyme PoyC catalyzes valine C β-methylation during polytheonamide biosynthesis. *Journal of the American Chemical Society*, *138*(48), 15515–15518. https://doi.org/10.1021/jacs.6b06697.

Prodromou, C., & Pearl, L. H. (1992). Recursive PCR: A novel technique for total gene synthesis. *Protein Engineering, Design and Selection*, *5*(8), 827–829. https://doi.org/10.1093/protein/5.8.827.

Schada von Borzyskowski, L., Remus-Emsermann, M., Weishaupt, R., Vorholt, J. A., & Erb, T. J. (2015). A set of versatile brick vectors and promoters for the assembly, expression, and integration of synthetic operons in *Methylobacterium extorquens* AM1 and other alphaproteobacteria. *ACS Synthetic Biology*, *4*(4), 430–443. https://doi.org/10.1021/sb500221v.

Schwämmle, V., Verano-Braga, T., & Roepstorff, P. (2015). Computational and statistical methods for high-throughput analysis of post-translational modifications of proteins. *Journal of Proteomics*, *129*, 3–15. https://doi.org/10.1016/j.jprot.2015.07.016.

Selvaraj, G., & Iyer, V. N. (1981). Genetic transformation of *Rhizobium meliloti* by plasmid DNA. *Gene*, *15*(2–3), 279–283. https://doi.org/10.1016/0378-1119(81)90137-2.

Shieh, Y.-W., Minguez, P., Bork, P., Auburger, J. J., Guilbride, D. L., Kramer, G., et al. (2015). Operon structure and cotranslational subunit association direct protein assembly in bacteria. *Science*, *350*(6261), 678–680. https://doi.org/10.1126/science.aac8171.

Simon, R., Priefer, U., & Pühler, A. (1983). A broad host range mobilization system for in vivo genetic engineering: Transposon mutagenesis in Gram negative bacteria. *Bio/Technology*, *1*(9), 784–791. https://doi.org/10.1038/nbt1183-784.

Smanski, M. J., Zhou, H., Claesen, J., Shen, B., Fischbach, M. A., & Voigt, C. A. (2016). Synthetic biology to access and expand nature's chemical diversity. *Nature Reviews Microbiology*, *14*(3), 135–149. https://doi.org/10.1038/nrmicro.2015.24.

Tett, A. J., Rudder, S. J., Bourdès, A., Karunakaran, R., & Poole, P. S. (2012). Regulatable vectors for environmental gene expression in alphaproteobacteria. *Applied and Environmental Microbiology*, *78*(19), 7137–7140. https://doi.org/10.1128/AEM.01188-12.

Vincze, E., & Bowra, S. (2006). Transformation of rhizobia with broad-host-range plasmids by using a freeze-thaw method. *Applied and Environmental Microbiology*, *72*(3), 2290–2293. https://doi.org/10.1128/AEM.72.3.2290-2293.2006.

Walsh, C. T. (2015). A chemocentric view of the natural product inventory. *Nature Chemical Biology*, *11*(9), 620–624. https://doi.org/10.1038/nchembio.1894.

Weber, E., Engler, C., Gruetzner, R., Werner, S., & Marillonnet, S. (2011). A modular cloning system for standardized assembly of multigene constructs. *PLoS One*, *6*(2), e16765 https://doi.org/10.1371/journal.pone.0016765.

Welander, P. V., Coleman, M. L., Sessions, A. L., Summons, R. E., & Newman, D. K. (2010). Identification of a methylase required for 2-methylhopanoid production and implications for the interpretation of sedimentary hopanes. *Proceedings of the National Academy of Sciences of the United States of America*, *107*(19), 8537–8542. https://doi.org/10.1073/pnas.0912949107.

Wells, J. N., Bergendahl, L. T., & Marsh, J. A. (2016). Operon gene order is optimized for ordered protein complex assembly. *Cell Reports*, *14*(4), 679–685. https://doi.org/10.1016/j.celrep.2015.12.085.

Werner, W. J., Allen, K. D., Hu, K., Helms, G. L., Chen, B. S., & Wang, S. C. (2011). In vitro phosphinate methylation by PhpK from *Kitasatospora phosalacinea*. *Biochemistry*, *50*(42), 8986–8988. https://doi.org/10.1021/bi201220r.

Wilson, M. C., Mori, T., Rückert, C., Uria, A. R., Helf, M. J., Takada, K., et al. (2014). An environmental bacterial taxon with a large and distinct metabolic repertoire. *Nature*, *506*(7486), 58–62. https://doi.org/10.1038/nature12959.

Zhang, H., Boghigian, B. A., Armando, J., & Pfeifer, B. A. (2011). Methods and options for the heterologous production of complex natural products. *Natural Product Reports*, *28*(1), 125–151. https://doi.org/10.1039/C0NP00037J.

Zhang, Q., van der Donk, W. A., & Liu, W. (2012). Radical-mediated enzymatic methylation: A tale of two SAMs. *Accounts of Chemical Research*, *45*(4), 555–564. https://doi.org/10.1021/ar200202c.

Zhou, S., Alkhalaf, L. M., de Los Santos, E. L., & Challis, G. L. (2016). Mechanistic insights into class B radical-*S*-adenosylmethionine methylases: Ubiquitous tailoring enzymes in natural product biosynthesis. *Current Opinion in Chemical Biology*, *35*, 73–79. https://doi.org/10.1016/j.cbpa.2016.08.021.

CHAPTER NINE

Biosynthetic Insights of Calyculin- and Misakinolide-Type Compounds in "*Candidatus* Entotheonella sp."

Agustinus R. Uria*[,1], Jörn Piel[†,1], Toshiyuki Wakimoto*[,1]
*Faculty of Pharmaceutical Sciences, Hokkaido University, Sapporo, Japan
[†]Institute of Microbiology, Eidgenössische Technische Hochschule (ETH) Zurich, Zurich, Switzerland
[1]Corresponding authors: e-mail address: agustinus.uria@gmail.com; jpiel@ethz.ch; wakimoto@pharm.hokudai.ac.jp

Contents

Abstract

Microbial symbionts are recognized as the important sources of numerous sponge-derived metabolites with potent biological activities. The limitation to cultivate the majority of potential symbionts has hampered attempts to explore and exploit their natural products for further development toward medical applications. Metagenomics-guided approaches have enabled cloning of natural product biosynthesis genes from uncultured microbial symbionts. Subsequent activation of biosynthesis genes in easily culturable bacteria could lead to the sustainable production of rare sponge-derived compounds. In this chapter, we highlight metagenomic strategies to reveal natural product biosynthetic pathways in sponge metagenomes based on the calyculin and

Methods in Enzymology, Volume 604
ISSN 0076-6879
https://doi.org/10.1016/bs.mie.2018.02.017

misakinolide polyketides. Techniques to identify the compound producer are briefly discussed. We further describe examples of functional studies of the biosynthetic pathways of these two compound types with a special emphasis on the general experimental protocols for the activity assays of key proteins involved in their biosynthesis.

ABBREVIATIONS

A domain adenylation domain
ACP acyl carrier protein
AT acyltransferase
CARD-FISH catalyzed reporter deposition-fluorescence in situ hybridization
ECH enoyl-CoA hydratases
FACS fluorescence-activated cell sorting
HMGS 3-hydroxy-3-methylglutaryl synthase
HPLC high-performance liquid chromatography
HPLC-DAD HPLC diode array detector
KR ketoreductase
KS ketosynthase
LCMS liquid chromatography–mass spectrometry
MALDI-IMS matrix-assisted desorption–ionization imaging mass spectrometry
MDA multiple displacement amplification
NGS next-generation sequencing
NRPS nonribosomal peptide synthetase
PCR polymerase chain reaction
PKS polyketide synthase

1. INTRODUCTION

Among marine organisms, sponges are the dominant sources of potent natural products with 283 new structures reported in 2014 (Blunt, Copp, Keyzers, Munro, & Prinsep, 2016). Sponges that belong to the order Tetractinellida (previously called Lithistida) (Morrow & Cárdenas, 2015) are especially well known as the important source of highly complex bioactive secondary metabolites with more than 300 structurally diverse compounds reported from this order, including polyketides, peptides, alkaloids, pigments, lipids, and sterols (Bewley & Faulkner, 1998; D'Auria, Zampella, & Zollo, 2002; Fusetani & Matsunaga, 1993; Winder, Pomponi, & Wright, 2011). Many natural products isolated from marine sponges are proposed to be produced by the associated bacterial symbionts due to the frequently observed phenomena that they are structurally similar to typical bacterial compounds (Bewley & Faulkner, 1998; Piel, 2004). However, the limitation to cultivate the majority of the microbial symbionts outside their host (Friedrich, Fischer, Proksch, Hacker, & Hentschel, 2001; Webster & Hill, 2001) has hampered attempts to harness their potential natural products using cultivation-based approaches. Metagenomics has emerged as a powerful tool to investigate the vast majority of hidden uncultured microbial diversity from any environmental sample for the discovery of natural product biosynthetic pathways and other applications. Using metagenomics-guided approaches, a number of sponge-derived natural product biosynthetic pathways were discovered (Agarwal et al., 2017; Fisch et al., 2009; Freeman et al., 2012; Piel et al., 2004; Wakimoto et al., 2014; Wilson et al., 2014; Ueoka et al., 2015).

Two notable examples of sponge-derived complex compounds, in which their biosynthetic pathways, have been characterized using metagenomic approaches are calyculin A from *Discodermia calyx* (Wakimoto et al., 2014) and misakinolide A from *Theonella swinhoei* (Ueoka et al., 2015). These two compound types exhibit interesting modes of action and promising therapeutic potential (Bubb, Spector, Bershadsky, & Korn, 1995; Ishihara et al., 1989; Klenchin, King, Tanaka, Marriott, & Rayment, 2005; Saito et al., 1998; Terry, Spector, Higa, & Bubb, 1997); and therefore, discovery of their biosynthetic pathways not only provides convincing proof for the bacterial origin of sponge-derived natural products but also facilitates the sustainable supply of rare bioactive compounds through heterologous production in easily culturable bacteria. Moreover, with the genes in hand, pathways could be genetically altered to produce structurally novel analogues with improved pharmacological profiles (Uria & Piel, 2009). In this chapter, we describe metagenomics-guided approaches that we have developed to gain access to the biosynthetic pathways of sponge natural products, exemplified by calyculin A and misakinolide A. Furthermore, we specifically describe strategies and protocols about functional studies of the discovered biosynthetic pathways, which can generally be applied and developed for biosynthetic pathways of other natural products.

1.1 Calyculin-Type Compounds

Calyculin A was first reported in the marine sponge *D. calyx* by Fusetani, Matsunaga, and coworkers. It exhibited potent activity based on the starfish egg assay (Kato et al., 1986). Other analogues of calyculin A discovered in *D. calyx* include calyculin B, C, and D (Kato, Fusetani, Matsunaga,

Hashimoto, & Koseki, 1988), E, F, G, and H (Matsunaga, Fujiki, Sakata, & Fusetani, 1991). It was proposed that calyculins B, E, and F are photochemically derived from calyculin A, whereas calyculins D, G, and H are presumably derived from calyculin C (Egami, Wakimoto, & Abe, 2014). Calyculin-type compounds were also reported from other species of marine sponges, including calyculinamides in the Epipolasid sponge *Lamellomorpha strongylata* from New Zealand deepwater (Dumdei, Blunt, Munro, & Pannell, 1997), *Luffariella geometrica* from Australia (Kehraus, König, & Wright, 2002), swinhoeiamide A in *T. swinhoei* from the Papua New Guinea (Edrada et al., 2002), and clavosines in the Astrophorid sponge *Myriastra clavosa* from Palau (Fu, Schmitz, Kelly-Borges, McCready, & Holmes, 1998) (Fig. 1). The occurrence of calyculin-type compounds among different sponge species and in geographically different locations suggests the microbial origin of their biosynthesis.

Calyculin-type compounds showed pronounced inhibitory activity toward a series of human cancer cell lines. Calyculin A showed inhibitory activity against several cancer cell lines (Suganuma et al., 1990; Wakimoto, Matsunaga, Takai, & Fusetani, 2002). They are also potent natural insecticides (Okada, Watanabe, Umeda, & Miyakado, 1991). Swinhoeiamide A has been shown to exhibit insecticidal and antifungal activities (Edrada et al., 2002). Calyculin-type compounds specifically inhibit protein phosphatases 1 and 2A,

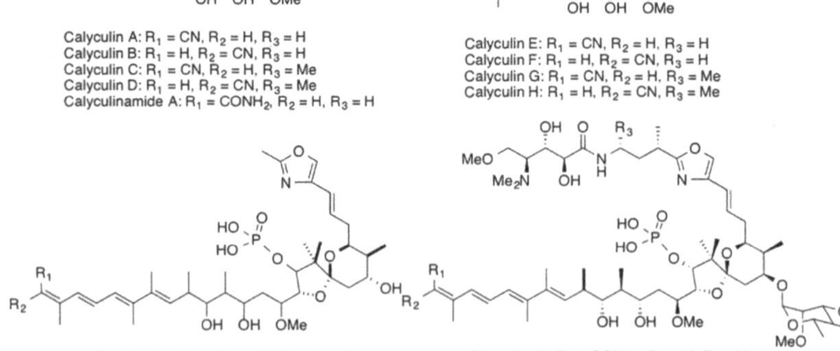

Fig. 1 Calyculin A and its derivatives.

in which the phosphate group plays a vital role for binding with the target proteins (Ishihara et al., 1989; Kita et al., 2002). Owing to their interesting biological activities, they are commercially marketed as molecular probes for studying intracellular signal transduction, as are okadaic acid and microcystin (Lindvall, Pihko, & Koskinen, 1997). Wakimoto, Egami, Abe, Piel, and coworkers have recently identified a biosynthetic gene cluster of calyculin A from the Japanese *D. calyx*. This exceptionally large gene cluster spanned over 150 kb (Wakimoto et al., 2014). In Section 2, we described experimental strategies to target the biosynthetic pathway of a complex polyketide/peptide hybrid, exemplified by calyculin A, followed with the protocol to functionally study one of the pathway components in Section 3.

1.2 Misakinolide-Type Compounds

Misakinolide A was initially identified as a monomeric 20-membered macrolide in *Theonella* sp. (Sakai, Higa, & Kashman, 1986), which was then revised as a symmetric cyclic 40-membered dimer (Kato et al., 1987). It is structurally related to a 44-membered swinholide A (Carmely & Kashman, 1985; Doi, Ishida, Kobayashi, & Kitagawa, 1991; Kitagawa et al., 1990) and hurghadolide A (Youssef & Mooberry, 2006). Although these three misakinolide-type compounds are potent actin polymerization inhibitors (Carmely & Kashman, 1985; Kitagawa et al., 1990; Youssef & Mooberry, 2006), they show different mechanisms of action. Misakinolide caps the growing barbed end of filaments, while swinholide A severs actin filaments (Bubb et al., 1995; Bubb & Spector, 1998; Terry et al., 1997). Hurghadolide A was 10 times more potent than swinholide A at disrupting microfilaments (Youssef & Mooberry, 2006). Misakinolide-type compounds are structurally similar to some macrolides reported from other organisms (Ueoka et al., 2015), such as luminaolide from the crustose coralline algae *Hydrolithon reinboldii* (Kitamura, Schupp, Nakano, & Uemura, 2009), tolytoxin from the terrestrial cyanobacterium *Scytonema* (Carmeli, Moore, & Patterson, 1990; Ishibashi, Moore, Patterson, Xu, & Clardy, 1986), and lobophorolide from the seaweed *Lobophora variegata* (Kubanek et al., 2003) (Fig. 2). These macrolides share core skeletons (indicated by red color) with various extended moieties (indicated by blue and green colors) shared between members of different families (Helfrich & Piel, 2016) (Fig. 2). The occurrence of these structurally related macrolides in other organisms suggests that misakinolide-type compounds are produced by the associated bacterial symbionts. This symbiosis hypothesis was experimentally supported by Faulkner and coworkers that showed the cellular localization of swinholide

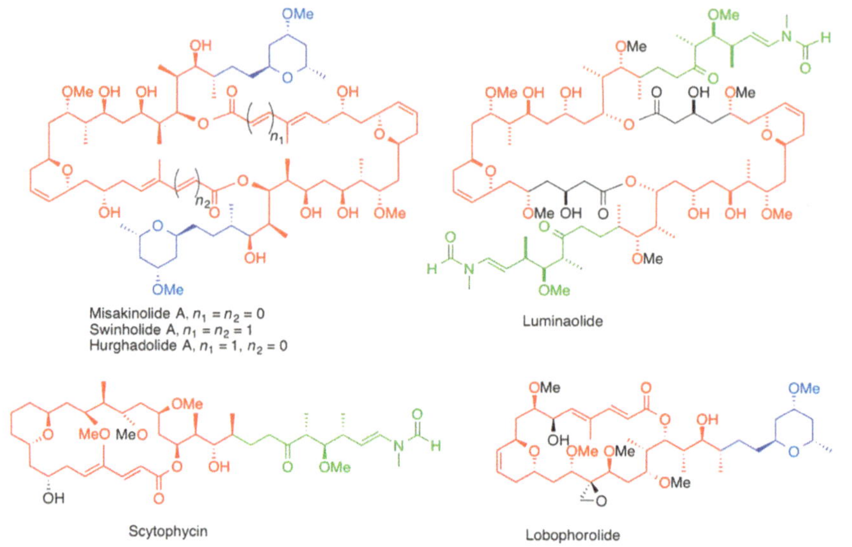

Fig. 2 Misakinolide-type compounds and structurally closely macrolides reported from diverse sources. The *same color* regions share structural features (Ueoka et al., 2015).

A in the mixed unicellular bacterial fraction of the Palauan *T. swinhoei*, while the antifungal theopalaumide mainly in the filamentous bacteria (Bewley, Holland, & Faulkner, 1996). A further separate study by Gerwick and coworkers indicated the ability of free-living cyanobacteria collected from two different locations to produce swinholide A (Andrianasolo et al., 2005). More recently, Sivonen and coworkers identified swinholide A in the terrestrial cyanobacterium *Nostoc* sp. UHCC 0450 (Humisto et al., 2017).

From bioactivity point of view, misakinolide-type compounds and the structurally related macrolides, except luminaolide, have been shown to target actin with similar binding characteristics. The cyclic core skeleton binds to the actin-binding site surface, while an extended "tail" region incorporates into a narrow cleft, leading to sequestration of globular actin (G-actin), severing of filamentous actin (F-actin), and capping of filaments ends (Allingham, Klenchin, & Rayment, 2006; Allingham, Zampella, D'Auria, & Rayment, 2005). Owing to the interesting mode of action, these actin-targeting macrolides have not only been used to study the organization, dynamics, and function of actin cytoskeleton (Braet, De Zanger, Jans, Spector, & Wisse, 1996; Bubb, Spector, Beyer, & Fosen, 2000; Rothkegel et al., 1996) but also can potentially be developed in the treatment of cancer and other human diseases (Senderowicz et al., 1995; Stingl, Andersen, & Emerman, 1992).

More recently, Piel and coworkers cloned a gene cluster encoding misakinolide biosynthesis from the metagenome of the Japanese *T. swinhoei* containing misakinolide A (Ueoka et al., 2015). Interestingly, this misakinolide biosynthetic pathway contains an additional component (Ueoka et al., 2015) that is a perfect match with the structures of swinholide A and hurghadolide A (Carmely & Kashman, 1985; Doi et al., 1991; Kitagawa et al., 1990; Youssef & Mooberry, 2006). Comparative study of the biosynthetic pathways of misakinolide-type compounds and other structurally related macrolides (scytophycin-like compounds and luminaolide) showed that the similar pathways evolved from their ancestral pathways by recombination process via loss exchange of the modular components (Ueoka et al., 2015). Strategies to target the biosynthetic pathway of a complex polyketide, exemplified by misakinolide A, followed with its functional study are discussed in the next sections.

1.3 Polyketide and Nonribosomal Peptide Biosynthesis

Structures of misakinolide-type compounds belong to polyketide class, a large group of structurally diverse secondary metabolites assembled by successive decarboxylative Claisen condensations between a thioesterified malonate derivate and an acyl thioester (Chan, Podevels, Kevany, & Thomas, 2009). Structures of calyculin-type compounds consist of peptide segment and polyketide segment connected by an oxazole ring. The peptide part harbors two λ-amino acids, while a polyketide part contains a 5,6-spiroacetal, phosphate, tetraene, and a terminal nitrile. These structural properties suggest a nonribosomal peptide and polyketide hybrid biosynthetic pathway mediates calyculin biosynthesis (Wakimoto, Egami, & Abe, 2016).

Complex bacterial polyketides and nonribosomal peptides are synthesized on large polyketide synthase (PKS) and nonribosomal peptide synthetase (NRPS) enzymes, respectively. To date, PKSs are grouped into three types that are different in the catalytic domain organization: types I, II, and III (Shen, 2003). Type I PKSs are multifunctional enzymes organized into modules. Each module consists of catalytic domains that perform one cycle of polyketide elongation and modification. One module minimally consists of two domains: an acyl carrier protein (ACP) domain that plays a role as an anchor for the building block and a ketosynthase (KS) domain that elongates the polyketide chain. An acyltransferase (AT) domain, responsible for selecting an appropriate acyl-CoA building block, is an additional domain either integrated into each module or external. Optional domains can be present to perform various functional modifications on the β-position (Piel, 2010; Rawlings, 2001). Since the PKS module architecture and function often correspond to the resulting intermediates, a polyketide structure can be predicted based on PKS domain organizations. This correlation, called the colinearity rule, is usually applied for a subclass of modular type I PKSs called "*cis*-AT PKSs" with an AT domain integrated in each module (Hertweck, 2009). However, the subclass *trans*-AT PKSs lacks integrated AT domains and contains enzymatic features to generate diverse complex polyketides (Piel, 2010). A free-standing AT domain usually acts in *trans* to load acyl building blocks into the *trans*-AT PKS assembly line (Cheng, Tang, & Shen, 2002).

Among domains present in *trans*-AT PKS system, the KS domain is ideally used as a basis for predicting structures of intermediates and even final biosynthetic products for two main reasons: (i) KS sequences are the most conserved sequences among PKS domains (Aparicio et al., 1996) and (ii) phylogenetic analysis of KS domains allows not only classifying them regarding whether they belong to fatty acid, *cis*-AT, or *trans*-AT systems, but also predicting which type of intermediate that the corresponding *trans*-AT PKS module elongates (Nguyen et al., 2008). Based on the modular architecture combined with the functional and phylogenetic studies of KS domains, structures of polyketides encoded on *trans*-AT PKS genes can accurately be predicted as has been shown for many polyketides (Helfrich & Piel, 2016), including thailandamide (Nguyen et al., 2008), psymberin (Fisch et al., 2009), elansolid (Teta et al., 2010), and sorangicin (Irschik et al., 2010). If the metagenome sequences are not available, this targeting approach is particularly useful for the isolation of rare *trans*-AT PKS gene clusters from marine sponges containing a large number of PKS-like fatty acid synthases (FASs) genes as frequently found in marine sponge metagenomes (Fieseler et al., 2007; Hochmuth & Piel, 2009).

Biosynthetic formation of nonribosomal peptides derives from 20 proteinogenic amino acids and a larger variety of nonproteinogenic amino and other acids as the building blocks during chain elongation and postassembly, resulting in a remarkable structural and functional diversity (Fischbach & Walsh, 2006). A NRPS is composed of a series of modules, in which each module incorporates and modifies an amino acid unit into the growing peptide chain. Therefore, the module order and number as well as domain composition in a NRPS pathway reflect the sequence and number of building blocks in the peptide product generated (Cane & Walsh, 1999; Konz & Marahiel, 1999; Schwarzer & Marahiel, 2001). A typical NRPS module contains at least three domains: an adenylation (A) domain for amino acid activation and incorporation, a thiolation (T) domain for thioesterification of the activated amino acid residue, and a condensation (C) domain for elongation of the growing peptide chain through transpeptidation. Among the three domains, A domain belong to the best-characterized NRPS domains. The first crystal structure of phenylalanine-specific A domain (Conti, Stachelhaus, Marahiel, & Brick, 1997) allowed subsequent determination of 10 amino acid residues generally responsible for substrate specificity (Stachelhaus, Mootz, & Marahiel, 1999). Such 10 residues, known as nonribosomal codes, along with the refinements (Challis, Ravel, & Townsend, 2000; Stachelhaus et al., 1999) are subsequently very useful as a basis for predicting the amino acid building blocks of a nonribosomal peptide (Röttig et al., 2011).

2. TARGETING BIOSYNTHETIC PATHWAYS

Many marine sponges harbor highly complex microbial symbionts that may account up to 60% of their total biomass with the bacterial density between 6.4×10^8 and 1.5×10^9 cells per mL of sponge extract (Scheuermayer et al., 2006). The sponge-associated microbial symbionts may consist of hundreds to thousands of microbial species (Taylor, Radax, Steger, & Wagner, 2007). Such high microbial abundance and diversity result in high complexity of many sponge metagenomes encoding numerous homologous genes from highly diverse biosynthetic pathways (Piel, 2010). Thus, searching for biosynthetic pathways in marine sponges and isolating their gene clusters can be highly challenging without effective targeting strategies. In this section, we describe metagenomic strategies that were developed to isolate biosynthetic gene clusters from sponges, using calyculin- and misakinolide-type compounds as examples. The strategies include detection of key DNA regions involved in the biosynthesis, generation, and efficient high-throughput screening of complex metagenomic libraries, and accurate bioinformatic prediction of biosynthetic pathways. Similar strategies are applicable for the biosynthetic pathways of other polyketides that belong to the subclass *trans*-AT PKSs, which play an important role in the production of polyketide in bacteria, such as sorangicin (Irschik et al., 2010), nosperin (Kampa et al., 2013), and toblerols (Ueoka, Bortfeld-Miller, Morinaka, Vorholt, & Piel, 2017).

2.1 Key Biosynthesis Gene Detection

Detection of key gene regions involved in the biosynthesis of a natural product requires a prior general information on its possible biosynthetic origin. As mentioned before, the calyculin structure contains a peptide and a polyketide portion. Misakinolide-type compounds can be predicted to be entirely polyketide derived. Calyculin is particularly characterized by the presence of two unusual β-branched methyl groups at C3 and C7 on the tetraene moiety, suggesting its biosynthesis involves a *trans*-AT PKS system. Unusual branching at the β-position of acyl-ACP intermediates in a process resembling isoprenoid-like methylation (Calderone, Iwig, Dorrestein, Kelleher, & Walsh, 2007; Calderone, Kowtoniuk, Kelleher, Walsh, & Dorrestein, 2006; Gu et al., 2006) is performed by many *trans*-AT PKS systems (Helfrich & Piel, 2016), such as pederin and onnamide A (Piel, 2002; Piel et al., 2004), bryostatins (Sudek et al., 2007), bacillaene (Moldenhauer, Chen, Borriss, & Piel, 2007), except curacin A (Calderone et al., 2006; Chang et al., 2004) (Fig. 3). For this reason, the diagnostic detection of the calyculin pathway was directed to one of the *trans*-AT PKS components.

Detection of key regions involved in calyculin biosynthesis can be based on phylogenetic analysis of KS domains derived from the *D. calyx* metagenome to identify KS sequences that belong to *trans*-AT PKSs. This was initially conducted with polymerase chain reaction (PCR) amplification of KS domains using degenerate primers (Piel, 2002), followed by generating

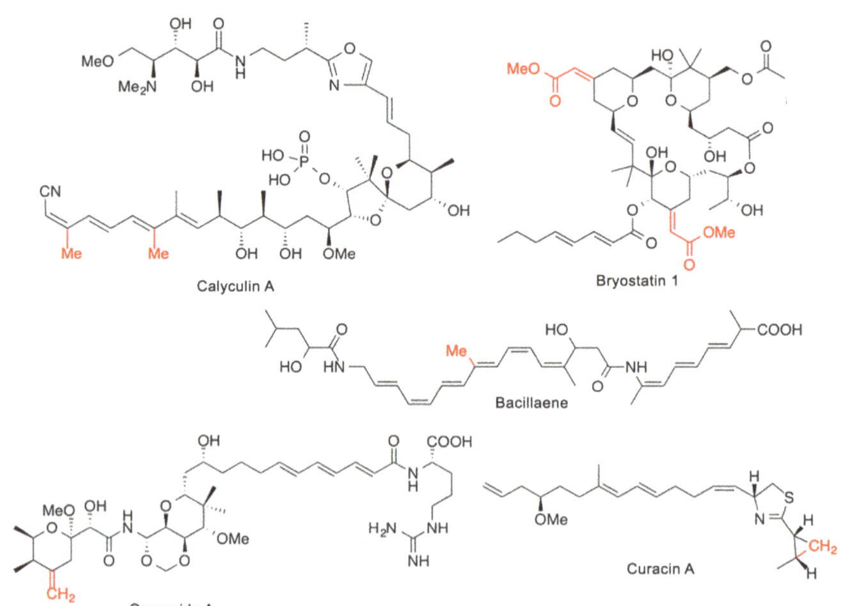

Fig. 3 β-Branching in calyculin A and other polyketides showing various β-branched moieties (*red*).

a KS-amplicon library and subsequent clone sequencing. Of the 58 cloned amplicons chosen for sequencing, 31 belonged to *trans*-AT PKSs, which subsequently became a basis for designing probes to screen a 250,000-membered metagenomics fosmid library. The screening resulted in isolating 12 fosmids that covered the majority of calyculin biosynthetic genes (Wakimoto et al., 2014). To obtain more evidence about the involvement of the isolated fosmids in coding for calyculin biosynthesis, such fosmids were subjected to further screening to detect gene regions responsible for the incorporation of β-branching as well as for the formation of the peptide portion containing an oxazole ring and a hydroxylated trimethylseryl unit (Wakimoto et al., 2014). It was assumed that the β-branch methyl incorporation in the biosynthesis of calyculins may occur through a 3-hydroxy-3-methylglutaryl synthase (HMGS)-like enzyme and two enoyl-CoA hydratases (ECHs) (Calderone et al., 2006; Gulder, Freeman, & Piel, 2011) (Fig. 4). Therefore, detecting the gene region encoding β-branching was carried out using degenerate primers based on one of the β-branching components, in this term HMGS motifs.

Biosynthetic formation of an oxazole ring usually begins with serine activation and incorporation into a NRPS assembly line by A domain followed by cyclodehydration to an oxazoline ring and subsequent oxidation to the oxazole by an FMN-dependent dehydrogenase, as has recently

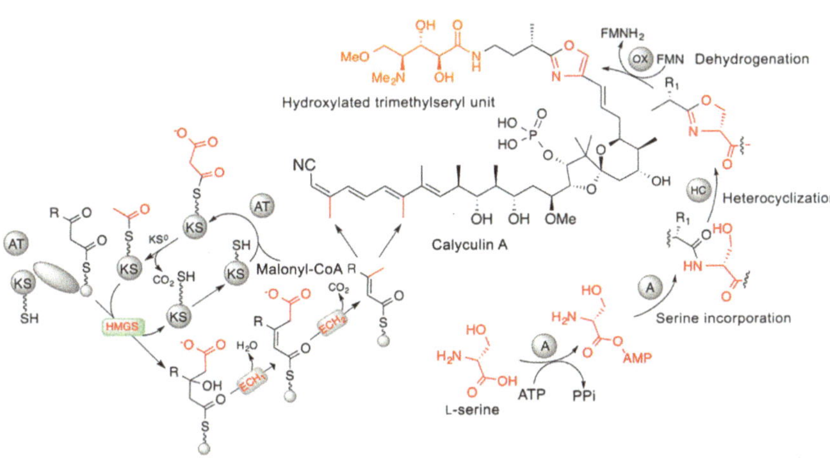

Fig. 4 The proposed biosynthetic origin of some moieties in calyculin A, which became a basis for identifying the biosynthesis genes. Formation of two β-branching, an oxazole ring, and a hydroxylated trimethylseryl unit suggests a hybrid of *trans*-AT PKS and NRPS catalyzes calyculin biosynthesis; and therefore, the diagnostic primers can be based on key components, such as KS domain, HMGS, and A domain.

been reported by Leadlay and coworkers for the C2-symmetrical macrodiolide conglobatin (Zhou et al., 2015). The formation of a hydroxylated trimethylseryl unit corresponding to the distal end of calyculin A may occur through serine activation onto an NRPS module by an A domain, and subsequent methylation of both the hydroxyl and amino groups of the uploaded serine. Based on this prediction, degenerate primers based on A domains were used for detecting the gene regions coding for the peptide portion formation (Fig. 4). Employing degenerate primers based on the conserved A3 and A7 motifs of A domains (Ayuso-Sacido & Genilloud, 2005) as well as the HMGS-like motifs (Calderone et al., 2006) led to the identification of a fosmid that harbor DNA regions coding for an expected A domain and HMGS (Wakimoto et al., 2014). This information suggested that detection of key biosynthetic regions involved in calyculin biosynthesis can be based on KS domains, β-branching proteins, and A domains (Wakimoto et al., 2014).

In terms of misakinolide-type compounds, the presence of tetrahydropyran (THP) rings and the suspected symbiotic origin suggested that a *trans*-AT PKS catalyzes their biosynthesis. In *trans*-AT PKSs, THP and furan rings are often formed by pyran synthase (PS) domains that are usually integrated in the minimal module architecture KS-DH-KR-DH-PS-ACP. A PS resembles DH domains, but it contains a modified active site in the catalytic motif HxxxGxxxxP (Pöplau, Frank, Morinaka, & Piel, 2013). Some examples of compounds containing THP rings, in which their biosynthetic pathways belong to *trans*-AT PKSs that harbor such minimal architecture, include pederin (Piel, 2002), bryostatin (Sudek et al., 2007), and sorangicin (Irschik et al., 2010). The function of the PS domain in the pederin biosynthetic pathway was experimentally demonstrated using open-chain synthetic substrates (Pöplau et al., 2013) (Fig. 5B). Based on the hypothesis that biosynthesis of misakinolide-type compounds occurs via a *trans*-AT PKS, degenerate primers based on KS domain motifs DPQQ and HTGTxLGD were used for the initial diagnostic detection of misakinolide biosynthetic pathway in the Japanese sponge *T. swinhoei* chemotype white (WA) (Wilson et al., 2014). This chemotype was characterized with white interior and contains misakinolide A (also known as bistheonellide A) (Kato et al., 1987) and theonellamides (30–35) (Matsunaga, Fusetani, Hashimoto, & Wälchli, 1989).

PCR amplification and cloning of DNA regions coding for KS domains using the sponge metagenome as a template resulted only in diverse KS sequences that belonged to the Sup (sponge ubiquitous PKS) group. Sup-type PKSs, highly abundant in many sponge microbiomes, contain a

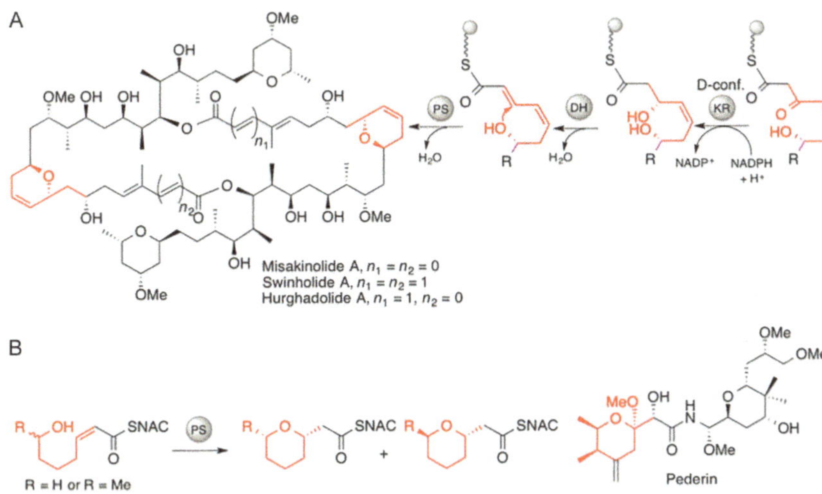

Fig. 5 Misakinolide-type compounds containing pyran rings. Pyran ring formation often involves a PS domain integrated in a *trans*-AT module. The function of PS domains in pyran formation has been studied in the biosynthetic pathways of misakinolide (A) and pederin (B).

single or more PKS modules that architecturally resemble animal FASs (KS-AT-DH-MT-ER-KR-ACP) and are possibly involved in biosynthesis of methyl-branched fatty acids (Fieseler et al., 2007). Therefore, none of these PKSs were expected to catalyze the biosynthesis of complex polyketides. To identify rare DNA species encoding polyketide biosynthesis, the sequence complexity of the sponge microbiome was reduced by differential centrifugation to separate symbiotic microbial consortia from sponge cells and to divide the consortia into less complex bacterial fractions (Piel et al., 2004). Two types of bacterial fractions were obtained: (i) a filamentous bacterial fraction dominated by "*Candidatus* Entotheonella" and (ii) a unicellular bacterial fraction (Wilson et al., 2014). Based on the earlier localization studies by Bewley et al. (1996), we hypothesized that misakinolide A (structurally similar to swinholide A) was present in the unicellular bacteria fraction, while theonellamide F (structurally similar to theopalaumide) was in the filamentous bacterial fraction. Our initial study was focused on the filamentous bacterial fraction to reveal the biosynthetic potential of "*Candidatus* Entotheonella" because it has unusual large morphology that is taxonomically different from known cultivated bacteria; and therefore, it is an interesting model to study the metabolic capacity of uncultured-yet bacteria (Brück et al., 2008; Schmidt, Obraztsova, Davidson, Faulkner, & Haygood, 2000; Wilson et al., 2014). To obtain preliminary insight into the polyketide biosynthetic potential of

"Entotheonella," PCR amplification with degenerate KS primers was carried out using the filamentous fraction as a template. Surprisingly, of the six different KS amplicons obtained, four of them were phylogenetically assigned to *trans*-AT PKS genes (Wilson et al., 2014). Four different KS sequences belong to *trans*-AT PKS were chosen as a basis to isolate PKS gene clusters for two main reasons: (i) biosynthesis of misakinolide A is most likely catalyzed by a *trans*-AT PKS due to the presence of two pyran rings and (ii) *trans*-AT PKS systems are associated with biosynthesis of pharmacologically potential complex polyketides, exemplified by onnamides (Nguyen et al., 2008; Piel et al., 2004) and psymberin (Fisch et al., 2009). Using such KS sequences as probes to screen a metagenomic library from the sponge *T. swinhoei* led to the isolation of misakinolide biosynthetic gene cluster. Furthermore, this gene cluster was attributed to "*Candidatus* Entotheonella" (Ueoka et al., 2015), as described in Section 2.3.

2.2 Biosynthetic Gene Cluster Cloning

Identification of natural product biosynthesis gene clusters in a highly complex microbial community such as sponge-associated microbial symbionts can be done by total metagenome sequencing. However, isolation of gene clusters encoding natural product biosynthesis is the issue here, which requires two things: (i) a sufficiently large-sized metagenomic library (typically consisting of more than 250,000 clones with an insert size range of 30–40 kb) and (ii) effective high-throughput screening of the complex library. In construction of a complex metagenomic library, three parameters that need to be considered in metagenome preparation are DNA quantity, size, and purity. Typically, at least 500 ng of sponge metagenomic DNA is needed for downstream steps. Obtaining large DNA fragments of around 40 kb is important for generating large-insert libraries based on fosmid vectors. The resulting libraries subsequently need to be screened to isolate clones that cover entire biosynthetic gene clusters of interest. The number of fosmid clones to be isolated from the libraries depends on the size of a biosynthetic gene cluster. Positive fosmid clones are subjected to DNA sequencing followed with bioinformatic analysis to predict if they correspond to a target biosynthetic gene cluster. The detailed procedures of how to construct and screen a fosmid-based metagenomic library are described by Hrvatin and Piel (2007) and Gurgui and Piel (2010). We described later how the procedures were applied to isolate gene clusters encoding the biosynthesis of calyculin A and misakinolide A (Fig. 6).

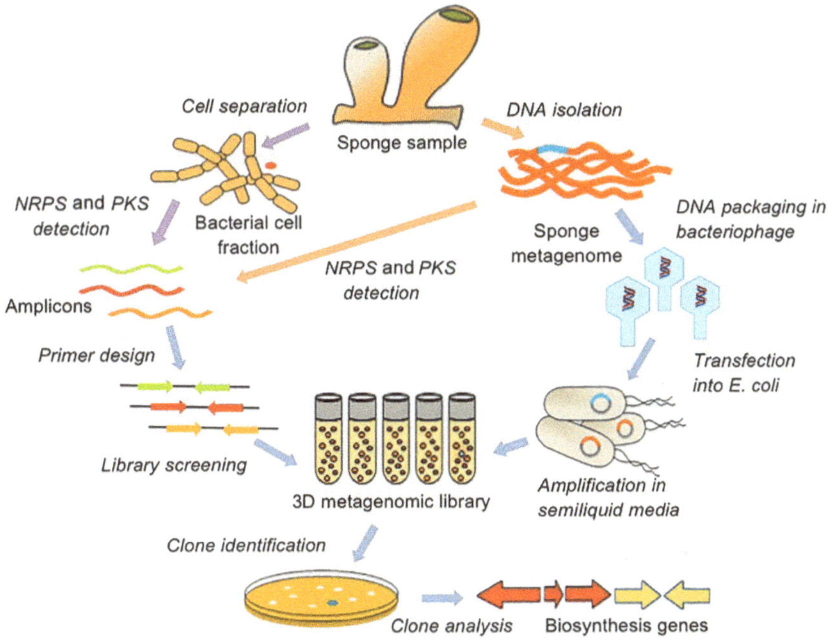

Fig. 6 General strategies for the construction and screening of 3D metagenomic libraries from marine sponges, developed for isolating onnamide biosynthesis genes (Piel et al., 2004), which were subsequently applied for obtaining calyculin biosynthesis genes (Wakimoto et al., 2014) and misakinolide biosynthesis genes (Ueoka et al., 2015).

2.2.1 Complex Metagenomic Library Construction and Screening

There are a number of methods to prepare high molecular weight (HMW) DNA from environmental samples (Uria & Zilda, 2016). For sponge samples, we usually employ a cetyltrimethylammonium bromide (CTAB)-based DNA isolation method (Piel et al., 2004) modified from Murray and Thompson (1980). This method begins with grinding a small amount of sponge sample in liquid nitrogen, followed with cell suspension in a lysis buffer containing lauryl sarcosyl and urea at high concentration and subsequent incubation at high temperature to allow efficient breakdown. The lysate is then subjected to phenol and chloroform extractions to remove proteins, lipids, and other organic interferences, followed with CTAB treatment to remove polysaccharides. Finally, the DNA is recovered and concentrated by isopropanol and ethanol precipitation (Gurgui & Piel, 2010; Hrvatin & Piel, 2007). As an alternative, microbial cells can initially be separated from the sponge tissue by differential centrifugation prior to cell lysis, as described later.

The following step-by-step procedure for the isolation of metagenomic DNA from a sponge sample (Fig. 7) is based on the protocol described by Fisch et al. (2009) and Gurgui and Piel (2010) with slight modification.

Steps 1a–3a for a whole sponge sample

1a. Place a sponge sample (~1 g) in a sterile mortar and cut it into small pieces.

2a. Pulverize the sample pieces in the mortar using a pestle under the liquid nitrogen.

3a. After the nitrogen is evaporated, transfer the homogenized sample immediately into a 50-mL falcon tube, gently suspend with 10 mL of lysis buffer (50 m*M* Tris–Cl, pH 7.5, 50 m*M* EDTA, 700 m*M* NaCl, 2% sodium lauroyl sarcosinate, 8 *M* urea), and incubate at 60°C for 20 min.

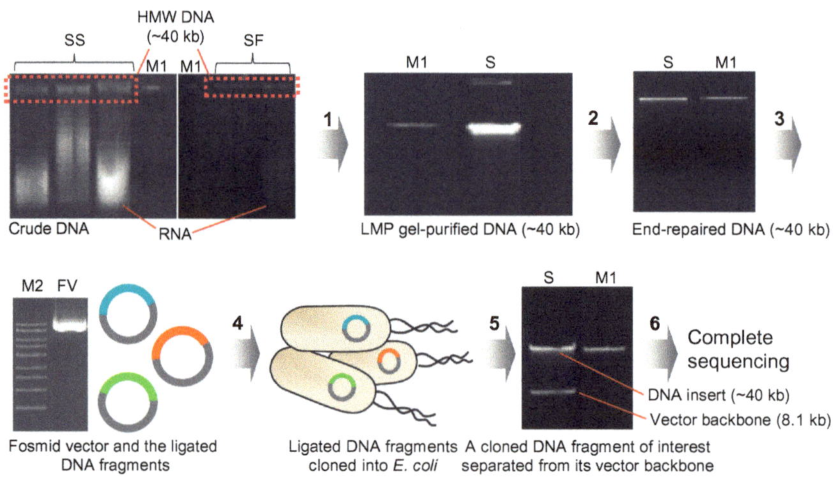

Fig. 7 Isolation and cloning of HMW metagenomic DNA from a marine sponge sample for metagenomic library construction. Total DNA obtained from the whole sponge sample or a microbial cell fraction was purified by LMP-gel extraction or chromatography column-based gel extraction (step 1). The resulting pure DNA (~40 kb) was subjected to end repairmen (step 2). The end-repaired DNA fragments were ligated with a blunt-ended and dephosphorylated fosmid vector of 8.1 kb (Epicentre) (step 3), and cloned into *E. coli* via phage-based transfection (step 4), resulting in a metagenomic library. After metagenomic library screening, a cloned DNA construct was prepared from a positive recombinant *E. coli* clone, treated with ATP-dependent DNase to remove the host chromosomal DNA, digested with the rare-cutting enzyme *Not*I to remove the vector backbone, and sent for insert sequencing. *Notes*: *S*, sample; *SS*, whole sponge-derived total DNA sample; *SF*, microbial cell fraction-derived total DNA sample; *M1*, 36-kb linear fosmid control marker (10 ng); *M2*, 1-kb DNA ladder; and *FV*, 8.1-kb fosmid vector.

Steps 1b–3b for bacterial cells prepared from a sponge sample

1b. Place a sponge sample (~4 g) in a 50-mL falcon tube.

2b. Add 10 mL sterile MQ water and squeeze the sample using a sterile stick (such as magnetic stirring bar retriever) to release microbial cells. Perform differential centrifugation of the supernatant containing microbial cells as follow:

- Transfer the supernatant into a 15-mL falcon tube and centrifuge at 200 × *g* for 5 min at 4°C.
- Transfer the supernatant into a 15-mL falcon tube and centrifuge at 600 × *g* for 5 min at 4°C.
- Transfer the supernatant into a 15-mL falcon tube and centrifuge at 1500 × *g* for 5 min at 4°C.
- Transfer the supernatant into a 15-mL falcon tube and centrifuge at 20,000 × *g* for 20 min at 4°C.

3b. Suspend the cell pellet obtained from each level of differential centrifugation with 2.5–5 mL lysis, depending on cell biomass, immerse it in liquid nitrogen for 5 min, and incubate at 60°C for 20 min, while mixing the suspension slowly by inverting 5–7 times.

4. Extract the resulting lysate (either from a whole sponge sample or from a bacterial cell suspension) with two times with an equal volume of phenol/chloroform/isoamyl alcohol mixture (in ratio 25:24:1) and centrifuge at 13,000 rpm at 5 min for each extraction time.

5. Transfer the upper layer into an empty falcon tube, add with alkaline proteinase K (Promega) at the final concentration of 50 μg/mL, and incubate at room temperature for approximately 1 h.

6. Add with an equal volume of chloroform, centrifuge at 13,000 rpm at 5 min, take the upper layer and transfer to a new falcon tube.

7. (Optional) This step is usually applied for a whole sponge sample, not for bacterial cells prepared from a sponge sample. Add the upper layer with CTAB (previously warmed at 55°C for 15 min) at the final concentration of 0.5%, mixed carefully, and incubated at 60°C for 10 min. Centrifuge the CTAB-treated extract at 13,000 rpm to precipitate the polysaccharides.

8. Add with 2 volumes of 100% cold ethanol or 0.7 volume of isopropanol, and with 0.1 volume of 3 *M* sodium acetate pH 7. Mix it gently by inverting the tube several times, store at room temperature for 1 h at room temperature or in the refrigerator overnight to improve the DNA precipitation, and centrifuge at the maximal speed (4°C) for 30 min.

9. Wash the resulting DNA pellet on the tube bottom with two times with cold 70% ethanol to remove the remaining salts and CTAB traces.
10. Dry the DNA pellet through air flow, and dissolve it in 200 μL of ddH_2O or storage buffer (EDTA 10 m*M*, Tris–Cl 50 m*M*, pH 7.0). To improve the DNA solubility, keep the wet DNA sample overnight in the refrigerator.
11. (Optional) RNA amount can be removed by adding RNAse at the final concentration of 10 μg/mL and subsequent 1-h incubation at 37°C. The RNA-free DNA can be recovered by isopropanol and ethanol precipitation as described earlier (steps 8–10) or chloroform extraction (step 6). Alternatively, RNA removal can be done by gel extraction on low melting point (LMP) gel.
12. Check the dissolved DNA (1–2 μL) by electrophoresis on 1% agarose gel. Fig. 7 showed HMW DNA isolated from a sponge sample on 1% agarose gel both in crude extract form and LMP gel-purified form.

For fosmid-based metagenomic library construction, it is necessary to use pure HMW DNA in the size range of 30–40 kb (Fig. 7). We usually obtain pure HMW DNA through electrophoretic separation on LMP agarose gel and subsequent gel extraction using a β-agarose-digesting enzyme such as GELase (Epicentre). This enzyme is able to break the carbohydrate backbone in the LMP gel into small and soluble oligosaccharides, allowing HMW DNA purification (http://www.epibio.com; Gurgui & Piel, 2010). During DNA isolation, the resulting DNA fragments may have frayed ends that are not blunt and without phosphate residues at 5′-termini, and therefore they require further enzymatic processing prior to ligation with a fosmid vector. The ready-to-use fosmid vector pCC1FOS™ (Epicentre) used in our previous metagenomic work is in the linear form with dephosphorylated blunt ends. To allow effective ligation with this fosmid vector, the frayed ends for DNA fragments should be made blunt using T4 DNA polymerase that fills in the recessed 5′-termini and remove the protruding 3′-termini, while the 5′-OH ends should be converted to the 5′-phosphate group using terminal phosphotransferase (Brown, 2006). The end-repaired DNA fragments are subsequently ligated with the pCC1FOS vector and packaged in lambda phage particles. The phage particles are subsequently used to transfect *Escherichia coli*, thereby resulting in recombinant clones harboring metagenomic DNA fragments derived from a sponge sample (Fig. 7). To create growing well-separated clones at high densities between 1000 and 4000 clones/mL (Ueoka et al., 2015), the clones can be grown in semiliquid medium, thereby creating a 3D library format (Hrvatin & Piel, 2007). The resulting 3D library can conveniently be preserved in glycerol and stored at −80°C for several years.

Generating the 3D format of a metagenomic library as described earlier allows further screening to identify clones that harbor portions of a target biosynthesis genes (Fig. 6). To rapidly screen the library, the pooling strategy reported by Hrvatin and Piel (2007) can be used. This screening strategy in principle consists of three steps: clone pooling, PCR detection, and clone dilution. Positive clone pools are diluted and regrown at lower densities and screened further until individual positive clones are detected. Once a positive clone is isolated from a metagenomic library, primer walking strategy can be performed in order to identify other clones harboring an entire biosynthetic gene cluster. For the misakinolide pathway, in total five positive clones were isolated and subjected to complete sequencing. It is often necessary to get general information inside an isolated clone by subcloning prior to complete sequencing (Ueoka et al., 2015). Another way to obtain insight into the genetic content of a clone before complete sequencing is by PCR detection using degenerate primers, as has been shown for screening of clones encoding the upstream part of calyculin biosynthetic pathway. In total, 12 clones were isolated from the fosmid library of the *D. calyx* metagenome, some of which were screened further using degenerate primers based on conserved motifs of HMGS and A domains, leading to identification of a positive clone proposed to encode the formation of the β-branching and peptide portion of calyculin A (Wakimoto et al., 2014). Obtaining positive clones from metagenomic libraries which are verified by subcloning and/or PCR detection provides a basis for further complete sequencing of all clones that cover the entire biosynthesis gene cluster of a target compound. Prior to complete sequencing of positive clones, the contaminating linear host chromosomal DNA can selectively be removed from the recombinant fosmid preparation using ATP-dependent DNase, leaving circular fosmid DNA intact (Epicentre; www.epibio.com/). A rare-cutting enzyme (*Not*I in case of using the pCC1FOS vector) can subsequently be used to remove the vector backbone. Large inserts without vector backbone are completely sequenced through a de novo sequencing approach (Fig. 7).

2.2.2 Bioinformatic Analysis

Using illumina-based method, one of the next-generation sequencing (NGS) approaches, Wakimoto et al. (2014) sequenced all 12 fosmid clones isolated from the *D. calyx* metagenomic library. The assembly and another round sequencing to close sequence gaps between contigs generated an assembled sequence ranged over 200 kb. Subsequent annotation of the entire assembled sequence revealed that it contained an over 150-kb biosynthesis gene cluster that encodes NRPS and *trans*-AT PKS modules architecturally corresponding to a calyculin biosynthetic pathway (Wakimoto et al., 2014). For the misakinolide pathway, complete sequencing, assembly, and annotation of five overlapping fosmid clones derived from *T. swinhoei* metagenomic library revealed a continuous sequence of ~103 kb that encodes *trans*-AT PKS modules corresponding to misakinolide-type compounds (Ueoka et al., 2015).

Bioinformatic analyses to predict the intermediates and final products generated by PKS and/or NRPS pathways generally involve two following steps: (i) predicting the module architecture and domain composition and (ii) sequence analysis to get preliminary insight into the function of each domain by identifying key motifs that might define the installed intermediates and stereochemistry. Some well known online softwares that can be helpful to predict the module architecture and domain composition rapidly include Basic Alignment Search Tool (BLAST) (Altschul, Gish, Miller, Myers, & Lipman, 1990; McGinnis & Madden, 2004) that can be run in the Geneious program (Drummond et al., 2011), PKS/NRPS analysis website (Bachmann & Ravel, 2009), Natural Product Domain Seeker (NaPDoS) (Ziemert et al., 2012), Antibiotics and Secondary Metabolite Shell (antiSMASH) (Blin et al., 2017; Medema et al., 2011), and NP.searcher (Li, Ung, Zajkowski, Garneau-Tsodikova, & Sherman, 2009). Furthermore, analysis of key domains in the NRPS and PKS pathways can provide more insight into the resulting intermediates and even stereochemistry. For example, NRPSpredictor2 (Röttig et al., 2011) as a part of antiSMASH and NRPSsp (Prieto, García-Estrada, Lorenzana, & Martín, 2012) is used to predict nonribosomal codes, thereby revealing substrate specificity of A domains in NRPS modules.

For *trans*-AT PKS modules, characterized by the lack of ATs integrated in each module and often with noncanonical modifications, accurate prediction of intermediates is based on the phylogenetic analysis of KS domains (Piel, 2010). For PKS modules harboring KR domains that reduce β-carbonyl group on their intermediates, the stereochemistry of the resulting chiral center having hydroxyl group can be predicted based on the KR motifs, LDD and PxxxN (Baerga-Ortiz et al., 2006; Caffrey, 2003; Reid et al., 2003). It is important to analyze the amino acid sequences of individual domains for the presence or absence of important catalytic conserved motifs or essential conserved residues. For example, a functional DH domain usually contains the conserved HxxxGxxxxP motif near to its N-terminus (Bevitt, Cortes, Haydock, & Leadlay, 1992) as well as the conserved Dxxx(Q/H) motif. The histidine (H) and gylcine (G) residues in the first motif form the catalytic dyad for the activity (Cortes et al., 1995; Tsai, 2009), while the catalytic aspartic acid (D) in the second motif is essential for the activity (Keatinge-Clay, 2008; Valenzano et al., 2010). Functional KS domains contain a conserved HGTGT motif, of which the His residue is important for decarboxylative condensation (Davies, Heath, White, & Rock, 2000). Deletion or degradation of such typical catalytic motifs may indicate that domains are nonfunctional.

The biosynthetic enzymes for calyculin A and misakinolide A contain unusual PKS modules with noncanonical features. Some noncanonical modifications found in calyculin PKS modules, as suggested through bioinformatic analysis, include nonelongating KS domains that account for one-third of total 29 KS domains (modules 2, 4, 8, 10, 13, 17, 20, 22, and 32), *N*-MT and *O*-MT domains (1, 2, and 22), and module splits between two proteins (module 32). Sequence analysis of KR domains indicated that the generated β-hydroxyl groups were in perfect match with the stereochemistry in the calyculin A structure, except for the OH group that links to a phosphate moiety. The NRPS/PKS genes are interrupted by a monooxygenase gene that is most likely involved in a chain shortening mechanism. The domain order, module architecture, predicted substrate specificity of A domains, and stereochemical motifs of KS domains strongly suggested that the discovered *trans*-AT PKS pathway is in good agreement with the structure of calyculin A (Fig. 8) (Wakimoto et al., 2014).

For misakinolide biosynthesis, bioinformatic analysis of a *trans*-AT PKS modules that mediate misakinolide biosynthesis revealed some unusual features, such as three modules split between two distinct proteins (modules 5, 10, and 14), some presumed inactive domains exemplified by nonelongating KSs (modules 3, 11, and 19), multiple MTs and O-MTs that introduce methyl groups (modules 3, 5, 6, 7, 9, 11, and 17), and a PS domain generating a pyran ring (module 15). Furthermore, phylogenetic analysis of KS domains and subsequent analysis of a diagnostic motif in KR domains indicated that the pathway perfectly matched to the structure and stereochemistry of misakinolide A (Fig. 9). Interestingly, the module architecture contains a penultimate module (module 18) that would rather match to the extended congeners swinholide A and hurghadolide A. However, chemical analysis of the sponge extract through LC-HRMS showed the presence of misakinolide A and other closely related analogues (didesmethylmisakinolide A, bistheonellide B, hydroxymisakinolide A, and dihydroxymisakinolide A), but not the extended congeners (Ueoka et al., 2015.)

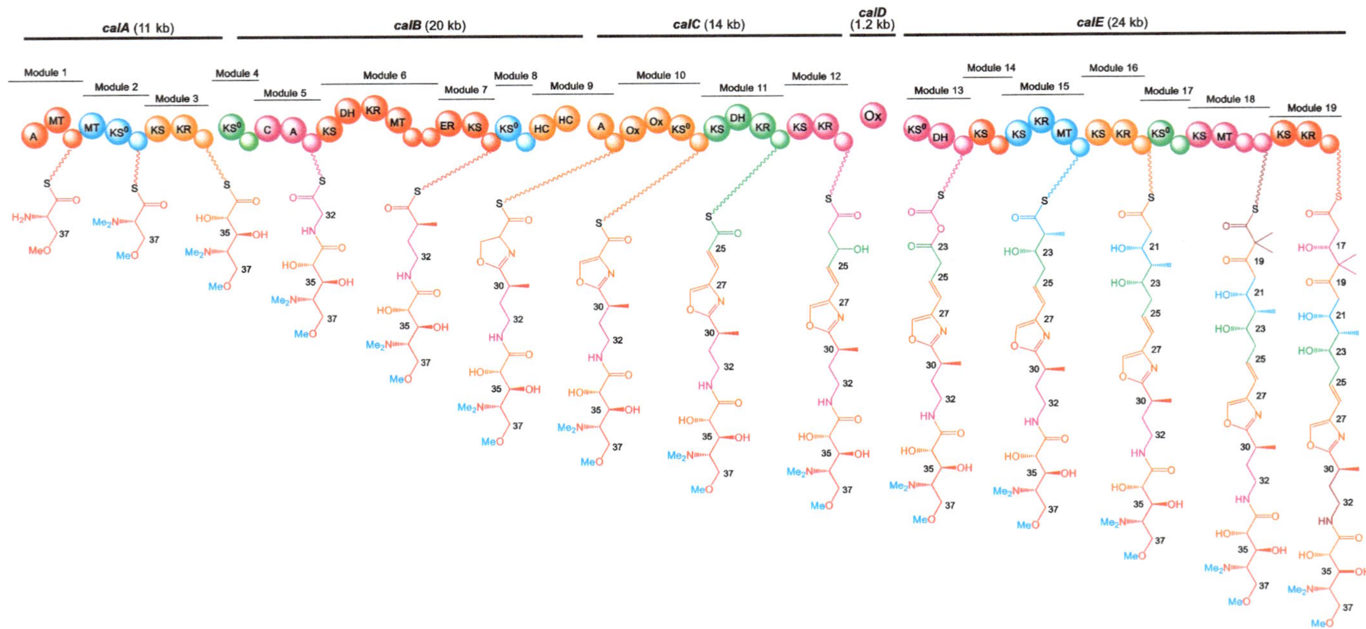

Fig. 8 The proposed biosynthetic pathway for calyculin A based on the bioinformatic analysis of the gene cluster isolated from the metagenome of the Japanese sponge *D. calyx*. The isolated gene cluster encodes multifunctional enzymes (CalA–CalI). The upstream cluster region outside of the calyculin assembly line encodes AT (CalY), KS^0 (CalW), HMGS (CalT), and two ECHs (CalS and R), which could be responsible for the introduction of β-branched methyl groups on C3 and 7 of the tetraene moiety in cooperation with modules 26 and 28, respectively (Fig. 4). Other modification enzymes encoded on the upstream region are presumably involved in the tailoring reactions to transform nascent PKS and NRPS product into mature calyculin A, which is finally phosphorylated by CalQ to generate phosphocalyculin A (Fig. 10).

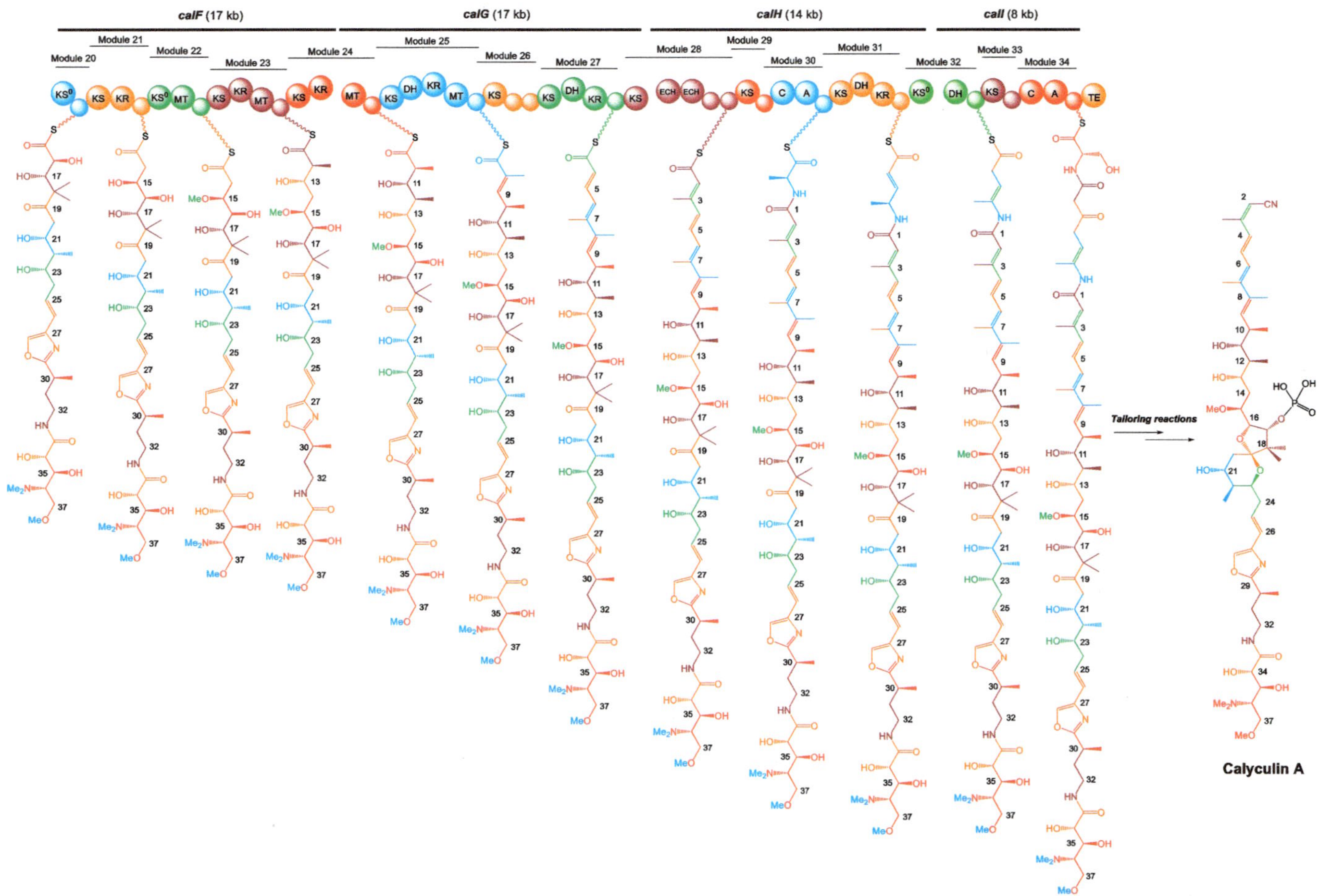

calF (17 kb)
calG (17 kb)
calH (14 kb)
calI (8 kb)
Module 20
Module 21
Module 22
Module 23
Module 24
Module 25
Module 26
Module 27
Module 28
Module 29
Module 30
Module 31
Module 32
Module 33
Module 34
Tailoring reactions
Calyculin A

Fig. 8—Cont'd

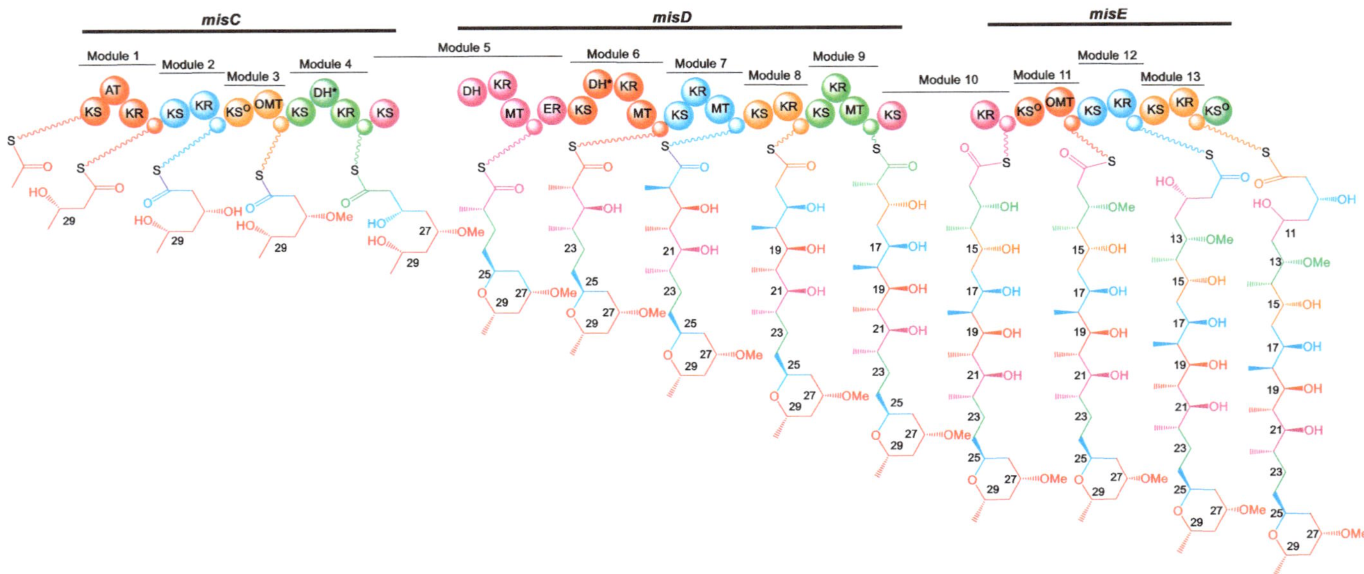

Fig. 9 The proposed biosynthetic pathway for misakinolide A based on the bioinformatic analysis of a gene cluster cloned from the metagenome of the Japanese sponge *T. swinhoei*. The module architecture, intermediate specificity, and stereochemical configuration were predicted based on the analyses of KS and KR sequences. It consists of multifunctional enzymes (MisC–MisG) with 15 elongation modules that correspond with the incorporation of 15 malonyl extender units. MisF contains module 18 that matches to the structures of swinholide A and hurghadolide A. Since misakinolide A instead of swinholide A was detected in the sponge, it was proposed that module 18 is nonfunctional and may facilitate the transfer of the misakinolide A monomer from ACP17 to ACP18 without elongation. The TE domain at the downstream end of the misakinolide assembly line likely ligates two misakinolide monomers (one that binds to TE and another one to the terminal ACP), resulting in cyclization to form the dimeric misakinolide A.

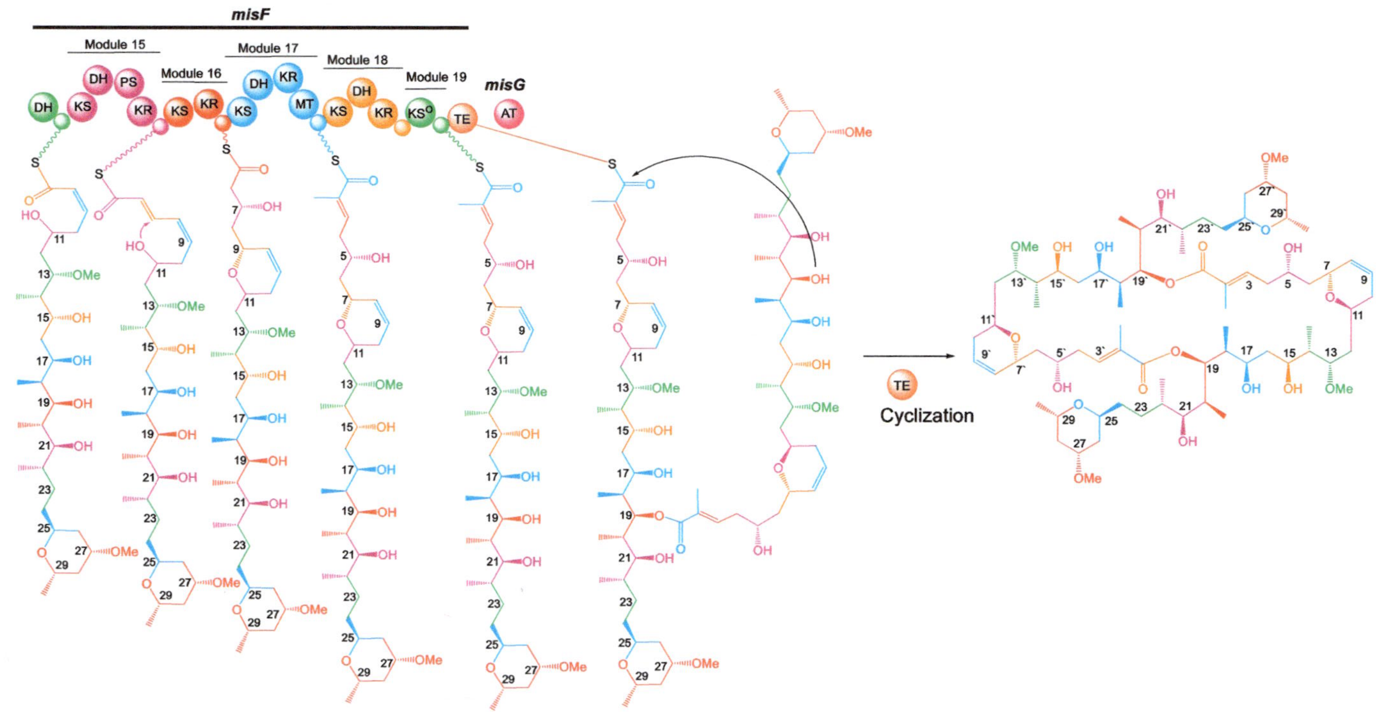
misF
Module 15
Module 16
Module 17
Module 18
Module 19
misG
DH
KS
PS
KR
MT
TE
AT
Cyclization

Fig. 9—Cont'd

2.3 Compound Producer Identification

The isolated calyculin and misakinolide biosynthesis clusters showed a typical bacterial architecture, such as tightly packed genes, lack of introns and polyadenylation sites, and preceding Shine–Dalgarno, suggesting bacterial origin of calyculin-type and misakinolide-type compounds (Ueoka et al., 2015; Wakimoto et al., 2014). To connect metabolites or biosynthetic genes to uncultured bacteria, a number of methods are available, such as differential centrifugation (Bewley et al., 1996) and percoll density gradient (Ridley et al., 2005) for cell separation followed by chemical analysis of extracts, catalyzed reporter deposition-fluorescence in situ hybridization (CARD-FISH) (Pernthaler, Pernthaler, & Amann, 2002), fluorescence-activated cell sorting (FACS) (Unson, Holland, & Faulkner, 1994; Zoetendal et al., 2002), laser microdissection for the isolation of bacterial cells, multiple displacement amplification (MDA) for whole-genome amplification (Binga, Lasken, & Neufeld, 2008; Chen et al., 2008), and matrix-assisted desorption–ionization imaging mass spectrometry (MALDI-IMS) for compound localization (Yang et al., 2012).

To identify the producer of calyculin A, microscopic observation of the homogenate of the calyculin-containing sponge *D. calyx* was initially carried out, showing the presence of two filamentous bacterial population types that are different in colors and morphologies (large size and small size with bright color) (Wakimoto et al., 2014). CARD-FISH analysis (Amann & Fuchs, 2008; Pernthaler et al., 2002) revealed the location of the calyculin biosynthesis gene cluster only in the large filamentous bacteria (Wakimoto et al., 2014). Furthermore, using laser microdissection and subsequent PCR detection (Gloess, Grossart, Allgaier, Ratering, & Hupfer, 2008), single filaments of the large filamentous bacteria were individually isolated and identified as a member of the candidate genus "Entotheonella" (Kimura et al., 2012; Wakimoto et al., 2014), an environmental taxon belonging to the proposed candidate phylum "Tectomicrobia" that was previously also shown to harbor "Entotheonella factor" TSY1, the producer of almost all bioactive polyketides and peptides known from the Japanese *T. swinhoei* chemotype Y (Wilson et al., 2014).

To identify the producer of misakinolide, extended regions outside the misakinolide biosynthesis gene cluster were initially analyzed. These exhibited >80% nucleotide identity to the previously reported sequenced genomes of "E. factor" and an accompanying variant in *T. swinhoei* Y, "*Candidatus* Entotheonella gemina" TSY2 (Wilson et al., 2014). Mechanical separation of the filamentous cell fraction from the misakinolide chemotype *T. swinhoei* WA by differential centrifugation and subsequent pyrosequencing of 16S rRNA gene amplicons suggested accumulation (approximately 70%) of an "Entotheonella" phylotype designated as "Entotheonella serta" TSWA1 (98.3% identity of 16S rRNA to that of "E. factor") (Ueoka et al., 2015). Furthermore, single-bacterial filament separation by FACS, MDA (Binga et al., 2008), and PCR analysis indicated the presence of misakinolide biosynthesis genes as well as the "Entotheonella" 16S gene in the amplified DNA of some single filaments. Additional evidence for "Entotheonella" as producers was obtained by MALDI-IMS and CARD-FISH analysis, revealing a highly localized distribution of misakinolide and "Entotheonella" filaments in thin tissue regions at the same confined locations (Ueoka et al., 2015). In conclusion, the integrated techniques of differential centrifugation, FACS or laser microdissection, MDA, PCR, MALDI-IMS, CARD-FISH, and gene homologies all support "Entotheonella" are responsible for the biosynthesis of the polyketides.

3. FUNCTIONAL STUDIES OF KEY PATHWAY COMPONENTS

To experimentally verify the function of the cloned genes in the putative calyculin biosynthetic pathway, three putative phosphotransferase genes (*calM*, *calP*, and *calQ*) clustered at the upstream region were investigated as candidates for phosphate processing through phosphorylation and dephosphorylation (Wakimoto et al., 2014). Such functional study was based on the finding that phosphocalyculin A was accumulated in the freeze-dried and liquid N_2-frozen fresh sponge tissue, while calyculin A was the main metabolite in the sponge extract prepared in the conventional way, suggesting the presence of enzyme-catalyzed reactions that perform phosphorylation and dephosphorylation. Coexpression of the three genes in *E. coli* with the chaperons GroES and GroEL resulted in soluble forms. Further activity assays with calyculin A and calyculinamide A as the substrates revealed that CalQ catalyzed monophosphorylation reaction that converted calyculin A into phosphocalyculin A, as verified by HPLC, LCMS, and NMR analyses (Wakimoto et al., 2014). This finding suggested the ecological role of CalQ in generating the less active diphosphate form to protect the host sponge from strong toxicity of calyculin A. A dephosphorylation reaction that converted the less toxic form phosphocalyculin A into the much more toxic

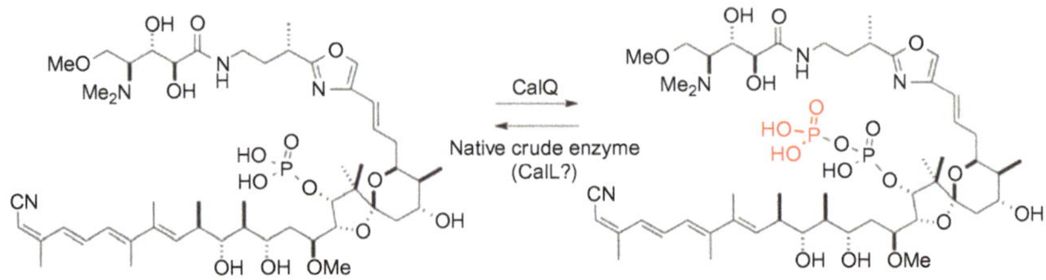

Fig. 10 Enzymatic phosphorylation and dephosphorylation in calyculin biosynthesis.

form calyculin A was observed using crude enzyme prepared from the fresh sponge, suggesting the presence of an enzyme that plays an ecological role to activate chemical defense upon tissue wounding (Wakimoto et al., 2014) (Fig. 10).

In the functional study of misakinolide biosynthetic pathway, the PS domain was analyzed, which is located in module 15 of MisF (Fig. 9). This PS was predicted to form a dihydropyran ring (Fig. 10A) that is similar to the mechanism previously reported for the PS domain of the pederin biosynthetic pathway (Pöplau et al., 2013). The function of the putative *mis* PS domain was experimentally verified by heterologous expression of the encoded gene region and subsequent activity assay of the recombinant domain using a synthetic substrate (Fig. 11B). The products (THP) were characterized by NMR (Ueoka et al., 2015) (Fig. 11B).

3.1 General Concept for Protein Activity Assays

There are a number of methods or assays to test the activity of proteins or enzymes involved in polyketide and nonribosomal peptide biosynthesis, depending on protein types or kinds. The domain activities can be evaluated in vitro using putative intermediates required by polyketide and nonribosomal peptide assemblies (Ehmann, Trauger, Stachelhaus, & Walsh, 2000; Gilbert et al., 1995). Synthetic intermediates in the form of SNACs, such as β-ketoacyl-SNACs and aminoacyl-SNACs with various lengths, are good mimics of the phosphopantetheine arm of ACP; and therefore, they can be recognized and processed by domains in PKS and NRPS (Cane, Prabhakaran, Tan, & Ott, 1991; Ehmann et al., 2000; Gilbert et al., 1995; Gokhale, Hunziker, Cane, & Khosla, 1998). Specificity for an A domain toward synthetic SNACs, for example, can be tested using mass spectrometry (MS)-based pyrophosphate (PPi) exchange assay (Phelan, Du, McLean, & Bachmann, 2009), as has been applied for the A domain regions encoded on the pathways of hormaomycin (Crüsemann, Kohlhaas, & Piel, 2013; Höfer et al., 2011), and keramamide D

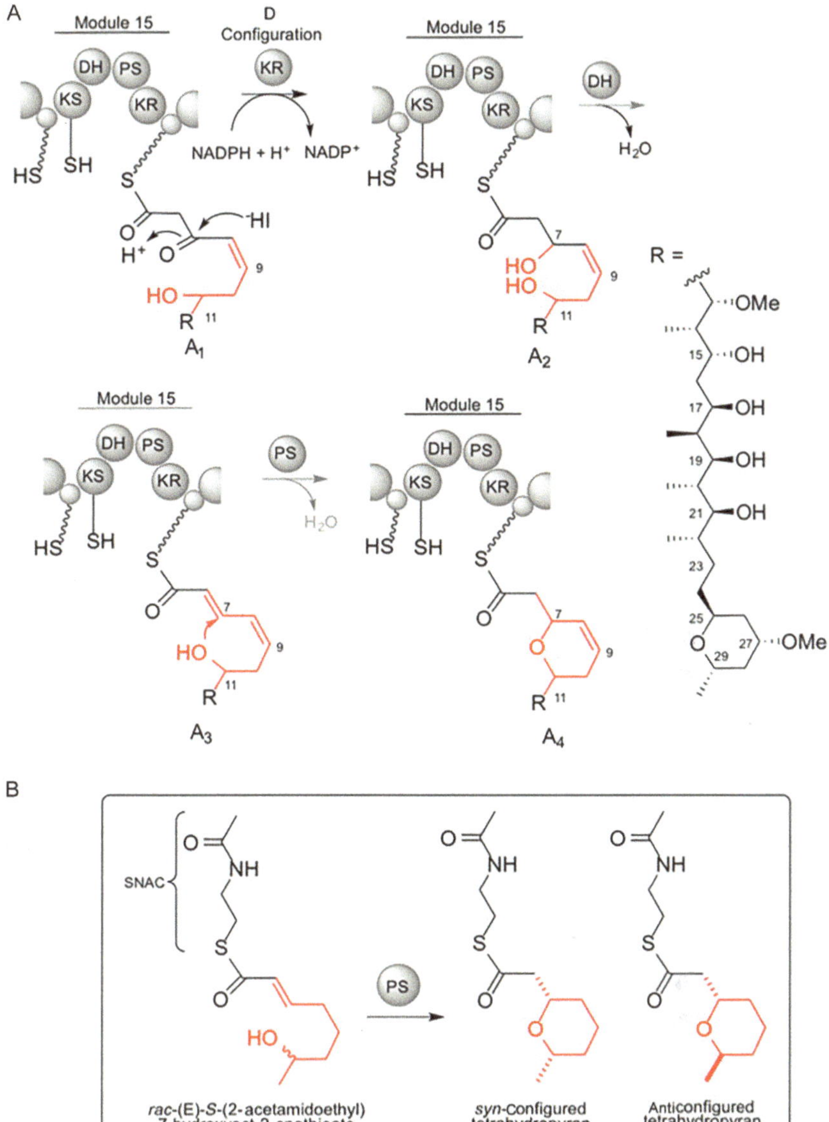

Fig. 11 Formation of the intermediate bearing a pyran ring in module 15 of misakinolide PKS pathway. (A) The proposed mechanism of the pyran ring formation. KS15 would catalyze the decarboxylative condensation between the upstream *cis*-2,3-enoyl-*S*-ACP intermediate and the downstream malonyl-*S*-ACP, generating a β-keto-*S*-ACP (A1). Further β-ketoreduction by KR15 would generate a β-hydroxylated intermediate (A2). The cognate DH would dehydrate the C9 β-OH group of A2 into an α,β-*E*-olefinic moiety (C6–C7), resulting in A3. Subsequent PS-catalyzed reduction of the C_{11} hydroxyl group to the C_7 olefinic moiety (A3) would give an intermediate bearing the C_7–C_{11} pyran ring (A4). (B) Observed pyran ring formation catalyzed by heterologously produced PS from module 15. SNAC, *N*-acetyl cysteamine thioester.

and cyclotheonamide A (Wilson et al., 2014). Testing the activity of a KS domain toward synthetic substrates, for instance, can be based on the MS-based method developed by Piel, Oldham, and coworkers to address substrate specificity of KS domains in *trans*-AT PKSs (Jenner et al., 2013; Kohlhaas et al., 2013).

Activity of a PS domain in pyran ring formation can be assayed using HPLC and LCMS analyses and subsequent NMR, as has been shown for the PS region encoded on the pathways of pederin (Pöplau et al., 2013) and misakinolide A (Ueoka et al., 2015). For phosphorylation or dephosphorylation activity, the reaction products can be detected by HPLC followed by testing their biological activity, as shown by Wakimoto et al. (2014) for phosphotransferases encoded on calyculin pathway. As the initial step to perform the functional study, the gene region (e.g., PS-encoding region) or an entire gene (e.g., phosphotransferase gene) is first PCR-amplified using a high-fidelity DNA polymerase and subsequently cloned into a heterologous expression system, such as pET expression system (Novagen) or pCold DNA system (TaKaRa). The expression constructs should be designed to allow His-tag purification. In case that the expressed proteins are insoluble forms, coexpression with one of chaperon plasmid series is often necessary to improve the solubility, as shown for the expressed phosphotransferase (CalQ) (Wakimoto et al., 2014). Experimental steps in measuring the activity of a PS domain for the ability to form a pyran ring are described later. Subsequently, assays for testing phosphorylation and dephosphorylation activities are also described.

3.2 Activity Assay for PS

Activity assay of the expressed PS domain for pyran formation described later is based on the protocol reported by Pöplau et al. (2013) and Ueoka et al. (2015). Prior to the protein activity assays, it is necessary to desalt and exchange the elution buffer with the storage buffer, as well as to concentrate the expressed protein dissolved in the storage buffer.

3.2.1 Instrument, Buffers, and Reagents

- Storage buffer (25 m*M* HEPES, 100 m*M* NaCl, 10% glycerol, pH 7.5)
- GE Healthcare PD column
- Sartorius Vivaspin 500 column (MWCO 10,000)
- Dimethyl sulfoxide (DMSO)
- Synthetic SNAC substrate: rac-(*E*)-*S*-(2-acetamidoethyl) 7-hydroxyoct-2-enethioate, prepared according to Pöplau et al. (2013)
- Synthetic SNAC standard: *syn*-configured THP (Pöplau et al., 2013)
- Ethyl acetate (EtOAc)

- Methanol
- HPLC instruments: analytical reversed phase HPLC-MS and semi-preparative reversed-phase HPLC
- HPLC columns: Knauer Eurospher II 100-5, Phenomenex column Luna C18
- $CDCl_3$
- NMR 600 MHz

3.2.2 Procedure (*Pöplau et al., 2013*; *Ueoka et al., 2015*)

1. Desalt and exchange the elution buffer for the purified PS protein with the storage buffer using GE Healthcare PD column.
2. Concentrate the expressed PS protein dissolved in the storage buffer using a Sartorius Vivaspin 500 column (MWCO 10,000).
3. Prepare the SNAC substrate dissolved in ddH_2O containing 2% DMSO ($10\times$ stock).
4. Prepare a reaction mixture of total volume of 200 μL containing c.4 mg/mL of PS protein (180 μL) and 12 m*M* SNAC substrate (20 μL), and incubate at 37°C for 16 h.
5. Extract the reaction with EtOAc (2×200 μL) and evaporate it. Resuspend the dried sample precipitate with 750 μL methanol for further HPLC and LCMS analyses.
6. Analyze the reaction assays using analytical reversed phase HPLC–MS. The following condition recommended:
   - Knauer column Eurospher II 100-5 (5 μL, $250\times 4\,mm^2$)
   - Flow rate 1 mL/min, mobile phase 50% methanol in H_2O
   - Detection $\lambda = 230$ nm

   Notes: HPLC–MS analysis can be conducted on an Agilent 1200 Series HPLC system coupled to an ESI-microQ-TOF mass spectrometer (Bruker Daltonik) or Dionex Ultimate 3000 UHPLC coupled to a Thermo Scientific Q Exactive mass spectrometer.
7. Purify the assay reactions by semipreparative reversed-phase HPLC. The following condition recommended:
   - Phenomenex column Luna C18 (5 μL, $250\times 10\,mm^2$)
   - Flow rate 3 mL/min
   - Mobile phase 30% CH_3CN in H_2O

   Notes: The reaction products (anticonfigured and *syn*-configured) were detected at different retention times (17.1 and 22.3 min, respectively).
8. Resuspended the precipitate in $CDCl_3$ for NMR measurement to verify the reaction products.

3.3 Activity Assays of Phosphotransferase

Phosphorylation activity assay of the expressed phosphotransferase and dephosphorylation activity assay of the crude enzyme as described later are based on the protocols reported by Wakimoto et al. (2014). In HPLC analysis of the reaction products, chromatographic conditions, such as mobile phase, flow rate, injection volume, column temperature, detection wavelength, need to be optimized to obtain well-separated peaks, and minimal analysis time. Gradient elution also requires optimization (Zhao, Wen, Upur, & Tian, 2017).

3.3.1 Instrument, Buffers, and Reagents

- Storage buffer (25 m*M* HEPES, 100 m*M* NaCl, 10% glycerol, pH 7.5)
- Sartorius Vivaspin 500 column (MWCO 10,000)
- DMSO
- Reaction substrates for phosphorylation and dephosphorylation activity assays
- Phenylmethylsulfonyl fluoride (PMSF)
- EtOAc
- Methanol
- HPLC instrument: DAD RP HPLC
- HPLC columns: Cosmosil MS-II C18
- $CDCl_3$
- NMR 600 MHz

3.3.2 Procedures (*Wakimoto et al., 2014*)

Phosphorylation activity assay

1. Prepare compounds as the reaction substrates, either purified from the sponge sample or provided by chemical synthesis, or both. In term of calyculin biosynthetic pathway, calyculin A and hemycalyculin A were prepared from the sponge *D. calyx*, while semisynthetic derivatives include calyculinamide A, dephospohnocalyculin A, and 11,13-*O*-isopropylidene-calyculin A (Wakimoto et al., 2002).
2. Make a reaction mixture of a total volume of 50 μL that consists of 100 m*M* Tris–HCl (pH 8.0), 10 m*M* $MgCl_2$, 1 m*M* ATP, 0.001% Triton X-100, 10 μg of the substrates, and a purified recombinant protein. Incubate the mixture at 20°C for 1 h.
3. Analyze the reaction mixture by diode array detector (DAD) RP HPLC using Cosmosil MS-II C18 ($4.6 \times 250\,mm^2$, Nacalai Tesque) with the following conditions:

- Mobile phase (A): 0.05% trifluoroacetic acid in acetonitrile (CH_3CN)
- Linear gradient elution: 0–20 min, 30%–100% A; 20–50 min, 100% A
- Flow rate: 0.8 mL/min
- Detection (λ_{max}): 340 nm

Note: Cosmosil MS-II is a monomeric type of C18 stationary phase with the pH range of 2–10, particularly recommended for LMW basic organic compounds (Nacalai Tesque).

4. Test the biological activity of the reaction product (phosphorylated compounds) against P388 mouse leukemia cells and HeLa cells according to a standard method (Wakimoto et al., 2014).

Dephosphorylation activity assay

1. Fresh sponge samples should be kept at 4°C during transportation from the sampling site to the laboratory, which is then immediately frozen in liquid nitrogen.
2. Lyophilize the flash-frozen sample and then keep it in a freezer (−30°C) until use.
3. Soak the lyophilized sample in methanol for a day and extract it. This methanol extract contains a dominant phosphorylated compound (in this case phosphocalyculin A), which is used as the substrate for the enzyme reaction.
4. Homogenize a frozen sponge sample (10 g) by grinding it in mortar and pestle on ice.
5. Extract the homogenate with 10 mL of 50 m*M* Tris–Cl pH 8.0 containing 1 m*M* PMSF on ice for 2 h.
6. Centrifuge the resulting suspension at 20,000 × *g* for 1 h. The resulting supernatant is used as the crude enzyme solution.
7. Prepare a reaction mixture consisting of 50 μL of the crude enzyme and approximately 10 μg phosphocalyculin A in 50 m*M* Tris–Cl pH 8.0. Incubated the mixture at 20°C for 3 h.
8. Analyze the reaction product using DAD RP HPLC (λ_{max} = 340 nm).
9. Test the biological activity of the reaction product (dephosphorylated compounds) against P388 mouse leukemia cells and HeLa cells according to a standard method.

4. SUMMARY AND CONCLUSION

Finding the biosynthesis genes for sponge-derived natural products could become an important basis for generating sustainable production systems of rare pharmacologically important compounds. Metagenomic strategies to search for natural product biosynthesis genes from highly complex microbial consortia, such as symbiotic bacteria in marine sponges, include identification of key gene or regions involved in biosynthesis, cloning of entire biosynthetic gene clusters, and functional studies. Preliminary detection of a potential biosynthetic gene cluster requires prior information or hypothesis on the enzymatic steps or proteins involved in biosynthesis. This may become a basis on deciding what types of degenerate primers to be used or designed to target a gene or gene regions coding for key protein. For example, two β-branched methyl moieties in calyculins and two pyran rings in misakinolide-type compounds indicate the involvement of *trans*-AT PKS systems in their biosynthesis. One of the important components in *trans*-AT PKS modules is the KS domain that can be chosen as diagnostic feature to identify biosynthesis genes. Cloning of biosynthetic gene clusters from sponge metagenomes includes construction of highly complex metagenomic libraries, efficient library screening, and powerful bioinformatic analysis for the accurate prediction of the pathway's intermediates and final products. Using the integrated cellular, molecular, and chemical techniques, such as differential centrifugation, percoll density gradient, CARD-FISH, FACS, MDA, and MALDI-IMS, have recently enabled identification of the uncultivated compound-producing bacterial symbionts. The biosynthetic pathways and final product predicted based on bioinformatic sequence analysis need to be experimentally proven by functional studies on one or more proteins involved in the biosynthesis of a target compound. Functional studies require heterologous expression of key proteins and activity assays, since gene knockout experiments are presently not applicable to the as-yet-uncultured bacteria. The heterologous expression of "Entotheonella"-derived proteins needs careful manipulations in *E. coli* expression system. It is also difficult to expect the intermediate for post-PKS modification without the disruption of biosynthetic pathway. Nevertheless, functional study with heterologously expressed protein is the only promising method to confirm the biosynthetic pathway of interest in "Entotheonella." With the technical improvements of heterologous expression system, unique biosynthetic gene clusters with novel processes will continue to emerge from uncultivated sponge symbionts.

ACKNOWLEDGMENTS

A.R.U. is an International Research Fellow of the Japan Society for the Promotion of Science (JSPS). T.W. appreciates funding by the Takeda Science Foundation, the Astellas Foundation for Research on Metabolic Disorders, the Naito Foundation and Grants-in-Aid from the Ministry of Education, Culture, Sports, Science and Technology (MEXT), Japan (JSPS KAKENHI Grant Numbers JP16703511, JP15547389, JP15597834, and JP14506930). J.P. is grateful for funding by the European Research Council (ERC Advanced Grant "SynPlex"), the Swiss National Science Foundation (205321_165695), and the Helmut Horten Foundation.

REFERENCES

Agarwal, V., Blanton, J. M., Podell, S., Taton, A., Schorn, M. A., Busch, J., et al. (2017). Metagenomic discovery of polybrominated diphenyl ether biosynthesis by marine sponges. *Nature Chemical Biology, 13*, 537–543. https://doi.org/10.1038/nchembio.2330.

Allingham, J. S., Klenchin, V. A., & Rayment, I. (2006). Actin-targeting natural products: Structures, properties, and mechanisms of action. *Cellular and Molecular Life Sciences, 63*(18), 2119–2134. https://doi.org/10.1007/s00018-006-6157-9.

Allingham, J. S., Zampella, A., D'Auria, M. V., & Rayment, I. (2005). Structures of microfilament destabilizing toxins bound to actin provide insight into toxin design and activity. *Proceedings of the National Academy of Sciences of the United States of America, 102*, 14527–14532. https://doi.org/10.1073/pnas.0502089102.

Altschul, S. F., Gish, W., Miller, W., Myers, E. W., & Lipman, D. J. (1990). Basic local alignment search tool. *Journal of Molecular Biology, 215*, 403–410. https://doi.org/10.1016/S0022-2836(05)80360-2.

Amann, R., & Fuchs, B. M. (2008). Single-cell identification in microbial communities by improved fluorescence in situ hybridization techniques. *Nature Reviews Microbiology, 6*, 339–348. https://doi.org/10.1038/nrmicro1888.

Andrianasolo, E. H., Gross, H., Goeger, D., Musafija-Girt, M., McPhail, K., Leal, R. M., et al. (2005). Isolation of swinholide A and related glycosylated derivatives from two field collections of marine cyanobacteria. *Organic Letters*, 7(7), 1375–1378. https://doi.org/10.1021/ol050188x.

Aparicio, J. F., Molnar, I., Schwecke, T., König, A., Haydock, S. F., Khaw, L. E., et al. (1996). Organization of the biosynthetic gene cluster for rapamycin in *Streptomyces hygroscopicus*: Analysis of the enzymatic domains in the modular polyketide synthase. *Gene, 169*, 9–16. https://doi.org/10.1016/0378-1119(95)00800-4.

Ayuso-Sacido, A., & Genilloud, O. (2005). New primers for the screening of NRPS and PKS-I systems in actinomycetes: Detection and distribution of these biosynthetic gene sequences in major taxonomic groups. *Microbial Ecology, 49*(1), 10–24. https://doi.org/10.1007/s00248-004-0249-6.

Bachmann, B. O., & Ravel, J. (2009). Methods for in silico prediction of microbial polyketide and nonribosomal peptide biosynthetic pathways from DNA sequence data. *Methods in Enzymology, 458*, 181–217. https://doi.org/10.1016/S0076-6879(09)04808-3.

Baerga-Ortiz, A., Popovic, B., Siskos, A. P., O' Hare, H. M., Spiteller, D., Williams, M. G., et al. (2006). Directed mutagenesis alters the stereochemistry of catalysis by isolated ketoreductase domains from the erythromycin polyketide synthase. *Chemistry & Biology, 13*, 277–285. https://doi.org/10.1016/j.chembiol.2006.01.004.

Bevitt, D. J., Cortes, J., Haydock, S. F., & Leadlay, P. F. (1992). 6-Deoxyerythronolide-B synthase 2 from *Saccharopolyspora erythraea*. Cloning of the structural gene, sequence analysis and inferred domain structure of the multifunctional enzyme. *European Journal of Biochemistry, 204*, 39–49. https://doi.org/10.1111/j.1432-1033.1992.tb16603.x.

Bewley, C. A., & Faulkner, D. J. (1998). Lithistid sponges: Star performers or hosts to the stars. *Angewandte Chemie International Edition, 37*, 2162–2178. DOI: 10.1002/(SICI)1521-3773(19980904)37:16<2162::AID-ANIE2162>3.0.CO;2-2.

Bewley, C. A., Holland, N. D., & Faulkner, D. J. (1996). Two classes of metabolites from *Theonella swinhoei* are localized in distinct populations of bacterial symbionts. *Experientia, 52*, 716–722.

Binga, E. K., Lasken, R. S., & Neufeld, J. D. (2008). Something from (almost) nothing: The impact of multiple displacement amplification on microbial ecology. *The International Society for Microbial Ecology Journal, 2*, 233–241. https://doi.org/10.1038/ismej.2008.10.

Blin, K., Wolf, T., Chevrette, M. G., Lu, X., Schwalen, C. J., Kautsar, S. A., et al. (2017). AntiSMASH 4.0—Improvements in chemistry prediction and gene cluster boundary identification. *Nucleic Acids Research, 45*, W36–W41. https://doi.org/10.1093/nar/gkx319.

Blunt, J. W., Copp, B. R., Keyzers, R. A., Munro, M. H. G., & Prinsep, M. R. (2016). Marine natural products. *Natural Product Reports*, *33*, 382–431. https://doi.org/10.1039/c5np00156k.

Braet, F., De Zanger, R., Jans, D., Spector, I., & Wisse, E. (1996). Microfilament-disrupting agent latrunculin A induces and increased number of fenestrae in rat liver sinusoidal endothelial cells: Comparison with cytochalasin B. *Hepatology*, *24*, 627–635. https://doi.org/10.1053/jhep.1996.v24.pm0008781335.

Brown, T. A. (2006). *Gene cloning & DNA analysis, an introduction* (5th ed.). Blackwell Publishing Ltd.

Bubb, M. R., & Spector, I. (1998). Use of the F-actin-binding drugs, misakinolide A and swinholide A. Methods in enzymology. In R. B. Valee (Ed.), *Analysis of actomyosin-related systems: vol. 298. Molecular motors and the cytoskeleton—Part B* (pp. 26–32): Elsevier. https://doi.org/10.1016/S0076-6879(98)98005-3.

Bubb, M. R., Spector, I., Bershadsky, A. D., & Korn, E. D. (1995). Swinholide A is a microfilament disrupting marine toxin that stabilizes actin dimers and severs actin filaments. *Journal of Biological Chemistry*, *270*, 3463–3466. https://doi.org/10.1074/jbc.270.8.3463.

Bubb, M. R., Spector, I., Beyer, B. B., & Fosen, K. M. (2000). Effects ofjasplakinolide on the kinetics of actin polymerization: An explanation for certain in vivo observations. *Journal of Biological Chemistry*, *275*, 5163–5170. https://doi.org/10.1074/jbc.275.7.5163.

Brück, W. M., Sennett, S. H., Pomponi, S. A., Willenz, P., & McCarthy, P. J. (2008). Identification of the bacterial symbiont Entotheonella sp. in the mesohyl of the marine sponge Discodermia sp. *The ISME Journal*, *2*, 335–339. https://doi.org/10.1038/ismej.2007.91.

Caffrey, P. (2003). Conserved amino acid residues correlating with ketoreductase stereospecificity in modular polyketide synthases. *Chembiochem*, *4*(7), 654–657. https://doi.org/10.1002/cbic.200300581.

Calderone, C. T., Iwig, D. F., Dorrestein, P. C., Kelleher, N. L., & Walsh, C. T. (2007). Incorporation of nonnethyl branches by isoprenoid-like logic: Multiple β-alkylation events in the biosynthesis of myxovirescin A1. *Chemistry & Biology*, *14*, 835–846. https://doi.org/10.1016/j.chembiol.2007.06.008.

Calderone, C. T., Kowtoniuk, W. E., Kelleher, N. L., Walsh, C. T., & Dorrestein, P. C. (2006). Convergence of isoprene and polyketide biosynthetic machinery: Isoprenyl-S-carrier proteins in the pksX pathway of *Bacillus subtilis*. *Proceedings of the National Academy of Sciences of the United States of America*, *103*, 8977–8982. https://doi.org/10.1073/pnas.0603148103.

Cane, D. E., Prabhakaran, P. C., Tan, W., & Ott, W. R. (1991). Macrolide biosynthesis. 6 mechanism of polyketide chain elongation. *Tetrahedron Letter*, *32*(40), 5457–5460. https://doi.org/10.1016/0040-4039(91)80057-D.

Cane, D. E., & Walsh, C. T. (1999). The parallel and convergent universes of polyketide synthases and nonribosomal peptide synthetases. *Chemistry & Biology*, *6*(12), R319–25. https://doi.org/10.1016/S1074-5521(00)80001-0.

Carmeli, S., Moore, R. E., & Patterson, G. M. L. (1990). Tolytoxin and new scytophycins from three species of *Scytonema*. *Journal of Natural Products*, *53*(6), 1533–1542. https://doi.org/10.1021/np50072a021.

Carmely, S., & Kashman, Y. (1985). Structure of swinholide A, a new macrolide from the marine sponge, *Theonella swinhoei*. *Tetrahedron Letter*, *26*(4), 511–514. https://doi.org/10.1016/S0040-4039(00)61925-1.

Challis, G. L., Ravel, J., & Townsend, C. A. (2000). Predictive, structure-based model of amino acid recognition by nonribosomal peptide synthase adenylation domains. *Chemistry & Biology*, 7, 211–224. https://doi.org/10.1016/S1074-5521(00)00091-0.

Chan, Y. A., Podevels, A. M., Kevany, B. M., & Thomas, M. G. (2009). Biosynthesis of polyketide synthase extender units. *Natural Product Reports*, *26*(1), 90–114. https://doi.org/10.1039/b801658p.

Chang, Z., Sitachitta, N., Rossi, J. V., Roberts, M. A., Flatt, P. M., Jia, J., et al. (2004). Biosynthetic pathway investigations and gene cluster of curacin A, an antitubulin natural product from the tropical marine cyanobacterium *Lyngbya majuscula*. *Gene*, *67*(1), 1356–1367. https://doi.org/10.1021/np0499261.

Chen, Y., Dumont, M. G., Neufeld, J. D., Bodrossy, L., Stralis-Pavese, N., McNamara, N. P., et al. (2008). Revealing the uncultivated majority: Combining DNA stable-isotope probing, multiple displacement amplification and metagenomic analyses of uncultivated *Methylocystis* in acidic peatlands. *Environmental Microbiology*, *10*(10), 2609–2622. https://doi.org/10.1111/j.1462-2920.2008.01683.x.

Cheng, Y.-Q., Tang, G.-L., & Shen, B. (2002). Type I polyketide synthase requiring a discrete acyltransferase for polyketide biosynthesis. *Proceedings of the National Academy of Sciences of the United States of America*, *100*(6), 3149–3154. https://doi.org/10.1073/pnas.0537286100.

Conti, E., Stachelhaus, T., Marahiel, M. A., & Brick, P. (1997). Structural basis for the activation of phenylalanine in the non-ribosomal biosynthesis of gramicidin S. *EMBO Journal*, *16*, 4174–4183. https://doi.org/10.1093/emboj/16.14.4174.

Cortes, J., Wiesmann, K. E., Roberts, G. A., Brown, M. J., Staunton, J., & Leadley, P. F. (1995). Repositioning of a domain in a modular polyketide synthase to promote specific chain cleavage. *Science*, *268*(5216), 1487–1489. https://doi.org/10.1126/science.7770773.

Crüsemann, M., Kohlhaas, C., & Piel, J. (2013). Evolution-guided engineering of nonribosomal peptide synthetase adenylation domains. *Chemical Science*, *4*, 1041–1045. https://doi.org/10.1039/C2SC21722H.

D'Auria, M. V., Zampella, A., & Zollo, F. (2002). The chemistry of lithistid sponge: A spectacular source of new metabolites. *Studies in Natural Products Chemistry*, *26*(part G), 1175–1258. https://doi.org/10.1016/S1572-5995(02)80027-8.

Davies, C., Heath, R. J., White, S. W., & Rock, C. O. (2000). The 1.8 angstrom crystal structure and active-site architecture of beta-ketoacyl-acyl carrier protein synthase III (FabH) from *Escherichia coli*. *Structure*, *8*, 185–195. https://doi.org/10.1016/S0969-2126(00)00094-0.

Doi, M., Ishida, T., Kobayashi, M., & Kitagawa, I. (1991). Molecular conformation of swinholide A, a potent cytotoxic dimeric macrolide from the Okinawan marine sponge *Theonella swinhoei*: X-ray crystal structure of its diketone derivative. *Journal of Organic Chemistry*, *56*, 3629–3632. https://doi.org/10.1021/jo00011a033.

Drummond, A. J., Ashton, B., Cheung, M., Heled, J., Kearse, M., Moir, R., et al. (2011). *Geneious v5.5*. https://www.geneious.com/.

Dumdei, E., Blunt, J., Munro, M., & Pannell, L. (1997). Isolation of calyculins, calyculinamides, and swinholide H from the New Zealand deep-water marine sponge *Lamellomorpha strongylata*. *Journal of Organic Chemistry*, *62*, 2636–2639. https://doi.org/10.1021/jo961745j.

Edrada, R., Ebel, R., Supriyono, A., Wray, V., Schupp, P., Steube, K., et al. (2002). Swinhoeiamide A, a new highly active calyculin derivative from the marine sponge *Theonella swinhoei*. *Journal of Natural Products*, *65*, 1168–1172. https://doi.org/10.1021/np020049d.

Egami, Y., Wakimoto, T., & Abe, I. (2014). Phosphocalyculin C as a pyrophosphate protoxin of calyculin C in the marine sponge *Discodermia calyx*. *Bioorganic & Medical Chemistry Letter*, *24*, 5150–5153. https://doi.org/10.1016/j.bmcl.2014.

Ehmann, D. E., Trauger, J. W., Stachelhaus, T., & Walsh, C. T. (2000). Aminoacyl-SNACS as small-molecules substrates for the condensation domains of nonribosomal peptide synthases. *Chemistry & Biology*, 7(10), 765–772. https://doi.org/10.1016/S1074-5521(00)00022-3.

Fieseler, L., Hentschel, U., Grozdanov, L., Schirmer, A., Wen, G., Platzer, M., et al. (2007). Widespread occurrence and genomic context of unusually small polyketide synthase genes in microbial consortia associated with marine sponges. *Applied Environmental Microbiology*, *73*, 2144–2155. https://doi.org/10.1128/AEM.02260-06.

Fisch, K. M., Gurgui, C., Heyke, N., van der Sar, S. A., Anderson, S. A., Webb, V. L., et al. (2009). Polyketide assembly lines of uncultivated sponge symbionts from structure-based gene targeting. *Nature Chemical Biology*, *5*(7), 494–501. https://doi.org/10.1038/nchembio.176.

Fischbach, M. A., & Walsh, C. T. (2006). Assembly-line enzymology for polyketide and nonribosomal peptide antibiotics: Logic, machinery, and mechanisms. *Chemical Reviews*, *106*(8), 3468–3496. https://doi.org/10.1021/cr0503097.

Freeman, M. F., Gurgui, C., Helf, M. J., Morinaka, B. I., Uria, A. R., Oldham, N. J., et al. (2012). Metagenome mining reveals polytheonamides as posttranslationally modified ribosomal peptides. *Science*, *338*, 387–390. https://doi.org/10.1126/science.

Friedrich, A. B., Fischer, I., Proksch, P., Hacker, J., & Hentschel, U. (2001). Temporal variation of the microbial community associated with the Mediterranean sponge *Aplysina aerophoba*. *FEMS Microbiology Ecology*, *38*(2–3), 105–113. https://doi.org/10.1016/S0168-6496(01)00185-4.

Fu, X., Schmitz, F. J., Kelly-Borges, M., McCready, T. L., & Holmes, C. F. B. (1998). Clavosines A–C from the marine sponge *Myriastra clavosa*: Potent cytotoxins and inhibitors of protein phosphatases 1 and 2A. *Journal of Organic Chemistry*, *63*, 7957–7963. https://doi.org/10.1021/jo981249q.

Fusetani, N., & Matsunaga, S. (1993). Bioactive sponge peptides. *Chemical Reviews*, *93*, 1793–1806. https://doi.org/10.1021/cr00021a007.

Gilbert, I. H., Ginty, M., O'Neill, J. A., Simpson, T. J., Staunton, J., & Willis, C. L. (1995). Synthesis of β-keto and α,β-unsaturated *N*-acetylcysteamine thioesters. *Bioorganic & Medical Chemistry*, *5*(15), 1587–1590. https://doi.org/10.1016/0960-894X(95)00276-Y.

Gloess, S., Grossart, H. P., Allgaier, M., Ratering, S., & Hupfer, M. (2008). Use of laser microdissection for phylogenetic characterization of polyphosphate-accumulating bacteria. *Applied and Environmental Microbiology*, *74*, 4231–4235. https://doi.org/10.1128/AEM.02545-07.

Gokhale, R. S., Hunziker, D., Cane, D. E., & Khosla, C. (1998). Mechanism and specificity of the terminal thioesterase domain from the erythromycin polyketide synthase. *Chemistry & Biology*, *6*, 117–125. https://doi.org/10.1016/S1074-5521(99)80008-8.

Gu, L. C., Jia, J., Liu, H., Håkansson, K., Gerwick, W. H., & Sherman, D. H. (2006). Metabolic coupling of dehydration and decarboxylation in the curacin A pathway: Functional identification of a mechanistically diverse enzyme pair. *Journal of American Chemical Society*, *128*, 9014–9015. https://doi.org/10.1021/ja0626382.

Gulder, T. M., Freeman, M. F., & Piel, J. (2011). The catalytic diversity of multimodular polyketide synthases: Natural product biosynthesis beyond textbook assembly rules. *Topics in Current Chemistry*, 1–53. https://doi.org/10.1007/128_2010_113.

Gurgui, C., & Piel, J. (2010). Metagenomic approaches to identify and isolate bioactive natural products from microbiota of marine sponges. *Methods in Molecular Biology*, *668*, 247–264. Springer Science + Business Media LLC. https://doi.org/10.1007/978-1-60761-823-2_17.

Helfrich, E. J. N., & Piel, J. (2016). Biosynthesis of polyketides by *trans*-AT polyketide synthases. *Natural Product Reports*, *33*(1), 231–316. https://doi.org/10.1039/c5np00125k.

Hertweck, C. (2009). The biosynthetic logic of polyketide diversity. *Angewandte Chemie International Edition*, *48*, 4688–4716. https://doi.org/10.1002/anie.200806121.

Hochmuth, T., & Piel, J. (2009). Polyketide synthases of bacterial symbionts in sponges—Evolution-based applications in natural products research. *Phytochemistry*, *70*(15–16), 1841–1849. https://doi.org/10.1016/j.phytochem.2009.04.010.

Höfer, I., Crüsemann, M., Radzom, M., Geers, B., Flachshaar, D., Cai, X., et al. (2011). Insights into the biosynthesis of hormaomycin, an exceptionally complex bacterial signaling metabolite. *Chemistry & Biology*, *18*(3), 381–391. https://doi.org/10.1016/j.chembiol.2010.12.018.

Hrvatin, S., & Piel, J. (2007). Rapid isolation of rare clones from highly complex DNA libraries by PCR analysis of liquid gel pools. *Journal of Microbiological Methods*, *68*(2), 434–436. https://doi.org/10.1016/j.mimet.2006.09.009.

Humisto, A., Jokela, J., Liu, L., Wahlsten, M., Wang, H., Permi, P., et al. (2017). The swinholide biosynthetic gene cluster from a terrestrial cyanobacterium *Nostoc* sp. UHCC 0450. *Applied and Environmental Microbiology*, *84*(3), e02321–17. https://doi.org/10.1128/AEM.02321-17.

Irschik, H., Kopp, M., Weissman, K. J., Buntin, K., Piel, J., & Müller, R. (2010). Analysis of the sorangicin gene cluster reinforces the utility of a combined phylogenetic/retrobiosynthetic analysis for deciphering natural product assembly by *trans*-AT PKS. *ChemBioChem*, *11*(13), 1840–1849. https://doi.org/10.1002/cbic.201000313.

Ishibashi, M., Moore, R. E., Patterson, G. M. L., Xu, C., & Clardy, J. (1986). Scytophycins, cytotoxic and antimycotic agents from the cyanophyte *Scytonema pseudohofmanni*. *Journal of Organic Chemistry*, *51*(26), 5300–5306. https://doi.org/10.1021/jo00376a047.

Ishihara, H., Martin, B. L., Brautigan, D. L., Karaki, H., Ozaki, H., Kato, Y., et al. (1989). Calyculin A and okadaic acid: Inhibitors of protein phosphatase activity. *Biochemical and Biophysical Research Communications*, *159*, 871–877. https://doi.org/10.1016/0006-291X(89)92189-X.

Jenner, M., Frank, S., Kampa, A., Kohlhaas, Pöplau, P., Briggs, G. S., et al. (2013). Substrate specificity in ketosynthase domains for *trans*-AT polyketide synthases. *Angewandte Chemie International Edition*, *52*(4), 1143–1147. https://doi.org/10.1002/anie.201207690.

Kampa, A., Gagunashvili, A. N., Gulder, T. A. M., Morinaka, B. I., Daolio, C., Godejohann, M., et al. (2013). Metagenomic natural product discovery in lichen provides evidence for a family of biosynthetic pathways in diverse symbioses. *Proceedings of the National Academy of Sciences of the United States of America*, *110*(33), E3129–37. https://doi.org/10.1073/pnas.1305867110.

Kato, Y., Fusetani, N., Matsunaga, S., Hashimoto, K., Fujita, S., & Furuya, T. (1986). Bioactive marine metabolites. Part 16. Calyculin A, a novel antitumor metabolite from the marine sponge *Discodermia calyx*. *Journal of American Chemical Society*, *108*, 2780–2781. https://doi.org/10.1021/ja00270a061.

Kato, Y., Fusetani, N., Matsunaga, S., Hashimoto, K., & Koseki, K. (1988). Bioactive marine metabolites. 24. Isolation and structure elucidation of calyculins B, C, and D, novel antitumor metabolites, from the marine sponge *Discodermia calyx*. *Journal of Organic Chemistry*, *53*, 3930–3932. https://doi.org/10.1021/jo00252a009.

Kato, Y., Fusetani, N., Matsunaga, S., Hashimoto, K., Sakai, R., Higa, T., et al. (1987). Antitumor macrodiolides isolated from a marine sponge *Theonella* sp.: Structure revision misakinolide A. *Tetrahedron Letters*, *28*(49), 6225–6228. https://doi.org/10.1016/S0040-4039(00)61853-1.

Keatinge-Clay, T. A. (2008). Crystal structure of the erythromycin polyketide synthase dehydratase. *Journal of Molecular Biology*, *384*(4), 941–953. https://doi.org/10.1016/j.jmb.2008.09.084.

Kehraus, S., König, G., & Wright, A. (2002). A new cytotoxic calyculinamide derivative, geometricin A, from the Australian sponge *Luffariella geometrica*. *Journal of Natural Products*, *65*, 1056–1058. https://doi.org/10.1021/np010544u.

Kimura, M., Wakimoto, T., Egami, Y., Tan, K. C., Ise, Y., & Abe, I. (2012). Calyxamides A and B, cytotoxic cyclic peptides from the marine sponge *Discodermia calyx*. *Journal of Natural Products*, *75*, 290–294. https://doi.org/10.1021/np2009187.

Kita, A., Matsunaga, S., Takai, A., Kataiwa, H., Wakimoto, T., Fusetani, N., et al. (2002). Crystal structure of the complex between calyculin A and the catalytic subunit of protein phosphatase 1. *Structure*, *10*(5), 715–724. https://doi.org/10.1016/S0969-2126(02)00764-5.

Kitagawa, I., Kobayashi, M., Katori, T., Yamashita, M., Tanaka, J., Doi, M., et al. (1990). Absolute stereostructure of swinholide A, a potent cytotoxic macrolide from the

Okinawan marine sponge *Theonella swinhoei*. *Journal of American Chemical Society*, *112*(9), 3710–3712. https://doi.org/10.1021/ja00165a094.

Kitamura, M., Schupp, P. J., Nakano, Y., & Uemura, D. (2009). Luminaolide, a novel metamorphosis-enhancing macrodiolide for scleractinian coral larvae from crustose coralline algae. *Tetrahendron Letters*, *50*(47), 6606–6609. https://doi.org/10.1016/j.tetlet.2009.09.065.

Klenchin, V. A., King, R., Tanaka, J., Marriott, G., & Rayment, I. (2005). Structural basis of swinholide A binding to actin. *Chemistry & Biology*, *12*(3), 287–291. https://doi.org/10.1016/j.chembiol.2005.02.011.

Kohlhaas, C., Jenner, M., Kampa, A., Briggs, G. S., Alfonso, J. P., Piel, J., et al. (2013). Amino acid accepting ketosynthase domain from a *trans*-AT polyketide synthase exhibits high selectivity for predicted intermediates. *Chemical Science*, *4*, 3212–3217. https://doi.org/10.1039/C3SC50540E.

Konz, D., & Marahiel, M. A. (1999). How do peptide synthetases generate structural diversity? *Chemistry & Biology*, *6*(2), R39–R48. https://doi.org/10.1016/S1074-5521(99)80002-7.

Kubanek, J., Jensen, P. R., Keifer, P. A., Sullards, M. C., Collins, D. O., & Fenical, W. (2003). Seaweed resistance to microbial attack: A targeted chemical defense against marine fungi. *Proceedings of the National Academy of Sciences of the United States of America*, *100*(12), 6916–6921. https://doi.org/10.1073/pnas.1131855100.

Li, M. H., Ung, P. M., Zajkowski, J., Garneau-Tsodikova, S., & Sherman, D. H. (2009). Automated genome mining for natural products. *BMC Bioinformatics*, *10*, 185. https://doi.org/10.1186/1471-2105-10-185.

Lindvall, M. K., Pihko, P. M., & Koskinen, A. M. (1997). The binding mode of calyculin A to protein phosphatase-1. A novel spiroketal vector model. *Journal of Biological Chemistry*, *272*(37), 23312–23316. https://doi.org/10.1074/jbc.272.37.23312.

Matsunaga, S., Fujiki, H., Sakata, D., & Fusetani, N. (1991). Calyculins E, F, G, and H, additional inhibitors of protein phosphatases 1 and 2A, from the marine sponge *Discodermia calyx*. *Tetrahedron*, *47*(18–19), 2999–3006. https://doi.org/10.1016/S0040-4020(01)96031-4.

Matsunaga, S., Fusetani, N., Hashimoto, K., & Wälchli, M. (1989). Theonellamide F. a novel antifungal bicyclic peptide from a marine sponge *Theonella* sp.. *Journal of American Chemical Society*, *111*(7), 2582–2588. https://doi.org/10.1021/ja00189a035.

McGinnis, S., & Madden, T. L. (2004). BLAST: At the core of a powerful and diverse set of sequence analysis tools. *Nucleic Acids Research*, *32*, W20–5. https://doi.org/10.1093/nar/gkh435.

Medema, M. H., Blin, K., Cimermancic, P., de Jager, V., Zakrzewski, P., Fischbach, M. A., et al. (2011). antiSMASH: Rapid identification, annotation and analysis of secondary metabolite biosynthesis gene clusters in bacterial and fungal genome sequences. *Nucleic Acids Research*, *39*, W339–46. https://doi.org/10.1093/nar/gkr466.

Moldenhauer, J., Chen, X.-H., Borriss, R., & Piel, J. (2007). Biosynthesis of the antibiotic bacillaene, the product of a giant polyketide synthase complex of the *trans*-AT family. *Angewandte Chemie International Edition*, *119*, 8343–8345. https://doi.org/10.1002/anie.200703386.

Morrow, C., & Cárdenas, P. (2015). Proposal for a revised classification of the Demospongiae (Porifera). *Frontiers in Zoology*, *12*, 7–33. https://doi.org/10.1186/s12983-015-0099-8.

Murray, M. G., & Thompson, W. F. (1980). Rapid isolation of high molecular weight plant DNA. *Nucleic Acids Research*, *8*, 4321–4325. https://doi.org/10.1093/nar/8.19.4321.

Nguyen, T., Ishida, K., Jenke-Kodama, H., Dittmann, E., Gurgui, C., Hochmuth, T., et al. (2008). Exploiting the mosaic structure of *trans*-acyltransferase polyketide synthases for natural product discovery and pathway dissection. *Nature Biotechnology*, *26*(2), 225–233. https://doi.org/10.1038/nbt1379.

Okada, A., Watanabe, K., Umeda, K., & Miyakado, M. (1991). Calyculin E and F, novel insecticidal metabolites from the marine sponge, *Discodermia* sp. *Agricultural and Biological Chemistry, 55*, 2765–2771. https://doi.org/10.1271/bbb1961.55.2765.

Pernthaler, A., Pernthaler, J., & Amann, R. (2002). Fluorescence in situ hybridization and catalyzed reporter deposition for the identification of marine bacteria. *Applied and Environmental Microbiology, 68*, 3094–3101. https://doi.org/10.1128/AEM.68.6.3094-3101.2002.

Phelan, V. V., Du, Y., McLean, J. A., & Bachmann, B. O. (2009). Adenylation enzyme characterization using γ-$^{18}O_4$-ATP pyrophosphate exchange. *Chemistry & Biology, 16*, 473–478. https://doi.org/10.1016/j.chembiol.2009.04.007.

Piel, J. (2002). A polyketide synthase-peptide synthetase gene cluster from an uncultured bacterial symbiont of *Paederus beetles. Proceedings of the National Academy of Sciences of the United States of America, 99*, 14002–14007. https://doi.org/10.1073/pnas.222481399.

Piel, J. (2004). Metabolites from symbiotic bacteria. *Natural Product Reports, 21*, 519–538. https://doi.org/10.1039/b703499g.

Piel, J. (2010). Biosynthesis of polyketides by *trans*-AT polyketide synthases. *Natural Product Reports, 27*, 996–1047. https://doi.org/10.1039/b816430b.

Piel, J., Hui, D., Wen, G., Butzke, D., Platzer, M., Fusetani, N., et al. (2004). Antitumor polyketide biosynthesis by an uncultivated bacterial symbiont of the marine sponge *Theonella swinhoei. Proceedings of the National Academy of Sciences of the United States of America, 101*, 16222–16227. https://doi.org/10.1073/pnas.0405976101.

Pöplau, P., Frank, S., Morinaka, B. I., & Piel, J. (2013). An enzymatic domain for the formation of cyclic ethers in complex polyketides. *Angewandte Chemie International Edition, 52*, 13215–13218. https://doi.org/10.1002/anie.201307406.

Prieto, C., García-Estrada, C., Lorenzana, D., & Martín, J. F. (2012). NRPSsp: Nonribosomal peptide synthase substrate predictor. *Bioinformatics, 28*(3), 426–427. https://doi.org/10.1093/bioinformatics/btr659.

Rawlings, B. J. (2001). Type I polyketide biosynthesis in bacteria (Part B). *Natural Product Reports, 18*, 231–281. https://doi.org/10.1039/B100191O.

Reid, R., Piagentini, M., Rodriguez, E., Ashley, G., Viswanathan, N., Carney, J., et al. (2003). A model of structure and catalysis of ketoreductase domains in modular polyketide synthases. *Biochemistry, 42*(1), 72–79. https://doi.org/10.1021/bi0268706.

Ridley, C. P., Bergquist, P. R., Harper, M. K., Faulkner, D. J., Hooper, J. N., & Haygood, M. G. (2005). Speciation and biosynthetic variation in four dictyoceratid sponges and their cyanobacterial symbiont, *Oscillatoria spongeliae. Chemistry & Biology, 12*(3), 397–406. https://doi.org/10.1016/j.chembiol.2005.02.003.

Rothkegel, M., Mayboroda, O., Rohde, M., Wucherpfennig, C., Valenta, R., & Jockusch, B. M. (1996). Plant and animal profilins are functionally equivalent and stabilize microfilaments in living animal cells. *Journal of Cell Science, 109*, 83–90.

Röttig, M., Medema, M. H., Blin, K., Weber, T., Rausch, C., & Kohlbacher, O. (2011). NRPSpredictor2—A web server for predicting NRPS adenylation domain specificity. *Nucleic Acids Research, 39*. W362–7 https://doi.org/10.1093/nar/gkr323.

Saito, S., Watabe, S., Ozaki, H., Kobayashi, M., Suzuki, T., Kobayashi, H., et al. (1998). Actin-depolymerizing effect of dimeric macrolides, bistheonellide A and swinholide A. *Journal of Biochemistry, 123*(4), 571–578.

Sakai, R., Higa, T., & Kashman, Y. (1986). Misakinolide-A, an antitumor macrolide from the marine sponge *Theonella* sp. *Chemistry Letters, 15*(9), 1499–1502. https://doi.org/10.1246/cl.1986.1499.

Scheuermayer, M., Pimentel-Elardo, S., Fieseler, L., Grozdanov, L., & Hentschel, U. (2006). Microorganisms of sponges: Phylogenetic diversity and biotechnological potential. In P. Proksch & W. E. G. Müller (Eds.), *Frontiers in Marine Biotechnology* (pp. 289–312). London, UK: Horizon Scientific Press. ISBN: 978-1-904933-18-2.

Schmidt, E. W., Obraztsova, A. Y., Davidson, S. K., Faulkner, D. J., & Haygood, M. G. (2000). Identification of the antifungal peptide-containing symbiont of the marine sponge *Theonella swinhoei* as a novel δ-proteobacterium, "Candidatus *Entotheonella palauensis*" *Marine Biology*, *136*(6), 969–977. https://doi.org/10.1007/s0022700002.

Schwarzer, D., & Marahiel, M. A. (2001). Multimodular biocatalysts for natural product assembly. *Naturwissenschaften*, *88*(3), 93–101. https://doi.org/10.1007/s0011401002.

Senderowicz, A. M. J., Kaur, G., Sainz, E., Laing, C., Inman, W. D., Rodriquez, J., et al. (1995). Jasplakinolide's inhibition of the growth of prostate carcinoma cells in vitro with disruption of the actin cytoskeleton. *Journal of the National Cancer Institute*, *87*(1), 46–51. https://doi.org/10.1093/jnci/87.1.46.

Shen, B. (2003). Polyketide biosynthesis beyond the type I, II and III polyketide synthase paradigms. *Current Opinion in Chemical Biology*, 7(2), 285–295. https://doi.org/10.1016/S1367-5931(03)00020-6.

Stachelhaus, T., Mootz, H. D., & Marahiel, M. A. (1999). The specificity-conferring code of adenylation domains in nonribosomal peptide synthetases. *Chemistry & Biology*, *6*, 493–505. https://doi.org/10.1016/S1074-5521(99)80082-9.

Stingl, J., Andersen, R. J., & Emerman, J. T. (1992). In vitro screening of crude extracts and pure metabolites obtained from marine invertebrates for the treatment of breast cancer. *Cancer Chemotheraphy and Pharmacology*, *30*(5), 401–406. https://doi.org/10.1007/BF00689969.

Sudek, S., Lopanik, N. B., Waggoner, L. E., Hildebrand, M., Anderson, C., Liu, H., et al. (2007). Identification of the putative bryostatin polyketide synthase gene cluster from "*Candidatus* Endobugula sertula", the uncultivated microbial symbiont of the marine bryozoan *Bugula neritina*. *Journal of Natural Products*, *70*(1), 67–74. https://doi.org/10.1021/np060361d.

Suganuma, M., Fujiki, H., Furuya-Suguri, H., Yoshizawa, S., Yamamoto, S., Kato, Y., et al. (1990). Calyculin A, and inhibitor of protein phosphatases, a potent tumor promoter on CD-1 mouse skin. *Cancer Research*, *50*, 3521–3525.

Taylor, M. W., Radax, R., Steger, D., & Wagner, M. (2007). Sponge-associated microorganisms: Evolution, ecological, and biotechnological potential. *Microbiology and Molecular Biology Reviews*, *71*(2), 295–347. https://doi.org/10.1128/MMBR.00040-06.

Terry, D. R., Spector, I., Higa, T., & Bubb, M. R. (1997). Misakinolide A is a marine macrolide that caps but does not sever filamentous actin. *Journal of Biological Chemistry*, *272*(12), 7841–7845. https://doi.org/10.1074/jbc.272.12.7841.

Teta, R., Gurgui, M., Helfrich, E. J. N., Künne, S., Schneider, A., van Echten-Deckert, G., et al. (2010). Genome mining reveals *trans*-AT polyketide synthase directed antibiotic biosynthesis in the bacterial phylum Bacteroidetes. *ChemBioChem*, *11*, 2506–2512. https://doi.org/10.1002/cbic.201000542.

Tsai, S.-C., & Ames, B. D. (2009). Chapter 2 Structural enzymology of polyketide synthases. In D. A. Hopwood (Ed.), *Methods in enzymology.* : *Vol. 459. Complex enzymes in microbial natural product biosynthesis, part B: Polyketides, aminocoumarins and carbohydrates* (pp. 17–47) Elsevier. https://doi.org/10.1016/S0076-6879(09)04602-3.

Ueoka, R., Bortfeld-Miller, M., Morinaka, B. I., Vorholt, J. A., & Piel, J. (2017). Toblerols, cyclopropanol-containing modulators of methylobacterial antibiosis generated by an unusual polyketide synthase. *Angewandte Chemie International Edition*, *56*, 1–6. https://doi.org/10.1002/anie.201709056.

Ueoka, R., Uria, A. R., Reiter, S., Mori, T., Karbaum, P., Peters, E. E., et al. (2015). Metabolic and evolutionary origin of actin-biding polyketides from diverse organisms. *Nature Chemical Biology*, *11*, 705–712. https://doi.org/10.1038/nchembio.1870.

Unson, M. D., Holland, N. D., & Faulkner, D. J. (1994). A brominated secondary metabolite synthesized by the cyanobacterial symbiont of a marine sponge and accumulation of the crystalline metabolite in the sponge tissue. *Marine Biology*, *119*(1), 1–11. https://doi.org/10.1007/BF00350100.

Uria, A. R., & Piel, J. (2009). Cultivation-independent approaches to investigate the chemistry of marine symbiotic bacteria. *Phytochemistry Reviews*, *8*(2), 401–414. https://doi.org/10.1007/s11101-009-9127-7.

Uria, A. R., & Zilda, D. S. (2016). Metagenomics-guided mining of commercially useful biocatalysts from marine microorganisms. In S.-K. Kim & F. Toldrá (Eds.), *Advances in food and nutrition research. Marine enzymes biotechnology: Production and industrial applications, part I—Vol. 78 Production of enzymes* (pp. 1–26). Elsevier. https://doi.org/10.1016/bs.afnr.2016.05.001.

Valenzano, C. R., You, Y.-O., Garg, A., Keatinge-Clay, A., Khosla, C., & Cane, D. E. (2010). Stereospecificity of the dehydratase domain of the erythromycin polyketide synthase. *Journal of American Chemical Society*, *132*, 14697–14699. https://doi.org/10.1021/ja107344h.

Wakimoto, T., Egami, Y., & Abe, I. (2016). Calyculin: Nature's way of making the sponge-derived cytotoxin. *Natural Product Reports*, *33*, 751–760. https://doi.org/10.1039/C5NP00123D.

Wakimoto, T., Egami, Y., Nakashima, Y., Wakimoto, Y., Mori, T., Awakawa, T., et al. (2014). Calyculin biogenesis from a pyrophosphate protoxin produced by a sponge symbiont. *Nature Chemical Biology*, *10*, 648–655. https://doi.org/10.1038/nchembio.1573.

Wakimoto, T., Matsunaga, S., Takai, A., & Fusetani, N. (2002). Insight into binding of calyculin A to protein phosphatase 1: Isolation of hemicalyculin a and chemical transformation of calyculin A. *Chemistry & Biology*, *9*(3), 309–319. https://doi.org/10.1016/S1074-5521(02)00118-7.

Webster, N. S., & Hill, R. T. (2001). The culturable microbial community of the great barrier reef sponge *Rhopaloeides odorabile* is dominated by an α-proteobacterium. *Marine Biology (Berlin)*, *138*, 843–851. https://doi.org/10.1007/s0022700005.

Wilson, M. C., Mori, T., Ruckert, C., Uria, A. R., Helf, M. J., Takada, K., et al. (2014). An environmental bacterial taxon with a large and distinct metabolic repertoire. *Nature*, *506*, 58–62. https://doi.org/10.1038/nature12959.

Winder, P. L., Pomponi, S. A., & Wright, A. E. (2011). Natural products from the lithistida: A review of the literature since 2000. *Marine Drugs*, *9*(12), 2643–2682. https://doi.org/10.3390/md9122643.

Yang, J. Y., Phelanb, V. V., Simkovskyd, R., Watrousa, J. D., Triale, R. M., Fleminge, T. C., et al. (2012). Primer on agar-based microbial imaging mass spectrometry. *Journal of Bacteriology*, *194*, 6023–6028. https://doi.org/10.1128/JB.00823-12.

Youssef, T. A., & Mooberry, S. L. (2006). Hurghadolide A and swinholide I, potent actin-microfilament disrupters from the Red Sea sponge *Theonella swinhoei*. *Journal of Natural Products*, *69*(1), 154–157. https://doi.org/10.1021/np050404a.

Zhao, L., Wen, E., Upur, H., & Tian, S. (2017). High performance liquid chromatography-diode array detector method for the simultaneous determination of five compounds in the pulp and seed of sea buckthorn. *Pharmacognosy Magazine*, *13*(49), 136–140. https://doi.org/10.4103/0973-1296.197656.

Zhou, Y., Murphy, A. C., Samborskyy, M., Prediger, P., Dias, L. C., & Leadlay, P. F. (2015). Iterative mechanism of macrodiolide formation in the anticancer compound conglobatin. *Chemistry & Biology*, *22*(6), 745–754. https://doi.org/10.1016/j.chembiol.2015.05.010.

Ziemert, N., Podell, S., Penn, K., Badger, J. H., Allen, E., & Jensen, P. R. (2012). The natural product domain seeker NaPDoS: A phylogeny based bioinformatic tool to classify secondary metabolite gene diversity. *PLoS One*, 7(3). e34064. https://doi.org/10.1371/journal.pone.0034064.

Zoetendal, E. G., Ben-Amor, K., Harmsen, H. J. M., Schut, F., Akkermans, A. D. L., & de Vos, W. M. (2002). Quantification of uncultured *Ruminococcus obeum*-like bacteria in human fecal samples by fluorescent in situ hybridization and flow cytometry using 16S rRNA-targeted probes. *Applied and Environmental Microbiology*, *68*(9), 4225–4232. https://doi.org/10.1128/AEM.68.9.4225-4232.2002.

FURTHER READING

Saito, S., Watabe, S., Ozaki, H., Fusetani, N., & Karaki, H. (1994). Mycalolide B, a novel actin depolymerizing agent. *Journal of Biological Chemistry*, *69*, 29710–29714.

Smith, D. R. M., Uria, A. R., Helfrich, E. J. N., Milbredt, D., van Pée, K.-H., Piel, J., et al. (2017). An unusual flavin-dependent halogenase from the metagenome of the marine sponge *Theonella swinhoei* WA. *ACS Chemical Biology*, *12*(5), 1281–1287. https://doi.org/10.1021/acschembio.6b01115.

Weber, T., Blin, K., Duddela, S., Krug, D., Kim, H. U., Bruccoleri, R., et al. (2015). antiSMASH 3.0—A comprehensive resource for the genome mining of biosynthetic gene clusters. *Nucleic Acids Research*, *43*, W237–43. https://doi.org/10.1093/nar/gkv437.

SECTION III

Natural Product Biosynthetic Enzymology of Marine Bacteria and Fungi

CHAPTER TEN

Chemoenzymatic Synthesis of Starting Materials and Characterization of Halogenases Requiring Acyl Carrier Protein-Tethered Substrates

Hem R. Thapa*, Andrew J. Lail*, Neha Garg*, Vinayak Agarwal*,†,1
*School of Chemistry and Biochemistry, Georgia Institute of Technology, Atlanta, GA, United States
†School of Biological Sciences, Georgia Institute of Technology, Atlanta, GA, United States
[1]Corresponding author: e-mail address: vagarwal@gatech.edu

Contents

Abstract

Flavin-adenine dinucleotide (FAD)-dependent halogenases are widespread in natural product biosynthetic gene clusters and have been demonstrated to employ small organic molecules as substrates for halogenation, as well as substrates that are tethered to carrier proteins (CPs). Despite numerous reports of FAD-dependent halogenases utilizing CP-tethered substrates, only a few have been biochemically characterized due to limited accessibility to the physiological substrates. Here, we describe a method for the preparation of acyl-*S*-CP substrates and their use in biochemical assays to query the activity of FAD-dependent halogenases. Furthermore, we describe a mass spectrometry-based method for the characterization of acyl-*S*-CP substrates and the corresponding halogenated products generated by the halogenases. Finally, we test

Methods in Enzymology, Volume 604
ISSN 0076-6879
https://doi.org/10.1016/bs.mie.2018.01.028

the substrate specificity of a physiological chlorinase and a physiological brominase from marine bacteria, and, for the first time, demonstrate the distinct halide specificity of halogenases. The methodology described here will enable characterization of new halogenases employing CP-tethered substrates.

1. INTRODUCTION

Halogenation is a modification that is routinely afforded upon natural products and often confers unique biological activities to these organic molecules, including, but not limited to desirable pharmaceutical antibiotic and cytotoxic activities. Bioactivities of small molecules can be critically dependent on halogenation, such as for the antibiotics vancomycin (**1**, Fig. 1) and chloramphenicol (**2**). Four principal classes of halogenating enzymes have been described to date, their mechanisms, and activities tailored to halogenate diverse substrates varying from electron-rich aromatic rings to unactivated alkyl centers (Agarwal et al., 2017). Of the four classes, this manuscript describes procedures to assay the activity of halogenases that employ flavin-adenine dinucleotide (FAD) as a cofactor. FAD-dependent halogenases oxidize halides (X^-, X=Cl, Br, and I) to their corresponding halenium (X^+) ions that are regiospecifically delivered either as a haloamines or as hypohalites to electron-rich centers to catalyze electrophilic substitution reactions (Flecks et al., 2008; Yeh, Blasiak, Koglin, Drennan, & Walsh, 2007). FAD-dependent halogenases require a partner enzyme, flavin

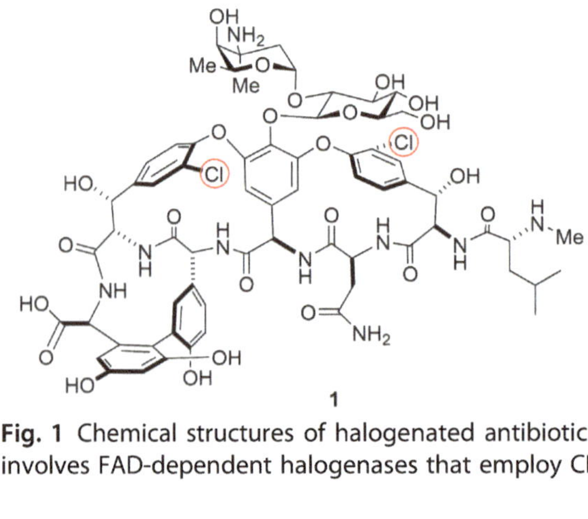

1

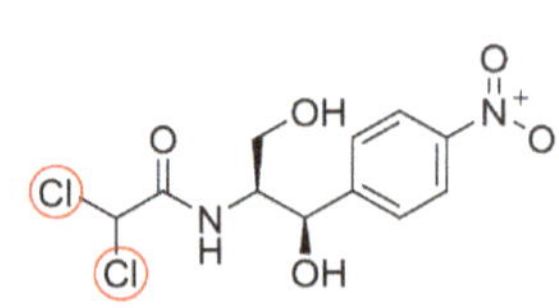

2

Fig. 1 Chemical structures of halogenated antibiotics **1** and **2**, biosynthesis of which involves FAD-dependent halogenases that employ CP-tethered substrates.

reductase, to deliver reduced FAD ($FADH_2$) that is required for halogenase activity. Electrons to generate $FADH_2$ from FAD are derived from NAD(P)H. Molecular oxygen acts as the final sink for these electrons, together with the electrons derived from the oxidation of the halide. The catalytic scheme for FAD-dependent halogenases is shown in Fig. 2.

The inventory of halogenated natural products involving the participation of FAD-dependent halogenases in their respective biosynthetic schemes is already impressive and continues to expand. Furthermore, characterization of their biochemical activities has advanced FAD-dependent halogenases to a stage where they can be successfully used as diagnostic elements to identify natural product biosynthetic gene clusters in genomic and metagenomics datasets (Agarwal, Blanton, et al., 2017; Agarwal, Miles, et al., 2017; Jordan & Moore, 2016). Enzymatic functionalization of carbon centers by halogenation has attracted attention toward the engineering of FAD-dependent halogenases to improve their thermal and solvent tolerance, to expand their substrate scope, and to use these enzymes in a combinatorial fashion to engineer natural product scaffolds (Brown & O'Connor, 2015; Latham, Brandenburger, Shepherd, Menon, & Micklefield, 2018; Menon et al., 2017; Payne, Andorfer, & Lewis, 2016;

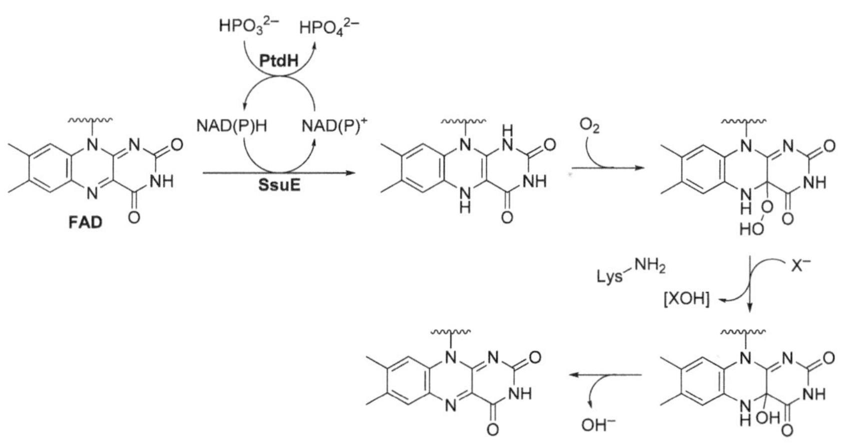

Fig. 2 Catalytic mechanism for FAD-dependent halogenases. In this study, the *Escherichia coli* flavin reductase SsuE is used to deliver $FADH_2$ to the halogenases while the phosphite dehydrogenase PtdH is used to regenerate NADPH (see Section 4.1 for experimental details). The identity of the electrophilic halogenating species generated by FAD-dependent halogenases, hypohalite ([XOH]), or a haloamine generated by the capture of the halenium by the side chain amine of a conserved Lys residue in the halogenase active site is presently not clear.

Schnepel, Minges, Frese, & Sewald, 2016). The efforts alluded to above are principally directed toward FAD-dependent halogenases that use small-molecule organic substrates, such as L-Trp. Much of our current mechanistic understanding for FAD-dependent halogenases relies on the crystal structures of and biochemical studies performed using L-Trp chlorinases (Agarwal, Miles, et al., 2017).

In contrast, the pace of characterization of FAD-dependent halogenases that employ substrates tethered to carrier proteins (CPs) has been sluggish. For instance, despite the identification of the biosynthetic gene cluster and establishment of aryl-halogenation as being critical to antibiotic efficacy, reconstitution of the bioactivity of the FAD-dependent halogenase VhaA participating in the biosynthesis of **1** has only recently been achieved (Pinchman & Boger, 2013a, 2013b; Schmartz, Zerbe, Abou-Hadeed, & Robinson, 2014; van Wageningen et al., 1998). As a second example, while the biosynthetic gene cluster has been identified and the crystal structure of the enzyme determined, a firm biochemical characterization of the FAD-dependent halogenase CmlS involved in the biosynthesis of **2** is still elusive (He, Magarvey, Piraee, & Vining, 2001; Podzelinska et al., 2010). Two challenges precipitate this knowledge gap. First, the timing of halogenation relative to other biochemical modifications in multimodular schemes involving CP-tethered substrates can be hard to decipher. Second, synthesis of CP-tethered substrates in itself can be challenging as is described in detail later.

CPs are small proteins with a conserved four-helical bundle structure. The side chain hydroxyl of a conserved serine residue in apo-CPs is post-translationally esterified to a phosphopantetheine "arm" derived from coenzyme A (CoA-SH), a transformation afforded by phosphopantetheinyl transferases to yield holo-CPs (Fig. 3) (Beld, Sonnenschein, Vickery, Noel, & Burkart, 2014). Following esterification with phosphopantetheine, the terminal thiol of holo-CPs is condensed with adenylated carboxylic acids in an ATP-dependent ligation reaction to yield acyl-*S*-CPs. Acyl-*S*-CPs serve to carry carboxylic acids through reaction schemes while the carboxylic acid is activated by the formation of the high-energy thioester bond with the phosphopantetheine arm. Following modifications of the acyl moiety, the thioester bond can be displaced by various nucleophiles, by water to yield carboxylic acids, by alcohols to yield esters and lactones, by amines to yield amides and lactams, by carbanions to yield Claisen condensation products, or transthioesterified. These modifications are extensively represented in primary metabolism and in natural product biosynthetic pathways.

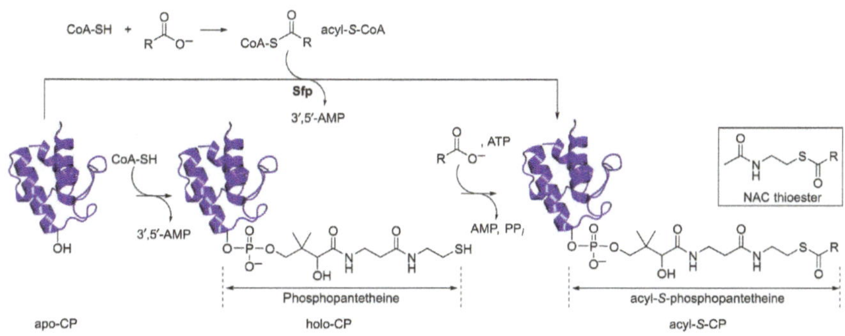

Fig. 3 Posttranslational maturation of apo-CP by transfer of the phosphopantetheine arm from CoA-SH to the side chain hydroxyl of a conserved serine residue to yield holo-CP, followed by formation of the thioester bond with a carboxylic acid to generate acyl-*S*-CPs. In some cases, acyl-*S*-CP substrates can be substituted by abbreviated NAC thioesters. An alternate route to generating acyl-*S*-CPs relies on the transfer of the acyl-*S*-phosphopantetheine from acyl-S-CoAs, chemically synthesized by the ligation of CoA-SH with carboxylic acids, by the phosphopantetheinyl transferase Sfp. The *E. coli* CP (PDB: 4ETW) is shown in cartoon representation as a representative example of the tertiary structure of CPs and the site for the conserved Ser residue (Agarwal, Lin, Lukk, Nair, & Cronan, 2012).

CPs, when expressed as recombinant proteins in heterologous hosts such as *Escherichia coli*, are posttranslationally modified in situ to their corresponding holo forms by phosphopantetheinyl transferases that are present in the heterologous host's proteome. However, these conversions usually do not reach completion and heterologously expressed CPs are purified as a mixture of apo and holo states. The discovery and characterization of the broadly substrate tolerant phosphopantetheinyl transferase Sfp has provided a straightforward, and now widely employed, route for the conversion of apo-CPs to holo-CPs (Quadri et al., 1998). However, the conversion of holo-CPs to acyl-*S*-CPs can be complicated by the nonavailability of an appropriate ligase, specifically when the carboxylic acid to be ligated to the holo-CP is an advanced biosynthetic intermediate for which a cognate ligase does not exist.

Strategies to address this challenge have been devised. Numerous biosynthetic enzymes that employ acyl-*S*-CoA and acyl-*S*-CP substrates have been reported to use abbreviated substrates, such as *N*-acetylcysteamine thioesters (NAC thioesters, Fig. 3). Strategies for the elaboration of NAC thioesters were recently reviewed (Franke & Hertweck, 2016). An alternate strategy is based on the substrate tolerance of Sfp. The crystal structure of Sfp in complex with CoA-SH reveals that the enzyme makes extensive contacts with the 3′-phosphoadenosine and the pyrophosphate moieties of CoA-SH. However, the cysteamine moiety is not actively engaged by the enzyme (Reuter, Mofid, Marahiel, & Ficner, 1999). Thus, Sfp can directly catalyze the production of acyl-*S*-CPs starting from apo-CPs by the transfer of the acyl-*S*-phosphopantetheine moiety from acyl-*S*-CoAs (Fig. 3). This strategy requires the elaboration of acyl-*S*-CoAs by the synthetic ligation of carboxylic acids to CoA-SH, a reaction that proceeds in low yields while using typically high-priced CoA-SH as a substrate, and requires a chromatographic purification of the acyl-*S*-CoA product (Mishra & Drueckhammer, 2000). Hence, alternate routes for the elaboration of acyl-*S*-CoAs have been sought (Peter, Vogeli, Cortina, & Erb, 2016). By adopting the substrate promiscuity of the CoA-SH biosynthetic enzymes together with the broad substrate tolerance of Sfp, we have recently reported a chemoenzymatic scheme for the elaboration of acyl-*S*-CoAs and acyl-*S*-CPs at a multimilligram scale (Agarwal et al., 2015; Nazi, Koteva, & Wright, 2004; Strauss & Begley, 2002). These procedures are outlined in Section 3 together with mass spectrometric assays to query the acylation states of CPs. In Section 4, we describe procedures to assay for the activity of FAD-dependent halogenases using acyl-*S*-CP substrates together with the mass spectrometric characterization of the assay products. Using these procedures, for the first time, we experimentally validate the distinction between physiological chlorinases and brominases.

2. EQUIPMENTS AND CONSUMABLES

Equipments

1. Temperature controlled incubators and shakers for bacterial growth
2. Refrigerated centrifuges
3. Sonicator
4. SDS-PAGE gel casting accessories and electrophoresis chamber
5. GE Healthcare Amersham Imager 600
6. White light transilluminator
7. Spectrophotometer
8. GE Healthcare Aktaprime plus chromatography system
9. Agilent 1260 infinity quaternary LC coupled to impact II ToF mass spectrometer (Bruker Daltonics)

Consumables

1. Chromatography: His-Trap FF 5 mL (GE Healthcare); Hi-Trap Q 5 mL (GE Healthcare); Hi-Load 16/600 Superdex 75pg (GE Healthcare);

Aeris 3.6 μm WIDEPORE XB-C18 200 Å 250 × 4.6 mm^2 (Phenomenex).
2. Antibiotics: Kanamycin monosulfate and ampicillin trihydrate (Fisher).
3. Chemicals and reagents: All chemicals and reagents were purchased from Fisher Scientific unless noted otherwise. Tris base, sodium chloride, imidazole, glycerol, HEPES, potassium chloride, potassium bromide, sodium phosphite dibasic pentahydrate, magnesium chloride tetrahydrate, isopropyl β-D-1-thiogalactopyranoside (IPTG), adenosine 5′-triphosphate disodium salt dihydrate (ATP) (Sigma-Aldrich), FAD (Sigma-Aldrich), nicotinamide adenine dinucleotide phosphate hydrate ($NADP^+$) (Sigma-Aldrich), and tris(2-carboxyethyl)phosphine hydrochloride (Sigma-Aldrich), protein assay dye reagent (Bio-Rad), Brilliant Blue R-250 (Bio-Rad).
4. Solvents: Chromatography grade acetonitrile, water, and formic acid (Fisher).
5. Media: Luria-Bertani (LB) broth and terrific broth (TB) (Fisher).
6. Centrifugal filters: Amicon ultra-15 centrifugal filters with MWCO of 3, 10, and 30 kDa (EMD Millipore).

3. PREPARATION OF Acyl-*S*-CPs

3.1 Discussion

The biosynthesis of the halogenated pyrrole containing marine natural products pentabromopseudilin (**3**, Fig. 4), marinopyrrole A (**4**), and chlorizidine (**5**) each involves halogenation of a pyrrolyl-*S*-CP intermediate by the FAD-dependent halogenases Bmp2, Mpy16, and Clz5, respectively (Agarwal et al., 2014; Mantovani & Moore, 2013; Yamanaka, Ryan, Gulder, Hughes, & Moore, 2012). As described above, ligases that append pyrrole-2-carboxylic acid to the corresponding holo-*S*-CPs (Bmp1, Mpy15, and Clz18, respectively) do not exist, since pyrrolyl-*S*-CPs are advanced biosynthetic intermediates that are derived by the enzymatic oxidation of prolyl-*S*-CPs (Fig. 4) (Agarwal et al., 2014; Thomas, Burkart, & Walsh, 2002; Walsh, Garneau-Tsodikova, & Howard-Jones, 2006). It should be noted that the synthesis and halogenation of pyrrolyl-*S*-CPs are not restricted to marine natural products. The discovery of gene clusters encoding production of **3–5** was inspired by the prior characterization of the molecular logic for the construction of the dichlorinated pyrrole moiety in the terrestrial natural product pyoluteorin (**6**) (Dorrestein, Yeh, Garneau-Tsodikova, Kelleher, & Walsh, 2005; Thomas et al., 2002).

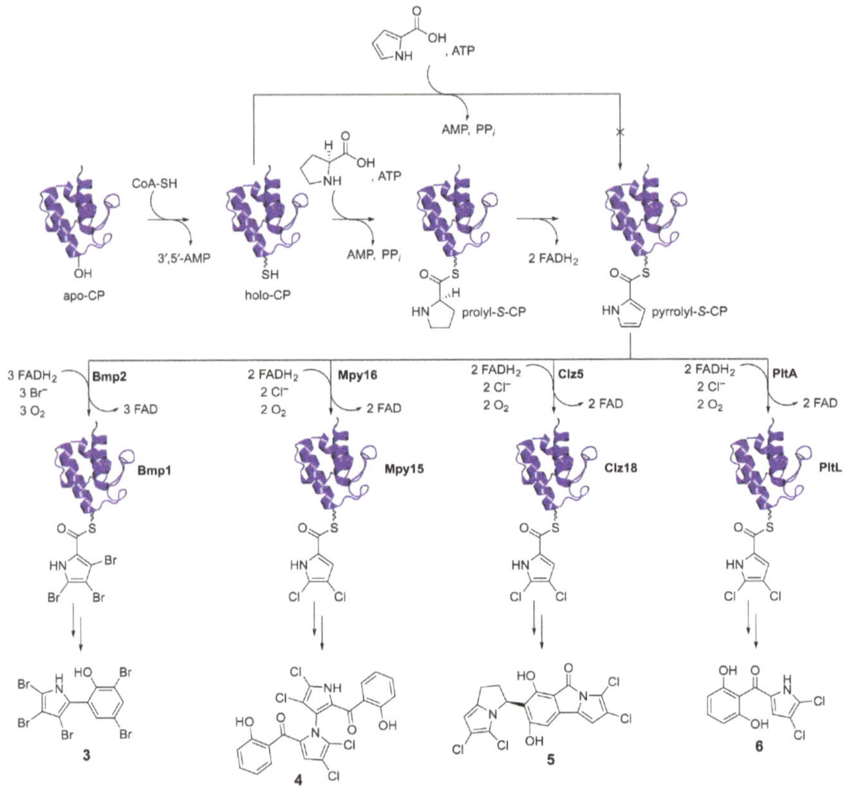

Fig. 4 Molecular logic for the construction of halogenated pyrrole moiety in **3–6**. The phosphopantetheine arm of holo-CPs is abbreviated for clarity.

As ATP-dependent ligases that ligate pyrrole-2-carboxylic acid to holo-CPs are not available, alternate strategies have been employed for the synthesis of pyrrolyl-*S*-CP substrates in order to assay the activities of the FAD-dependent halogenases as illustrated in Fig. 4. In a strategy that mimics the natural product biosynthetic scheme, ATP-dependent ligases are used to ligate L-Pro to holo-CPs to generate prolyl-*S*-CPs, followed by oxidation of the prolyl heterocycle to the aromatic pyrrolyl ring by FAD-dependent dehydrogenases (Dorrestein et al., 2005; Thomas et al., 2002). This approach is complicated by the FAD-dependent dehydrogenase being difficult to purify, and which does not yield stoichiometric conversion of the prolyl-*S*-CP substrate to the pyrrolyl-*S*-CP product (Agarwal et al., 2014; Dorrestein et al., 2005; Thomas et al., 2002). Based on the prior demonstration of the broad substrate tolerance for the coenzyme A biosynthetic enzymes CoaA, CoaD, and CoaE, and the phosphopantetheinyl transferase

Sfp, we have described a chemoenzymatic scheme to generate pyrrolyl-*S*-CPs, the procedures for which are described here (Agarwal et al., 2015; El Gamal et al., 2016; Nazi et al., 2004; Quadri et al., 1998; Strauss & Begley, 2002; Worthington & Burkart, 2006). The strategy described affords multimilligram quantities of highly pure pyrrolyl-*S*-CPs and does not employ the FAD-dependent prolyl-*S*-CP dehydrogenases. The chemoenzymatic scheme, as illustrated in Fig. 5 involves pyrrolyl-*S*-pantetheine (**7**), the synthesis of which has been described in detail previously (Agarwal et al., 2015).

3.2 Procedure

Buffers and solvents

TB media prepared according to manufacturer's directions and sterile autoclaved prior to use.

LB media prepared according to manufacturer's directions and sterile autoclaved prior to use.

Resuspension buffer: 20 m*M* Tris–HCl (pH 8.0), 500 m*M* NaCl, 10% glycerol.

Wash buffer: 20 m*M* Tris–HCl (pH 8.0), 1 *M* NaCl, 30 m*M* imidazole.

Elution buffer: 20 m*M* Tris–HCl (pH 8.0), 1 *M* NaCl, 250 m*M* imidazole.

Elution buffer (#2): 20 m*M* Tris–HCl (pH 8.0), 1 *M* NaCl, 250 m*M* imidazole, 10% glycerol.

Dialysis buffer: 20 m*M* Tris–HCl (pH 8.9), 50 m*M* KCl.

Ion-exchange buffer A: 20 m*M* Tris–HCl (pH 8.9).

Ion-exchange buffer B: 20 m*M* Tris–HCl (pH 8.9), 1 *M* KCl.

Storage buffer: 20 m*M* HEPES (pH 7.5), 50 m*M* KCl, 10% glycerol.

Assay quenching buffer: MeCN + 0.1% formic acid.

HPLC solvent A: Water + 0.1% formic acid.

HPLC solvent B: MeCN + 0.1% formic acid.

Bacterial strains

E. coli DH5α for plasmid maintenance and cloning.

E. coli BL21Gold(DE3) for overexpression of recombinant proteins.

3.2.1 Preparation of Plasmid DNA

Table 1 lists the plasmids that are used in the procedures described in this study. All recombinant proteins are appended with fusion tags to assist in purification of recombinant proteins.

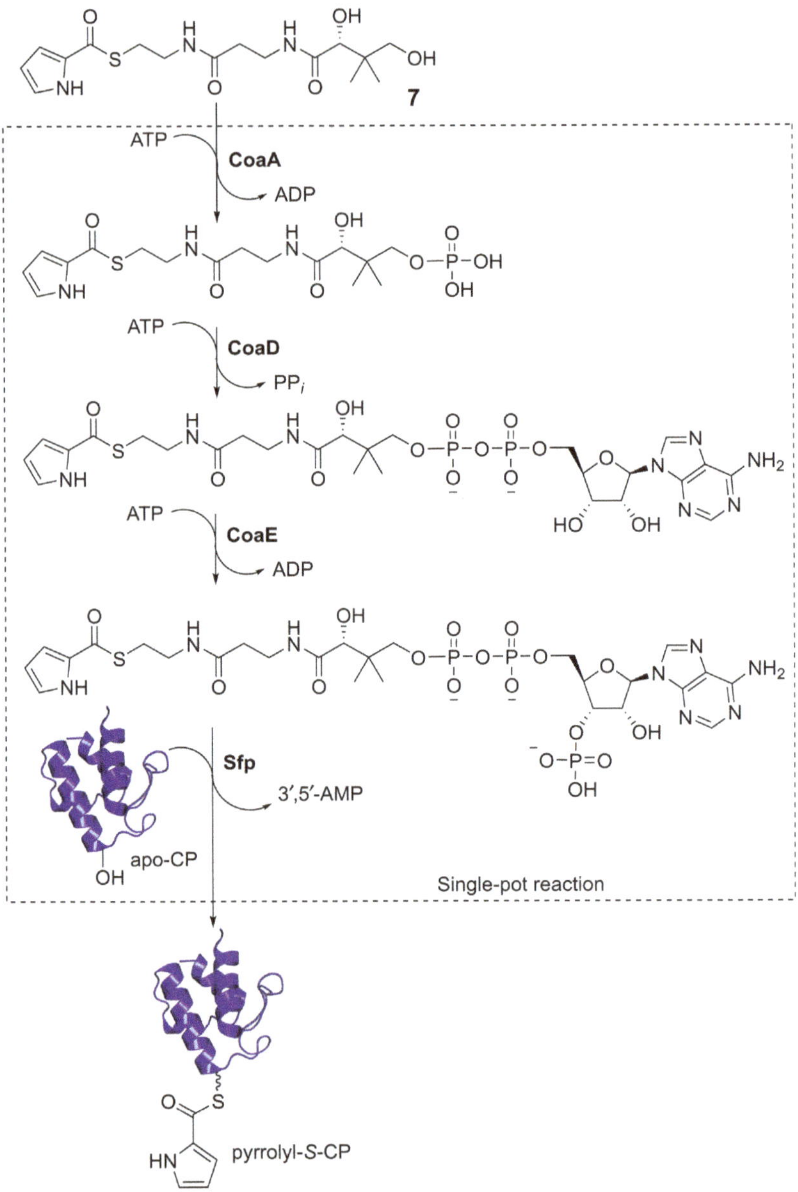

Fig. 5 Route for the enzymatic synthesis of pyrrolyl-*S*-CPs starting from **7** with a sequential action of CoaA, CoaD, CoaE, and Sfp enzymes. The reaction can be performed as a "single-pot" reaction in which all four enzymes can be incubated with **7**, apo-CPs, and ATP to yield pyrrolyl-*S*-CPs, as is described.

Table 1 Plasmids Used in This Study

Construct	Fusion Tag	Plasmid Backbone	Source Organism	Reference
Bmp2	N-terminal His_6	pET-28b(+)	*Marinomonas mediterranea* MMB-1	Agarwal et al. (2014)
Mpy16	N-terminal His_6 followed by maltose-binding protein (MBP)	pET-28b(+)	*Streptomyces* sp. CNQ-418	El Gamal et al. (2016)
Bmp1[a]	N-terminal His_6	pET-28b(+)	*M. mediterranea* MMB-1	Agarwal et al. (2014)
Mpy15	N-terminal His_6	pET-28b(+)	*Streptomyces* sp. CNQ-418	El Gamal et al. (2016)
SsuE	N-terminal His_6	pET-28b(+)	*E. coli*	Agarwal et al. (2014)
PtdH	N-terminal His_6	pET-15b(+)	*Pseudomonas stutzeri*	Johannes, Woodyer, and Zhao (2007)
Sfp	C-terminal His_6	pET-24b(+)	*Bacillus subtilis*	Agarwal et al. (2014)
CoaA	N-terminal His_6	pET-28b(+)	*E. coli*	Agarwal et al. (2015)
CoaD	N-terminal His_6 followed by MBP	pET-28b(+)	*E. coli*	Agarwal et al. (2015)
CoaE	N-terminal His_6 followed by MBP	pET-28b(+)	*E. coli*	Agarwal et al. (2015)

[a]Bmp1 is a didomain protein with an N-terminal CP domain and a C-terminal thioesterase domain. In this study, for clarity, Bmp1 refers to the Bmp1 CP domain (Agarwal et al., 2014; El Gamal et al., 2016).

3.2.2 Expression and Purification of Recombinant Proteins

All proteins described in this study were recombinantly expressed and purified in an identical manner unless otherwise noted.

Experimental protocol

1. Transform plasmid DNA into competent BL21Gold(DE3) cells and grow transformants on LB agar media with antibiotic selection for 14–16h at 37°C.
2. Pick individual colonies and grow for 14–16h in 5mL TB media with antibiotic selection with vigorous shaking at 37°C.
3. Use 5mL of the starter culture to innoculate 1L of TB media with antibiotic selection and grow with vigorous shaking at 37°C till optical density (measured at 600nm) reaches 0.6.

4. Reduce the incubation temperature to 18°C and induce protein expression by the addition of 0.3 m*M* IPTG (isopropyl-β-D-1-thiogalactopyranoside). Grow for additional 18 h with vigorous shaking at 18°C.
5. Collect biomass by centrifugation at 5000 × *g* for 15 min at 4°C. Resuspend the pellet in 50 mL resuspension buffer that is prechilled to 4°C.
6. Lyse bacterial cells by sonication. Care should be exercised to maintain the temperature close to 4°C during sonication.
7. Clarify the lysate by centrifugation at 30,000 × *g* for 1 h at 4°C. Collect the supernatant. The pellet can be discarded.
8. Apply the supernatant to a 5 mL His-Trap chromatography column using an AKTA Prime FPLC system operating at 4°C.
9. After the supernatant has been applied to the column, wash the column extensively with wash buffer.
10. Elute the bound protein by a continuous linear gradient to 100% elution buffer across 20 column volumes and collect the eluate in 5 mL fractions. Analyze the fractions by SDS-PAGE with Coomassie staining. *Note*: Elute SsuE using elution buffer (#2). Collect fractions with pure SsuE protein and proceed directly to step 18. Do not concentrate the SsuE protein as it is prone to aggregation.
11. Analyze the fractions by SDS-PAGE with Coomassie staining.
12. Pool the fractions with pure protein and dialyze the protein sample overnight using dialysis buffer.
13. Apply the dialyzed protein sample to a 5 mL Hi-Trap Q chromatography column using an AKTA Prime FPLC system operating at 4°C.
14. Wash the column with two column volumes of 95% ion-exchange buffer A and 5% ion-exchange buffer B.
15. Elute the bound protein by a continuous linear gradient to 100% ion-exchange buffer B across 20 column volumes and collect the eluate in 5 mL fractions.
16. Analyze the fractions by SDS-PAGE with Coomassie staining.
17. Pool the fractions with pure protein and concentrate to 2.5 mL final volume using centrifugal filters with 10 kDa molecular weight cut-off (MWCO). For CPs, use centrifugal filters with 3 kDa MWCO.
18. Desalt the concentrated protein into storage buffer using gravity flow desalting columns.

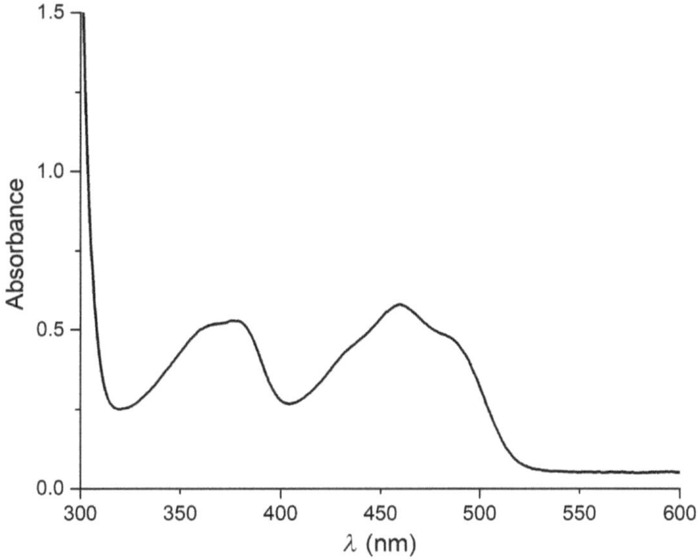

Fig. 6 UV–vis absorbance spectra for Bmp2 demonstrating the presence of FAD cofactor.

19. For halogenases, record an absorbance spectra from 300 to 600 nm. The absorbance spectra should demonstrate the presence of FAD. Typical absorbance spectra for Bmp2 halogenase are shown in Fig. 6.
20. Measure protein concentration using Bradford assay. Store purified proteins at −80°C in 100 μL aliquots. Store SsuE protein at 4°C and use without further modifications. *Note*: Procedures described here typically yield greater than 30 mg of purified CPs and greater than 50 mg of all other proteins. SsuE protein, as purified above, retains activity for ~2 days at 4°C. A fresh purification of SsuE is recommended after that.

3.2.3 Chemoenzymatic Synthesis of Pyrrolyl-S-CPs

The scheme for the synthesis of pyrrolyl-*S*-CPs is illustrated in Fig. 5. Briefly, the four catalysts—CoaA, CoaD, CoaE, and Sfp—are incubated together with the substrates **7**, ATP/Mg^{2+}, and purified apo-CPs as listed in Table 2. The reaction is conducted in a "single-pot" without the requirement of purifying the intermediates. After incubation at 30°C, the reaction is purified by size exclusion chromatography and the pyrrolyl-*S*-CP is collected. Size exclusion chromatography removes the residual ATP and other

Table 2 Final Concentration of Assay Components for Synthesis of Pyrrolyl-*S*-CPs

Component	Concentration in Assay
apo-CP	250 μ*M*
7	1 m*M*
CoaA[a]	1 μ*M*
CoaD[a]	1 μ*M*
CoaE[a]	1 μ*M*
Sfp[a]	1 μ*M*
ATP	9 m*M*
$MgCl_2$	10 m*M*
HEPES-Na (pH 7.9)	50 m*M*
Water	

[a]CoaA, CoaD, CoaE, and Sfp can be premixed as a single "enzyme cocktail" if multiple assays are being set up in parallel.

small-molecule products generated during the course of the reaction and yields pure pyrrolyl-*S*-CP in a defined buffer. At this stage, the pyrrolyl-*S*-CPs can be concentrated and stored at −80°C in small aliquots for use in halogenation assays described below. As shown in Fig. 5, this procedure generates pyrrolyl-*S*-CoA in situ which then acts as a substrate for the Sfp enzyme for transfer of the pyrrolyl-*S*-phosphopantetheine moiety to the side chain hydroxyl of the apo-CP. As Sfp requires the CP to be in the apo state for the transfer of the pyrrolyl-*S*-phosphopantetheine moiety, holo-CPs are not responsive to this methodology. The procedure illustrated in Fig. 5 is fundamentally different from the biosynthetic scheme as shown in Fig. 4 in which a prolyl ligase is used to ligate L-pro to the phosphopantetheine thiol of a holo-CP and the prolyl side chain is subsequently oxidized to generate pyrrolyl-*S*-CP.

Experimental protocol

1. Thaw aliquots of apo-CPs (Bmp1 and Mpy15), Sfp, PtdH, CoaA, CoaD, and CoaE on ice. Centrifuge briefly using a table-top centrifuge.
2. Prepare a fresh aliquot of 200 m*M* ATP in prechilled 50 m*M* HEPES-Na (pH 7.9) buffer.
3. For each CP, in duplicate, add the reaction components to the final concentrations delineated in Table 2. The reactions should be set up

in duplicate, with one of the reactions serving as the negative control (see below).

4. Incubate the reaction at 30°C for 5 min.
5. Initiate one of the two reactions by adding ATP. Adjust the final reaction volume to 2.5 mL with water.
6. For the other reaction, add water in place of ATP. This reaction is henceforth referred to as "negative control."
7. Incubate the reactions at 30°C for 3 h.
8. Centrifuge the reaction tubes at 4°C to remove any aggregated proteins.
9. Inject the reactions on a Superdex 75 16/200 size exclusion chromatography column equilibrated in 20 m*M* Tris–HCl (pH 8.0) using an AKTA Prime FPLC system operating at 4°C. Monitor the absorbance of the eluate at 280 nm.
10. For reactions in which ATP is added, two major peaks will be observed on the absorbance chromatogram. The first peak corresponds to the modified CPs, and the second peak corresponds to the residual ATP, and ADP and 3′,5′-AMP that are generated during the course of the reaction. The second peak is not observed for the negative control reaction.
11. Collect the eluate corresponding the first peak of the absorbance chromatogram. The presence of CP can be verified by SDS-PAGE and Coomassie staining as shown in Fig. 7.
12. Withdraw 100 μL aliquot of the eluate. Inject 10 μL of this sample on a HPLC–MS/MS system and elute using the following solvent gradient

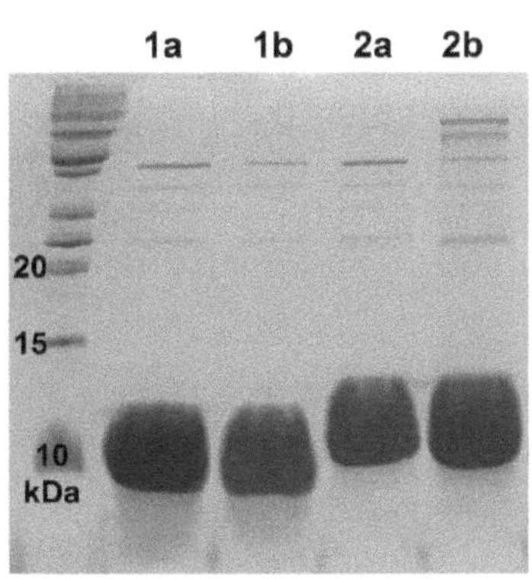

Fig. 7 SDS-PAGE demonstrating 1a: 20 μg apo-Bmp1; 1b: 20 μg pyrrolyl-*S*-Bmp1; 2a: 20 μg apo-Mpy15; and 2b: 20 μg pyrrolyl-*S*-Mpy15.

at a flow rate of 0.5 mL/min using an Aeris 3.6 μm widepore XB-C18 200 Å 250 × 4.6 mm^2 column:

a. 0–5 min: 5% HPLC solvent B
b. 5–30 min: linear gradient to 70% HPLC solvent B
c. 30–31 min: linear gradient to 95% HPLC solvent B
d. 31–35 min: 95% HPLC solvent B
e. 35–36 min: linear gradient to 5% HPLC solvent B
f. 36–38 min: 5% HPLC solvent B
g. 38–39 min: linear gradient to 95% HPLC solvent B
h. 39–42 min: 95% HPLC solvent B
i. 42–43 min: linear gradient to 5% HPLC solvent B
j. 43–46 min: 5% HPLC solvent B

Note: The elution gradient described here includes reintroduction of high concentration of the organic solvent in the mobile phase which affords a rigorous regeneration of the stationary phase.

13. Acquire mass spectrometry data between 5 and 40 min in the positive ionization mode in the range of 100–2000 *m/z*. Perform an external calibration with ESI-L Low Concentration Tuning Mix (Agilent Technologies) prior to data collection and use internal calibrant hexakis (1H,1H,3H–tetrafluoropropoxy)phosphazene (SynQuest Laboratories) throughout the runs. Use the following instrument parameters: capillary voltage 4500 V; nebulizer gas (nitrogen) pressure 2 bar; ion source temperature 200°C; dry gas flow 9 L/min, spectral rate 3 Hz for MS^1, and 6 Hz for MS^2. Select the three most intense ions per MS^1 for acquiring MS/MS fragmentation and use collision-induced dissociation (CID) energies listed in Table 3. Use basic stepping function to fragment ions at 50% and 125% of the CID calculated for each *m/z* from Table 3 with timing of 50% for each step. Similarly, employ basic stepping of collision RF of 550 and 800 Vpp with a timing of 50% for each step and transfer time stepping of 57 and 90 μs with a timing of 50% for each step. Set MS/MS active exclusion parameter to "3" and release after 30 s. Exclude the mass of internal calibrant from the MS/MS list using a mass range of *m/z* 921.5–924.5.
14. The data analysis procedure is described in Section 3.3.
15. Concentrate the eluate from the size exclusion chromatography column using a 3 kDa MWCO centrifugal filter and measure the protein concentration using the Bradford assay. Store the protein in small aliquots at −80°C.

Table 3 Collision-Induced Energies for Tandem MS Data Collection

Type	Mass	Width	Collision	Charge State
Base	100	4	22	1
Base	100	4	22	2
Base	300	5	27	1
Base	300	5	27	2
Base	500	6	35	1
Base	500	6	30	2
Base	1000	8	45	1
Base	1000	8	35	2
Base	2000	10	50	1
Base	2000	10	50	2

3.3 Data Analysis

The acylation states of CPs can be differentiated based on the mass observed in the MS^1 mass spectra as well as characteristic phosphopantetheine and acyl-*S*-phosphopantetheine ejection ions that is observed in the MS^2 spectra for holo- and acyl-*S*-CPs, respectively, as illustrated in Fig. 8 (Dorrestein et al., 2006; Dorrestein & Kelleher, 2006). Under positive ionization mode, mining for the masses of these ejection ions in the MS^2 spectra allow for deconvolution of the acylation states of the CPs which can be further corroborated by calculating the difference in the MS^1 masses as illustrated below for apo-Bmp1 and pyrrolyl-*S*-Bmp1 (Figs. 9 and 10). As mentioned previously, expression of CPs in heterologous hosts, *E. coli* in this study, will lead to partial conversion of apo-CPs to holo-CPs, which will also be deconvoluted as described below.

Data analysis protocol

1. Generate MS^2 extracted ion chromatograms (EICs) for $[M+H]^{1+}=261.13$ and $[M+H]^{1+}=354.15$ corresponding to the phosphopantetheine and pyrrolyl-*S*-phosphopantetheine ejection ions, respectively, as shown in Fig. 8. The EICs for the reactions with and without ATP as described in Section 3.2.3 for Bmp1 are shown in Fig. 9.
2. Using the retention times of the MS^2 ejection ions shown in Fig. 9, determine the retention time of MS^1 spectra corresponding to holo- and pyrrolyl-*S*-Bmp1 and generate their corresponding MS^1 spectra.

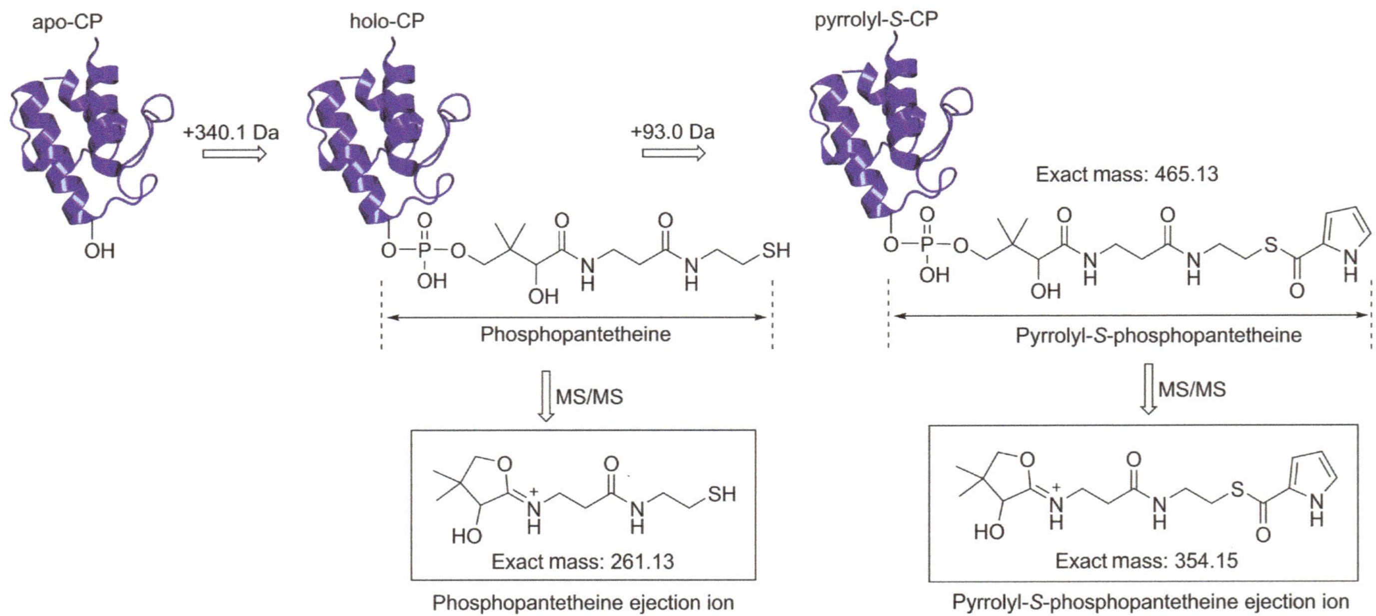

Fig. 8 Under MS/MS conditions in a mass spectrometer, holo-CPs generate a characteristic pantetheine ejection ion of $[M+H]^{1+}=261.13$ that can be detected in the mass spectrometry dataset to signify the presence of holo-CPs. Similarly, acyl-*S*-CPs, such as pyrrolyl-*S*-CPs, also generate the corresponding ejection ions that can be used to identify acyl-*S*-CPs in mass spectrometry datasets. Apo-CPs do not generate these characteristic ejection ions. The theoretical parent mass difference between the apo-CPs, holo-CPs, and pyrrolyl-*S*-CPs is shown.

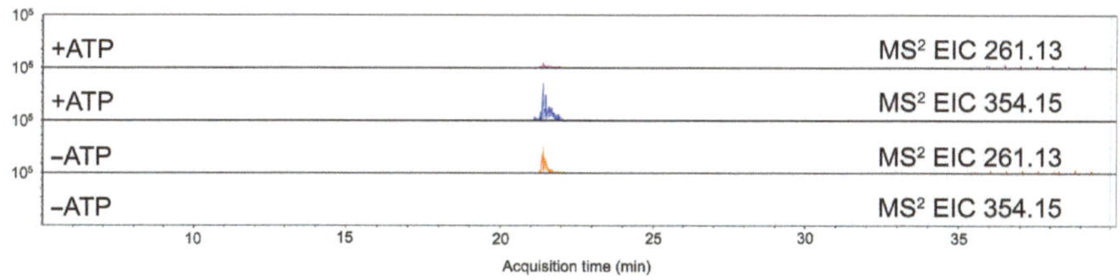

Fig. 9 EICs for Bmp1 reactions with and without ATP as described in Section 3.2.3. The MS^2 pyrrolyl-*S*-phosphopantetheine ejection ion (354.15 Da) is observed only when ATP is present in the reaction. However the phosphopantetheine ejection ion (261.13 Da), which corresponds to the holo-Bmp1 is observed in both reactions. As mentioned before, during recombinant expression in heterologous hosts, apo-CPs are partially converted to their holo state by the activity of the host encoded phosphopantetheinyl transferase(s).

These MS^1 spectra are shown in Fig. 10. The MS^1 spectra corresponding to the apo state has to be identified empirically.

3. Deconvolute the mass of the apo-, holo-, and pyrrolyl-*S*-Bmp1 using the formula:

   Deconvoluted mass = (monoisotopic mass × *z*) − (*z* × mass of H^+).

   Thus:

   For $[M+13H]^{13+}$ species for apo-Bmp1 as shown in Fig. 10, the calculated deconvoluted mass is = (817.35 × 13) − (13 × 1.01) = 10,612.42 Da

   For $[M+13H]^{13+}$ species for holo-Bmp1 as shown in Fig. 10, the calculated deconvoluted mass is = (843.50 × 13) − (13 × 1.01) = 10,952.37 Da

   For $[M+13H]^{13+}$ species for pyrrolyl-*S*-Bmp1 as shown in Fig. 10, the calculated deconvoluted mass is = (850.66 × 13) − (13 × 1.01) = 11,045.45 Da.

4. Mass difference between holo-Bmp1 and apo-Bmp1 (10,952.37–10,612.42 = 339.95 Da) and between pyrrolyl-*S*-Bmp1 and holo-Bmp1 (11,045.45–10,952.37 = 93.08 Da) is in good agreement with the theoretical mass differences as shown in Fig. 8.

Note: It is not necessary that identical $[M+nH]^{n+}$ species be used for parent mass deconvolution across all MS^1 spectra as shown in Fig. 10. Clusters of ions corresponding to different charge states should be manually examined, and the charge state that offers the best resolution between the isotopic peaks should be selected for parent mass deconvolution. Different charge states should lead to identical, or closely matching parent ion masses (depending on the resolution of the mass spectrometer, see below). The resolution of the isotopic peaks will decrease as the charge state increases. However, the

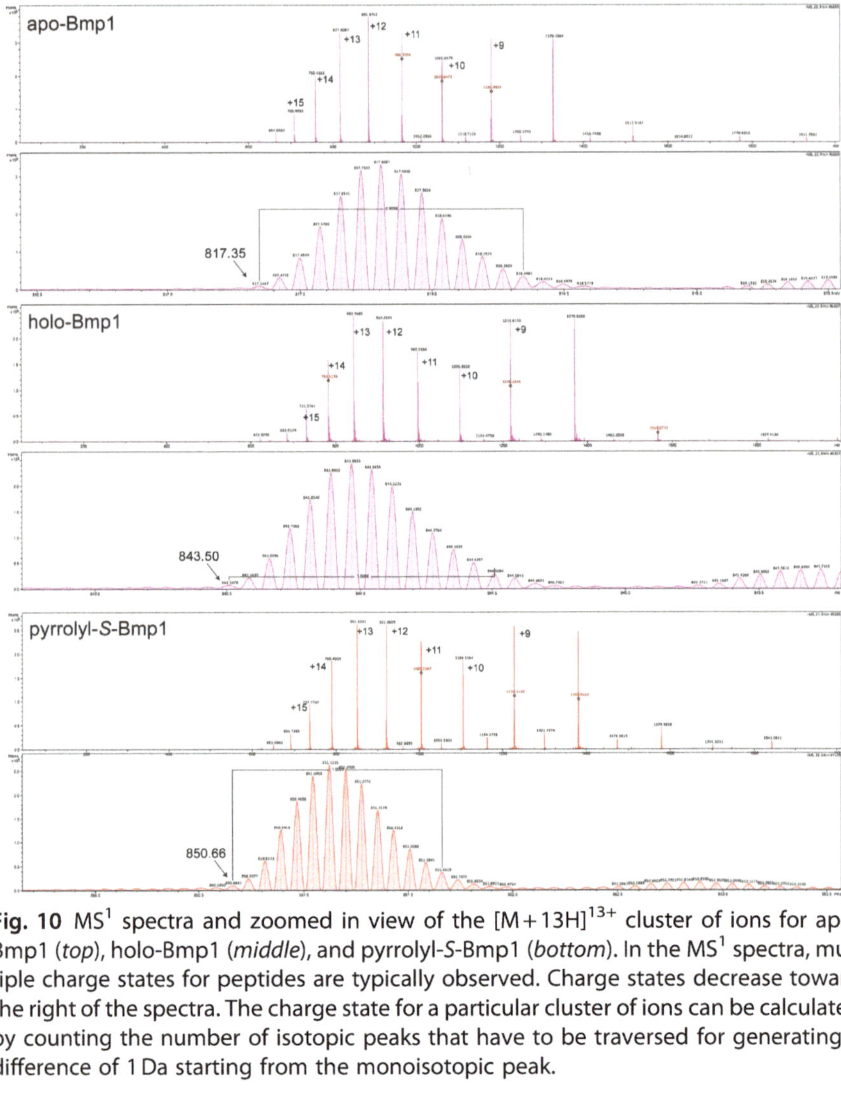

Fig. 10 MS[1] spectra and zoomed in view of the $[M+13H]^{13+}$ cluster of ions for apo-Bmp1 (*top*), holo-Bmp1 (*middle*), and pyrrolyl-*S*-Bmp1 (*bottom*). In the MS[1] spectra, multiple charge states for peptides are typically observed. Charge states decrease toward the right of the spectra. The charge state for a particular cluster of ions can be calculated by counting the number of isotopic peaks that have to be traversed for generating a difference of 1 Da starting from the monoisotopic peak.

intensity of lower charge states might be too low for the monoisotopic peak to be identified accurately. This tradeoff between resolution and abundance must be empirically balanced for each MS[1] spectra. For relatively high resolution mass spectrometers, such as the Bruker Impact II Q-ToF (resolution ~40,000) and the Agilent 6530 Q-ToF (resolution ~20,000), the isotopic resolution is usually sufficient that monoisotopic peaks can be accurately identified even for low abundance charge states. For lower resolution mass spectrometers, such as the Bruker amaZon SL ion-trap (resolution ~2000), only higher abundance charge states can be used to accurately deconvolute parent ion peptide masses. In our experience, ToF and ion-trap instruments both yield identical ejection ions for acyl-*S*-CP species. As has been reported previously, attention must be paid to fragmentation energies that are employed to generate the MS^2 spectra (Agarwal et al., 2015). At low fragmentation energies, ejection ions will not be observed in sufficient abundance to generate MS^2 EICs. At high fragmentation energies, acyl-*S*-pantetheine ejection ions can degrade to pantetheine ejection ions due to cleavage of the labile thioester bond. The appropriate fragmentation energy must also be empirically determined.

4. HALOGENATION OF Acyl-*S*-CPs

4.1 Discussion

FAD-dependent halogenases acting on substrates tethered to CPs are mechanistically identical to FAD-dependent halogenases acting on small-molecule substrates such as L-tryptophan (Fig. 2). Halogenases acting on CP-tethered substrates routinely catalyze multiple halogenations on their CP-tethered aromatic substrates. For instance, the halogenase Bmp2 brominates all four carbon atoms of a pyrrole ring while Mpy16, Clz5, and PltA dichlorinate the pyrrole ring (Dorrestein et al., 2005; El Gamal et al., 2016; Mantovani & Moore, 2013) (Fig. 4). Note, while Bmp2 tetrabrominates the pyrrole ring, a tetrabrominated product will not be produced with the absence of a partner thioesterase enzyme (El Gamal et al., 2016). In the absence of the thioesterase, a tribrominated product, as shown in Fig. 4, will predominate. The halogenase HrmQ, involved in the biosynthesis of hormaomycin, only monochlorinates the pyrrole ring (Hofer et al., 2011). Similarly, the FAD-dependent halogenase AltN likely mono- and dibrominates the phenyl ring en route production of the alterochromides (Ross, Gulland, Dorrestein, & Moore, 2015), while the chlorinase SgcC3 only monochlorinates the phenyl ring during the biosynthesis of the enediyne C-1027 (Lin, Van Lanen, & Shen, 2007). FAD-halogenases that work on small organic molecules exclusively catalyze monohalogenations in a regiospecific fashion which is perhaps a result of the highly ordered binding of the substrate in the halogenase active site (Agarwal, Miles, et al., 2017; Bitto et al., 2008; Dong et al., 2005; Zhu et al., 2009). While a crystal structure of FAD-dependent halogenase with a CP-tethered substrate is not available yet, the active sites of Bmp2, Mpy16, and PltA seem to be more accommodating of substrates of different steric sizes (El Gamal et al., 2016; Pang, Garneau-Tsodikova, & Tsodikov, 2015). Recent discovery of the bifunctional methyltransferase-halogenase AoiQ and the MalA/MalA' indole halogenases, together with the fungal halogenase CazI, however challenge this notion and is likely to further expand the mechanistic repertoire of FAD-dependent halogenases (Chankhamjon et al., 2016; Fraley et al., 2017; Sato et al., 2016).

In this section, we focus on the description of the unique activities of FAD-dependent halogenases that affect multiple halogenations on their aromatic substrates. We describe the procedures to assay for the activity of the pyrrolyl-*S*-Bmp1 tribrominase Bmp2 and the pyrrolyl-*S*-dichlorinase Mpy16. FAD-dependent halogenases require a partner flavin reductase that provides $FADH_2$ to the halogenase (Fig. 2). We employ the *E. coli* flavin reductase SsuE for this purpose (Eichhorn, van der Ploeg, & Leisinger, 1999). Electrons required for the reduction of FAD by SsuE are derived from NADPH. In the assay described below, NADPH is regenerated by the thermostable phosphite dehydrogenase PtdH in situ (Fig. 2) (Johannes et al., 2007). Using these procedures, we experimentally demonstrate, for the first time, that physiological chlorinases as defined by Drennan and coworkers can chlorinate as well as brominate (Blasiak & Drennan, 2009). However, the brominase Bmp2 can only brominate and is incapable of chlorination activity, thus providing a firm experimentally validated distinction between physiological chlorinases and brominases.

4.2 Procedure

The purification of halogenases followed the protein purification protocol described in Section 3.2.2, and synthesis of pyrrolyl-*S*-CP substrates is described in Section 3.

Experimental protocol

1. Thaw aliquots of halogenases, pyrrolyl-*S*-CPs, and PtdH on ice. Centrifuge briefly using a table-top centrifuge.
2. The pyrrolyl-*S*-CPs can be analyzed by HPLC–MS/MS as described in Section 3.2.3 to ensure that no degradation has occurred. In our experience, pyrrolyl-*S*-Bmp1 and pyrrolyl-*S*-Mpy15 do not undergo degradation during long-term storage at −80°C.
3. For halogenation of pyrrolyl-*S*-Mpy15 by Mpy16, keep the chloride concentration in the assay constant at 100 m*M* and vary the concentration of bromide as 0, 25, 50, 100, and 200 m*M*. Add assays components

Table 4 Concentration of Assay Components for Halogenation of Pyrrolyl-*S*-Mpy15 by Mpy16

Component	Concentration in Assay
Pyrrolyl-*S*-Mpy15	50 μ*M*
Mpy16	20 μ*M*
TCEP[a]	5 m*M*
Na-phosphite	10 m*M*
FAD	20 μ*M*
$NADP^+$	1 m*M*
PtdH	10 μ*M*
SsuE	50 n*M*
KCl	100 m*M*
KBr	0, 5, 25, 50, 100, and 200 m*M*
HEPES-Na (pH 7.9)	20 m*M*
10% glycerol[b]	

[a]Tris(2-carboxyethyl)phosphine.
[b]10% glycerol was used in place of water to adjust the final volume of the assay to 100 μL.

according to Table 4 to a final volume of 100 μL. In case of halogenation of pyrrolyl-*S*-Bmp1 by Bmp2, keep the bromide concentration in the assay constant at 100 m*M* and vary the concentration of chloride as 0, 25, 50, 100, and 200 m*M*. Add assays components according to Table 5 to a final volume of 100 μL.

4. Incubate the reactions at 30°C for 3 h.
5. After 3 h incubation, add 8 μL of trypsin (1 mg/mL) and incubate at 30°C for 1 h.
6. Add 108 μL of assay quenching buffer and centrifuge at 16,000 × *g* for 30 min. Carefully withdraw the supernatant into an appropriate vial.
7. Inject 10 μL of this sample on a HPLC–MS/MS system and elute using the following solvent gradient at a flow rate of 0.5 mL/min using an Aeris 3.6 μm widepore XB-C18 200 Å 250 × 4.6 mm^2 column:
   a. 0–5 min: 5% HPLC solvent B
   b. 5–30 min: linear gradient to 70% HPLC solvent B
   c. 30–31 min: linear gradient to 95% HPLC solvent B
   d. 31–35 min: 95% HPLC solvent B
   e. 35–36 min: linear gradient to 5% HPLC solvent B

Table 5 Concentration of Assay Components for Halogenation of Pyrrolyl-*S*-Bmp1 by Bmp2

Component	Concentration in Assay
Pyrrolyl-*S*-Bmp1	50 μ*M*
Bmp2	20 μ*M*
TCEP	5 m*M*
Na-phosphite	10 m*M*
FAD	20 μ*M*
$NADP^+$	1 m*M*
PtdH	10 μ*M*
SsuE	50 n*M*
KBr	100 m*M*
KCl	0, 5, 25, 50, 100, and 200 m*M*
HEPES-Na (pH 7.9)	20 m*M*
10% glycerol	

   f. 36–38 min: 5% HPLC solvent B
   g. 38–39 min: linear gradient to 95% HPLC solvent B
   h. 39–42 min: 95% HPLC solvent B
   i. 42–43 min: linear gradient to 5% HPLC solvent B
   j. 43–46 min: 5% HPLC solvent B

8. Acquire mass spectrometry data between 5 and 40 min in the positive ionization mode. The data analysis procedure is described below.

4.3 Data Analysis

Data analysis for the halogenation of pyrrolyl-*S*-Mpy15 by the FAD-dependent halogenase Mpy16 is described in detail here using Figs. 11 and 12. The data analysis procedures for the halogenation of pyrrolyl-*S*-Bmp1 by Bmp2 are identical.

1. Generate MS^2 EICs, as described in Section 3.3, for $[M+H]^{1+} = 388.11$ and $[M+H]^{1+} = 422.07$ corresponding to the monochloropyrrolyl-*S*-phosphopantetheine and dichloropyrrolyl-*S*-phosphopantetheine ejection ions, respectively. These EICs correspond to the monochlorinated and dichlorinated products generated by Mpy16 using pyrrolyl-*S*-Mpy15 as substrate (Fig. 11). As Mpy16 can also accept bromide in

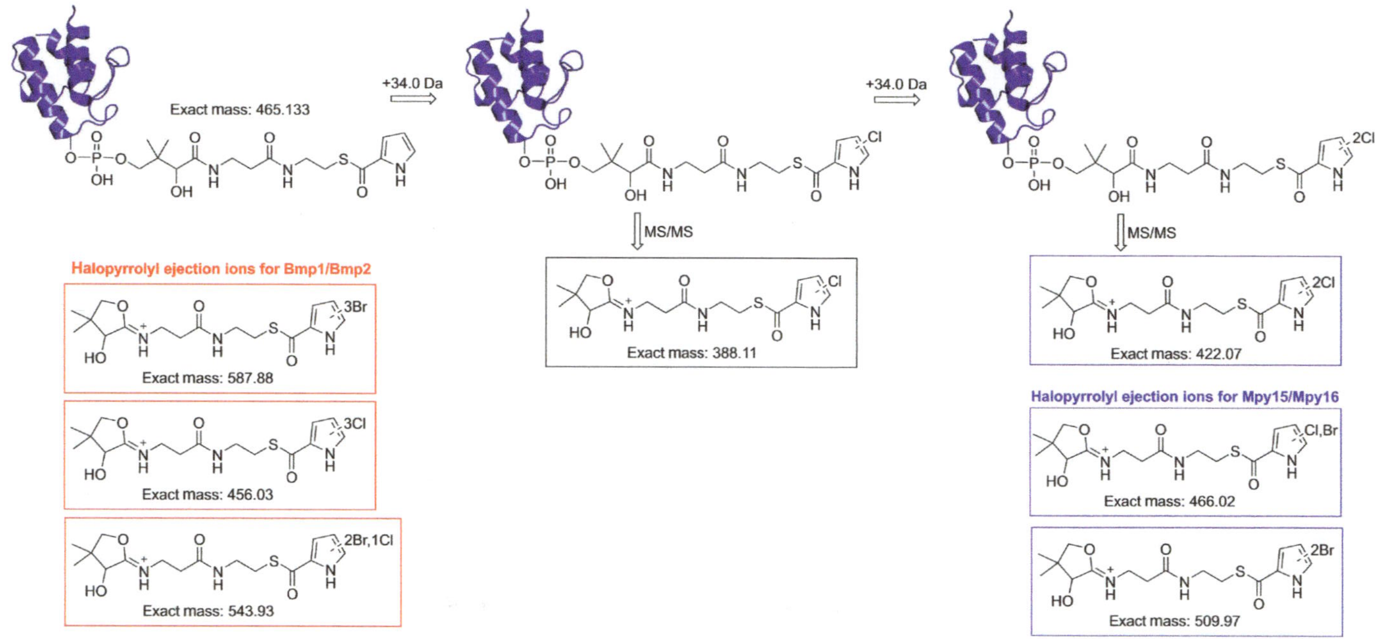

Fig. 11 MS^2 ejection ions for hypothetical dihalogenated products generated by halogenation of pyrrolyl-*S*-Mpy15 by Mpy16 (in *blue boxes*) and by trihalogenated products generated by the halogenation of pyrrolyl-*S*-Bmp1 by Bmp2 (in *red boxes*).

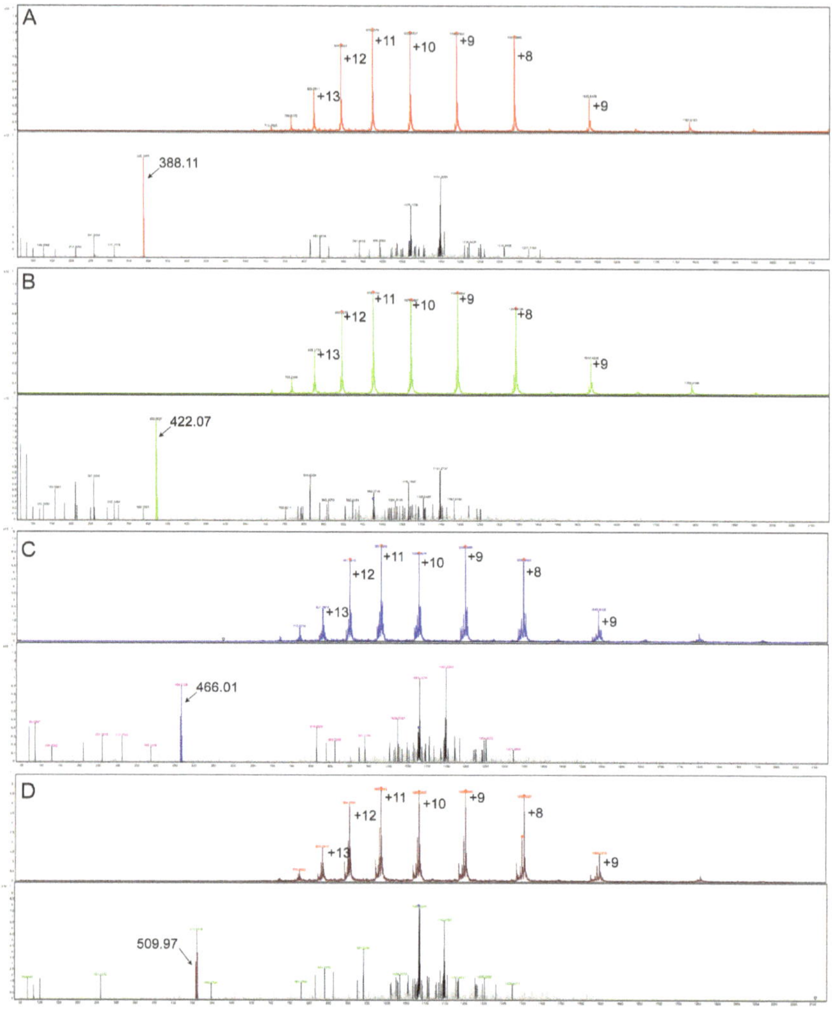

Fig. 12 MS^1 (*top*) and MS^2 (*bottom*) spectra corresponding to (A) monochloropyrrolyl-*S*-Mpy15, (B) dichloropyrrolyl-*S*-Mpy15, (C) mixed monochloro,monobromopyrrolyl-*S*-Mpy15, and (D) dibromopyrrolyl-*S*-Mpy15 products generated by Mpy16. The different peptide charge states in the MS^1 spectra and the ejection ions in the MS^2 spectra are illustrated.

place of chloride (vide infra), generate MS^2 EICs for $[M+H]^{1+} = 466.02$ and $[M+H]^{1+} = 509.97$ corresponding to the monochloro,monobromopyrrolyl-*S*-phosphopantetheine and dibromopyrrolyl-*S*-phosphopantetheine ejection ions, respectively. These EICs correspond to the mixed monochlorinated–monobrominated and dibromintaed products generated by Mpy16 (Fig. 11).

2. Using the retention times of the four MS^2 ejection ions illustrated in Fig. 11, determine the MS^1 spectra corresponding to monochlorinated-, dichlorinated-, mixed monochlorinated–monobrominated-, and dibromintaed-pyrrolyl-*S*-Mpy15. The MS^1 spectra for the four possible products generated by Mpy16, along with the MS^2 spectra showing the corresponding ejection ions are illustrated in Fig. 12.
3. Using the procedures described in Section 3.3, select an appropriate ionization state of the acyl-*S*-Mpy15 peptide to deconvolute the parent masses. The thus-calculated mass differences should be in good agreement with the theoretical mass differences as calculated in Fig. 11.
4. To determine the halide specificity of Mpy16, extract MS^2 EICs for all reactions listed in Table 4 for the following ejection ions: $[M+H]^{1+}=422.07$, $[M+H]^{1+}=466.02$, and $[M+H]^{1+}=509.97$. As illustrated in Fig. 11 (blue boxes), these MS^2 ions correspond, respectively, to the dichlorinated, dibrominated, and mixed monobrominated–monochlorinated products that can possibly be generated by Mpy16.
5. Similarly, to determine the halide specificity of Bmp2, extract MS^2 EICs for all reactions listed in Table 5 for the following ejection ions: $[M+H]^{1+}=587.88$, $[M+H]^{1+}=456.03$, and $[M+H]^{1+}=543.93$. As illustrated in Fig. 11 (red boxes), these MS^2 ions correspond, respectively, to the tribrominated, trichlorinated, and mixed dibrominated–monochlorinated products that can hypothetically be generated by Bmp2.

   Note: It has been demonstrated that Mpy16 specifically catalyzes dihalogenation of the pyrrole ring, while in the absence of the thioesterase, Bmp2 catalyzes trihalogenation of the pyrrole ring (El Gamal et al., 2016).
6. The thus-generated EICs for all reactions listed in Table 4 are plotted in Fig. 13A, and EICs for all reactions listed in Table 5 are plotted in Fig. 13B.

Note: For the halogenation of pyrrolyl-*S*-Bmp1 by Bmp2, under no relative concentration of chloride and bromide in the assay can a chlorinated product be observed (Fig. 13B). A different trend is observed for Mpy16. At high relative concentration of chloride in the assay, Mpy16 generates a predominantly dichlorinated product (Fig. 13A). However, as the relative

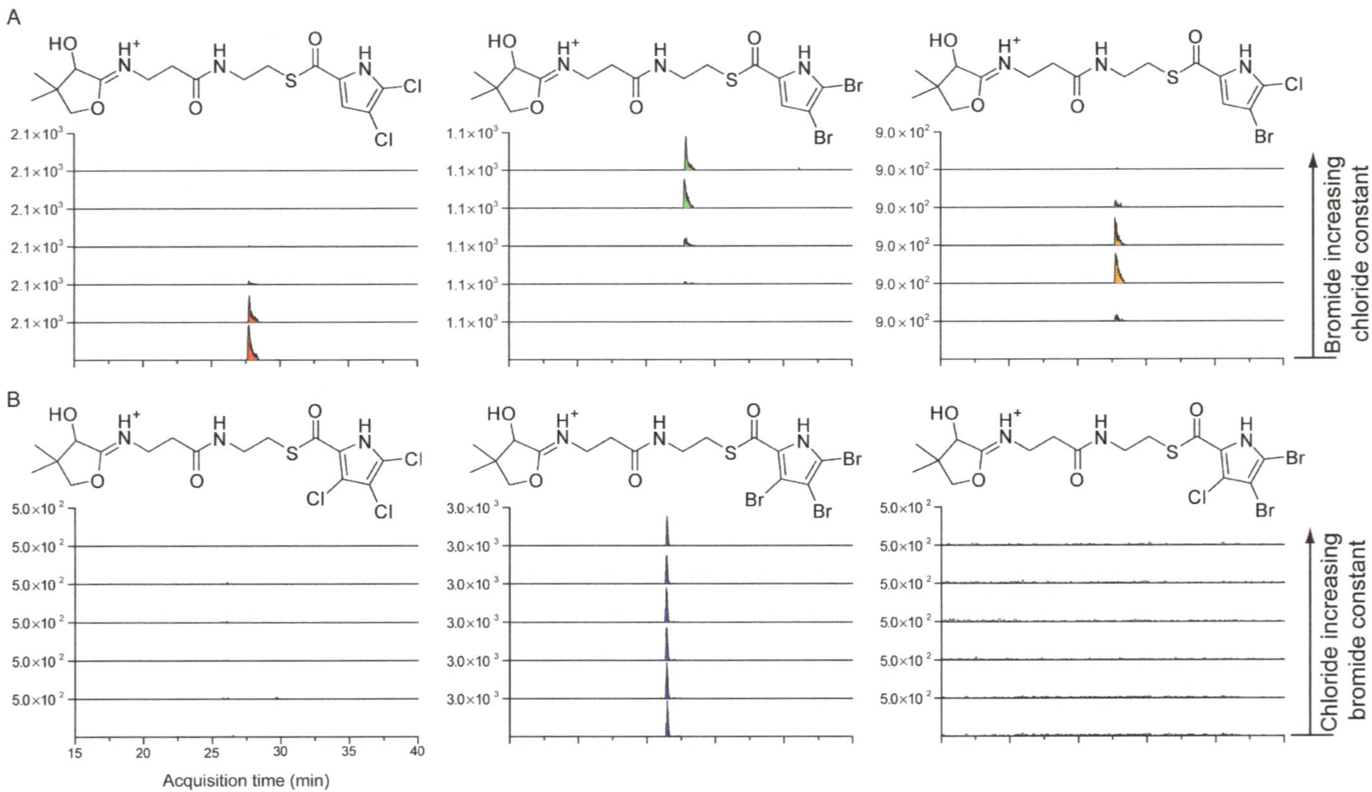

Fig. 13 MS^2 EICs (A) for all reactions listed in Table 4 for ejection ions corresponding to the dichlorinated, dibrominated, and mixed monobrominated–monochlorinated products that can possibly be generated by Mpy16. (B) MS^2 EICs for all reactions listed in Table 5 for ejection ions corresponding to trichlorinated, tribrominated, and mixed dibrominated–monochlorinated products that can hypothetically be generated by Bmp2.

concentration of bromide in the assay increases, ejection ion corresponding to a mixed brominated–chlorinated product is observed. At high relative concentration of bromide in the assay, the chlorinating activity of Mpy16 is completely abolished and only the ejection ion corresponding to dibrominated product is observed. Hence, the activity of a chlorinase is fundamentally different from a brominase, in that, a chlorinase (such as Mpy16) is an effective brominating catalyst depending on the relative concentration of halides in the assay. However, a brominase (such as Bmp2) is specific for incorporating bromide. To the best of our knowledge, this is the first experimental demonstration of the halide specificity for a chlorinase and a brominase and is aligned with the theoretical distinction between chlorinases and brominases (Blasiak & Drennan, 2009).

5. CONCLUSIONS

Access to starting material in sufficient quantity and purity is a major obstacle to investigating the reactions catalyzed by halogenase that require CP-tethered substrates. Crystal structures of halogenases in complex with their small-molecule organic substrates and products, such as for the Trp halogenases, have been critical in advancing our mechanistic understanding of FAD-dependent halogenases and their engineering for novel applications (Agarwal, Miles, et al., 2017; Latham et al., 2018). However, a cocrystal structure of a FAD-dependent halogenase in complex with a CP-tethered substrate has so far been elusive. In addition to the halogenase, cocrystallization requires access to large quantities of highly pure substrates, and the lack of the above-mentioned crystal structure is perhaps reflective of lack of access to these substrates needed to drive these efforts forward. The procedures describe in this study will enable these efforts in the future.

Mechanistically, FAD-dependent halogenases that employ CP-tethered substrates are presumed to be identical to FAD-dependent halogenases that employ small-molecule substrates such as L-Trp. The structural engagement of the FAD cofactor and the positioning of key catalytic resides for the two subclasses of FAD-dependent halogenases is highly similar. However, some key biochemical differences remain to be addressed. Primary among these is the absence of a catalytic Glu reside in acyl-*S*-CP halogenases (Fig. 14). To facilitate a electrophilic aromatic substitution by the halenium, the side chain of this Glu residue is postulated to serve as the catalytic base to abstract a proton from the substrate molecule, a role that is supported by kinetic and structural studies (Dong et al., 2005; Flecks et al., 2008). Thus, the

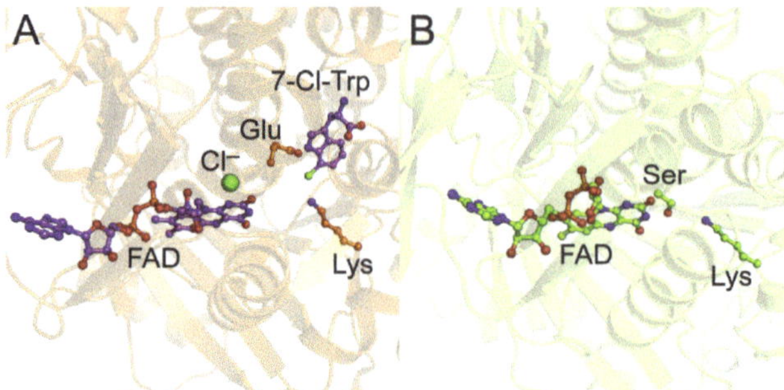

Fig. 14 Comparison of active sites of (A) L-Trp chlorinase PrnA (Dong et al., 2005) and (B) pyrrolyl-*S*-Mpy15 chlorinase Mpy16 (El Gamal et al., 2016). While the cofactor FAD and the catalytic Lys side chain are positioned analogously for the two structures, Mpy16, like other acyl-*S*-CP halogenases, is missing the key Glu residue which is replaced by a Ser. PrnA is complexed with a Cl^- anion and the product, 7-chlorotryptophan.

identity of the analogous catalytic base is still elusive for acyl-*S*-CP halogenases. Addressing these mechanistic questions will undoubtedly be aided by availability of a cocrystal structure of a FAD-dependent halogenase in complex with its cognate acyl-*S*-CP substrate. While this study specifically deals with FAD-dependent halogenases, these procedures described herein can be extended to other classes of halogenases as well. Most pertinently, numerous α-ketoglutarate-dependent nonheme Fe(II) (NHFe) halogenases have been described in literature that require substrates that are tethered to CPs (Agarwal, Miles, et al., 2017). However, the activities of several of these enzymes have not been reconstituted in vitro, perhaps due to lack of access to CP-tethered substrates.

Owing to high abundance of halides in seawater, the marine environment is an exceptionally rich source of halogenated natural products, specifically, of brominated natural products that are exclusively identified from marine sources (Blunt, Copp, Keyzers, Munro, & Prinsep, 2017). The existence of brominated natural products implies that brominases, such as Bmp2, have devised a "molecular filter" that rejects chloride being used as a substrate by the enzyme. This filter is extraordinarily selective given the ~600-fold higher concentration of chloride in seawater as compared to bromide. The chlorinases such as Mpy16, on the other hand, do not exert a similar halide selectivity filter. Chlorinases are equally adept at incorporating chloride and bromide. Hence, it is likely that the presence of chlorinated natural products in the marine environment, as by extension in the terrestrial environment as well, is a consequence of greater abundance and availability of chloride as compared to bromide.

ACKNOWLEDGMENTS

We gratefully acknowledge financial support from the NIH (R00ES026620) and Georgia Institute of Technology.

REFERENCES

Agarwal, V., Blanton, J. M., Podell, S., Taton, A., Schorn, M. A., Busch, J., et al. (2017). Metagenomic discovery of polybrominated diphenyl ether biosynthesis by marine sponges. *Nature Chemical Biology*, *13*(5), 537–543. https://doi.org/10.1038/nchembio.2330.

Agarwal, V., Diethelm, S., Ray, L., Garg, N., Awakawa, T., Dorrestein, P. C., et al. (2015). Chemoenzymatic synthesis of acyl coenzyme A substrates enables *in situ* labeling of small molecules and proteins. *Organic Letters*, *17*(18), 4452–4455. https://doi.org/10.1021/acs.orglett.5b02113.

Agarwal, V., El Gamal, A. A., Yamanaka, K., Poth, D., Kersten, R. D., Schorn, M., et al. (2014). Biosynthesis of polybrominated aromatic organic compounds by marine bacteria. *Nature Chemical Biology*, *10*(8), 640–647. https://doi.org/10.1038/nchembio.1564.

Agarwal, V., Lin, S., Lukk, T., Nair, S. K., & Cronan, J. E. (2012). Structure of the enzyme-acyl carrier protein (ACP) substrate gatekeeper complex required for biotin synthesis. *Proceedings of the National Academy of Sciences of the United States of America*, *109*(43), 17406–17411. https://doi.org/10.1073/pnas.1207028109.

Agarwal, V., Miles, Z. D., Winter, J. M., Eustaquio, A. S., El Gamal, A. A., & Moore, B. S. (2017). Enzymatic halogenation and dehalogenation reactions: Pervasive and mechanistically diverse. *Chemical Reviews*, *117*(8), 5619–5674. https://doi.org/10.1021/acs.chemrev.6b00571.

Beld, J., Sonnenschein, E. C., Vickery, C. R., Noel, J. P., & Burkart, M. D. (2014). The phosphopantetheinyl transferases: Catalysis of a post-translational modification crucial for life. *Natural Product Reports*, *31*(1), 61–108. https://doi.org/10.1039/c3np70054b.

Bitto, E., Huang, Y., Bingman, C. A., Singh, S., Thorson, J. S., & Phillips, G. N., Jr. (2008). The structure of flavin-dependent tryptophan 7-halogenase RebH. *Proteins*, *70*(1), 289–293. https://doi.org/10.1002/prot.21627.

Blasiak, L. C., & Drennan, C. L. (2009). Structural perspective on enzymatic halogenation. *Accounts of Chemical Research*, *42*(1), 147–155. https://doi.org/10.1021/ar800088r.

Blunt, J. W., Copp, B. R., Keyzers, R. A., Munro, M. H. G., & Prinsep, M. R. (2017). Marine natural products. *Natural Product Reports*, *34*(3), 235–294. https://doi.org/10.1039/c6np00124f.

Brown, S., & O'Connor, S. E. (2015). Halogenase engineering for the generation of new natural product analogues. *ChemBioChem*, *16*(15), 2129–2135. https://doi.org/10.1002/cbic.201500338.

Chankhamjon, P., Tsunematsu, Y., Ishida-Ito, M., Sasa, Y., Meyer, F., Boettger-Schmidt, D., et al. (2016). Regioselective dichlorination of a non-activated aliphatic carbon atom and phenolic bismethylation by a multifunctional fungal flavoenzyme. *Angewandte Chemie (International ed. in English)*, *55*(39), 11955–11959. https://doi.org/10.1002/anie.201604516.

Dong, C., Flecks, S., Unversucht, S., Haupt, C., van Pee, K. H., & Naismith, J. H. (2005). Tryptophan 7-halogenase (PrnA) structure suggests a mechanism for regioselective chlorination. *Science (New York, NY)*, *309*(5744), 2216–2219. https://doi.org/10.1126/science.1116510.

Dorrestein, P. C., Bumpus, S. B., Calderone, C. T., Garneau-Tsodikova, S., Aron, Z. D., Straight, P. D., et al. (2006). Facile detection of acyl and peptidyl intermediates on thiotemplate carrier domains via phosphopantetheinyl elimination reactions during tandem mass spectrometry. *Biochemistry*, *45*(42), 12756–12766. https://doi.org/10.1021/bi061169d.

Dorrestein, P. C., & Kelleher, N. L. (2006). Dissecting non-ribosomal and polyketide biosynthetic machineries using electrospray ionization Fourier-Transform mass spectrometry. *Natural Product Reports*, *23*(6), 893–918. https://doi.org/10.1039/b511400b.

Dorrestein, P. C., Yeh, E., Garneau-Tsodikova, S., Kelleher, N. L., & Walsh, C. T. (2005). Dichlorination of a pyrrolyl-S-carrier protein by FADH2-dependent halogenase PltA during pyoluteorin biosynthesis. *Proceedings of the National Academy of Sciences of the United States of America*, *102*(39), 13843–13848. https://doi.org/10.1073/pnas.0506964102.

Eichhorn, E., van der Ploeg, J. R., & Leisinger, T. (1999). Characterization of a two-component alkanesulfonate monooxygenase from *Escherichia coli*. *Journal of Biological Chemistry*, *274*(38), 26639–26646.

El Gamal, A., Agarwal, V., Diethelm, S., Rahman, I., Schorn, M. A., Sneed, J. M., et al. (2016). Biosynthesis of coral settlement cue tetrabromopyrrole in marine bacteria by a uniquely adapted brominase–thioesterase enzyme pair. *Proceedings of the National Academy of Sciences of the United States of America*, *113*(14), 3797–3802. https://doi.org/10.1073/pnas.15196951131519695113.

Flecks, S., Patallo, E. P., Zhu, X., Ernyei, A. J., Seifert, G., Schneider, A., et al. (2008). New insights into the mechanism of enzymatic chlorination of tryptophan. *Angewandte Chemie (International ed. in English)*, *47*(49), 9533–9536. https://doi.org/10.1002/anie.200802466.

Fraley, A. E., Garcia-Borras, M., Tripathi, A., Khare, D., Mercado-Marin, E. V., Tran, H., et al. (2017). Function and structure of MalA/MalA', iterative halogenases for late-stage C–H functionalization of indole alkaloids. *Journal of the American Chemical Society*, *139*(34), 12060–12068. https://doi.org/10.1021/jacs.7b06773.

Franke, J., & Hertweck, C. (2016). Biomimetic thioesters as probes for enzymatic assembly lines: Synthesis, applications, and challenges. *Cell Chemical Biology*, *23*(10), 1179–1192. https://doi.org/10.1016/j.chembiol.2016.08.014.

He, J., Magarvey, N., Piraee, M., & Vining, L. C. (2001). The gene cluster for chloramphenicol biosynthesis in *Streptomyces venezuelae* ISP5230 includes novel shikimate pathway homologues and a monomodular non-ribosomal peptide synthetase gene. *Microbiology*, *147*(Pt. 10), 2817–2829. https://doi.org/10.1099/00221287-147-10-2817.

Hofer, I., Crusemann, M., Radzom, M., Geers, B., Flachshaar, D., Cai, X., et al. (2011). Insights into the biosynthesis of hormaomycin, an exceptionally complex bacterial signaling metabolite. *Chemistry and Biology*, *18*(3), 381–391. https://doi.org/10.1016/j.chembiol.2010.12.018.

Johannes, T. W., Woodyer, R. D., & Zhao, H. (2007). Efficient regeneration of NADPH using an engineered phosphite dehydrogenase. *Biotechnology and Bioengineering*, *96*(1), 18–26. https://doi.org/10.1002/bit.21168.

Jordan, P. A., & Moore, B. S. (2016). Biosynthetic pathway connects cryptic ribosomally synthesized posttranslationally modified peptide genes with pyrroloquinoline alkaloids. *Cell Chemical Biology*, *23*(12), 1504–1514. https://doi.org/10.1016/j.chembiol.2016.10.009.

Latham, J., Brandenburger, E., Shepherd, S. A., Menon, B. R. K., & Micklefield, J. (2018). Development of halogenase enzymes for use in synthesis. *Chemical Reviews*, *118*(1), 232–269. https://doi.org/10.1021/acs.chemrev.7b00032.

Lin, S., Van Lanen, S. G., & Shen, B. (2007). Regiospecific chlorination of (S)-beta-tyrosyl-S-carrier protein catalyzed by SgcC3 in the biosynthesis of the enediyne antitumor antibiotic C-1027. *Journal of the American Chemical Society*, *129*(41), 12432–12438. https://doi.org/10.1021/ja072311g.

Mantovani, S. M., & Moore, B. S. (2013). Flavin-linked oxidase catalyzes pyrrolizine formation of dichloropyrrole-containing polyketide extender unit in chlorizidine A. *Journal of the American Chemical Society*, *135*(48), 18032–18035. https://doi.org/10.1021/ja409520v.

Menon, B. R. K., Brandenburger, E., Sharif, H. H., Klemstein, U., Shepherd, S. A., Greaney, M. F., et al. (2017). RadH: A versatile halogenase for integration into synthetic pathways. *Angewandte Chemie (International ed. in English)*, *56*(39), 11841–11845. https://doi.org/10.1002/anie.201706342.

Mishra, P. K., & Drueckhammer, D. G. (2000). Coenzyme A analogues and derivatives: Synthesis and applications as mechanistic probes of coenzyme A ester-utilizing enzymes. *Chemical Reviews*, *100*(9), 3283–3310.

Nazi, I., Koteva, K. P., & Wright, G. D. (2004). One-pot chemoenzymatic preparation of coenzyme A analogues. *Analytical Biochemistry*, *324*(1), 100–105.

Pang, A. H., Garneau-Tsodikova, S., & Tsodikov, O. V. (2015). Crystal structure of halogenase PltA from the pyoluteorin biosynthetic pathway. *Journal of Structural Biology*, *192*(3), 349–357. https://doi.org/10.1016/j.jsb.2015.09.013.

Payne, J. T., Andorfer, M. C., & Lewis, J. C. (2016). Engineering flavin-dependent halogenases. *Methods in Enzymology*, *575*, 93–126. https://doi.org/10.1016/bs.mie.2016.03.024.

Peter, D. M., Vogeli, B., Cortina, N. S., & Erb, T. J. (2016). A chemo-enzymatic road map to the synthesis of CoA esters. *Molecules*, *21*(4), 517. https://doi.org/10.3390/molecules21040517.

Pinchman, J. R., & Boger, D. L. (2013a). Investigation into the functional impact of the vancomycin C-ring aryl chloride. *Bioorganic and Medicinal Chemistry Letters*, *23*(17), 4817–4819. https://doi.org/10.1016/j.bmcl.2013.06.080.

Pinchman, J. R., & Boger, D. L. (2013b). Probing the role of the vancomycin e-ring aryl chloride: Selective divergent synthesis and evaluation of alternatively substituted E-ring analogues. *Journal of Medicinal Chemistry*, *56*(10), 4116–4124. https://doi.org/10.1021/jm4004494.

Podzelinska, K., Latimer, R., Bhattacharya, A., Vining, L. C., Zechel, D. L., & Jia, Z. (2010). Chloramphenicol biosynthesis: The structure of CmlS, a flavin-dependent halogenase showing a covalent flavin-aspartate bond. *Journal of Molecular Biology*, *397*(1), 316–331. https://doi.org/10.1016/j.jmb.2010.01.020.

Quadri, L. E., Weinreb, P. H., Lei, M., Nakano, M. M., Zuber, P., & Walsh, C. T. (1998). Characterization of Sfp, a *Bacillus subtilis* phosphopantetheinyl transferase for peptidyl carrier protein domains in peptide synthetases. *Biochemistry*, *37*(6), 1585–1595. https://doi.org/10.1021/bi9719861bi9719861.

Reuter, K., Mofid, M. R., Marahiel, M. A., & Ficner, R. (1999). Crystal structure of the surfactin synthetase-activating enzyme sfp: A prototype of the 4'-phosphopantetheinyl transferase superfamily. *EMBO Journal*, *18*(23), 6823–6831. https://doi.org/10.1093/emboj/18.23.6823.

Ross, A. C., Gulland, L. E., Dorrestein, P. C., & Moore, B. S. (2015). Targeted capture and heterologous expression of the *Pseudoalteromonas* alterochromide gene cluster in *Escherichia coli* represents a promising natural product exploratory platform. *ACS Synthetic Biology*, *4*(4), 414–420. https://doi.org/10.1021/sb500280q.

Sato, M., Winter, J. M., Kishimoto, S., Noguchi, H., Tang, Y., & Watanabe, K. (2016). Combinatorial generation of chemical diversity by redox enzymes in chaetoviridin biosynthesis. *Organic Letters*, *18*(6), 1446–1449. https://doi.org/10.1021/acs.orglett.6b00380.

Schmartz, P. C., Zerbe, K., Abou-Hadeed, K., & Robinson, J. A. (2014). Bis-chlorination of a hexapeptide-PCP conjugate by the halogenase involved in vancomycin biosynthesis. *Organic & Biomolecular Chemistry*, *12*(30), 5574–5577. https://doi.org/10.1039/c4ob00474d.

Schnepel, C., Minges, H., Frese, M., & Sewald, N. (2016). A high-throughput fluorescence assay to determine the activity of tryptophan halogenases. *Angewandte Chemie (International ed. in English)*, *55*(45), 14159–14163. https://doi.org/10.1002/anie.201605635.

Strauss, E., & Begley, T. P. (2002). The antibiotic activity of N-pentylpantothenamide results from its conversion to ethyldethia-coenzyme A, a coenzyme A antimetabolite. *Journal of Biological Chemistry*, *277*(50), 48205–48209. https://doi.org/10.1074/jbc.M204560200M204560200.

Thomas, M. G., Burkart, M. D., & Walsh, C. T. (2002). Conversion of L-proline to pyrrolyl-2-carboxyl-S-PCP during undecylprodigiosin and pyoluteorin biosynthesis. *Chemistry and Biology*, *9*(2), 171–184. https://doi.org/10.1016/S1074-5521(02)00100-X.

van Wageningen, A. M., Kirkpatrick, P. N., Williams, D. H., Harris, B. R., Kershaw, J. K., Lennard, N. J., et al. (1998). Sequencing and analysis of genes involved in the biosynthesis of a vancomycin group antibiotic. *Chemistry and Biology*, *5*(3), 155–162.

Walsh, C. T., Garneau-Tsodikova, S., & Howard-Jones, A. R. (2006). Biological formation of pyrroles: Nature's logic and enzymatic machinery. *Natural Product Reports*, *23*(4), 517–531. https://doi.org/10.1039/b605245m.

Worthington, A. S., & Burkart, M. D. (2006). One-pot chemo-enzymatic synthesis of reporter-modified proteins. *Organic & Biomolecular Chemistry*, *4*(1), 44–46. https://doi.org/10.1039/b512735a.

Yamanaka, K., Ryan, K. S., Gulder, T. A., Hughes, C. C., & Moore, B. S. (2012). Flavoenzyme-catalyzed atropo-selective N,C-bipyrrole homocoupling in marinopyrrole biosynthesis. *Journal of the American Chemical Society*, *134*(30), 12434–12437. https://doi.org/10.1021/ja305670f.

Yeh, E., Blasiak, L. C., Koglin, A., Drennan, C. L., & Walsh, C. T. (2007). Chlorination by a long-lived intermediate in the mechanism of flavin-dependent halogenases. *Biochemistry*, *46*(5), 1284–1292. https://doi.org/10.1021/bi0621213.

Zhu, X., De Laurentis, W., Leang, K., Herrmann, J., Ihlefeld, K., van Pee, K. H., et al. (2009). Structural insights into regioselectivity in the enzymatic chlorination of tryptophan. *Journal of Molecular Biology*, *391*(1), 74–85. https://doi.org/10.1016/j.jmb.2009.06.008.

CHAPTER ELEVEN

Preparation, Assay, and Application of Chlorinase SalL for the Chemoenzymatic Synthesis of *S*-Adenosyl-L-Methionine and Analogs

Tony D. Davis*, Sylvia Kunakom†, Michael D. Burkart*,[1], Alessandra S. Eustaquio†,[1]

*University of California San Diego, San Diego, CA, United States
†University of Illinois at Chicago, College of Pharmacy, and Center for Biomolecular Sciences, Chicago, IL, United States
[1]Corresponding authors: e-mail address: mburkart@ucsd.edu; ase@uic.edu

Contents

Abstract

S-adenosyl-L-methionine (SAM) is universal in biology, serving as the second most common cofactor in a variety of enzymatic reactions. One of the main roles of SAM is the methylation of nucleic acids, proteins, and metabolites. Methylation often imparts

Methods in Enzymology, Volume 604
ISSN 0076-6879
https://doi.org/10.1016/bs.mie.2018.02.012

regulatory control to DNA and proteins, and leads to an increase in the activity of specialized metabolites such as those developed as pharmaceuticals. There has been increased interest in using SAM analogs in methyltransferase-catalyzed modification of biomolecules. However, SAM and its analogs are expensive and unstable, degrading rapidly under physiological conditions. Thus, the availability of methods to prepare SAM in situ is desirable. In addition, synthetic methods to generate SAM analogs suffer from low yields and poor diastereoselectivity. The chlorinase SalL from the marine bacterium *Salinispora tropica* catalyzes the reversible, nucleophilic attack of chloride at the C5′ ribosyl carbon of SAM leading to the formation of 5′-chloro-5′-deoxyadenosine (ClDA) with concomitant displacement of L-methionine. It has been demonstrated that the in vitro equilibrium of the SalL-catalyzed reaction favors the synthesis of SAM. In this chapter, we describe methods for the preparation of SalL, and the chemoenzymatic synthesis of SAM and SAM analogs from ClDA and L-methionine congeners using SalL. In addition, we describe procedures for the in situ chemoenzymatic synthesis of SAM coupled to DNA, peptide, and metabolite methylation, and to the incorporation of isotopes into alkylated products.

1. INTRODUCTION

S-adenosyl-L-methionine (SAM or AdoMet) is ubiquitous in nature and the second most utilized enzyme substrate after adenosine triphosphate (ATP). SAM is not only a source of methyl groups to methylate a variety of substrates, including nucleic acid, proteins, and small molecules (Cantoni, 1975; Thomsen, Vogensen, Buchardt, Burkart, & Clausen, 2013), but also a source of methylene, amino, ribosyl, and aminopropyl groups, in addition to 5′-deoxyadenosyl radicals (Fontecave, Atta, & Mulliez, 2004). Remarkably, SAM has also recently been shown to serve as a cofactor in coordinating pericyclic reactions via electrostatic catalysis (Ohashi et al., 2017). Another role for SAM is to serve as a substrate in reactions catalyzed by SAM-dependent halogenases of the 5′-halo-5′-deoxyadenosine (XDA) synthase class (Agarwal et al., 2017; Dong et al., 2004; Eustáquio, Pojer, Noel, & Moore, 2008).

Halogenated compounds are pervasive across all domains of life. Over 5000 organohalogens are known. Organochlorines and organobromines are the most abundant with >2000 known examples of each, whereas organoiodines are more scarce (<200) and organofluorines are very rare with only five rigorously identified compounds (Chan & O'Hagan, 2012; Gribble, 2012; Wang, Zhou, Fredimoses, Liao, & Liu, 2014). Halogenating enzymes employ various mechanistic strategies depending on the properties of the halogen atom and the reactivity of the organic substrate. Most halogenases activate chloride, bromide, and iodide via oxidation by utilizing cofactors such as heme-iron, vanadium, flavin, and Fe(II)/α-ketoglutarate (Agarwal et al., 2017). For fluoride, on the other hand, which evades oxidation due to its extreme electronegativity, an orthogonal strategy has been described that uses SAM as a cosubstrate in an S_N2-type, nucleophilic substitution reaction (Fig. 1). The only fluorinase described thus far (FlA) catalyzes the first committed step in the biosynthesis of fluoroacetate and fluorothreonine in the soil bacterium *Streptomyces cattleya* in which fluoride attacks the C5′ electrophilic carbon of SAM, displacing L-methionine and generating 5′-fluoro-5′-deoxyadenosine (FDA) (Dong et al., 2004). FlA was later identified in other bacteria in addition to *S. cattleya* (Deng et al., 2014; Huang et al., 2014).

Four years after the structural characterization of the fluorinase, the chlorinase SalL was reported from the marine bacterium *Salinispora tropica* (Eustáquio et al., 2008). SalL is analogous to fluorinase and the only other structurally characterized SAM-dependent halogenase of the XDA synthase class (Fig. 1). SalL catalyzes the first committed step in the biosynthesis of proteasome inhibitor salinosporamide A (Eustáquio et al., 2009, 2008; Feling et al., 2003). Both FlA and SalL can also catalyze the reverse reaction, i.e., synthesis of SAM from XDA. For FlA, the in vitro equilibrium

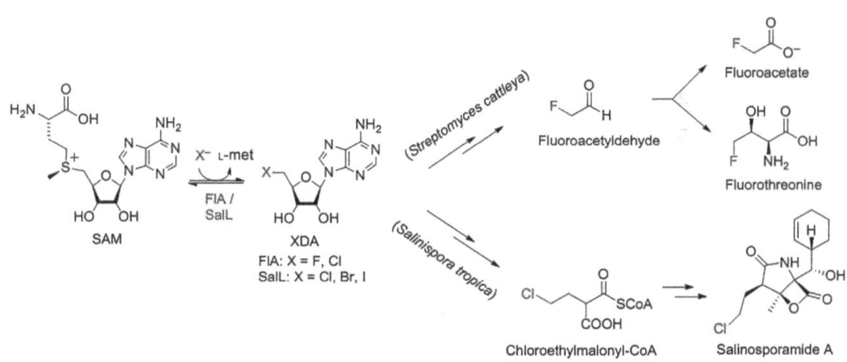

Fig. 1 Reaction catalyzed by SAM-dependent halogenases. Fluorinase FlA from soil bacterium *Streptomyces cattleya* (Dong et al., 2004) and chlorinase SalL from marine bacterium *Salinispora tropica* (Eustáquio et al., 2008) catalyze reversible S_N2-type, nucleophilic substitution reactions in which the halide attacks the C5′ ribosyl carbon of SAM, generating 5′-halo-5′-deoxyadenosine (XDA) and releasing L-methionine. FlA and SalL catalyze the first committed step in the biosynthesis of fluorometabolites and salinosporamide A, respectively.

lies in favor of FDA by a factor of three (Dong et al., 2004). That the forward reaction is favored by FlA is unsurprising given that the C–F bond is the strongest bond in organic chemistry. For chlorinase SalL, on the other hand, synthesis of SAM from 5′-chloro-5′-deoxyadenosine (ClDA) and L-methionine (L-met) is favored, with the relative enzyme efficiency (k_{cat}/K_m) for the reverse reaction (SAM synthesis) being several orders of magnitude higher than the forward halogenation reaction (Eustáquio et al., 2008). The fluorinase was also reported to favor the reverse reaction in vitro when using chloride as the halide (Deng et al., 2006).

As a versatile methyl donor to biomolecules, there has been considerable interest in exploring the synthesis of SAM analogs of chemical and biological utility. For instance, nonnatural SAM analogs could be used to modify cellular targets such as DNA, RNA, proteins, and metabolites. Moreover, SAM itself is an expensive and unstable cofactor, degrading rapidly under physiological or alkaline conditions. Thus, the availability of methods to prepare SAM in situ is desirable.

Traditionally, SAM analogs have been chemically synthesized from *S*-adenosyl-L-homocysteine. However, this synthesis suffers from low yields and lack of diastereoselectivity (Dalhoff, Lukinavicius, Klimasauskas, & Weinhold, 2006a, 2006b). Furthermore, SAM synthesis is challenging due to its inherent lability and tendency to racemize (Hoffman, 1986). Chemoenzymatic synthesis using chlorinase SalL and ClDA offers an alternative to chemical synthesis. In addition to being stable and commercially available, ClDA can also be synthesized economically from adenosine and $SOCl_2$ (Sinhababu, Bartel, Pochopin, & Borchardt, 1985). Chemoenzymatic synthesis of SAM and analogs in situ could also be coupled with downstream applications such as methylation or alkylation by methyltransferases. This strategy can also be used for the facile incorporation of isotopes into alkylated products.

We have demonstrated the use of SalL for the diastereoselective synthesis of SAM and SAM congeners from ClDA and L-met analogs (see Section 3). We have also coupled the SalL-catalyzed production of SAM with DNA and with metabolite methylation (Lipson et al., 2013). Importantly, we have demonstrated that in situ generated SAM analogs were recognized by rat protein arginine methyltransferase 1 (rPRMT1) which alkylated its peptide substrate (Thomsen et al., 2013).

In this chapter, we describe procedures for the preparation and assay of SalL, the SalL-mediated chemoenzymatic synthesis of SAM analogs from

ClDA and L-met congeners, and the in situ chemoenzymatic synthesis of SAM coupled to DNA, peptide, and metabolite methylation, and to the incorporation of isotopes into alkylated products.

2. PREPARATION AND ASSAY OF CHLORINASE SalL

2.1 Purification of Recombinant SalL From *Escherichia coli*

Recombinant SalL containing a N-terminal His_8 tag was obtained by gene expression in and protein purification from *E. coli* BL21(DE3) (Eustáquio et al., 2008).

2.1.1 Cloning of the salL *Gene Into an Expression Vector*

The *salL* gene can be amplified by PCR using total genomic DNA isolated from *S. tropica* CNB-440 as a template and appropriate primers. The following are suggested primers that were used to clone *salL* into pHIS8 (Jez, Ferrer, Bowman, Dixon, & Noel, 2000) by restriction/ligation, yielding plasmid pAEM7 (Eustáquio et al., 2008):

for: 5′-CGTGGTTCCCATGG*C**ATG**CAGCACAATCTCATTGC*-3′ (*Nco*I site underlined)

rev: 5′-GCTCGAATTCAAGCTT*GTCAGCTACCCGAGCACCG*-3′ (*Hin*dIII site underlined)

The primers were designed to ensure that the ATG (Met start codon, bold) was in-frame with the His_8 tag. The priming sites are in italics. Bases upstream of the restriction sites are random sequences to facilitate cutting of the PCR product with the respective restriction enzyme. *salL* was amplified by PCR using the Expand High Fidelity PCR system (Roche) following the manufacturer's instructions and using 5% DMSO (v/v) and annealing temperature of 58°C. Alternative, proofreading DNA polymerases include Q5® High Fidelity polymerase and Phusion® High Fidelity polymerase (New England Biolabs). The expected size of the PCR product when using the primers above is 888 bp. Alternatively, synthetic DNA can be ordered from GenScript. The accession code for the *salL* gene (Strop_1026) of *S. tropica* CNB-440 is NC_009380.1 (position 1,162,700 to 1,163,551, complement). If using the primers above, digest the PCR product with *Nco*I and *Hin*dIII and ligate it into the same sites of pHIS8 to yield the pHIS8-*salL* expression vector (pAEM7). While the PCR product can be digested and cloned directly into pHIS8, we subcloned the PCR product into the pGEM-T Easy vector (Promega) by following the manufacturer's instructions. The insert of the obtained plasmid (pAEM5) was sequenced and confirmed to contain no mistakes. Following digestion of pAEM5 with *Hind*III, *Nco*I, and *Dra*I (DraI cuts only the vector backbone), and isopropanol precipitation, the *Hind*III–*Nco*I *salL* fragment (868 bp) was cloned into the same sites of pHIS8. pHIS8 contains an N-terminal, 8 × His tag that will facilitate protein purification. Note that the pAEM7 *salL* expression plasmid used here contains native DNA from *S. tropica* rather than DNA which is codon-optimized for expression in *E. coli*.

2.1.2 Expression of salL *in* E. coli *BL21(DE3) and Purification of Recombinant SalL Enzyme*

2.1.2.1 Materials

- Competent *E. coli* BL-21 (DE3) cells (Novagen)
- *salL* expression vector (pAEM7)
- Luria–Bertani (LB) agar (Becton Dickinson)
- LB broth (Becton Dickinson)
- Terrific broth (TB) medium (Becton Dickinson)
- Ni^{2+}-NTA resin (Qiagen)
- Lysis buffer: 50 m*M* sodium phosphate buffer (pH 8.0), 500 m*M* NaCl, 20 m*M* imidazole (pH 8.0), 20 m*M* β-mercaptoethanol, 10% (v/v) glycerol, and 1% (v/v) Tween-20, and 1 mg/mL lysozyme
- Wash buffer: 50 m*M* sodium phosphate buffer (pH 8.0), 500 m*M* NaCl, 20 m*M* imidazole (pH 8.0), 20 m*M* β-mercaptoethanol, and 10% (v/v) glycerol
- Elution buffer: 50 m*M* sodium phosphate buffer (pH 8.0), 500 m*M* NaCl, 250 m*M* imidazole (pH 8.0), 20 m*M* β-mercaptoethanol, and 10% (v/v) glycerol
- Storage buffer: 50 m*M* sodium phosphate buffer (pH 7.9), 20% glycerol
- Disposable polypropylene column (from Qiagen)
- Disposable PD-10 desalting columns (Sephadex G-25, GE Healthcare)
- SDS-PAGE (10% Bis-Tris NuPAGE gel with MOPS running buffer, ThermoFisher)

2.1.2.2 Procedure

1. Competent *E. coli* BL-21 (DE3) cells are commercially available (e.g., Novagen, cat. No. 69450-3). Follow the manufacturer's instructions to introduce the *salL*-containing expression vector into *E. coli* BL-21 (DE3) by chemical transformation. Plate the transformation reaction

on LB agar containing 50 mg/L kanamycin (pHIS8 contains a kanamycin-resistance gene) and incubate the plates overnight at 37°C.

2. Inoculate a single colony onto 50 mL LB containing 50 mg/L kanamycin (use a 250-mL flask) and incubate the flask overnight at 37°C, 200 rpm.
3. Prepare cryo stocks for long-term storage by mixing the *E. coli* BL21 (DE3) culture containing the *salL* plasmid 1:1 with sterile glycerol 40%, giving a final glycerol concentration of 20%. Store cryo stocks at −80°C.
4. Use the overnight LB culture to inoculate 1 L of TB media containing 50 mg/L kanamycin so that the starting $A_{600}=0.1$ (10–20 mL of the overnight *E. coli* seed culture are expected to be necessary). Incubate the flask at 25°C, 225 rpm until $A_{600}=0.8$. Once an A_{600} of 0.8 is reached, induce cells with 0.25 m*M* IPTG (final concentration) and incubate overnight at 20°C, 225 rpm. A 2.8-L Fernbach flask is recommended for cultivation of 1 L cultures.
5. Harvest cells by centrifugation (e.g., 15 min at 5000 × *g*) and resuspend the cell pellet in ~20 mL of cold lysis buffer. Incubate on ice for 30 min.
6. *Note*: take care to keep the preparation cold by cooling centrifuges, buffers, and keeping the sample on ice.
7. Lyse cells by sonication on ice (e.g., 2 s on, 2 s off for 3 min total). Alternatively, a French press can be used for cell lysis.
8. Centrifuge the lysate at high speed (>10,000 × *g*) for 30 min at 4°C.
9. The clear lysate is then purified using Ni^{2+}-NTA resin (Qiagen) and a modified batch purification method as briefly described below (refer to the manufacturer's instructions "The QIAexpressionist" for more details).
10. Transfer the clear lysate to a clean, autoclaved beaker on ice, and add 5 mL Ni^{2+}-NTA to the clear lysate. Mix gently using a magnetic stirrer for 1 h. Alternatively, use a rotary shaker for mixing the sample contained in a conical tube.
11. If using a beaker for mixing, transfer the lysate/Ni-NTA suspension to a conical tube and centrifuge at 3000 × *g* for 5 min at 4°C to collect the Ni-NTA resin.
12. Discard the supernatant and wash the resin twice with 25 mL wash buffer each as following. Add 25 mL wash buffer to the Ni-NTA pellet, resuspend gently, and centrifuge at 3000 × *g* for 5 min at 4°C. Discard supernatant and repeat.

13. After the last wash, resuspend the Ni-NTA pellet in ~5 mL wash buffer and transfer the resin to an empty polypropylene column with the bottom outlet capped. Remove bottom cap and discard flow-through. Wash once with 25 mL cold wash buffer. Discard flow-through.
14. Elute the His_8-tagged SalL with 2 × 5 mL elution buffer.
15. Assess the quality of the protein preparation using SDS-PAGE (e.g., 10% Bis-Tris NuPAGE gel with MOPS running buffer, ThermoFisher). The expected size of the His_8-tagged protein is 32.4 kDa (Fig. 2).
16. The expected yield is >60 mg recombinant, soluble SalL per liter of culture. Protein yield can be calculated using the Bradford assay (Bradford, 1976).
17. Desalting can be performed using PD-10 columns (Sephadex G-25): cut the bottom tip off the column and let the storage liquid flow (discard); equilibrate the column with 25 mL storage buffer; load sample in a total volume of 2.5 mL per PD-10 column; discard the flow-through; elute with 3.5 mL storage buffer. Alternatively, dialysis can be used for desalting.

2.2 SalL Activity Assay

The reaction catalyzed by chlorinase SalL is reversible. Chloride, bromide, and iodide are accepted as the halide cosubstrates in the reaction with SAM (forward reaction) to yield XDA. Kinetic studies demonstrated that the

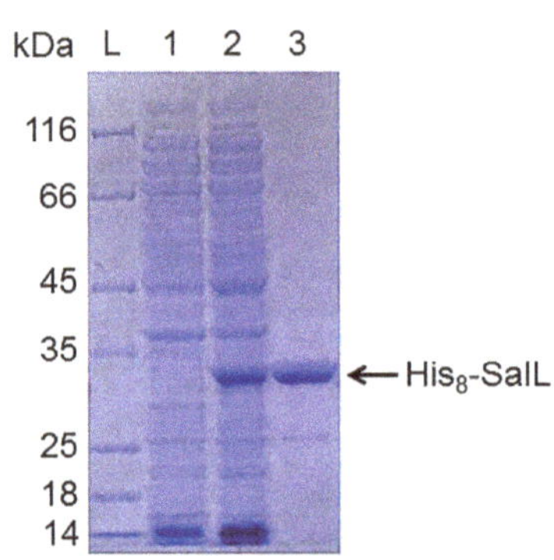

Fig. 2 SDS-PAGE analysis of purified, recombinant SalL. 10% Bis-Tris NuPAGE gel with MOPS running buffer (ThermoFisher). L, Fisherbrand EZ-Run protein marker; (1) soluble protein before induction; (2) soluble protein after induction; and (3) His_8-SalL purified by Ni-NTA affinity chromatography.

reverse reaction (formation of SAM from XDA and L-met) is favored in vitro, with the relative enzyme efficiency (k_{cat}/K_m) of the reverse reaction being orders of magnitude higher than the forward reaction (Eustáquio et al., 2008). The activity of recombinant SalL can be investigated by a HPLC-based method as described below.

2.2.1 Reagents

- *S*-adenosyl-L-methionine (Sigma-Aldrich)
- KCl (analytical grade)
- L-Methionine (Sigma-Aldrich)
- 5′-Chloro-5′-deoxyadenosine (Sigma-Aldrich)
- 50 m*M* sodium phosphate buffer pH 7.9
- Recombinant SalL (Section 2.1)

2.2.2 Instrumentation and Solvents

- Agilent HPLC 1200
- C18 column (250 × 4.60 mm, 5 μm particle size)
- Mobile phase A: 95% 50 m*M* KH_2PO_4; 5% acetonitrile
- Mobile phase B: 80% 50 m*M* KH_2PO_4; 20% acetonitrile
- *Note*: 50 m*M* KH_2PO_4 gives pH ~4.5; there is no need to adjust the pH.

2.2.3 Procedure

1. In order to test if your SalL preparation is active, set up two different reactions by incubating SalL (400 n*M*) with
   (i) SAM (0.5 m*M*) and KCl (400 m*M*)
   (ii) L-Methionine (10 m*M*) and ClDA (0.5 m*M*)
   in 50 m*M* sodium phosphate buffer (pH 7.9) at 37°C for 1 h. In addition, set up following controls: reactions with inactivated enzyme (the enzyme is first boiled for 5 min and then, the reagents are added) and boiled enzyme only (the enzyme is boiled for 5 min in buffer).
2. Boil the reaction mixture for 2 min at 95°C and centrifuge for 30 min, 10,000 × *g* to remove precipitated protein. 10 μL of clear supernatant is then analyzed by HPLC using a C18 column (250 × 4.60 mm, 5 μm particle size) and 1 mL/min flow rate. After equilibrating the system for at least 15 min using 100% mobile phase A, run following gradient: 100% A for 1 min, 0%–100% B for 29 min, 100%–0% B over 5 min, 100% A for 5 min. Inject also SAM and ClDA standards for comparison.

The retention times obtained under the conditions described were 16.4 min (ClDA) and 18.1 min (SAM).

3. *Notes*:
   - SAM is unstable and SAM solutions should be freshly prepared.
   - Recombinant SalL purified from *E. coli* copurifies with adenosine and ClDA. Thus, the boiled enzyme negative control may have a small peak corresponding to ClDA.

3. CHEMOENZYMATIC SYNTHESIS OF SAM ANALOGS

3.1 Overview of Chemoenzymatic Methods to Prepare SAM and SAM Analogs

SAM is biosynthesized by methionine adenosyltransferases (MAT) from methionine and ATP (Cantoni, 1975; Fig. 3). SAM is an essential cofactor for SAM-dependent methyltransferases to methylate various substrates including nucleic acids, proteins, lipids, and small molecules (Bauerle, Schwalm, & Booker, 2015; Lin, 2011; Martin & McMillan, 2002; Struck, Thompson, Wong, & Micklefield, 2012). While transmethylation produces small and modest chemical changes, these drastically affect molecular recognition of substrates by cellular targets to control processes such as

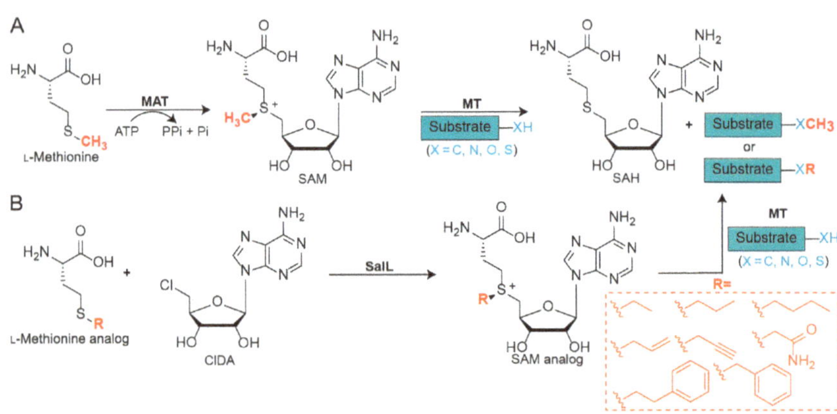

Fig. 3 Biosynthesis, utilization, and chemoenzymatic synthesis of *S*-adenosyl-L-methionine (SAM) analogs. (A) Biosynthesis of SAM in vivo is catalyzed by L-methionine-adenosyltransferases (MAT). Subsequently, nucleic acid, protein, or small molecule substrates are methylated by SAM-dependent methyltransferases (MT). *SAH*, *S*-adenosyl-L-homocysteine. (B) Chemoenzymatic synthesis of SAM analogs in vitro using chlorinase SalL from *Salinispora tropica*. *ClDA*, 5′-chloro-5′-deoxyadenosine.

epigenetic regulation of protein expression, neurotransmission, and natural products biosynthesis. The prevalence of SAM in cellular processes has inspired the development of analogs as chemical reporters for nucleic acid (Dalhoff et al., 2006a, 2006b; Lukinavicius et al., 2007; Motorin et al., 2011; Vranken et al., 2016), protein (Guo et al., 2014; Islam et al., 2012, 2013; Islam, Zheng, Yu, Deng, & Luo, 2011; Peters et al., 2010; Wang et al., 2013; Wang & Luo, 2013; Wang, Zheng, Yu, Deng, & Luo, 2011), small molecule (Lee et al., 2010; Singh et al., 2014; Stecher et al., 2009; Thomsen et al., 2013; Zhang, Weller, Thorson, & Rajski, 2006), and natural products methyltransferases (Winter et al., 2013; Zhang et al., 2006).

The preparation of SAM and its analogs is especially challenging due to inherent lability, tendency to racemize under experimental conditions, and rapid degradation at physiological pH (Hoffman, 1986). Chemical synthesis of SAM and SAM derivatives involves nucleophilic substitution between *S*-adenosyl-L-homocysteine (SAH or AdoCy) and an alkyl halide to produce a mixture of two sulfonium diastereomers, usually in variable yields (Dalhoff et al., 2006a, 2006b; de la Haba, Jamieson, Mudd, & Richards, 1959; Stecher et al., 2009). The *S,S* diastereomer serves as the methyl donor, while the *S,R* diastereomer inhibits the methyltransferase (Borchardt & Wu, 1976; Khani-Oskouee, Jones, & Woodard, 1984). Additionally, analogs of SAM where sulfur is replaced with selenium are more reactive methyl donors due to the longer, weaker Se–C bonds and do not racemize (Iwig & Booker, 2004). Notably, select Se-adenosyl-L-selenomethionines (SeAM), i.e., propargyl SeAM, are more stable than their SAM analogs (Bothwell et al., 2012; Bothwell & Luo, 2014). Overall, these early synthetic efforts ignited investigations in which nonnatural SAM analogs could be used by endogenous methyltransferases to modify and profile various cellular targets.

To complement chemical synthesis of SAM analogs, we pioneered chemoenzymatic methods to expand the range of SAM analogs available for labeling methyltransferase substrates (Lipson et al., 2013; Thomsen et al., 2013; Vranken et al., 2016). We employed two homologous halogenases, SalL and FlA, to produce SAM and SAM analogs. As discussed earlier, SalL catalyzes the reaction of SAM with chloride to produce ClDA and L-methionine via nucleophilic attack in *S. tropica* (Fig. 1). Similarly, *S. cattleya* uses fluorinase FlA to react fluoride with SAM to afford FDA and L-methionine. Based on Le Chatelier's Principle, we hypothesized that under low salt concentrations and excess L-methionine, these halogenases could catalyze the biologically reversed reaction to produce SAM in situ (Fig. 3) from easily prepared ClDA. Indeed, we observed SAM production by both SalL and FlA in various buffers, including Tris, HEPES, and phosphate buffers at pH 7–8, although SalL produced SAM in substantially higher yields than FlA. Furthermore, we generated a W190A point mutant of SalL that could accommodate larger alkyl substrates (Thomsen et al., 2013). Both wild type (WT) and mutant SalL generated alkyl, allylic, and benzylic SAM analogs, although the WT was most effective at generating small SAM analogs (methyl, ethyl), whereas both the WT and W190A mutant were equally effective at producing larger SAM analogs (propyl, butyl, allyl, and benzyl). Importantly, we demonstrated that our in situ-generated SAM congeners were recognized by nucleic acid, peptide, and small molecule methyltransferases and alkylated their respective substrates (Thomsen et al., 2013).

3.2 SalL-Mediated Chemoenzymatic Synthesis of SAM and SAM Analogs

3.2.1 Reagents

- DMSO (Sigma-Aldrich)
- HPLC grade acetonitrile (Fisher Scientific)
- 5′-chloro-5′-deoxyadenosine (Sigma-Aldrich)
- L-methionine (Sigma-Aldrich)
- *S*-ethyl-L-homocysteine (Sigma-Aldrich)
- *S*-propyl-L-homocysteine (prepared as described by Thomsen et al., 2013)
- *S*-butyl-L-homocysteine (Toronto Research Chemicals, Inc.)
- *S*-allyl-L-homocysteine (prepared as described by Thomsen et al., 2013)
- *S*-propargyl-L-homocysteine (prepared as described by Thomsen et al., 2013)
- *S*-benzyl-L-homocysteine (prepared as described by Thomsen et al., 2013)
- *S*-phenylethyl-L-homocysteine (prepared as described by Thomsen et al., 2013)
- *S*-(2-amino-2-carboxylethyl)-L-homocysteine (Sigma-Aldrich)
- *S*-(2-amino-2-oxoethyl)-L-homocysteine (prepared as described by Thomsen et al., 2013)
- *S*-adenosyl-L-methionine (New England Biolabs)

3.2.2 Enzyme

- SalL (expressed and purified as described in Section 2.1.2)

3.2.3 Stock Solutions and Buffers

- 300 m*M* ClDA in DMSO
- 150 m*M* L-methionine in 250 m*M* NaOH
- 15–150 m*M* L-methionine analog in 250 m*M* NaOH
- 32 m*M* SAM in 5 m*M* H_2SO_4 and 10% ethanol
- 50 m*M* sodium phosphate buffer, pH 6.8
- 400 m*M* sodium formate, pH 3.0
- 50 m*M* HEPES, pH 7.3

3.2.4 Analytical Instrumentation

- Agilent 6230 ESI-TOF LC-MS
- Agilent HPLC 1200
- Agilent Poroshell 300SB (5 μm, 2.1 × 75 mm) column

3.2.5 Procedure

1. Equilibrate 200 μ*M* ClDA and 3 μ*M* SalL in 50 m*M* phosphate buffer, pH 6.8 at 37°C.
2. Add 1.5–15 m*M* L-methionine or analog from stock. Due to the poor solubility of the benzyl and phenylethyl analogs of L-methionine, these were added at a final concentration of 1.5 m*M* from a 15 m*M* stock solution in 250 m*M* NaOH. L-Methionine and all other analogs were added at a final concentration of 15 m*M* from a 150 m*M* stock solution in 250 m*M* NaOH. In all assays, the addition of L-methionine or analog was 10% of the total reaction volume, resulting in a final pH of 7.4.
3. Quench reactions at various time points (0–5 h) by the addition of equal volumes of 400 m*M* sodium formate. Store quenched aliquots at 4°C until HPLC analysis.
4. Analyze aliquots using analytical reversed phase HPLC using a Synergi™ Hydro RP column (4 μm, 4.6 × 150 mm, Phenomenex) using the following gradient:

   Buffer A: 10 m*M* NH_4HCO_2, pH 3.2
   Buffer B: 90% CH_3CN, 10 m*M* NH_4HCO_2, pH 3.2
   0 min: 100% buffer A, 1 mL/min
   5 min: 98% buffer A, 1.5 mL/min
   10 min: 70% buffer A, 2 mL/min
   12 min: 0% buffer A, 2 mL/min

Table 1 LC-MS Retention Times and Chromatographic Analyses of Select SAM Analogs

Analog	Retention Time	Observed Monoisotopic Mass
SAM	4 min	399.1555
Ethyl-SAM	5 min	413.1671
Propyl-SAM	8.5, 8.9 min[a]	427.1784
Butyl-SAM	9.5 min	441.2366
Allyl-SAM	7.2, 7.9 min[a]	425.1862
Benzyl-SAM	10.4 min	475.2242

[a]Mixture of diastereomers.

Note: Complete conversion of ClDA (retention time = 11 min) into SAM (retention time = 4 min) was observed within 30 min.

5. Verify the identity of ClDA and SAM by collecting 0.1 mL fractions from the HPLC and analyze by LC-MS. Table 1 provides a summary of LC-MS data for select SAM analogs. See Thomsen et al. (2013) for representative LC-MS traces and chromatograms.

3.3 In Situ Chemoenzymatic Synthesis of SAM Coupled to DNA Methylation by *Hha*I

*Hha*I is a DNA methyltransferase that methylates cytosine at C5 (Sankpal & Rao, 2002). In this assay, *Hha*I is coupled with in situ SAM production with methylation of lambda DNA.

3.3.1 Reagents

- N^6-methyladenine-free lambda DNA (New England Biolabs)

3.3.2 Enzymes

- SalL (expressed and purified as described in Section 2.1.2)
- *Hha*I methyltransferase (New England Biolabs)
- 0.2 U/μL HinP1I restriction enzyme (New England Biolabs)

3.3.3 Stock Solutions and Buffers

- 300 m*M* ClDA in DMSO
- 150 m*M* L-methionine in 250 m*M* NaOH
- *Hha*I buffer (5 m*M* 2-mercaptoethanol, 10 m*M* EDTA, 50 m*M* Tris–HCl, pH 7.5)

- NEB4 buffer (10 m*M* KOAc, 4 m*M* Tris–HOAc, 2 m*M* $MgOAc_2$, 0.2 m*M* dithiothreitol (DTT), pH 7.9)

3.3.4 Procedure

1. Equilibrate 50 ng/μL N^6-methyladenine-free lambda DNA, 4 U/μL *Hha*I, 200 μ*M* ClDA, and 15 m*M* L-methionine in *Hha*I buffer at 37°C.

   Note: The buffers used for DNA methyltransferase *Hha*I and SalL activity assays were incompatible. *Hha*I was inactive in 50 m*M* sodium phosphate buffer, pH 7.9; however, the activity of SalL was reduced by 30% in *Hha*I buffer. Thus, coupled assays were conducted in *Hha*I buffer.
2. Add SalL to a final concentration of 900 n*M*.
3. After 1 h, quench reaction by heating at 80°C for 5 min.
4. Dilute reaction in an equal volume of HinP1I in 2 × NEB4 buffer to afford a final concentration of 0.1 U/μL HinP1I.
5. Incubate 1 h at 37°C.
6. Analyze lambda DNA cleavage by gel electrophoresis on 2% agarose gel.

3.4 In Situ Chemoenzymatic Synthesis of SAM Coupled to Teicoplanin Methylation by MtfA

MtfA is a small molecule methyltransferase from the chloroeremomycin biosynthetic pathway in *Amycolatopsis orientalis* that catalyzes *N*-methylation of vancomycin-like glycopeptides (Shi et al., 2009). In this assay, MtfA is used to couple in situ SAM production with methylation of teicoplanin.

3.4.1 Reagents

- Teicoplanin (TargetMol)

3.4.2 Enzymes

- SalL (expressed and purified as described in Section 2.1.2)
- MtfA methyltransferase (expressed and purified as described below)

3.4.3 Stock Solutions and Buffers

- 300 m*M* ClDA in DMSO
- 150 m*M* L-methionine (or [*methyl*-^{13}C]-L-methionine) in 250 m*M* NaOH
- 50 m*M* HEPES, pH 7.3

3.4.4 Expression and Purification of MtfA Methyltransferase

One liter of 2× YT medium (Sigma-Aldrich) supplemented with 50 μg/mL kanamycin was inoculated with *E. coli* BL21 (DE3) cells transformed with *mtfA*/pET28a plasmid (obtained from Prof. Gerard D. Wright, McMaster University). The cells were cultured at 37°C to an OD_{600} of 0.6. The cells were cooled to 0°C on ice, induced with 100 μ*M* isopropyl β-D-1-thiogalactopyranoside (IPTG), and grown for 20 h at 16°C. Cells were harvested via centrifugation at 2000 × g, 30 min, 4°C, and the pellet was resuspended in lysis buffer (50 m*M* HEPES, 500 m*M* NaCl, 20 m*M* imidazole, and 1 m*M* DTT, pH 7.5) and lysed by passage through a French pressure cell. The lysate was clarified by centrifugation at 4150 × g, 1.5 h, 4°C. His_6-MtfA was purified using 2 mL Ni-NTA resin (GE Healthcare) that was equilibrated with lysis buffer. The column was washed with 40 mL of buffer A (50 m*M* HEPES, 1 M NaCl, 20 m*M* imidazole, 1 m*M* DTT, pH 7.5), followed by 40 mL of buffer B (50 m*M* HEPES, 400 m*M* NaCl, 30 m*M* imidazole, 1 m*M* DTT, pH 7.5). The protein was eluted with buffer C (50 m*M* HEPES, 400 m*M* NaCl, 200 m*M* imidazole, 1 m*M* DTT, pH 7.5). We obtained around 50 mg of MtfA/L of cell culture. Eluted proteins were visualized with 12% SDS-PAGE with Coomassie Brilliant Blue staining. Imidazole was removed using spin concentration into 20 m*M* HEPES, 20 m*M* NaCl, and 5 m*M* DTT, pH 7.5. Pure protein was stored at −20°C in 10 m*M* HEPES, 10 m*M* NaCl, 2.5 m*M* DTT, pH 7.5 containing 50% glycerol. The protein is stable as a glycerol stock at −20°C.

3.4.5 Procedure

1. Equilibrate a 50 μL mixture of 20 m*M* ClDA, 50 m*M* L-methionine, and 1.4 μ*M* SalL in 32 m*M* phosphate buffer, pH 7.9 at 37°C.
2. After 1–3 h, add 250 μL of a solution containing: 1 m*M* teicoplanin, 1 mg/mL BSA, and 54 μ*M* MtfA in 50 m*M* HEPES, pH 7.3.
3. Monitor the methylation of teicoplanin by removing aliquots at various times and analyzing by LC-MS using a C18 column (3 μm, 2.1 × 150 mm, Atlantis) and an isocratic mixture of 3:1 H_2O/CH_3CN (v/v) with 0.1% formic acid. Under these conditions, the retention times of teicoplanin and *N*-methyl-teicoplanin were 9 and 9.5 min, respectively.

3.5 In Situ Chemoenzymatic Synthesis of ^{13}C-Labeled SAM Coupled With Isotope Incorporation Into Teicoplanin via Methylation by MtfA

We envisioned SalL-mediated in situ chemoenzymatic synthesis of SAM as a compelling and facile way to incorporate isotopes into biomolecules. Here, we describe the use of this technology to incorporate ^{13}C into teicoplanin for sensitive NMR identification (Lipson et al., 2013).

3.5.1 Reagents

- d_6-DMSO (Cambridge Isotope Laboratories, Item No. DLM-10-10)
- [*Methyl*-^{13}C]-L-methionine (Cambridge Isotope Laboratories, Item No. CLM-206-PK)
- Teicoplanin (TargetMol)

3.5.2 Enzymes

- SalL (expressed and purified as described in Section 2.1.2)
- MtfA methyltransferase (expressed and purified as described in Section 3.4.4)

3.5.3 Stock Solutions and Buffers

- 300 m*M* ClDA in d_6-DMSO
- 300 m*M* [*methyl*-^{13}C]-L-methionine in d_6-DMSO
- 10 m*M* phosphate buffer, pH 7.5

3.5.4 Analytical Instrumentation

- Varian VX500 NMR spectrometer equipped with an X-Sens cold probe

3.5.5 Procedure

1. Prepare 800 μL of a mixture of 350 n*M* SalL and 350 n*M* MtfA in 10 m*M* phosphate buffer, pH 7.5.
2. Prepare 200 μL of a mixture of 7 m*M* teicoplanin, 142 m*M* [*methyl*-^{13}C]-L-methionine, and 14.2 m*M* ClDA in 10 m*M* phosphate buffer, pH 7.5 containing 50% d_6-DMSO.
3. Transfer the mixture of teicoplanin, [*methyl*-^{13}C]-L-methionine, and ClDA mixture to the mixture of SalL and MtfA.
4. Transfer the entire solution to a clean 5 mm NMR tube.

5. Collect ^{13}C-NMR spectra on a Varian VX500 NMR spectrometer at various time points. For each time point, we acquired data for ≥64 scans with a delay time (d1) of 5 s. In our hands, we observed the appearance of [methyl-^{13}C]-*N*-teicoplanin within 3–9 h, as evidenced by a ^{13}C-NMR signal at $\delta = 33.7$ ppm.

3.6 In Situ Chemoenzymatic Synthesis of SAM Coupled to RGG Peptide Methylation by rPRMT1

Protein arginine methyltransferase 1 (PRMT1) is responsible for the majority of arginine methylation in mammalian cells. In the assay described below, rPRMT1, a rat homolog that dimethylates arginine residues, is used to couple in situ SAM production with methylation of arginine residues in the RGG peptide. Transfer of ethyl, allyl, and benzyl congeners of L-methionine has also been observed (Thomsen et al., 2013). Overall, sterics likely influences transfer since we observed RGG allylation, but did not observe RGG propylation, although the corresponding SAM analogs have similar sizes and are formed in situ at similar rates. RGG allylation most likely occurs via an S_N2' mechanism.

3.6.1 Reagents

- RGG peptide (expressed and purified as described by Thomsen et al., 2013)

3.6.2 Enzymes

- SalL (expressed and purified as described in Section 2.1.2)
- rPRMT1 (expressed and purified as described by Thomsen et al., 2013)

3.6.3 Stock Solutions and Buffers

- 300 m*M* ClDA in DMSO
- 150 m*M* L-methionine in 250 m*M* NaOH
- 400 m*M* sodium formate, pH 3.0
- 50 m*M* phosphate buffer, pH 7.4

3.6.4 Procedure Peptide Methylation Using rPRMT1

1. Equilibrate a mixture containing 200 μ*M* ClDA, 15 m*M* L-methionine, 1 μ*M* SalL, 10 μ*M* RGG peptide, and 2 μ*M* rPRMT1 in 50 m*M* phosphate buffer, pH 7.4 at 37°C.
2. After 12 h, quench reaction by the addition of an equal volume of 400 m*M* sodium formate.
3. Assess the formation of methylated RGG peptide using LC-MS.

4. SUMMARY AND CONCLUSIONS

We described procedures for the chemoenzymatic synthesis of SAM and analogs that employ chlorinase SalL discovered from a marine actinomycete bacterium. Recombinant SalL can be purified from *E. coli* in relatively high yields, i.e., over 60 mg soluble enzyme per liter culture. The SalL-catalyzed synthesis of SAM analogs from L-methionine congeners and ClDA is diastereoselective. Moreover, we described methods for in situ synthesis of SAM coupled with DNA, protein, and metabolite methylation, and with the incorporation of isotopes into biomolecules.

In addition to SAM-dependent halogenases, numerous groups have used methionine adenosyltransferase wild type and mutants that accommodate structural changes in L-methionine and ATP to generate stable SAM analogs that alkylate a variety of acceptor substrates (Huber, Johnson, Zhang, & Thorson, 2016; Huber et al., 2016; Singh et al., 2014; Wang et al., 2013). Overall, the development of chemoenzymatic methods and protein engineering efforts has expanded the structural scope of SAM analogs that can be generated for biological studies.

ACKNOWLEDGMENTS

Financial support for this work was provided by the National Center for Advancing Translational Sciences, National Institutes of Health (NIH), under Grant KL2TR002002 (to A.S.E.), NIH GM095970 (to M.D.B), and startup funds from the Department of Medicinal Chemistry and Pharmacognosy and the Center for Biomolecular Sciences, University of Illinois at Chicago (to A.S.E.). T.D.D. was supported by the NIH/NIGMS IRACDA K12 GM068524 award. The content is solely the responsibility of the authors and does not necessarily represent the official views of the NIH. Plasmid pAEM7 for the expression of *salL* was constructed in Prof. Bradley S. Moore's laboratory (Scripps Institution of Oceanography). The authors also graciously thank Prof. Joseph P. Noel (Salk Institute for Biological Studies) for vector pHIS8, Prof. Gerard D. Wright (McMaster University) for the *mtfA* gene construct, Dr. Yongxuan Su (UC San Diego) for mass spectral analyses, and Drs. Anthony Mrse and Xuemei Huang (UC San Diego) for assistance with acquiring NMR spectral data.

REFERENCES

Agarwal, V., Miles, Z. D., Winter, J. M., Eustáquio, A. S., El Gamal, A. A., & Moore, B. S. (2017). Enzymatic halogenation and dehalogenation reactions: Pervasive and mechanistically diverse. *Chemical Reviews*, *117*, 5619–5674.

Bauerle, M. R., Schwalm, E. L., & Booker, S. J. (2015). Mechanistic diversity of radical *S*-adenosylmethionine (SAM)-dependent methylation. *Journal of Biological Chemistry*, *290*, 3995–4002.

Borchardt, R. T., & Wu, Y. S. (1976). Potential inhibitors of *S*-adenosylmethionine-dependent methyltransferases V: Role of the asymmetric

sulfonium pole in the enzymatic binding of *S*-adenosyl-L-methionine. *Journal of Medicinal Chemistry*, *19*, 1099–1103.
Bothwell, I. R., Islam, K., Chen, Y., Zheng, W., Blum, G., Deng, H., et al. (2012). Se-adeonsyl-L-selenomethionine cofactor analogue as a reporter of protein methylation. *Journal of the American Chemical Society*, *134*, 14905–14912.
Bothwell, I. R., & Luo, M. (2014). Large-scale, protection free synthesis of Se-adenosyl-L-selenomethionine analogues and their application as cofactor surrogates of methyltransferases. *Organic Letters*, *16*, 3056–3059.
Bradford, M. M. (1976). A rapid and sensitive method for the quantitation of microgram quantities of protein utilizing the principle of protein-dye binding. *Analytical Biochemistry*, *72*, 248–254.
Cantoni, G. L. (1975). Biological methylation: Selected aspects. *Annual Review of Biochemistry*, *44*, 435–451.
Chan, K. K., & O'Hagan, D. (2012). The rare fluorinated natural products and biotechnological prospects for fluorine enzymology. *Methods in Enzymology*, *516*, 219–235.
Dalhoff, C., Lukinavicius, G., Klimasauskas, S., & Weinhold, E. (2006a). Direct transfer of extended groups from synthetic cofactors by DNA methyltransferases. *Nature Chemical Biology*, *2*, 31–32.
Dalhoff, C., Lukinavicius, G., Klimasauskas, S., & Weinhold, E. (2006b). Synthesis of *S*-adenosyl-L-methionine analogs and their use for sequence-specific transalkylation of DNA by methyltransferases. *Nature Protocols*, *1*, 1879–1886.
de la Haba, G., Jamieson, G. A., Mudd, S. H., & Richards, H. H. (1959). *S*-adenosylmethionine: The relation of configuration at the sulfonium center to enzymatic reactivity. *Journal of the American Chemical Society*, *81*, 3975–3980.
Deng, H., Cobb, S. L., McEwan, A. R., McGlinchey, R. P., Naismith, J. H., O'Hagan, D., et al. (2006). The fluorinase from *Streptomyces cattleya* is also a chlorinase. *Angewandte Chemie*, *45*, 759–762.
Deng, H., Ma, L., Bandaranayaka, N., Qin, Z., Mann, G., Kyeremeh, K., et al. (2014). Identification of fluorinases from *Streptomyces* sp. MA37, *Norcardia brasiliensis*, and *Actinoplanes* sp. N902-109 by genome mining. *Chembiochem*, *15*, 364–368.
Dong, C., Huang, F., Deng, H., Schaffrath, C., Spencer, J. B., O'Hagan, D., et al. (2004). Crystal structure and mechanism of a bacterial fluorinating enzyme. *Nature*, *427*, 561–565.
Eustáquio, A. S., McGlinchey, R. P., Liu, Y., Hazzard, C., Beer, L. L., Florova, G., et al. (2009). Biosynthesis of the salinosporamide A polyketide synthase substrate chloroethylmalonyl-coenzyme A from *S*-adenosyl-L-methionine. *Proceedings of the National Academy of Sciences of the United States of America*, *106*, 12295–12300.
Eustáquio, A. S., Pojer, F., Noel, J. P., & Moore, B. S. (2008). Discovery and characterization of a marine bacterial SAM-dependent chlorinase. *Nature Chemical Biology*, *4*, 66–74.
Feling, R. H., Buchanan, G. O., Mincer, T. J., Kauffman, C. A., Jensen, P. R., & Fenical, W. (2003). Salinosporamide A: A highly cytotoxic proteasome inhibitor from a novel microbial source, a marine bacterium of the new genus *Salinospora*. *Angewandte Chemie*, *42*, 355–357.
Fontecave, M., Atta, M., & Mulliez, E. (2004). *S*-adenosylmethionine: Nothing goes to waste. *Trends in Biochemical Sciences*, *29*, 243–249.
Gribble, G. W. (2012). Recently discovered naturally occurring heterocyclic organohalogen compounds. *Heterocycles*, *84*, 157–207.
Guo, H., Wang, R., Zheng, W., Chen, Y., Blum, G., Deng, H., et al. (2014). Profiling substrates of protein arginine N-methyltransferase 3 with *S*-adenosyl-L-methionine analogues. *ACS Chemical Biology*, *9*, 476–484.
Hoffman, J. L. (1986). Chromatographic analysis of the chiral and covalent instability of *S*-adenosyl-L-methionine. *Biochemistry*, *25*, 4444–4449.

Huang, S., Ma, L., Tong, M. H., Yu, Y., O'Hagan, D., & Deng, H. (2014). Fluoroacetate biosynthesis from the marine-derived bacterium *Streptomyces xinghaiensis* NRRL B-24674. *Organic and Biomolecular Chemistry, 12*, 4828–4831.

Huber, T. D., Johnson, B. R., Zhang, J., & Thorson, J. S. (2016). AdoMet analog synthesis and utilization: Current state of the art. *Current Opinion in Biotechnology, 42*, 189–197.

Huber, T. D., Wang, F., Singh, S., Johnson, B. R., Zhang, J., Sunkara, M., et al. (2016). Functional AdoMet isosteres resistant to classical AdoMet degradation pathways. *ACS Chemical Biology, 11*, 2484–2491.

Islam, K., Bothwell, I., Chen, Y., Sengelaub, C., Wang, R., Deng, H., et al. (2012). Bioorthogonal profiling of protein methylation using azido derivative of *S*-adenosyl-L-methionine. *Journal of the American Chemical Society, 134*, 5909–5915.

Islam, K., Chen, Y., Wu, H., Bothwell, I. R., Blum, G. J., Zeng, H., et al. (2013). Defining efficient enzyme-cofactor pairs for bioorthogonal profiling of protein methylation. *Proceedings of the National Academy of Sciences of the United States of America, 110*, 16778–16783.

Islam, K., Zheng, W., Yu, H., Deng, H., & Luo, M. (2011). Expanding cofactor repertoire of protein lysine methyltransferase for substrate labeling. *ACS Chemical Biololgy, 6*, 679–684.

Iwig, D. F., & Booker, S. J. (2004). Insight into the polar ractivity of the onium chalcogen analogues of *S*-adenosyl-L-methionine. *Biochemistry, 43*, 13496–13509.

Jez, J. M., Ferrer, J. L., Bowman, M. E., Dixon, R. A., & Noel, J. P. (2000). Dissection of malonyl-coenzyme A decarboxylation from polyketide formation in the reaction mechanism of a plate polyketide synthase. *Biochemistry, 39*, 890–902.

Khani-Oskouee, S., Jones, J. P., & Woodard, R. W. (1984). Stereochemical course of the biosynthesis of L-aminocyclopropane-1-carboxylic acid. I. Role of the asymmetric sulfonium pole and the α-amino acid center. *Biochemical and Biophysical Research Communications, 121*, 181–187.

Lee, B. W., Sun, H. G., Zang, T., Kim, B. J., Alfaro, J. F., & Zhou, Z. S. (2010). Enzyme-catalyzed transfer of a ketone group from an *S*-adenosylmethionine analogue: A tool for the functional analysis of methyltransferases. *Journal of the American Chemical Society, 132*, 3642–3643.

Lin, H. (2011). *S*-adenosylmethionine-dependent alkylation reactions: When are radical reactions used? *Bioorganic Chemistry, 39*, 161–170.

Lipson, J. M., Thomsen, M., Moore, B. S., Clausen, R. P., La Clair, J. J., & Burkart, M. D. (2013). A tandem chemoenzymatic methylation of *S*-adenosyl-L-methionine. *Chembiochem, 14*, 950–953.

Lukinavicius, G., Lapiene, V., Stasevskij, Z., Dalhoff, C., Weinhold, E., & Klimasauskas, S. (2007). Targeted labeling of DNA by methyltransferase-directed transfer of activated groups (mTAG). *Journal of the American Chemical Society, 129*, 2758–2759.

Martin, J. L., & McMillan, F. M. (2002). SAM (dependent) I AM: The *S*-adenosylmethionine-dependent methyltransferase fold. *Current Opinion in Structural Biology, 12*, 783–793.

Motorin, Y., Burhenne, J., Teimer, R., Koynov, K., Willnow, S., Weinhold, E., et al. (2011). Expanding the chemical scope of RNA: Methyltransferases to site-specific alkynylation of RNA for click labeling. *Nucleic Acids Research, 39*, 1943–1952.

Ohashi, M., Liu, F., Hai, Y., Chen, M., Tang, M. C., Yang, Z., et al. (2017). SAM-dependent enzyme-catalysed pericyclic reactions in natural product biosynthesis. *Nature, 549*, 502–506.

Peters, W., Willnow, S., Duisken, M., Kleine, H., Macherey, T., Duncan, K. E., et al. (2010). Enzymatic site-specific functionalization of protein methyltransferase substrates with alkynes for click labeling. *Angewandte Chemie International Edition England, 49*, 5170–5173.

Sankpal, U. T., & Rao, D. N. (2002). Structure, function, and mechanism of HhaI DNA methyltransferases. *Critical Reviews in Biochemistry and Molecular Biology*, *37*, 167–197.

Shi, R., Lamb, S. S., Zakeri, B., Proteau, A., Cui, Q., Sulean, T., et al. (2009). Structure and function of the glycopeptide *N*-methyltransferase MtfA, a tool for the biosynthesis of modified glycopeptide antibiotics. *Chemistry & Biology*, *16*, 401–410.

Singh, S., Zhang, J., Huber, T. D., Sunkara, M., Hurley, K., Goff, R. D., et al. (2014). Facile chemoenzymatic strategies for the synthesis and utilization of *S*-adenosyl-L-methionine analogues. *Angewandte Chemie International Edition*, *53*, 3965–3969.

Sinhababu, A. K., Bartel, R. L., Pochopin, N., & Borchardt, R. T. (1985). Mechanism of action of *S*-adenosyl-L-homocysteine hydrolase. Measurement of kinetic isotope effects using adenosine-3′-d and *S*-adenosyl-L-homocysteine-3′-d as substrates. *Journal of American Chemical Society*, *107*, 7628–7632.

Stecher, H., Tengg, M., Ueberbacher, B. J., Remler, P., Schwab, H., Griengl, H., et al. (2009). Biocatalytic Friedel-Crafts alkylation using non-natural cofactors. *Angewandte Chemie International Edition England*, *48*, 9546–9548.

Struck, A. W., Thompson, M. L., Wong, L. S., & Micklefield, J. (2012). *S*-adenosyl-methionine-dependent methyltransferases: Highly versatile enzymes in biocatalysis, biosynthesis and other biotechnological applications. *Chembiochem*, *13*, 2642–2655.

Thomsen, M., Vogensen, S. B., Buchardt, J., Burkart, M. D., & Clausen, R. P. (2013). Chemoenzymatic synthesis and *in situ* application of *S*-adenosyl-L-methionine analogs. *Organic and Biomolecular Chemistry*, *11*, 7606–7610.

Vranken, C., Fin, A., Tufar, P., Hofkens, J., Burkart, M. D., & Tor, Y. (2016). Chemoenzymatic synthesis and utilization of a SAM analog with an isomorphic nucleobase. *Organic and Biomolecular Chemistry*, *14*, 6189–6192.

Wang, R., Islam, K., Liu, Y., Zheng, W., Tang, H., Lailler, N., et al. (2013). Profiling genome-wide chromatin methylation with engineered posttranslation apparatus within living cells. *Journal of the American Chemical Society*, *135*, 1048–1056.

Wang, R., & Luo, M. (2013). A journey toward bioorthogonal profiling of protein methylation inside living cells. *Current Opinion in Chemical Biology*, *17*, 729–737.

Wang, R., Zheng, W., Yu, H., Deng, H., & Luo, M. (2011). Labeling substrates of protein arginine methyltransferase with engineered enzymes and matched *S*-adenosyl-L-methionine analogues. *Journal of the American Chemical Society*, *133*, 7648–7651.

Wang, L., Zhou, X., Fredimoses, M., Liao, S., & Liu, Y. (2014). Naturally occurring organoiodines. *RSC Advances*, *4*, 57350–57376.

Winter, J. M., Chiou, G., Bothwell, I. R., Xu, W., Garg, N. K., Luo, M., et al. (2013). Expanding the structural diversity of polyketides by exploring the cofactor tolerance of an inline methyltransferase domain. *Organic Letters*, *15*, 3774–3777.

Zhang, C., Weller, R. L., Thorson, J. S., & Rajski, S. R. (2006). Natural product diversification using a non-natural cofactor analogue of *S*-adenosyl-L-methionine. *Journal of the American Chemical Society*, *128*, 2760–2761.

CHAPTER TWELVE

In Vitro Analysis of Cyanobacterial Nonheme Iron-Dependent Aliphatic Halogenases WelO5 and AmbO5

Xinyu Liu[1]
University of Pittsburgh, Pittsburgh, PA, United States
[1]Corresponding author: e-mail address: xinyuliu@pitt.edu

Contents

Abstract

Aliphatic carbon–halogen (C–X) bonds are prevalent in modern pharmaceuticals and bioactive natural products. Three distinct chemical strategies are known in Nature to generate these structural motifs. The first is via the nucleophilic substitution at a prefunctionalized electrophilic carbon center with a halide anion (X^-), known for the *S*-adenosyl-L-methionine-dependent halogenases. The second is via the electrophilic activation of an alkene or its equivalent by a halenium ion (X^+) donor, known for the haloperoxidases and flavin-dependent halogenases. The third is via the direct functionalization of an unactivated aliphatic C–H bond with a halogen radical ($X^{\bullet}$) equivalent, known for the 2-oxo-glutarate and nonheme iron-dependent halogenases. Due to the ubiquitous nature of aliphatic C–H groups in organic molecules, transformations that permit *chemo-*, *regio-*, and *stereo*-selective modification(s) at an unactivated sp^3-carbon center have been a long sought-after goal in chemical science. Two nonheme iron-dependent halogenases, WelO5 and AmbO5 involved in the biogenesis of cyanobacterial hapalindole-type alkaloids, have been recently shown able to perform this type of challenging transformation. In this chapter, experimental details for the in vitro reconstitution of WelO5 and AmbO5 enzymatic activities are presented.

Methods in Enzymology, Volume 604
ISSN 0076-6879
https://doi.org/10.1016/bs.mie.2018.02.015

1. INTRODUCTION

Carbon–halogen (C–X) bonds are prevalent structural motifs in modern pharmaceuticals and agrochemicals (Smith, Eastman, & Njardarson, 2014). Natural products containing this type of functional group, once considered as the "chance products of Nature" (Petty, 1961), are now known ubiquitous across all domains of life with nearly 5000 members isolated and characterized to date (Gribble, 2010). Over the past two decades, significant progresses have been made in delineating Nature's logics for the enzymatic C–X bond formations. These advances, along with their historical perspective, have been recently systematically reviewed (Agarwal et al., 2017; Vaillancourt, Yeh, Vosburg, Garneau-Tsodikova, & Walsh, 2006) and will not be discussed in detail in this chapter. Aliphatic C–X bonds, distinct from those with halogens attached to an *sp*^{2}- or *sp*-hybridized carbon center, are widely present in naturally occurring halometabolites (Fig. 1; Gribble, 2010). The attachment of a halogen atom to an *sp*^{3}-hybridized carbon center profoundly alters its reactivity, notably its electrophilicity. Thus, this type of modification often plays an important role in the bioactivity of halogenated natural products (Williams et al., 2005). The logics underlying the formation of an aliphatic C–X bond in natural product biosynthesis encompass all the key chemical principles established for the enzymatic C–X bond formation (Agarwal et al., 2017; Vaillancourt et al., 2006). These principles are briefly outlined here.

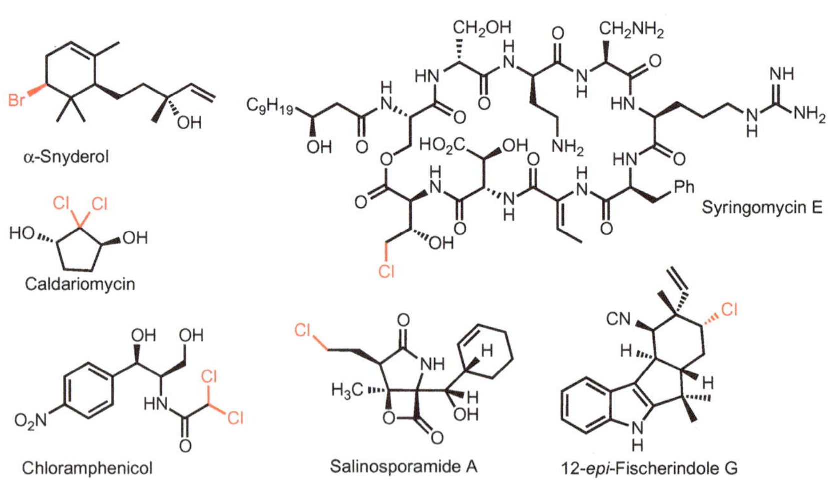

Fig. 1 Selected halogenated natural products with aliphatic carbon–halogen bonds.

Halogenated terpenoids of marine and terrestrial origins represent the largest family of naturally occurring halometabolites with aliphatic C–Cl and C–Br bonds (Gribble, 2010). This type of aliphatic C–X bonds originates primarily from an sp^2-hybridized carbon, amenable for activation by a halenium ion (X^+) donor that is generated oxidatively from a halide anion (X^-) by a haloperoxidase or a flavin-dependent halogenase. A notable example is the aliphatic C–Br bond formation in the biogenesis of α-snyderol from (*E*)-(+)-nerolidol by a vanadium-dependent bromoperoxidase (Fig. 2A; Carter-Franklin & Butler, 2004). Related mechanisms that involve the activation of an electron-rich sp^2-hybridized carbon with an enzymatically generated X^+ donor have been cited for the aliphatic C–Cl bond formations in the biosynthesis of caldariomycin and chloramphenicol, where the former requires a heme-dependent chloroperoxidase (Sundaramoorthy, Terner, & Poulos, 1995) and the latter calls for a flavin-dependent halogenase (Podzelinska et al., 2010).

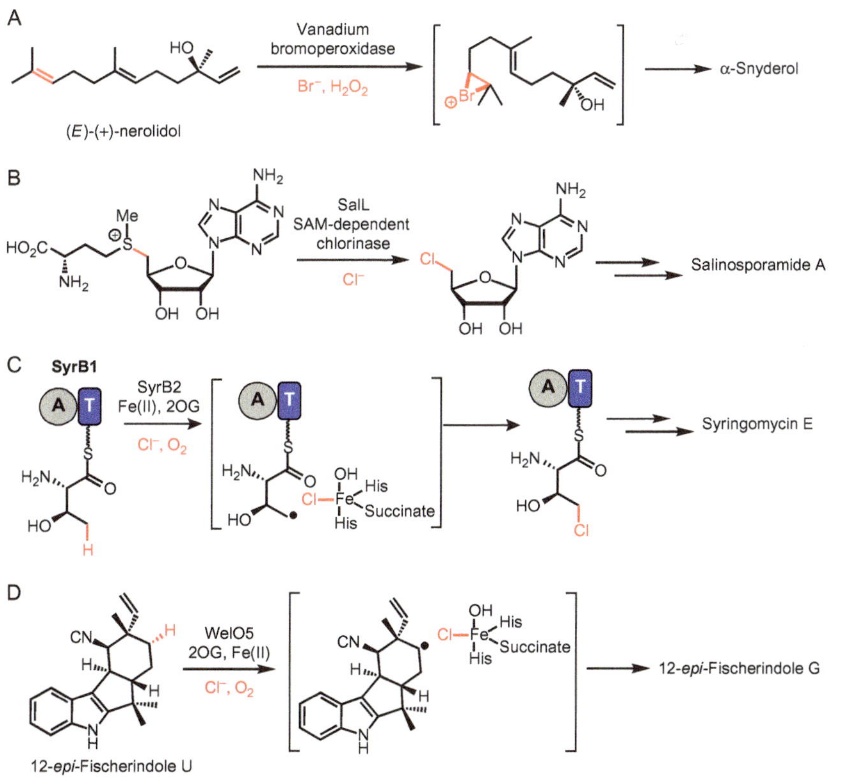

Fig. 2 Selected examples highlighting the key chemical principles underlying the formation of aliphatic carbon–halogen bonds in natural product biosynthesis.

Nucleophilic substitution by a halide anion (X^-) at an activated aliphatic carbon center of *S*-adenosyl-L-methionine (SAM), manifested by the SAM-dependent halogenases, constitutes the second major enzymatic strategy for introducing aliphatic C–X bonds to the halogenated natural products. When this substitution occurs on the activated methyl group of SAM, a halomethane, the simplest naturally occurring halometabolite, is generated and the corresponding enzyme is also known as the SAM-dependent halide methyltransferase (Wuosmaa & Hager, 1990). Alternatively, when the substitution occurs on the 5′ carbon of adenosine in SAM, a 5′-halo-5′-deoxyadenosine is generated as an intermediate for the downstream biosynthetic maturations. SalL, one of the two well-characterized SAM-dependent halogenases, catalyzes the formation of 5′-chloro-5′-deoxyadenosine en route to the biosynthesis of salinosporamide A (Fig. 2B; Eustaquio, Pojer, Noel, & Moore, 2008).

The third major enzymatic strategy for introducing aliphatic C–X bonds into halogenated natural product is via the direct modification of an aliphatic carbon–hydrogen (C–H) bond with a halogen radical ($X^\bullet$) equivalent, manifested by the nonheme iron-dependent halogenases. These halogenases constitute an important subset of a widespread family of oxygenases that combine reaction of O_2 at an H–X–D/E...H-coordinated mononuclear Fe(II) center with oxidative decarboxylation of a coordinated 2-oxo-glutarate (2OG) cofactor (Hausinger & Schofield, 2015). These enzymes generate an iron(IV)-oxo ($Fe^{IV}{=}O$) (ferryl) intermediate, a common mechanistic theme in all Fe/2OG enzymes, to initiate reaction with substrate by abstraction of a hydrogen atom ($H^\bullet$) (Krebs, Galonic Fujimori, Walsh, & Bollinger, 2007).

SyrB2, from the syringomycin biosynthetic pathway in *Pseudomonas syringae*, is the first biochemically characterized Fe/2OG halogenase (Vaillancourt, Yin, & Walsh, 2005). It monochlorinates the C4 methyl group in L-threonine covalently tethered as a thioester to the phosphopantetheine arm of its carrier protein, SyrB1 (Fig. 2C; Vaillancourt, Yin, et al., 2005). The X-ray crystal structure of Fe(II)·2OG SyrB2 with Cl^- (but in the absence of substrate, the threonine-appended carrier protein SyrB1) gave the first mechanistic insight into Fe/2OG-mediated halogenation (Blasiak, Vaillancourt, Walsh, & Drennan, 2006). Omission of the carboxylate ligand in the H-X-D/E...H motif via mutation to an Ala residue opens an Fe(II) coordination position that can be occupied by the halogen. Thus, the SyrB2 reaction proceeds by the same general pathway as Fe/2OG hydroxylases, involving a *cis*-halo-ferryl intermediate to abstract an $H^\bullet$ from C4 of the tethered threonine, then coupling the substrate carbon radical with the chlorine ligand instead of the hydroxyl ligand of the resultant Cl–Fe(III)–OH complex (Matthews, Krest, et al., 2009).

Due to the ubiquitous nature of aliphatic C–H groups in organic molecules, transformations that permit *chemo*-, *regio*-, and *stereo*-selective modification(s) at an unactivated *sp*^{3}-carbon center have been a long sought-after goal in chemical science (White, 2012). The discovery of SyrB2-type nonheme iron halogenases, widely present in the biosynthetic pathways for polyketide and nonribosomal peptide natural products (Galonic, Vaillancourt, & Walsh, 2006; Gu et al., 2009; Jiang et al., 2011; Neumann & Walsh, 2008; Ueki et al., 2006; Vaillancourt, Yeh, Vosburg, O'Connor, & Walsh, 2005), immediately raised the question whether they can be applied to biocatalytic processes. Moreover, the SyrB2 crystal structure seemingly suggests a minimal modification in the H-X-D/E...H Fe(II)-binding motif constitutes a key for the ferryl complex to promote chlorine transfer. This raised another intriguing question on whether this finding can be universally translated to the superfamily of Fe^{II}/2OG enzymes simply by deletion of the carboxylate ligand to accommodate a more diverse set of substrates. Unfortunately, neither was the case. SyrB2-type halogenases strictly require the small-molecular substrate tethered to a carried protein (Galonic et al., 2006; Gu et al., 2009; Jiang et al., 2011; Neumann & Walsh, 2008; Ueki et al., 2006; Vaillancourt, Yeh, et al., 2005; Vaillancourt, Yin, et al., 2005). Efforts to reprogram Fe/2OG hydroxylases by single-site substitution of the D/E in the H-X-D/E...H motif have not met with success (Gorres, Pua, & Raines, 2009; Grzyska, Muller, Campbell, & Hausinger, 2007). Even the restoration of the H-X-D/E...H motif in SyrB2 did not yield a competent enzyme for hydroxylation (Matthews, Neumann, et al., 2009). These data collectively suggest the challenging nature to utilize SyrB2-type halogenases for biocatalytic applications.

WelO5, from the welwitindolinone biosynthetic pathway in *Hapalosiphon welwitschii* (Hillwig, Fuhrman, et al., 2014; Hillwig, Zhu, & Liu, 2014), is the first biochemically characterized Fe/2OG halogenase that can act on free-standing small molecules independent of carrier proteins (Hillwig & Liu, 2014). WelO5 preferentially monochlorinates C13 of 12-*epi*-fischerindole U to yield 12-*epi*-fischerindole G (Fig. 2D) and is also able to convert 12-*epi*-hapalindole C to 12-*epi*-hapalindole E at a reduced efficiency (Fig. 3). A related halogenase AmbO5 (79% sequence identical to WelO5), from the ambiguine biosynthetic pathway in *Fischerella ambigua* (Hillwig, Zhu, et al., 2014), has been shown able to accommodate seven structurally distinct hapalindole-type alkaloids (Fig. 3; Hillwig, Zhu, Ittiamornkul, & Liu, 2016). In addition, the restoration of the H-X-D...H motif in WelO5 generates a catalytically competent hydroxylase WelO5(G166D) (Fig. 4A), suggesting a unique active site dynamics may be operant that is distinct from SyrB2. The cocrystal structure of Fe(II)·2OG WelO5 with Cl^- and 12-*epi*-fischerindole U (Fig. 4B;

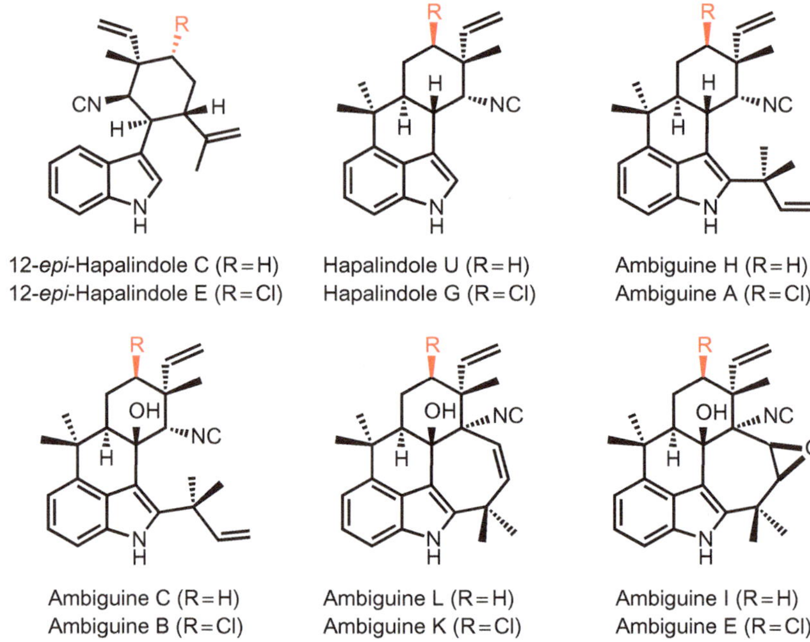

Fig. 3 Additional hapalindole-type alkaloids (R=H) that are validated substrates for WelO5 and/or AmbO5 aliphatic halogenases.

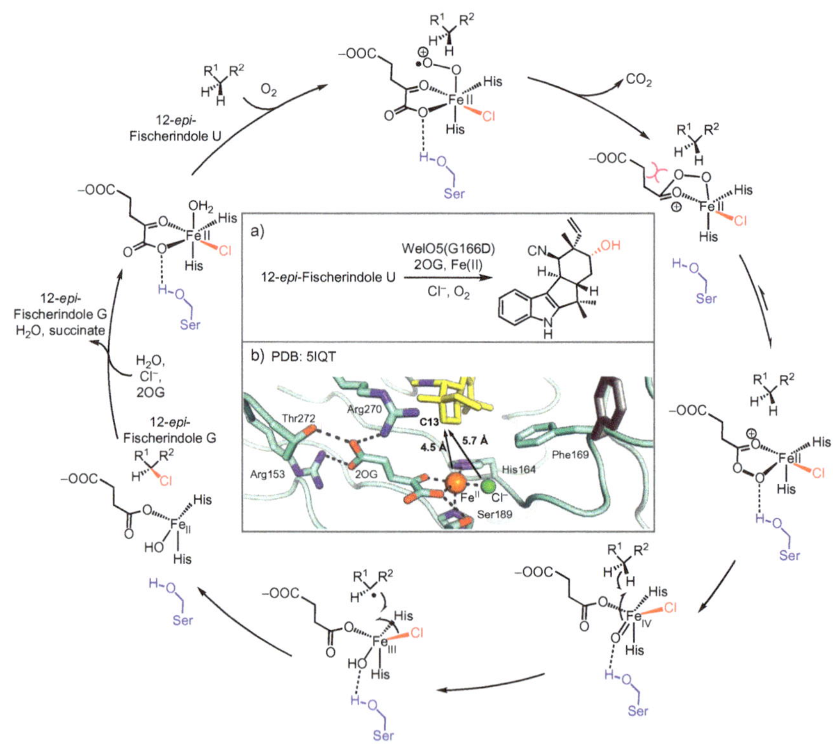

Fig. 4 Proposed reaction mechanism of WelO5 halogenase derived from its X-ray crystal structure in complex with 12-*epi*-fischerindole U, 2OG, Cl^-, and Fe(II) (inlet B) that accounts for the observed selective chlorination activity by the wild-type enzyme and the distinct hydroxylation activity by the WelO5(G166D) mutant (inlet A).

Mitchell et al., 2016) provided structural corroborations with the in vitro biochemical characteristics of WelO5 and AmbO5 (Hillwig & Liu, 2014; Hillwig et al., 2016; Zhu & Liu, 2017). This led to the identification of an external α-helical motif that defines the small-molecular substrate-binding pocket. Moreover, the WelO5 structure revealed that the target C13 of 12-*epi*-fischerindole U is naturally poised for hydroxylation in the active site (Fig. 4B), and the selective chlorine transfer is likely dictated by a previously unknown second sphere serine residue that helps control the geometry of the reactive Cl–Fe(IV)=O and Cl–Fe(III)–OH intermediates in the WelO5 catalytic cycle (Fig. 4). The structural characterization of WelO5 further led to the identification of an Fe/2OG hydroxylase (SadA) that can be readily engineered to gain halogenase activity (Mitchell et al., 2017). Together with the identification of additional WelO5-type halogenase that may mediate late-stage chlorinations beyond the hapalindole-type scaffolds (Moosmann et al., 2017), a promising future can be forecasted for engineering this family of enzymes for selective functionalization of aliphatic C–H groups with halogens in a diverse set of molecular scaffolds.

2. IN VITRO ANALYSIS OF WelO5 AND AmbO5 HALOGENASES

As for any given in vitro enzyme reconstitution, its success requires three key procedural components that are indispensable from each other: (1) the ability to procure sufficient quantity of the target enzyme, (2) the ability to procure sufficient quantity of the presumptive substrate(s) for the target enzyme, and (3) analytical procedures to assess the turnover of the substrate(s) by the target enzyme and to define the molecular nature of the enzymatic product. The following subsections detail my laboratory's experience on these three parts for WelO5 and AmbO5 halogenases.

2.1 Purification of WelO5 and AmbO5 Proteins

Genes (870 bp) encoding WelO5 and AmbO5 proteins can be readily amplified by PCR from the genomic DNA of *H. welwitschii* UTEX B 1830 and *F. ambigua* UTEX 1903, respectively. Both strains are available at the Culture Collection of Algae at the University of Texas at Austin (https://utex.org). Alternatively, these sequences (GenBank ID: AHI58816 for WelO5 and AKP23998 for AmbO5) can be synthesized by any reputable company offering a synthetic gene service. Codon optimization is not required, as the native sequences lead to an appreciable amount of soluble proteins in *Escherichia coli* (vide infra), sufficient for in vitro biochemical and structural studies. The pQTEV plasmid, available at Addgene (https://www.addgene.org/31291/), has been the vector of choice by my laboratory to clone the *wel*O5 and *amb*O5 genes. These constructs gave the N-His$_7$-tagged WelO5 and AmbO5 proteins upon overexpression in *E. coli*. Any other conventional cloning vectors that carry N-terminal His-Tag sequence, such as pET22 or pET28 plasmids, should perform equally well as the pQTEV. Site-directed mutagenesis can be routinely performed on the WelO5 and AmbO5 proteins readily once their expression constructs are generated using a PCR-based approach, of which details will not be included here. The following protocol describes a purification procedure for the WelO5 protein, which is equally applicable to the AmbO5 protein as well as their mutant proteins.

Day 1: Transform a pQTEV plasmid with the *wel*O5 gene cloned between the *Bam*HI and *Not*I sites to electro- or chemically competent C43 (DE3) or BL21(DE3) *E. coli* cells.

Day 2 evening: Pick a single colony of C43 (DE3) or BL21(DE3) *E. coli* cells with the pQTEV::*wel*O5 plasmid and prepare a starter culture by inoculating the picked colony to a 5-mL LB culture overnight at 30°C.

Day 3 morning: Inoculate 1 mL of the starter culture to a 2.8-L Fernbach flask containing 1 L of LB medium with 100 μg/mL ampicillin. Grow the cultures at 37°C to an optical density of approximately 0.6, then drop the temperature to 16°C. Add isopropyl β-D-thiogalactopyranoside (IPTG) (1 m*M*, final concentration) to induce protein expression.

Day 4 morning: Pellet the cells from the 1-L overnight culture by centrifugation at 8800 × *g* for 20 min. Resuspend the cell pellets in a 10-mL protein lysis buffer (50 m*M* Tris/pH 7.4, 500 m*M* NaCl, 20 m*M* imidazole, 10 m*M* β-mercaptoethanol, and 0.1% Tween-20) and perform cell lysis by sonicating on ice in a 10-s interval for 3 min and then pellet the cell debris by centrifugation at 20,000 × *g* and 4°C for 30 min. Transfer the cleared lysate to a 2-mL prewashed Ni-NTA bead (Qiagen) slurry and incubate for 1 h at 4°C. Thereafter, load the Ni-NTA beads on to a column and wash with 60 mL of binding buffer (50 m*M* Tris/pH 7.4, 500 m*M* NaCl, 0.1% Tween-20, 20 m*M* imidazole, 10 m*M* β-mercaptoethanol). The His-tagged WelO5 protein will be eluted by adding 5 mL of elution buffer (50 m*M* Tris/pH 7.4, 500 m*M* NaCl, 100 then 250 m*M* imidazole, β-mercaptoethanol) to the bead bed. The eluted protein will be subjected to a dialysis at 4°C using a 30-kDa molecular weight cutoff membrane (Spectrum Laboratory Products) with a buffer containing 50 m*M* HEPES (pH 7.4), 10 m*M* EDTA, 10 m*M* NaCl, 10% glycerol, and 0.5 m*M* DTT. This initial dialysis can be carried overnight and ensures the complete removal of metal ions in the eluted protein sample post-Ni-NTA column.

Day 5 morning: The WelO5 protein sample postdialysis against EDTA will be subjected to a second round of dialysis with a buffer containing 50 m*M* HEPES (pH 7.4), 10 m*M* NaCl, 10% glycerol, and 0.5 m*M* DTT to remove EDTA. This dialysis step to remove EDTA is recommend to be repeated one more time before the sample is concentrated with an Amicon MWCO 10 centrifugal filter. The WelO5 protein purified by the procedure described earlier is sufficiently pure for in vitro reconstitution experiment. Its purity can be judged by routine SDS-PAGE analysis, and its concentration can be estimated either using a UV–vis spectrometer or the colormetric Bradford assay.

Note: Literature describing the in vitro reconstitution of several carrier protein-dependent nonheme iron halogenases (Galonic et al., 2006; Vaillancourt, Yeh, et al., 2005; Vaillancourt, Yin, et al., 2005), including a protocol chapter in this book series (van Pee, 2012), has indicated that these proteins need to be purified in the *absence* of air. These observations are counterintuitive to the facts that these proteins are oxygenases and require molecular oxygen for their activities. My laboratory and others have independently validated (Hillwig & Liu, 2014; Matthews, Krest, et al., 2009), as the protocol described earlier, that these halogenases can be routinely purified in an atmospheric environment. The key to obtain consistently active proteins is to ensure the metal ions are completely stripped off from the protein initially by a dialysis against EDTA.

2.2 Procurement of Substrates for WelO5 and AmbO5 Proteins

Seven hapalindole-type alkaloids (Figs. 2D and 3) have been validated as the native substrates for WelO5 and/or AmbO5 halogenases (Hillwig & Liu, 2014; Hillwig et al., 2016). These alkaloids can be procured via three different routes. The first is via isolation from their native cyanobacterial producers (Smitka et al., 1992; Stratmann et al., 1994), which have been practiced by my laboratory for several years and a sample protocol will be provided here. The second is via de novo organic synthesis. Synthetic routes that consist of less than 10 steps from commercially available building blocks have been established for 12-*epi*-fischerindole U, 12-*epi*-hapalindole C, and hapalindole U (Baran, Maimone, & Richter, 2007; Baran & Richter, 2004). These synthetic advancements rendered the custom synthesis as a viable third option to procure these molecules. For example, 12-*epi*-hapalindole

C was known accessible in an estimated cost of 5000 USD/g via custom synthesis (2015 data, personal communication with the Biocatalysis group of Novartis AG, Basel, CH).

Described below is an exemplary protocol for the isolation of 12-*epi*-ficherindole U from *H. welwitschii* UTEX B 1830.

Step 1: Culture *H. welwitschii* UTEX B 1830 at a 20-L scale at room temperature in BG-11 medium (for recipe, see https://utex.org/products/bg-11-medium) with a 12h/12h light/dark cycle for approximately 5 weeks.

Step 2: Harvest the algal cells by filtration and then lyophilize. This typically yields 5–6g of freeze-dried material.

Step 3: Extract the lyophilized tissue with CH_2Cl_2/MeOH (v/v=1:1, 50mL) overnight with stirring at room temperature. The crude organic extract postfiltration will be dried under reduced pressure and redissolved in methanol (15mL). This methanol solution will be filtered through an organic solvent compatible syringe filter and injected on to a Phenomenex Luna C18 21.5×20mm reversed phase preparative HPLC column for initial crude separation of hapalindole-type compounds using a gradient from 50% to 100% methanol in water over 100min.

Step 4: The fractions that contain 12-*epi*-ficherindole U from the preparative HPLC run will need to be pooled, dried, and redissolved in methanol and subjected to another round of fine purification by HPLC. This second round HPLC is typically done via a C18 Luna Phenomenex 250×4.6mm column with a gradient from 50% to 80% acetonitrile in water over 35min and a UV detector wavelength set at 280nm. This procedure typically results in 0.2–1mg of 12-*epi*-ficherindole U from a 20-L cyanobacterial culture. The authenticity of 12-*epi*-ficherindole U will need to be verified by a ^1H NMR analysis in comparison with literature data (Stratmann et al., 1994). This NMR analysis step is critical, in particular for the first-time practitioner of this protocol, as *H. welwitschii* UTEX B 1830 produces several isomers of 12-*epi*-ficherindole U that cannot be readily distinguished by UV and mass spec analyses alone.

2.3 Activity Assay for WelO5 and AmbO5 Proteins

The protocol described here is a standard procedure for the WelO5 wild-type protein using 12-*epi*-ficherindole U as a substrate. This method can be readily applied to the AmbO5 protein as well as in combination with other substrates.

The analytical scale assay (100μL) is carried out in a 1.8-mL borosilicate glass vial, rather than an Eppendorf tube made of plastic. The reason for this is that my laboratory has found hapalindole-type compounds are inherently hydrophobic and tend to "stick" to the plastic surface of some Eppendorf tubes. This makes the subsequent enzymatic product recovery challenging. The metal cofactor Fe(II) is provided in the form of $(NH_4)_2Fe(SO_4)_2$. It is recommended the stock solution of $(NH_4)_2Fe(SO_4)_2$ to be made freshly to minimize its deterioration in the air, unless the practitioner's lab has a glove box to keep this solution.

Step 1: All assay components, except the WelO5 protein and Fe(II) salt, are premixed with water to a final volume of 90 μL with their final concentrations at 100 μL adjusted to the following values: 12-*epi*-ficherindole U (1 m*M*), 2OG (2 m*M*), NaCl (10 m*M*), and HEPES (pH 7.4, 20 m*M*). This mixture should be briefly purged with nitrogen gas.

Step 2: To this initially prepared solution, the WelO5 protein (20 μ*M*, final concentration) and $(NH_4)_2Fe(SO_4)_2$ (500 μ*M*, final concentration) are added with a small amount of water to adjust the final volume to 100 μL. The mixture will be briefly incubated under a nitrogen atmosphere before being purged with air and incubated aerobically at 30°C for 1 h.

Step 3: The enzymatic assay is stopped by extracting with ethyl acetate (0.5 mL × 2) and centrifuged. Ethyl acetate extracts are combined, dried under a stream of nitrogen gas, and redissolved in methanol (50 μL). A portion of the methanol solution is injected into a C18 Luna Phenomenex 250 × 4.6 mm HPLC column using a mobile phase gradient 50%–80% acetonitrile in water over 35 min with UV absorption monitoring set at 280 nm.

Step 4: 12-*epi*-ficherindole G is less polar than 12-*epi*-ficherindole U and can be readily identified by examining the HPLC chromatograph. For a first-time practitioner, the standard 12-*epi*-ficherindole G can be isolated from *H. welwitschii* UTEX B 1830 using the protocol described in Section 2.2, which can be used as a coelution standard in this assay. The HPLC analysis described in step 3 can be coupled to a MS detector to ascertain the chlorine transfer. To be absolutely certain on the structure of the chlorinated adduct, the enzymatic reaction has to be scaled up proportionally to at least 3 mL to allow for the enzymatic product to be isolated by semipreparative HPLC for NMR analysis.

3. OUTLOOK

Methods that enable selective intermolecular aliphatic C–H bond functionalization reactions have been long considered a Holy Grail of modern chemical science (Arndtsen, Bergman, Mobley, & Peterson, 1995). Despite the significant progress in this scientific frontier over the past two decades (Crabtree & Lei, 2017; Davies & Morton, 2017; Goldberg & Goldman, 2017; Hartwig, 2017; He, Wasa, Chan, Shao, & Yu, 2017; McIntosh, Farwell, & Arnold, 2014; Shang, Ilies, & Nakamura, 2017), broadly applicable chemical or enzymatic transformations that allow for the selective modification of an aliphatic C–H bond to a C–X bond in a diverse set of molecules have not yet emerged. The discovery of WelO5-type aliphatic halogenases provides a new platform and opportunity to explore this unmet need. While this statement is likely subject to rebuttals due to the limited currently known substrates of WelO5 and AmbO5 (Hillwig & Liu, 2014; Hillwig et al., 2016), several layers of evidence have accumulated over the past 3 years that augments this family of halogenases as a privileged system for biocatalysis development. These evidences address two important issues that often concern practitioners in this field: one is whether this family of halogenase is evolvable and can accept small-molecular substrates beyond hapalindole-type alkaloids; the other is if they are evolvable, then what are the rules? First of all, there is strong evidence that WelO5-type halogenases are naturally evolvable for alternating substrate preferences within the related structural class as well as for completely different structural scaffolds. By characterizing three WelO5-type halogenases related to the biosynthesis of hapalindole alkaloids (Hillwig & Liu, 2014; Hillwig et al., 2016; Zhu & Liu, 2017), my laboratory has identified a conserved C-terminal amino acid sequence that typically resides at 210–240 residues and is critically important for their substrate scope and tolerance of the identified hapalindole-type substrates. This observation corroborates with the substrate-bound X-ray crystal structure of WelO5 (Mitchell et al., 2016). More recently, a WelO5-type halogenase (MstM) was identified in the biosynthetic pathway of an *ent*-sterol-containing meroterpenoid in the cyanobacterium *Scytonema* sp. PCC10023 (Moosmann et al., 2017). Although the halogenation timing for MstM remains to be elucidated, the substrate for MstM is likely a steroid-like molecule, distinct from the hapalindole's indole monoterpene scaffold. Notably, the sequence of MstM diverges most notably from WelO5 and AmbO5 in the previously identified C-terminal motif that defines the substrate-binding pocket. Second, the X-ray crystal structure characterization of WelO5 discovered a novel structural element for controlling the selective halogen transfer by this family of 2OG/Fe(II) enzymes (Fig. 4) and laid the conceptual ground on how to engineer the broadly available 2OG/Fe(II) hydroxylases to halogenases. As a proof of principle, my laboratory has shown the *N*-suc-L-leucine hydroxylase SadA (Hibi et al., 2012) can readily gain chlorination activity by a rational engineering approach (Mitchell et al., 2017). This enzyme is now ready for lab-based directed evolution to further enhance its halogenation activity. It also offers a quick entry point for many other groups to pursue this goal simultaneously.

Overall, with growing microbial genomic data, in particular those related to cyanobacterial organisms (Shih et al., 2013), it is safe to envision more WelO5-type small-molecular aliphatic halogenases will be discovered. The comparative biochemical and structural characterizations of these new enzymes will provide us a better and more concrete picture on Nature's principle to evolve this family of enzymes for alternative structural scaffolds. This knowledge, in combination with those that will soon be learned from the directed evolution of SadA mutant protein to gain halogenation activity, will undoubtedly guide us synergistically toward a broadly applicable enzymatic system that enables late-stage aliphatic halogenations in small-molecular pharmaceuticals, agrichemicals, and bioactive natural products.

ACKNOWLEDGMENTS

The author is indebted to Dr. Matthew L. Hillwig, who established the protocols in the author's lab for assessing the enzymatic activity of WelO5 halogenase as described in this chapter. The initial discovery work associated with the WelO5 and AmbO5 halogenases that constitutes the basis for this chapter was supported by the unrestricted fund from Department of Chemistry, University of Pittsburgh.

REFERENCES

Agarwal, V., Miles, Z. D., Winter, J. M., Eustaquio, A. S., El Gamal, A. A., & Moore, B. S. (2017). Enzymatic halogenation and dehalogenation reactions: Pervasive and mechanistically diverse. *Chemical Reviews*, *117*(8), 5619–5674. https://doi.org/10.1021/acs.chemrev.6b00571.

Arndtsen, B. A., Bergman, R. G., Mobley, T. A., & Peterson, T. H. (1995). Selective intermolecular carbon-hydrogen bond activation by synthetic metal complexes in homogeneous solution. *Accounts of Chemical Research*, *28*(3), 154–162. https://doi.org/10.1021/ar00051a009.

Baran, P. S., Maimone, T. J., & Richter, J. M. (2007). Total synthesis of marine natural products without using protecting groups. *Nature*, *446*(7134), 404–408. https://doi.org/10.1038/nature05569.

Baran, P. S., & Richter, J. M. (2004). Direct coupling of indoles with carbonyl compounds: Short, enantioselective, gram-scale synthetic entry into the hapalindole and fischerindole alkaloid families. *Journal of the American Chemical Society*, *126*(24), 7450–7451. https://doi.org/10.1021/ja047874w.

Blasiak, L. C., Vaillancourt, F. H., Walsh, C. T., & Drennan, C. L. (2006). Crystal structure of the non-haem iron halogenase SyrB2 in syringomycin biosynthesis. *Nature*, *440*(7082), 368–371. https://doi.org/10.1038/nature04544.

Carter-Franklin, J. N., & Butler, A. (2004). Vanadium bromoperoxidase-catalyzed biosynthesis of halogenated marine natural products. *Journal of the American Chemical Society*, *126*(46), 15060–15066. https://doi.org/10.1021/ja047925p.

Crabtree, R. H., & Lei, A. (2017). Introduction: CH activation. *Chemical Reviews*, *117*(13), 8481–8482. https://doi.org/10.1021/acs.chemrev.7b00307.

Davies, H. M. L., & Morton, D. (2017). Collective approach to advancing C–H functionalization. *ACS Central Science*, *3*(9), 936–943. https://doi.org/10.1021/acscentsci.7b00329.

Eustaquio, A. S., Pojer, F., Noel, J. P., & Moore, B. S. (2008). Discovery and characterization of a marine bacterial SAM-dependent chlorinase. *Nature Chemical Biology*, *4*(1), 69–74. https://doi.org/10.1038/nchembio.2007.56.

Galonic, D. P., Vaillancourt, F. H., & Walsh, C. T. (2006). Halogenation of unactivated carbon centers in natural product biosynthesis: Trichlorination of leucine during barbamide biosynthesis. *Journal of the American Chemical Society*, *128*(12), 3900–3901. https://doi.org/10.1021/ja060151n.

Goldberg, K. I., & Goldman, A. S. (2017). Large-scale selective functionalization of alkanes. *Accounts of Chemical Research*, *50*(3), 620–626. https://doi.org/10.1021/acs.accounts.6b00621.

Gorres, K. L., Pua, K. H., & Raines, R. T. (2009). Stringency of the 2-His-1-Asp active-site motif in prolyl 4-hydroxylase. *PLoS One*, *4*(11), e7635. https://doi.org/10.1371/journal.pone.0007635.

Gribble, G. W. (2010). Naturally occurring organohalogen compounds—A comprehensive update. *Progress in the Chemistry of Organic Natural Products*, *91*, 1–613.

Grzyska, P. K., Muller, T. A., Campbell, M. G., & Hausinger, R. P. (2007). Metal ligand substitution and evidence for quinone formation in taurine/alpha-ketoglutarate dioxygenase. *Journal of Inorganic Biochemistry*, *101*(5), 797–808. https://doi.org/10.1016/j.jinorgbio.2007.01.011.

Gu, L., Wang, B., Kulkarni, A., Geders, T. W., Grindberg, R. V., Gerwick, L., et al. (2009). Metamorphic enzyme assembly in polyketide diversification. *Nature*, *459*(7247), 731–735. https://doi.org/10.1038/nature07870.

Hartwig, J. F. (2017). Catalyst-controlled site-selective bond activation. *Accounts of Chemical Research*, *50*(3), 549–555. https://doi.org/10.1021/acs.accounts.6b00546.

Hausinger, R., & Schofield, C. (2015). In R. Hausinger & C. Schofield (Eds.), *2-Oxoglutarate-dependent oxygenases*: The Royal Society of Chemistry.

He, J., Wasa, M., Chan, K. S. L., Shao, Q., & Yu, J.-Q. (2017). Palladium-catalyzed transformations of alkyl C–H bonds. *Chemical Reviews*, *117*(13), 8754–8786. https://doi.org/10.1021/acs.chemrev.6b00622.

Hibi, M., Kawashima, T., Kasahara, T., Sokolov, P. M., Smirnov, S. V., Kodera, T., et al. (2012). A novel Fe(II)/alpha-ketoglutarate-dependent dioxygenase from Burkholderia ambifaria has beta-hydroxylating activity of N-succinyl l-leucine. *Letters in Applied Microbiology*, *55*(6), 414–419. https://doi.org/10.1111/j.1472-765X.2012.03308.x.

Hillwig, M. L., Fuhrman, H. A., Ittiamornkul, K., Sevco, T. J., Kwak, D. H., & Liu, X. (2014). Identification and characterization of a welwitindolinone alkaloid biosynthetic gene cluster in the stigonematalean Cyanobacterium Hapalosiphon welwitschii. *Chembiochem*, *15*(5), 665–669. https://doi.org/10.1002/cbic.201300794.

Hillwig, M. L., & Liu, X. (2014). A new family of iron-dependent halogenases acts on free-standing substrates. *Nature Chemical Biology*, *10*(11), 921–923. https://doi.org/10.1038/nchembio.1625.

Hillwig, M. L., Zhu, Q., Ittiamornkul, K., & Liu, X. (2016). Discovery of a promiscuous non-heme iron halogenase in ambiguine alkaloid biogenesis: Implication for an evolvable enzyme family for late-stage halogenation of aliphatic carbons in small molecules. *Angewandte Chemie (International Ed. in English)*, *55*(19), 5780–5784. https://doi.org/10.1002/anie.201601447.

Hillwig, M. L., Zhu, Q., & Liu, X. (2014). Biosynthesis of ambiguine indole alkaloids in cyanobacterium Fischerella ambigua. *ACS Chemical Biology*, *9*(2), 372–377. https://doi.org/10.1021/cb400681n.

Jiang, W., Heemstra, J. R., Jr., Forseth, R. R., Neumann, C. S., Manaviazar, S., Schroeder, F. C., et al. (2011). Biosynthetic chlorination of the piperazate residue in kutzneride biosynthesis by KthP. *Biochemistry*, *50*(27), 6063–6072. https://doi.org/10.1021/bi200656k.

Krebs, C., Galonic Fujimori, D., Walsh, C. T., & Bollinger, J. M., Jr. (2007). Non-heme Fe (IV)-oxo intermediates. *Accounts of Chemical Research*, *40*(7), 484–492. https://doi.org/10.1021/ar700066p.

Matthews, M. L., Krest, C. M., Barr, E. W., Vaillancourt, F. H., Walsh, C. T., Green, M. T., et al. (2009). Substrate-triggered formation and remarkable stability of the C-H bond-cleaving chloroferryl intermediate in the aliphatic halogenase, SyrB2. *Biochemistry*, *48*(20), 4331–4343. https://doi.org/10.1021/bi900109z.

Matthews, M. L., Neumann, C. S., Miles, L. A., Grove, T. L., Booker, S. J., Krebs, C., et al. (2009). Substrate positioning controls the partition between halogenation and hydroxylation in the aliphatic halogenase, SyrB2. *Proceedings of the National Academy of Sciences of the United States of America*, *106*(42), 17723–17728. https://doi.org/10.1073/pnas.0909649106.

McIntosh, J. A., Farwell, C. C., & Arnold, F. H. (2014). Expanding P450 catalytic reaction space through evolution and engineering. *Current Opinion in Chemical Biology*, *19*, 126–134. https://doi.org/10.1016/j.cbpa.2014.02.001.

Mitchell, A. J., Dunham, N. P., Bergman, J. A., Wang, B., Zhu, Q., Chang, W. C., et al. (2017). Structure-guided reprogramming of a hydroxylase to halogenate its small molecule substrate. *Biochemistry*, *56*(3), 441–444. https://doi.org/10.1021/acs.biochem.6b01173.

Mitchell, A. J., Zhu, Q., Maggiolo, A. O., Ananth, N. R., Hillwig, M. L., Liu, X., et al. (2016). Structural basis for halogenation by iron- and 2-oxo-glutarate-dependent enzyme WelO5. *Nature Chemical Biology*, *12*(8), 636–640. https://doi.org/10.1038/nchembio.2112.

Moosmann, P., Ueoka, R., Grauso, L., Mangoni, A., Morinaka, B. I., Gugger, M., et al. (2017). Cyanobacterial ent-sterol-like natural products from a deviated ubiquinone pathway. *Angewandte Chemie (International Ed. in English)*, *56*(18), 4987–4990. https://doi.org/10.1002/anie.201611617.

Neumann, C. S., & Walsh, C. T. (2008). Biosynthesis of (-)-(1S,2R)-allocoronamic acyl thioester by an Fe(II)-dependent halogenase and a cyclopropane-forming flavoprotein. *Journal of the American Chemical Society*, *130*(43), 14022–14023. https://doi.org/10.1021/ja8064667.

Petty, M. A. (1961). An introduction to the origin and biochemistry of microbial halometabolites. *Bacteriological Reviews*, *25*, 111–130.

Podzelinska, K., Latimer, R., Bhattacharya, A., Vining, L. C., Zechel, D. L., & Jia, Z. (2010). Chloramphenicol biosynthesis: The structure of CmlS, a flavin-dependent halogenase showing a covalent flavin-aspartate bond. *Journal of Molecular Biology*, *397*(1), 316–331. https://doi.org/10.1016/j.jmb.2010.01.020.

Shang, R., Ilies, L., & Nakamura, E. (2017). Iron-catalyzed C–H bond activation. *Chemical Reviews*, *117*(13), 9086–9139. https://doi.org/10.1021/acs.chemrev.6b00772.

Shih, P. M., Wu, D., Latifi, A., Axen, S. D., Fewer, D. P., Talla, E., et al. (2013). Improving the coverage of the cyanobacterial phylum using diversity-driven genome sequencing. *Proceedings of the National Academy of Sciences of the United States of America*, *110*(3), 1053–1058. https://doi.org/10.1073/pnas.1217107110.

Smith, B. R., Eastman, C. M., & Njardarson, J. T. (2014). Beyond C, H, O, and N: Analysis of the elemental composition of U.S. FDA approved drug architectures. *Journal of Medicinal Chemistry*, *57*(23), 9764–9773. https://doi.org/10.1021/jm501105n.

Smitka, T. A., Bonjouklian, R., Doolin, L., Jones, N. D., Deeter, J. B., Yoshida, W. Y., et al. (1992). Ambiguine isonitriles, fungicidal hapalindole-type alkaloids from three genera of blue-green algae belonging to the Stigonemataceae. *Journal of Organic Chemistry*, *57*(3), 857–861. https://doi.org/10.1021/jo00029a014.

Stratmann, K., Moore, R. E., Bonjouklian, R., Deeter, J. B., Patterson, G. M. L., Shaffer, S., et al. (1994). Welwitindolinones, unusual alkaloids from the blue-green algae *Hapalosiphon welwitschii* and *Westiella intricata*. Relationship to fischerindoles and hapalinodoles. *Journal of the American Chemical Society*, *116*(22), 9935–9942. https://doi.org/10.1021/ja00101a015.

Sundaramoorthy, M., Terner, J., & Poulos, T. L. (1995). The crystal structure of chloroperoxidase: A heme peroxidase—Cytochrome P450 functional hybrid. *Structure*, *3*(12), 1367–1377.

Ueki, M., Galonic, D. P., Vaillancourt, F. H., Garneau-Tsodikova, S., Yeh, E., Vosburg, D. A., et al. (2006). Enzymatic generation of the antimetabolite gamma, gamma-dichloroaminobutyrate by NRPS and mononuclear iron halogenase action in a streptomycete. *Chemistry & Biology*, *13*(11), 1183–1191. https://doi.org/10.1016/j.chembiol.2006.09.012.

Vaillancourt, F. H., Yeh, E., Vosburg, D. A., Garneau-Tsodikova, S., & Walsh, C. T. (2006). Nature's inventory of halogenation catalysts: Oxidative strategies predominate. *Chemical Reviews*, *106*(8), 3364–3378. https://doi.org/10.1021/cr050313i.

Vaillancourt, F. H., Yeh, E., Vosburg, D. A., O'Connor, S. E., & Walsh, C. T. (2005). Cryptic chlorination by a non-haem iron enzyme during cyclopropyl amino acid biosynthesis. *Nature*, *436*(7054), 1191–1194. https://doi.org/10.1038/nature03797.

Vaillancourt, F. H., Yin, J., & Walsh, C. T. (2005). SyrB2 in syringomycin E biosynthesis is a nonheme Fe^{II} alpha-ketoglutarate- and O2-dependent halogenase. *Proceedings of the National Academy of Sciences of the United States of America*, *102*(29), 10111–10116. https://doi.org/10.1073/pnas.0504412102.

van Pee, K. H. (2012). Enzymatic chlorination and bromination. *Methods in Enzymology*, *516*, 237–257. https://doi.org/10.1016/b978-0-12-394291-3.00004-6.

White, M. C. (2012). Adding aliphatic C-H bond oxidations to synthesis. *Science*, *335*(6070), 807–809. https://doi.org/10.1126/science.1207661.

Williams, P. G., Buchanan, G. O., Feling, R. H., Kauffman, C. A., Jensen, P. R., & Fenical, W. (2005). New cytotoxic salinosporamides from the marine Actinomycete Salinispora tropica. *Journal of Organic Chemistry*, *70*(16), 6196–6203. https://doi.org/10.1021/jo050511+.

Wuosmaa, A. M., & Hager, L. P. (1990). Methyl chloride transferase: A carbocation route for biosynthesis of halometabolites. *Science*, *249*(4965), 160–162.

Zhu, Q., & Liu, X. (2017). Characterization of non-heme iron aliphatic halogenase WelO5* from Hapalosiphon welwitschiiIC-52-3: Identification of a minimal protein sequence motif that confers enzymatic chlorination specificity in the biosynthesis of welwitindolelinones. *Beilstein Journal of Organic Chemistry*, *13*, 1168–1173. https://doi.org/10.3762/bjoc.13.115.

CHAPTER THIRTEEN

Characterization and Biochemical Assays of *Streptomyces* Vanadium-Dependent Chloroperoxidases

Shaun M.K. McKinnie*, Zachary D. Miles*, Bradley S. Moore*,†,1

*Center for Marine Biotechnology and Biomedicine, Scripps Institution of Oceanography, University of California at San Diego, La Jolla, CA, United States
†Skaggs School of Pharmacy and Pharmaceutical Sciences, University of California at San Diego, La Jolla, CA, United States
[1]Corresponding author: e-mail address: bsmoore@ucsd.edu

Contents

Abstract

Vanadium-dependent haloperoxidases (VHPOs) are fascinating enzymes that facilitate electrophilic halogen incorporation into electron-rich substrates, simply requiring vanadate, a halide source, and cosubstrate hydrogen peroxide for activity. Initially characterized in fungi and red algae, VHPOs were long believed to have limited regio-, chemo-, and enantioselectivity in the production of halogenated metabolites. However, the recent discovery of homologues in the biosynthetic gene clusters of the stereoselectively halogenated meroterpenoids from marine-derived *Streptomyces* bacteria has revised this paradigm. Their intriguing transformations have both enhanced and contributed to the fields of synthetic organic and natural product chemistry. We, herein, describe the expression, purification, and chemical assays of two characterized vanadium-dependent chloroperoxidase enzymes (NapH1 and Mcl24), and one homologue devoid of chlorination activity (NapH3), involved in the biosyntheses of halogenated meroterpenoid products.

Methods in Enzymology, Volume 604
ISSN 0076-6879
https://doi.org/10.1016/bs.mie.2018.02.016

1. INTRODUCTION

Halogenated natural products are particularly intriguing due to their incorporation of electronegative halides into hydrocarbon backbones (Gribble, 2010). These metabolites are particularly prevalent in marine environments due to high concentrations of halides in sea water. The discovery of novel halogenated metabolites is intriguing from a chemical diversity perspective, in addition to the intricate enzymology responsible for the addition of these group 17 elements. A variety of biosynthetic approaches to incorporate halogens into molecules have been elucidated (Agarwal et al., 2017), with one of the most intriguing mechanisms exhibited by the vanadium-dependent haloperoxidase (VHPO) class of enzymes (Winter & Moore, 2009). Utilizing a vanadate prosthetic group to coordinate cosubstrate hydrogen peroxide, VHPOs perform a two-electron oxidation of electrophilic halides to generate a vanadium-bound hypohalous acid. This electrophilic halenium source can then react with electron-rich scaffolds, facilitating halogenation, cyclization, and hydration reactions. Intriguingly, VHPOs are not redox active at the vanadium metal center, remaining in the V^{5+} oxidation state, and thus do not require additional regeneration systems nor suffer oxidative inactivation during catalysis (Fig. 1). VHPO enzymes are more specifically categorized based on the most electronegative element they can oxidize; the vanadium-dependent chloroperoxidases (VCPOs) discussed in this chapter can therefore oxidize chloride, bromide, and iodide.

First identified in red algae and fungi, VHPO enzymes have been extensively studied for their roles in organohalogen production, and a thorough discussion of their history, characterization, biological potentials, and chemical applications are included within this series (Wever, Krenn, & Renirie, 2018). However, the discovery of VHPO genes within *Streptomyces* biosynthetic pathways of the regio- and stereospecifically halogenated meroterpenoid natural products napyradiomycin (Henkel & Zeeck, 1991; Shiomi et al., 1986) and merochlorin (Kaysser et al., 2012) was particularly intriguing. Heterologous expression of the *nap* (Winter et al., 2007) and *mcl* (Kaysser et al., 2012) gene clusters supported the hypothesis that VHPOs were responsible for halogenating the hybrid polyketide–terpenoid molecules. Subsequent in vitro characterization of recombinant VCPO enzymes NapH1 (Bernhardt, Okino, Winter, Miyanaga, & Moore, 2011), Mcl24 (Diethelm, Teufel, Kaysser, & Moore, 2014; Teufel et al., 2014), and homologue NapH3 (Miles et al., 2017) further validated this

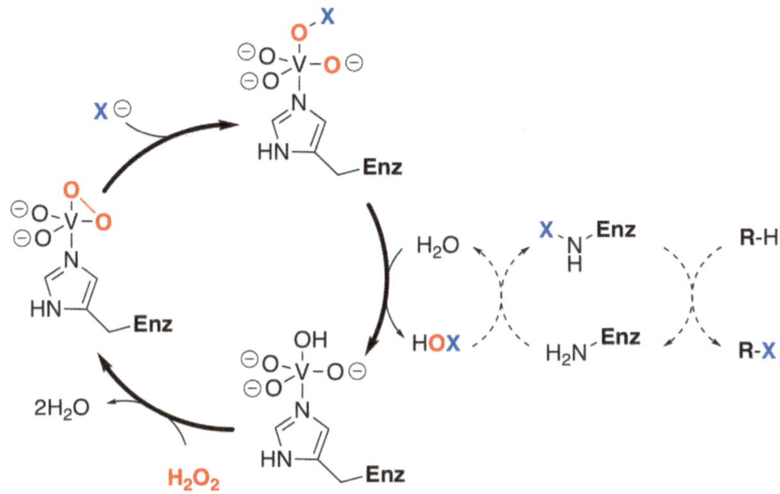

Fig. 1 Proposed catalytic mechanism for the two-electron oxidation of halide ions (X^-) by VHPO enzymes. In the *Streptomyces* VCPO enzymes, the oxidized HOX species is suspected to be intercepted by a neighboring lysine, generating an enzyme-bound halogen species believed to be responsible for the specificity. In the fungal and algal VHPOs, which do not exhibit specificity, the oxidized HOX species is released to rather react directly with electron-rich substrates.

hypothesis through elucidation of the distinct catalytic transformations. Although few examples have been characterized thus far, *Streptomyces* VCPO enzymes are capable of remarkable chemistry, inspiring biomimetic organic total syntheses, while also benefiting significantly from the contributions of the fields of organic and natural product chemistry (Moore, 2018).

This chapter will cover both the characterized VCPOs, NapH1 and Mcl24, involved in halogenated meroterpenoid biosynthesis and draw comparisons to homologue NapH3, which is not known to facilitate halogenating chemistry despite high sequence identity. Details, explanations, and challenges regarding their cloning, expression, purification, and enzymatic activity will be further explored.

2. MOLECULAR BIOLOGY OF *STREPTOMYCES* VCPO GENES

The identification of VCPO encoding genes in *Streptomyces* bacteria was facilitated by their coclustering within biosynthetic gene clusters associated with meroterpenoid natural products, thereby allowing for the context of their biosynthetic reactivity. All cloning of these VCPO enzymes were done using relatively GC-rich actinomycete gDNA; proper precautions should be taken including the proper choice of DNA polymerase and potential use of DMSO in the PCR reaction. The *napH1* and *napH3* genes were obtained by PCR amplification from *Streptomyces* sp. CNQ-525 gDNA. The *napH1* gene was ligated into a pHis8 vector, for expression of protein with an N-terminal octahistidine tag (Jez, Ferrer, Bowman, Dixon, & Noel, 2000) using the *Nco*I and *Not*I restriction sites using standard methodology (Bernhardt et al., 2011). In this same publication, the authors analogously ligated *napH3* into pHis8 that ultimately yielded soluble heterologous protein expression; however, no haloperoxidase activity was observed. Future examination of the annotated *napH3* gene using GeneMark (Besemer & Borodovsky, 2005) identified additional putative translational start sites. A larger open reading frame was subsequently amplified using PCR, and ligated into a pET28a vector (Novagen) again for expression of protein with an N-terminal hexahistidine tag, using the *Nde*I and *Hind*III restriction sites (Miles et al., 2017). This construct also resulted in soluble protein expression and is believed to be the full, functional gene. The *mcl24* gene was obtained by PCR amplification from *Streptomyces* sp. CNH-189 gDNA and was inserted into a pHis8 vector (Jez et al., 2000) using the *Bam*HI and *Hind*III restriction sites, generating an N-terminal octahistidine-tagged protein (Teufel et al., 2014).

To probe the roles of key active site residues in *Streptomyces* VCPO enzymes based on sequence alignment to red algal VHPO enzymes, site-directed mutagenesis was performed according to standard procedures. Owing to the high GC percentage of *Streptomyces* sp., the same precautions as detailed earlier should be taken, and with the expectation of a lower rate of success for mutations.

3. HETEROLOGOUS PROTEIN EXPRESSION AND PURIFICATION

To our satisfaction, these *Streptomyces* VCPO enzymes have thus far been readily expressed following transformation into *Escherichia coli* BL21-Gold (DE3) cells. However, this observation is not applicable to all VCPO enzymes, as recombinant protein expression of NapH4 from the napyradiomycin biosynthetic pathway (Bernhardt et al., 2011) and Mcl40 from the merochlorin biosynthetic pathway (unpublished results) have failed to yield soluble protein. Additional approaches to yield soluble and stable protein are currently under investigation within our lab. Although slight deviations exist in the literature procedures for the published VCPO enzymes, a general unified procedure for their recombinant overexpression and purification is outlined.

3.1 Recombinant VCPO Expression

3.1.1 Equipments

Erlenmeyer culture flasks (50 mL and 2.8 L)
Centrifuge tubes (500 mL, Thermo Scientific Nalgene)
Temperature-controlled centrifuge
Temperature-controlled shaker

3.1.2 Materials

E. coli BL21-Gold (DE3) competent cells
Lysogeny Broth (LB)
Terrific Broth (TB)
LB agar plates
50 mg/mL aqueous kanamycin solution
1 *M* isopropyl-β-D-thiogalactopyranoside (IPTG)
Buffer A (50 m*M* Tris pH 8.0, 500 m*M* NaCl, 25 m*M* imidazole)

3.1.3 Procedure

1. Transform *E. coli* BL21-Gold (DE3) competent cells with ~100 ng of vector according to the manufacturer's protocol. Plate each transformation on LB agar plates with 50 μg/mL kanamycin and incubate overnight at 37°C.
2. Select one positive colony from the plate and inoculate 25 mL of LB media with 50 μg/mL kanamycin. Incubate overnight at 37°C with shaking at ~200 rpm.
3. To 1.5 L of TB media supplemented with 50 μg/mL kanamycin add 15 mL of the overnight LB culture. Incubate at 37°C with shaking at ~200 rpm until the OD_{600} is approximately 0.6 (typically 3–4 h).
4. Cool the incubator to 18°C and maintain shaking at ~200 rpm for 1 h.
5. Add IPTG to a final concentration of 100 μ*M* to induce protein expression. Shake overnight (~16 h) at ~200 rpm.
6. Pellet cells by centrifugation (8000 × *g*, 5 min, 4°C), decant the supernatant media, and resuspend the cell pellets in 40 mL buffer A (50 m*M* Tris pH 8.0, 500 m*M* NaCl, 25 m*M* imidazole).
7. Store the resuspended cell pellets at −80°C or proceed directly to protein purification.

3.1.4 Notes

1. Overnight LB starter cultures can instead be initiated from a glycerol stock; however, we consistently observed better results from selecting a freshly transformed colony.
2. A final concentration of 10% glycerol was added to buffer A for NapH1 to aid in protein stability, or to all overexpressed VCPO cell pellets if they were to be stored at −80°C.
3. Minimal difference in yields or activities was observed between stored and freshly purified protein preparations.

3.2 Purification of Recombinant *Streptomyces* VCPO Enzymes

3.2.1 Equipments

Beaker (variable size)
Magnetic stir bar and stir plate
Sonifier
Temperature-controlled centrifuge
Centrifuge tubes (50 mL, Oak Ridge)
HisTrap FF column (GE Healthcare)
ÄKTA FPLC system (GE Healthcare)
Fraction collection tubes
Polyacrylamide gel apparatus
Amicon Ultra-15 10,000 MWCO spin concentrator (EMD Millipore)
PD-10 desalting column (GE Healthcare)

3.2.2 Materials

Buffer A (50 m*M* Tris pH 8.0, 500 m*M* NaCl, 25 m*M* imidazole)
Buffer B (50 m*M* Tris pH 8.0, 500 m*M* NaCl, 500 m*M* imidazole)
12% SDS-polyacrylamide gels
NapH1 and NapH3 storage buffer (25 m*M* HEPES-KOH pH 8.0, 300 m*M* KCl, 10% glycerol)
Mcl24 storage buffer (25 m*M* HEPES-NaOH pH 7.6, 200 m*M* NaCl, 10% glycerol)

3.2.3 Procedure

1. The resuspended cell pellet in buffer A was transferred to a beaker equipped with a small magnetic stir bar and kept on ice. Cells were lysed by sonication using a Branson digital sonifier (40% intensity) with magnetic stirring using a 15 s on/45 s off cycle for 40 min.
2. Cell debris was pelleted by centrifugation (16,000 × *g*, 30 min, 4°C).

3. The clarified supernatant was loaded onto a 5 mL HisTrap FF column (GE Healthcare), previously equilibrated with buffer A, at a flow rate of 2 mL/min using an ÄKTA FPLC system (GE Healthcare).
4. After loading the supernatant, the HisTrap column was washed with 5–10 column volumes (25–50 mL) of buffer A to remove unbound proteins. A gradient from 0% to 10% buffer B (50 m*M* Tris pH 8.0, 500 m*M* NaCl, 500 m*M* imidazole) was initiated over 10 mL, followed by five column volumes (25 mL) of 10% buffer B to elute weakly bound proteins.
5. Polyhistidine-tagged proteins were eluted using a gradient of 10%–100% buffer B over 60 mL, collecting 5 mL fractions.
6. Fractions containing purified recombinant VCPO were identified using SDS-PAGE (12% acrylamide gel), pooled, and concentrated by centrifugation (3500 × *g*, 4°C) using an Amicon Ultra-15 10,000 MWCO spin concentrator (EMD Millipore) to under 2.5 mL.
7. A PD-10 desalting column (GE Healthcare) was equilibrated following manufacturer instructions with >5 column volumes of cold storage buffer (NapH1 and NapH3: 25 m*M* HEPES-KOH pH 8.0, 300 m*M* KCl, 10% glycerol; Mcl24: 25 m*M* HEPES-NaOH pH 7.6, 200 m*M* NaCl, 10% glycerol).
8. The concentrated VCPO enzyme was applied to the PD-10 column, noting the volume added. Cold storage buffer was then added to bring the final volume applied to the column to precisely 2.5 mL. The buffer-exchanged enzyme was eluted using 3.5 mL of cold storage buffer, collecting a single fraction on ice. The PD-10 column can be regenerated through extensive washing (>10 column volumes) with storage buffer.
9. The recombinant VCPO enzyme was again concentrated (3500 × *g*, 4°C) using an Amicon Ultra-15 10,000 MWCO spin concentrator (EMD Millipore) to approximately 1 mL.
10. Next, the protein concentration was determined via the Bradford method using bovine serum albumin as a standard.
11. Finally, the protein was aliquoted into smaller fractions (50–100 μL), frozen on dry ice, and stored at −80°C until needed.

3.2.4 Notes

1. This procedure is also amenable to manual batch protein purification using a stepwise imidazole elution gradient and Ni-NTA resin.

2. Removing the N-terminal polyhistidine tag, possible through the use of a thrombin cleavage site, has little impact on protein activity or stability and was forgone for all published meroterpenoid biochemical assays.
3. To investigate their unique chemistry, the PD-10 desalting procedure described earlier is suitable for the removal of imidazole and retaining active VCPO enzymes in >90% purity (Fig. 2). Sephadex G-200 size-exclusion chromatography can also be used to additionally purify VCPO enzymes to greater purity with negligible impact on protein activity.
4. When thawing VCPO aliquots for activity assays, quickly thaw them using your fingers, and then subsequently place in ice until needed. NapH3 shows stability to successive freeze–thaw cycles; however, both NapH1 and Mcl24 will precipitate over time and should be used shortly after thawing (<1 h).
5. Typical production yields of recombinant VCPO proteins are as follows:

   NapH1: 5–10 mg/L; NapH3: 70–100 mg/L; Mcl24: 100–140 mg/L.

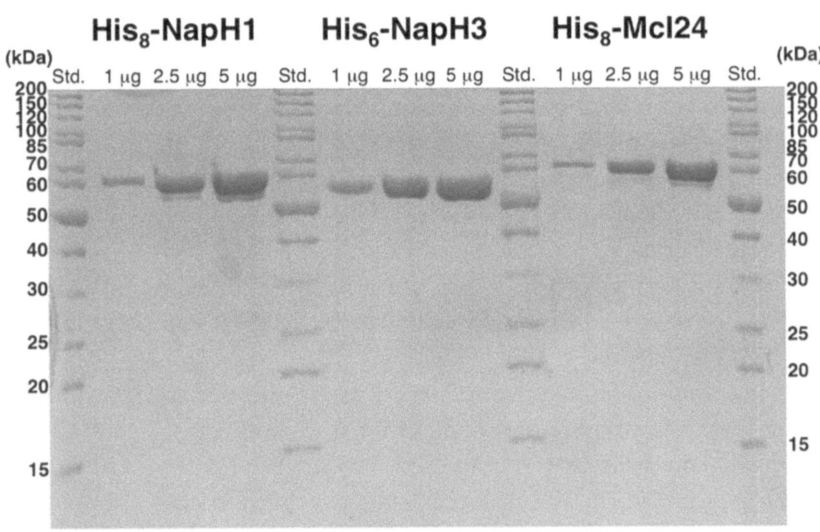

Fig. 2 SDS-PAGE gel (12%) of purified His_8-NapH1 (*left*), His_6-NapH3 (*middle*), and His_8-Mcl24 (*right*). The bands are consistent with the predicted molecular weights of 58.2, 53.8, and 57.6 kDa, respectively.

4. ENZYMATIC ASSAYS OF HETEROLOGOUSLY EXPRESSED *STREPTOMYCES* VCPO ENZYMES

4.1 Monochlorodimedone Assay

The classical way of investigating the halogenation potential of VHPO enzymes has involved the halogenation of monochlorodimedone (MCD). This assay is more thoroughly described in the algal VHPO article within the same series (Wever et al., 2018); however, a brief discussion of its application to marine *Streptomyces* VCPOs will be mentioned here. Algal and fungal VHPO enzymes generate a diffusible hypohalous acid (HOX species) that departs the active site. This can then halogenate the α-carbon between the two carbonyl moieties of MCD, disrupting the conjugation and diminishing the absorbance at 290 nm. However, a diffusible hypohalous acid would fail to provide the regio- and stereoselectivity synonymous with *Streptomyces* VCPO enzymatic transformations. Instead, the "Cl^{+}" species generated is proposed to be transiently intercepted by a neighboring lysine residue, generating a chloramine intermediate capable of facilitating this specific chemistry (Diethelm et al., 2014). When applied to the *Streptomyces* VCPO enzymes capable of chlorination reactions, NapH1 and Mcl24, the MCD assay fails to show any activity in the presence of chloride anions; instead exhibiting a decrease in absorbance at 290 nm only in the presence of bromide. Although the literature sample size is limited to these two enzymes, this trend has held consistent for additional *Streptomyces* VCPO enzymes currently under investigation within our lab that catalyze meroterpenoid chlorination chemistry. Our best attempt to rationalize this result suggests the weak enzymatic capture of the oxidized bromine species, allowing a freely dissociable hypobromite ion to exit the enzyme and appear active via the MCD assay. This is not the only discrepancy in the application of the MCD assay to analyze haloperoxidase enzymes (Wagner, Molitor, & König, 2008). However, despite the contradictory halide specificity results of the MCD assays and their applications to specific bacterial VCPOs, these experiments are useful to validate the proper folding and general halogenation activity of these enzymes prior to thorough in vitro interrogation with substrates. They can also provide insight into the oxidation rates of these VCPOs at different pH, a significant factor when analyzing the chemistry facilitated by Mcl24 (Fig. 3A and B).

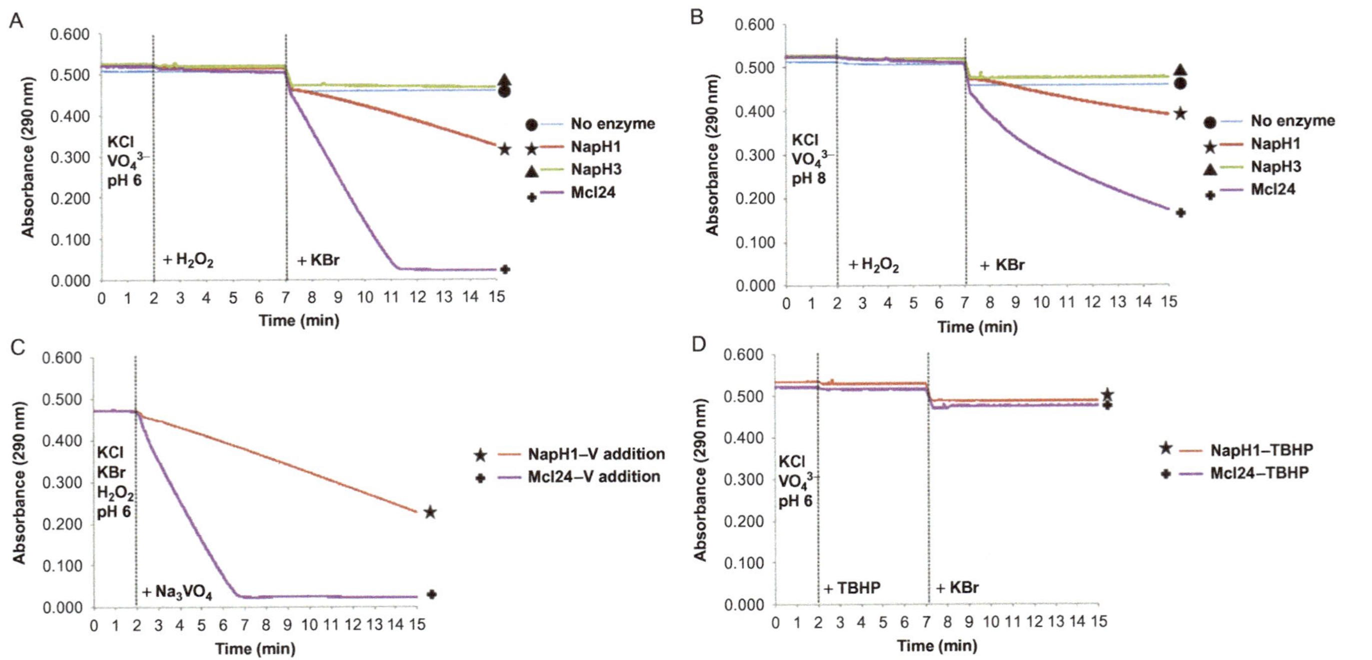

Fig. 3 Monochlorodimedone (MCD) assays of VCPO enzymes and no enzyme controls in: (A) MES buffer, pH 6.0; (B) HEPES buffer, pH 8.0; (C) initial VCPO preincubation with hydrogen peroxide followed by addition of exogenous sodium vanadate at 2 min (both chloride and bromide ions present); and (D) substitution of hydrogen peroxide with *tert*-butyl hydrogen peroxide (TBHP) in MES buffer, pH 6.0 to assess hydrogen peroxide dependency.

4.1.1 Equipments

Quartz cuvette
Spectrophotometer capable of kinetic scanning

4.1.2 Materials

0.5 *M* buffer at desired pH
2 *M* potassium chloride
2 *M* potassium bromide
500 μ*M* aqueous MCD solution
100 m*M* hydrogen peroxide
1 m*M* sodium orthovanadate (Na_3VO_4)
1 mg/mL enzyme solution

4.1.3 Procedure

1. Add all reaction components except MCD to a 1-mL quartz cuvette. Zero the UV spectrometer at 290 nm to this solution.
2. To a 1.5-mL microcentrifuge tube, add assay buffer (50 m*M*), MCD (50 μ*M*), potassium chloride (200 m*M*), and sodium orthovanadate (10 μ*M*) to the final concentrations shown in brackets. Add 10 μg of VCPO enzyme to this solution and transfer it to a quartz cuvette.
3. Record the absorbance at 290 nm of the solution every 1 s throughout the course of the assay using a kinetic scanning program. Monitor the initial 2 min to obtain a baseline MCD absorbance.
4. Add hydrogen peroxide to a final concentration of 1 m*M* with thorough pipette mixing. If no decrease in absorbance is observed 5 min following peroxide addition, add potassium bromide to a final concentration of 200 m*M*.

4.1.4 Notes

1. A quartz cuvette is needed to minimize UV absorbance/interference within the assay.
2. The two buffers used to assay these three VCPO enzymes at different pH were: 0.5 *M* 2-(*N*-morpholino)ethanesulfonic acid (MES), pH 6.0 and 0.5 *M* 4-(2-hydroxyethyl)-1-piperazineethanesulfonic acid (HEPES), pH 8.0.
3. Both hydrogen peroxide and vanadium salts must be exogenously supplied to initiate the oxidative activities of recombinant NapH1 and Mcl24. The assays as described earlier highlight the requirement of hydrogen peroxide for catalysis; to interrogate the necessity of exogenous

vanadate, an analogous assay can be established, adjusting the order of addition of MCD assay components. The addition of potassium bromide (200 m*M* final concentration) and substitution of sodium vanadate with hydrogen peroxide in step 2, establishing an MCD baseline absorbance for 2 min as in step 3, and adding sodium vanadate in step 4 can interrogate the presence of bound vanadium in the recombinant protein (Fig. 3C). These results show there is minimal time required to form the histidine-coordinated vanadate prosthetic group in both halogenases.

4. Hydrogen peroxide has yet to be conclusively demonstrated to be the native oxidant for VCPO enzymes in vivo. Although this remains to be thoroughly examined in vitro or in vivo, the substitution of H_2O_2 with small alkyl hydrogen peroxides (*tert*-butyl hydrogen peroxide, TBHP, Fig. 3D) fails to diminish the absorbance at 290 nm in the MCD assay with either halide ion. This is in contrast to the results discussed within Section 4.2.3, where Mcl24 appears to be capable of using TBHP as an oxidizing source, albeit at a significantly reduced result.

4.2 Chemical Assays

Inarguably, the most intriguing feature of these *Streptomyces* VCPO enzymes involves their capability to specifically halogenate meroterpenoid substrates in a variety of ways. While genetic evidence initially highlighted these enzymes and their roles in meroterpenoid halogenation and biosynthesis, in vitro characterization has uncovered a diverse suite of chemical activities deserving of individual discussion.

4.2.1 NapH1 Cyclization of Naphthomevalin to Napyradiomycin A1

In the napyradiomycin suite of molecules, VCPO NapH1 has been shown to catalyze the chloronium-induced cyclization of the dimethylallyl (prenyl) sidechain of naphthomevalin to generate the tricyclic compound napyradiomycin A1 (Bernhardt et al., 2011) (Fig. 4). This activity was originally demonstrated on methylated naphthomevalin analogue SF2415B1 isolated from *Streptomyces aculeolatus* NRRL 18442 (Winter et al., 2007), generating 7-methyl-napyradiomycin A1 (SF2415B3). In vitro assay conditions indicate NapH1 halogenation is diastereoselective at the newly formed stereocenter at the chlorine atom, implying a specific facial chloronium formation and cyclization by the adjacent hydroxy group. Incubation of NapH1 with SF2415B1 in the presence of bromide facilitated a product consistent with bromonium-induced cyclization, which is a

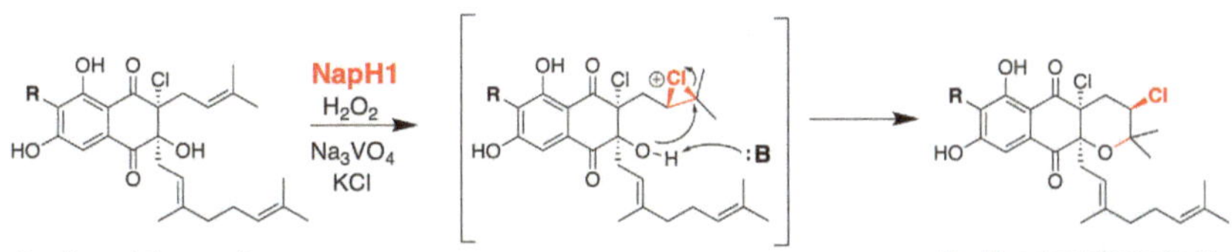

Fig. 4 NapH1 chlorination and cyclization of naphthomevalin (or methylated analogue SF2451B1) to tricyclic napyradiomycin A1 (or SF2451B3). (:B represents a general base in the reaction).

naturally produced napyradiomycin analogue (Farnaes et al., 2014). However, this reaction failed to go to completion in vitro due to probable self-bromination and inactivation of key residues in the VCPO required for activity and/or stability.

4.2.1.1 Procedure

1. For the in vitro preparative synthesis of SF2415B3 as reported (Bernhardt et al., 2011), starting material SF2415B1 (4.6 mg, 10 μmol) was added to an 100 mL aqueous solution of 50 m*M* MES buffer, pH 6.0, 50 m*M* potassium chloride, 10 μ*M* sodium vanadate, and 1 m*M* hydrogen peroxide in a round-bottom flask equipped with a stir bar. The reaction was initiated through the addition of recombinant NapH1 (50 μg) and stirred at 30°C for 8 h.
2. The reaction was quenched by extraction with ethyl acetate (3 × 100 mL). Pooled organic layers were dried over magnesium sulfate, filtered, and concentrated in vacuo. RP–HPLC with a semipreparative Phenomenex Synergi Polar-RP 4 μm, 250 × 10 mm column purified enzymatic SF2415B3 using a gradient of 60%–100% acetonitrile in water over 30 min.

4.2.1.2 Notes

1. The structural differences between the starting material and product are diagnostic enough to facilitate the application of thin-layer chromatography (TLC) to monitor the course of the reaction instead of solely liquid chromatography-based methods. Naphthomevalin R_f 0.35, napyradiomycin A1 R_f 0.40 (4:1 hexane:EtOAc).
2. As indicated in the procedure, this reaction is amenable to excess hydrogen peroxide. However, the stoichiometry of this cosubstrate can be decreased to 1 molar equivalent relative to the initial starting material

without significant loss of yield. Depending on the substrate used in the reaction, care must be taken to avoid oxidative degradation due to excess hydrogen peroxide.

3. NapH1 also catalyzes analogous chlorination and cyclization reactions on the des-methylated SF2415B1 substrate naphthomevalin, as highlighted in Fig. 4 (Miles et al., 2017). However, NapH1 failed to show any catalytic activity on the linear terpene alcohol precursor (±)-nerolidol or the related meroterpenoid-like compound lapachol (Bernhardt et al., 2011), highlighting the extreme selectivity NapH1 has for its requisite substrate. Reactions catalyzed by NapH1 are also stereoselective, evidenced by the single enantiomer of napyradiomycin A1 generated from the synthetic, racemic naphthomevalin substrate. The concomitant, unreacted naphthomevalin recovered from a NapH1 assay, shows an opposite circular dichroism spectrum to naphthomevalin naturally isolated from *Streptomyces* bacteria or that produced in vitro by NapH3 reaction (Miles et al., 2017).
4. Ethyl acetate was an optimal choice for the extraction of meroterpenoid products from the NapH1 reaction due to their high solubility in this solvent, facile phase separation of ethyl acetate and the aqueous reaction, and high volatility when concentrating in vacuo.

4.2.2 NapH3 α-Hydroxyketone Rearrangement to Generate Naphthomevalin

Although this reaction falls outside the scope of halogenation, the significance of NapH3 catalysis to the napyradiomycin biosynthetic pathway cannot be understated. In both the MCD assay and in vitro incubation with meroterpenoid substrates, NapH3 is devoid of halogenating activity with any halide ion. Rather, this VHPO homolog, which shares 57% sequence identity with the VCPO NapH1, facilitates an α-hydroxyketone rearrangement of the geranyl moiety of prenaphthomevalin to generate naphthomevalin, the substrate for NapH1 (Miles et al., 2017) (Fig. 5). Intriguingly, NapH3 does

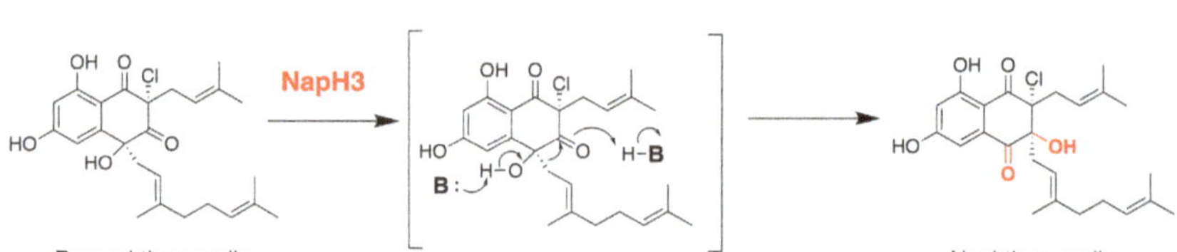

Fig. 5 NapH3 α-hydroxyketone rearrangement of synthetic "prenaphthomevalin" to generate naphthomevalin (B: represents a general base in the reaction).

not require exogenous vanadium or any other cofactor/cosubstrate to facilitate its prenyl migration reaction. Computational calculations have shown that this reaction is energetically favorable, and incubation of synthetic prenaphthomevalin in polar protic solvents can also facilitate this suprafacial 1,2-shift; however, the rate of this shift is dramatically accelerated in the presence of NapH3. This shift was also shown to be enantioselective when incubated with racemic prenaphthomevalin based on circular dichroism analyses.

4.2.2.1 Procedure

1. For the in vitro preparative scale enzymatic synthesis of naphthomevalin (Miles et al., 2017), prenaphthomevalin (7 mg, 18 μmol) dissolved in DMSO (200 μL) was added to a 10 mL solution of 50 m*M* HEPES-NaOH pH 8.0, 150 m*M* potassium chloride, and 10% glycerol. The reaction was initiated through the addition of recombinant NapH3 (20 mg) and stirred at room temperature for 2 h.
2. The reaction was quenched by extraction with ethyl acetate (3 × 10 mL). The solvent was removed in vacuo, and the crude mixture resuspended in methanol for reversed-phase HPLC purification on a semipreparative Phenomenex Luna C18(2), 5 μm, 250 × 10 mm column. Enzymatic naphthomevalin was purified at a flow rate of 2.5 mL/min using a gradient of 5%–100% B (A=0.1% aqueous trifluoroacetic acid, B=0.1% trifluoroacetic acid in acetonitrile) over 20 min.

4.2.2.2 Notes

To monitor the completion of this reaction, RP–HPLC is the most efficient approach and allows for clean separation of substrate and product. Both starting material and product have comparable TLC retention times and thorough NMR analyses can differentiate the conversion of the shift.

4.2.3 Multitasking Catalytic Activities of Mcl24 to Generate Merochlorins A, B, and X

VCPO Mcl24 from the merochlorin biosynthetic pathway shows an incredible ability to multitask, displaying all aforementioned VCPO functionalities (oxidative halogenation, terpene cyclization, and α-hydroxyketone rearrangement) within one enzyme (Diethelm et al., 2014; Miles et al., 2017; Teufel et al., 2014) (Fig. 6). At pH 6, Mcl24 catalyzes the halogenation, oxidative dearomatization, and terpene cyclization of premerochlorin to generate two dominant monochlorinated merochlorins. These structurally

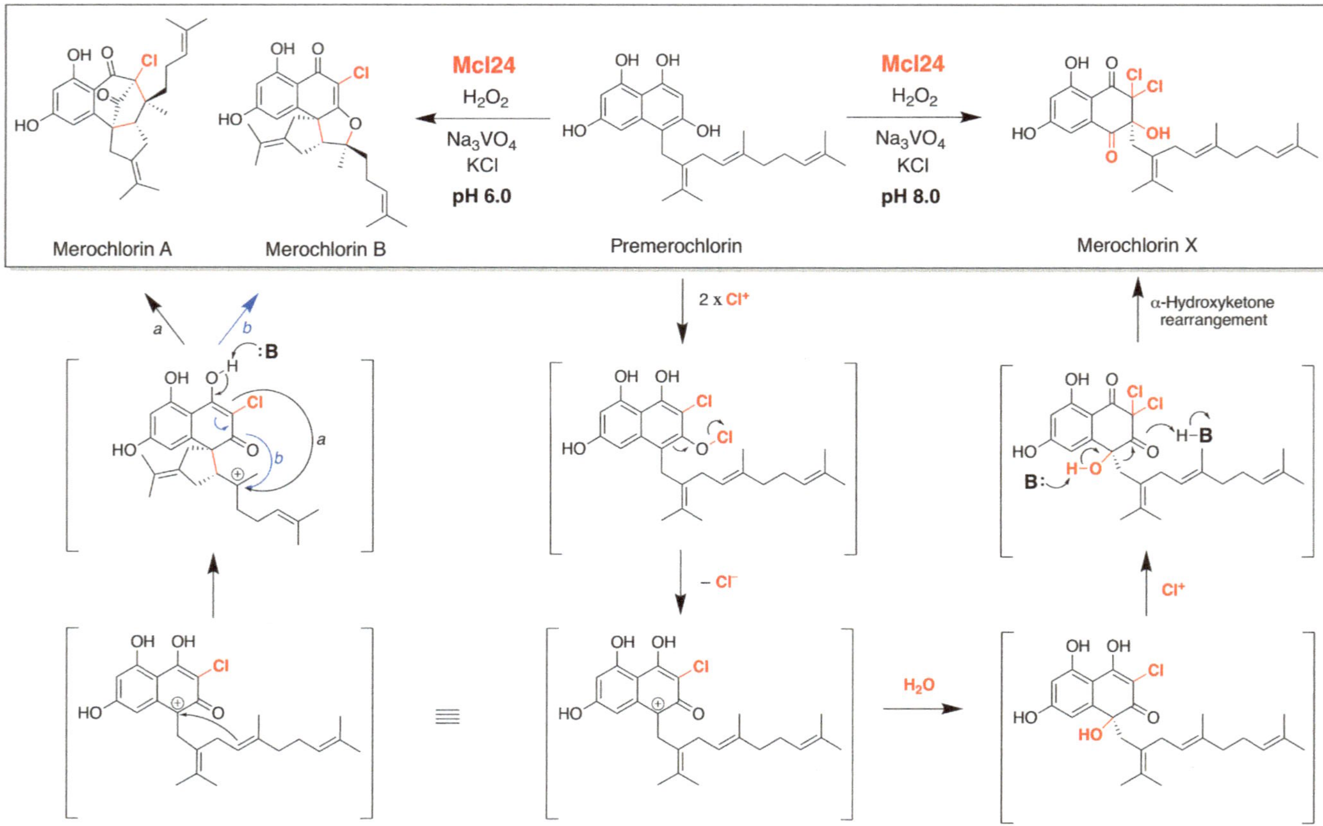

Fig. 6 Mcl24 reaction on "premerochlorin" substrate at different pH to synthesize merochlorins A, B, and X. Following dichlorination and oxidative dearomatization, Mcl24 branches in how the proposed benzylic cation (*bottom middle*) is quenched, either through terpene cyclizations to generate merochlorins A and B, or through hydration and an α-hydroxyketone rearrangement to generate merochlorin X. (B: represents a general base in the reaction.)

divergent molecules arise due to the final terpene cyclization cascade ending with a carbon–carbon bond, or oxygen–carbon bond in merochlorins A and B, respectively (Fig. 6). Biomimetic total syntheses have followed and supported this biosynthetic hypothesis for both merochlorin A (Pepper & George, 2013) and B (Meier, Strych, & Trauner, 2014) metabolites. However, under basic conditions (pH 8) the product distribution of Mcl24 is altered to predominantly catalyze the formation of merochlorin X, an α-hydroxyketone rearranged intermediate that may be further tailored by additional biosynthetic enzymes to generate merochlorins C and D (Fig. 6). The major mechanistic difference during the biosyntheses of these molecules involves the addition of water to the oxidatively dearomatized molecule, which Mcl24 catalyzes more favorably at a basic pH. Hydration of the benzylic cation facilitates dichlorination and the corresponding rearrangement.

When applied in vitro, Mcl24 generates a large number of by-products because of the substantial oxidative instability of the premerochlorin substrate. Minimizing the amount of hydrogen peroxide within the reaction to the stoichiometric equivalents required to facilitate the chemistry has improved yields substantially and should be explored when investigating the oxidation of sensitive naphthol-derived meroterpenoids with VCPO enzymes. Surprisingly, hydrogen peroxide can be substituted with the organic peroxide TBHP, generating small quantities of the major three merochlorins, although the rate of catalysis is dramatically reduced (Fig. 7). Investigation into appropriate hydrogen peroxide replacements and their substitution on VCPO activity is currently ongoing.

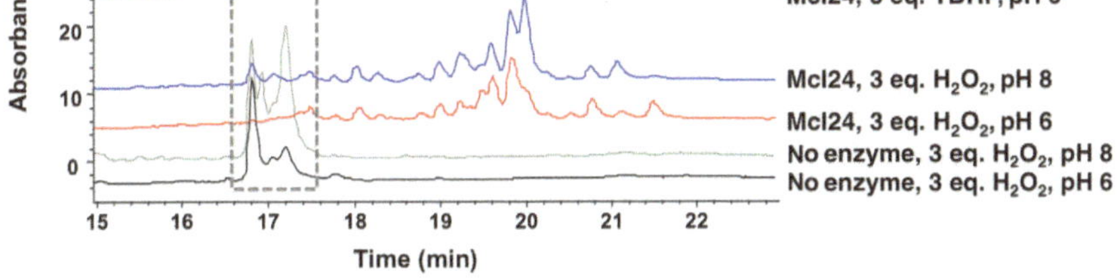

Fig. 7 RP-HPLC analyses of in vitro Mcl24 assays with synthetic "premerochlorin" under various conditions at both pH 6 and 8, and comparison to isolated merochlorin A, B, and X standards (UV absorbance at 340 nm).

4.2.3.1 Procedure

1. For the in vitro analytical scale production of merochlorin molecules, 1 m*M* premerochlorin (in a 100 m*M* DMSO stock solution) was added to a 0.5 mL solution of 50 m*M* buffer (pH 6.0—MES, pH 8.0—HEPES), 150 m*M* potassium chloride, 10% glycerol, 100 μ*M* sodium vanadate, and 1.4 mg Mcl24 (50 μ*M*) in a 1.5-mL microfuge tube and maintained under an inert atmosphere. To this solution, 3 m*M* hydrogen peroxide was added, mixed by pipetting, and left at room temperature (25°C) for 3 h.
2. A 50 μL aliquot of the reaction mixture was quenched into 50 μL of acetonitrile, mixed thoroughly, and centrifuged at 13,000 × *g* for 2 min to pellet precipitated protein. A 10 μL aliquot of the supernatant of the quenched reaction was analyzed by analytical HPLC (Phenomenex Luna C18(2) 5 μm, 100 × 4.6 mm column) using the following separation gradient: 5% B for 5 min, 5%–85% B over 13 min, 85%–100% B over 2 min, and 100% B over 3 min (A = 0.1% aqueous formic acid, B = 0.1% formic acid in acetonitrile).

4.2.3.2 Notes

1. The reported large-scale enzymatic preparations of merochlorins A, B (Diethelm et al., 2014), and X (Miles et al., 2017) differ slightly from the aforementioned procedure, using superstoichiometric quantities of hydrogen peroxide. Because of the oxidative instability of premerochlorin to these conditions, we have been able to decrease the amount of hydrogen peroxide to the minimum required for catalysis while retaining comparable merochlorin production. This approach is recommended for future large-scale enzymatic preparations of meroterpenoid products to minimize decomposition.
2. When preparing assays using oxygen-sensitive premerochlorin, maintain solutions under an inert atmosphere to minimize atmospheric oxidation of the unstable substrate prior to hydrogen peroxide addition.
3. As seen in Fig. 7, a substantial amount of other products are formed from an in vitro Mcl24 reaction with the premerochlorin substrate. Additional iso-, di-, and des-chlorinated merochlorin analogues have been isolated from in vitro reactions, prepared using biomimetic chemical chlorinating conditions, and produced in vivo by *Streptomyces* bacteria (Diethelm et al., 2014).

5. CONCLUSIONS AND FUTURE OUTLOOK

Streptomyces VCPO enzymes have shown remarkable abilities to facilitate specific and complex halogenation reactions in the biosyntheses of merochlorin and napyradiomycin molecules. Despite the novel and interesting chemistry facilitated by these enzymes, we have only begun to scratch the surface of understanding their potential as halogenating biocatalysts. Even within these two biosynthetic gene clusters, additional VHPOs remain to be functionally characterized. There is considerable genetic evidence to suggest the involvement of Mcl40 in the chloronium-mediated macrocyclization of merochlorin D to merochlorin C, generating an unprecedented chlorinated 15-membered ether. The activity of the last uncharacterized VCPO in the napyradiomycin biosynthetic gene cluster, NapH4, also remains to be explored. Similarly halogenated actinomycete meroterpenoids such as napyradiomycin B1 and azamerone (Cho, Kwon, Williams, Jensen, & Fenical, 2006) are anticipated to have VCPO enzymology implicated in their biosyntheses. Analogous α-hydroxyketone rearrangements may also be catalyzed by NapH3 homologues in the biosynthetic gene clusters of nonhalogenated meroterpenoids naphterpin (Shin-ya et al., 1990) and marinone (Pathirana, Jensen, & Fenical, 1992). Although the study of these enzymes in bacteria is still in its infancy, we hope that future work can continue to uncover additional applications of these specific VHPO enzymes and their unique roles in the enantioselective construction of complex natural product molecules.

ACKNOWLEDGMENTS

Research on meroterpenoid natural products has been generously supported by NIH Grant R01-AI047818 (B.S.M.) and funding from the Natural Sciences and Research Council of Canada (NSERC-PDF to S.M.K.M.).

REFERENCES

Agarwal, V., Miles, Z. D., Winter, J. M., Eustáquio, A. S., El Gamal, A. A., & Moore, B. S. (2017). Enzymatic halogenation and dehalogenation reactions: Pervasive and mechanistically diverse. *Chemical Reviews*, *117*, 5619–5674.

Bernhardt, P., Okino, T., Winter, J. M., Miyanaga, A., & Moore, B. S. (2011). Stereoselective vanadium-dependent chloroperoxidase in bacterial antibiotic biosynthesis. *Journal of the American Chemical Society*, *133*, 4268–4270.

Besemer, J., & Borodovsky, M. (2005). GeneMark: Web software for gene finding in prokaryotes, eukaryotes and virus. *Nucleic Acids Research*, *33*, W451–454.

Cho, J. Y., Kwon, H. C., Williams, P. G., Jensen, P. R., & Fenical, W. (2006). Azamerone, a terpenoid phthalazinone from a marine-derived bacterium related to the genus *Streptomyces* (Actinomycetales). *Organic Letters*, *8*, 2471–2474.

Diethelm, S., Teufel, R., Kaysser, L., & Moore, B. S. (2014). A multitasking vanadium-dependent chloroperoxidase as an inspiration for the chemical synthesis of the merochlorins. *Angewandte Chemie International Edition*, *53*, 11203–11206.

Farnaes, L., Coufal, N. G., Kauffman, C. A., Rheingold, A. L., DiPasquale, A. G., Jensen, P. R., et al. (2014). Napyradiomycin derivatives, produced by a marine-derived actinomycete, illustrate cytotoxicity by induction of apoptosis. *Journal of Natural Products*, *77*, 15–21.

Gribble, G. W. (2010). *Naturally occurring organohalogen compounds: A comprehensive update.* Vienna: Springer.

Henkel, T., & Zeeck, A. (1991). Secondary metabolites by chemical screening, 15. Structure and absolute configuration of naphthomevalin, a new dihydro-naphthoquinone antibiotic from *Streptomyces* sp. *The Journal of Antibiotics*, *44*, 665–669.

Jez, J. M., Ferrer, J. L., Bowman, M. E., Dixon, R. A., & Noel, J. P. (2000). Dissection of malonyl-coenzyme A decarboxylation from polyketide formation in the reaction mechanism of a plant polyketide synthase. *Biochemistry*, *39*, 890–902.

Kaysser, L., Bernhardt, P., Nam, S. J., Loesgen, S., Ruby, J. G., Skewes-Cox, P., et al. (2012). Merochlorins A–D, cyclic meroterpenoid antibiotics biosynthesized in divergent pathways with vanadium-dependent chloroperoxidases. *Journal of the American Chemical Society*, *134*, 11988–11991.

Meier, R., Strych, S., & Trauner, D. (2014). Biomimetic synthesis of (±)-merochlorin B. *Organic Letters*, *16*, 2634–2637.

Miles, Z. D., Diethelm, S., Pepper, H. P., Huang, D. M., George, J. H., & Moore, B. S. (2017). A unifying paradigm for naphthoquinone-based meroterpenoid (bio)synthesis. *Nature Chemistry*, *9*, 1235–1242.

Moore, B. S. (2018). Asymmetric alkene and arene halofunctionalization reactions in meroterpenoid biosynthesis. *Synlett*, *29*, 401–409. https://doi.org/10.1055/s-0036-1590919.

Pathirana, C., Jensen, P. R., & Fenical, W. (1992). Marinone and debromomarinone: Antibiotic sesquiterpenoid naphthoquinones of a new structure class from a marine bacterium. *Tetrahedron Letters*, *33*, 7663–7666.

Pepper, H. P., & George, J. H. (2013). Biomimetic total synthesis of (±)-merochlorin A. *Angewandte Chemie International Edition*, *52*, 12170–12173.

Shin-ya, K., Imai, S., Furihata, K., Hayakawa, Y., Kato, Y., Vanduyne, G. D., et al. (1990). Isolation and structural elucidation of an antioxidative agent, naphterpin. *The Journal of Antibiotics*, *43*, 444–447.

Shiomi, K., Iinuma, H., Hamada, M., Naganawa, H., Manabe, M., Matsuki, C., et al. (1986). Novel antibiotics napyradiomycins. Production, isolation, physico-chemical properties and biological activity. *The Journal of Antibiotics*, *39*, 487–493.

Teufel, R., Kaysser, L., Villaume, M. T., Diethelm, S., Carbullido, M. K., Baran, P. S., et al. (2014). One-pot enzymatic synthesis of merochlorin A and B. *Angewandte Chemie International Edition*, *53*, 11019–11022.

Wagner, C., Molitor, I. M., & König, G. M. (2008). Critical view on the monochlorodimedone assay utilized to detect haloperoxidase activity. *Phytochemistry*, *69*, 323–332.

Wever, R., Krenn, B. E., & Renirie, R. (2018). Marine vanadium-dependent haloperoxidases, their isolation, characterization, and application. *Methods in Enzymology*, *605*, in press.

Winter, J. M., Moffitt, M. C., Zazopoulos, E., McAlpine, J. B., Dorrestein, P. C., & Moore, B. S. (2007). Molecular basis for chloronium-mediated meroterpene cyclization—Cloning, sequencing, and heterologous expression of the napyradiomycin biosynthetic gene cluster. *Journal of Biological Chemistry*, *282*, 16362–16368.

Winter, J. M., & Moore, B. S. (2009). Exploring the chemistry and biology of vanadium-dependent haloperoxidases. *Journal of Biological Chemistry*, *284*, 18577–18581.

CHAPTER FOURTEEN

Unusual "Head-to-Torso" Coupling of Terpene Precursors as a New Strategy for the Structural Diversification of Natural Products

Robin Teufel[1]
ZBSA, Center for Biological Systems Analysis, University of Freiburg, Freiburg, Germany
Faculty of Biology, University of Freiburg, Freiburg, Germany
[1]Corresponding author: e-mail address: robin.teufel@zbsa.uni-freiburg.de

Contents

Abstract

Terpenoids are ubiquitous in nature and exhibit an immense structural diversity. Commonly, terpenoid biosynthesis involves prenyl diphosphate synthases (terpene synthases) that produce linear prenyl diphosphates with various chain lengths via conventional "head-to-tail" coupling of C5 units. Structural diversification, in contrast, is often mediated by terpene cyclases and additional tailoring enzymes. Recently, however, a few unusual prenyl diphosphate synthases were reported that catalyze noncanonical "head-to-torso" coupling reactions and thus formation of various branched prenyl diphosphates such as isosesquilavandulyl diphosphate. Here, I describe these

Methods in Enzymology, Volume 604
ISSN 0076-6879
https://doi.org/10.1016/bs.mie.2018.01.037

enzymes in detail and illustrate how these branching reactions are key to formation of various structurally distinct natural products such as the complex polycyclic merochlorin antibiotics from a marine bacterium.

1. INTRODUCTION

Many primary and secondary metabolites belong to the class of terpenoids (isoprenoids) that have numerous important functions, e.g., as visual pigments, defensive agents, or sex hormones. Generally, terpenoids derive from the C5 isoprene unit isopentenyl diphosphate (IPP) that can be condensed with allylic acceptor molecules (e.g., the isomeric dimethylallyl diphosphate (DMAPP) or larger prenyl diphosphates) by prenyl diphosphate synthases. Typically, the condensation reactions proceed through electrophilic substitution reactions and head-to-tail [C(1)′–C(4)] coupling of the two substrates, giving rise to linear, long-chain (≥C10) prenyl diphosphates (Fujihashi et al., 2001; Fig. 1A). According to the stereochemistry of their products, these enzymes can be classified as *cis*- and *trans*-prenyl diphosphate synthases, which, remarkably, share no sequential or structural homology despite similar substrate preferences and catalytic mechanisms (Fujihashi et al., 2001). Commonly, the linear prenyl diphosphates are further modified, e.g., by terpene cyclases that often catalyze complex cyclization cascades as well as additional tailoring enzymes (Dickschat, 2016).

More rarely, terpenoids feature non-head-to-tail connected isoprene units that are assembled, e.g., by cyclopropanation, cyclobutanation, or other branching reactions (Rivera et al., 2001; Thulasiram, Erickson, & Poulter, 2008). In addition, several noncanonical prenyl diphosphate synthases were recently discovered that catalyze unprecedented head-to-torso (also referred to as "head-to-middle") [C(1′)–C(2)] coupling reactions, thereby producing branched mono- and sesquiterpenes (Fig. 1B–D). In the following sections, I will highlight aspects of the discovery and characterization of a bacterial isosesquilavandulyl diphosphate synthase (ISLPPS), a homologous cyclolavandulyl diphosphate synthase (CLDS), as well as the plant enzyme lavandulyl diphosphate synthase (LPPS). The names of the enzymes and respective products derive from the monoterpenes lavandulol and lavandulyl acetate of lavender oil, where this unusual C(1′)–C(2) terpene branching was first observed and recently ascribed to the activity of LPPS.

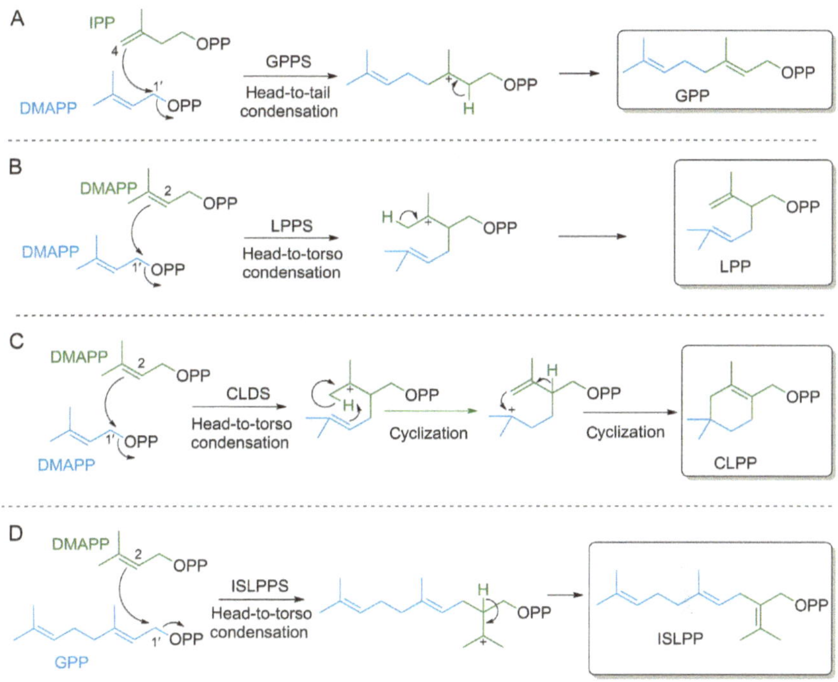

Fig. 1 Common head-to-tail and noncanonical head-to-torso (=head-to-middle) condensation of prenyl diphosphates and proposed catalytic mechanisms of the involved prenyl diphosphate synthases. For each reaction, prenyl diphosphates acting as nucleophiles and electrophiles are shown in *green* and *blue*, respectively. For simplicity, only protons involved in the corresponding catalytic steps are shown. (A) Geranyl diphosphate synthase (GPPS). (B) Lavandulyl diphosphate synthase (LPPS). (C) Cyclolavandulyl diphosphate synthase (CLDS). (D) Isosesquilavandulyl diphosphate synthase (ISLPPS) Mcl22.

2. THE BIOSYNTHESIS OF THE BACTERIAL MEROCHLORIN ANTIBIOTICS THAT FEATURE A UNIQUELY BRANCHED SESQUITERPENE MOIETY

Meroterpenoid natural products are polyketide–terpene hybrids with often potent bioactivities found in all domains of life. These compounds feature unusual molecular architectures, as exemplified by the recently discovered bacterial merochlorins from the marine *Streptomyces* sp. strain CNH-189 that exhibit high activity against clinically relevant methicillin-resistant *Staphylococcus aureus* strains (MICs of 2–4 μg/mL) (Sakoulas et al., 2012; Fig. 2A).

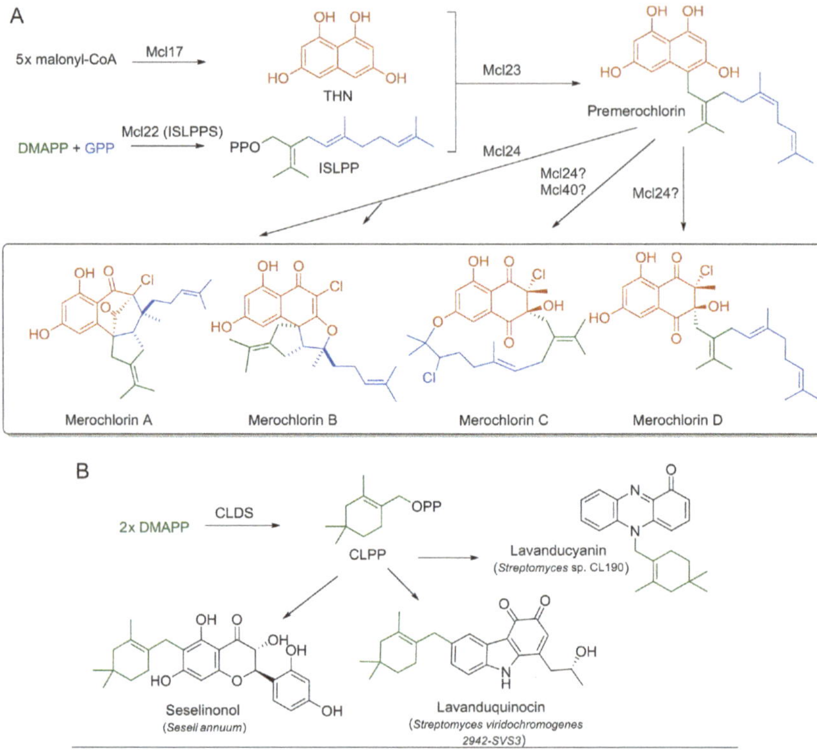

Fig. 2 (A) Merochlorin biosynthetic pathway. Malonyl-CoA is converted to THN by polyketide synthase Mcl17, whereas DMAPP and GPP are coupled by prenyl diphosphate synthase Mcl22 to afford ISLPP. The two precursors are then combined by prenyltransferase Mcl23 forming premerochlorin, which is further converted to merochlorins A and B by the vanadium-dependent haloperoxidase Mcl24. Mcl24 was furthermore recently shown to catalyze an unusual α-hydroketone rearrangement, which may occur en route to merochlorins C and D. Merochlorin C formation may require an additional vanadium-dependent haloperoxidase such as Mcl40. (B) Cyclolavandulyl diphosphate (CLPP) is formed by CLDS and then further incorporated into various bacterial products (e.g., lavanducyanin, lavanduquinocin) and possibly in a similar fashion into plant metabolites such as seselinonol.

The merochlorin biosynthetic gene cluster was identified and investigated shortly after and has a length of 57.6 kb, comprising a total of 41 open reading frames that encode the biosynthetic machinery, which allows for formation of various distinct merochlorins (A–D) (Kaysser et al., 2012). However, the function of many of the encoded proteins remains unclear. Aside from a few regulatory proteins, transporters, as well as putative enzymes for the biosynthesis of the isoprenoid precursors DMAPP and

GPP (geranyl diphosphate; C10), the gene cluster contained 13 genes with unknown function (Kaysser et al., 2012). Most likely, some are not required for merochlorin formation. This hypothesis was later substantiated by the finding that only the four enzymes Mcl17, Mcl22, Mcl23, and Mcl24 are required to convert the central metabolites DMAPP, GPP, and malonyl-CoA into the intricate and highly complex merochlorins A and B (Teufel et al., 2014).

Yet, how is such a remarkable series of biotransformation achieved that affords the structurally complex halogenated merochlorins A, B, C, and D? The naphthoquinone core of the merochlorins and other meroterpenoids derives from tetrahydroxynaphthalene (THN) that is assembled by a single enzyme, a chalcone synthase-like type III polyketide synthase (Mcl17 for merochlorins) that uses malonyl-coenzyme A as sole substrate. THN then becomes prenylated via regioselective electrophilic aromatic substitutions catalyzed by dedicated ABBA prenyltransferases, which utilize pathway-specific prenyl diphosphate precursors. In the case of the merochlorins, the transfer of a sesquiterpene is catalyzed by Mcl23 and gives rise to premerochlorin (Teufel et al., 2014). Further tailoring reactions eventually allow formation of the mature meroterpenoids. For merochlorins A and B, the vanadium-dependent haloperoxidase Mcl24 is remarkably responsible for exclusively converting premerochlorin into the mature compounds by catalyzing chlorination, oxidative dearomatization, and cyclization reactions (Fig. 2A; Diethelm, Teufel, Kaysser, & Moore, 2014).

While Mcl23 prenylates the nucleophilic C4 of the THN core en route to merochlorins A and B, the merochlorins C and D feature a prenyl moiety connected to the C3—a nonnucleophilic site that is not prone to undergo alkylation. Recently, a rationale for this unusual C3 prenylation pattern was provided by the discovery that Mcl24 not only catalyzes the conversion of premerochlorin into merochlorins A and B, but also an exceptional α-hydroxyketone rearrangement resulting in the migration of the C4 prenyl moiety to C3 that most likely sets up formation of merochlorins C and D (Fig. 2A; Diethelm et al., 2014; Miles et al., 2017).

Another salient feature of the merochlorins is the highly unusual C(1′)–C(2) branched sesquiterpene itself that has a unique double-bond arrangement. As will be described in more detail later, the prenyl diphosphate synthase Mcl22 generates this C15 prenyl diphosphate, which was named isosesquilavandulyl diphosphate and serves as substrate for Mcl23 (Fig. 1D; Teufel et al., 2014). The following sections present Mcl22 as a model enzyme for head-to-torso type prenyl diphosphate condensations.

3. Mcl22 IS AN ISLPPS

In the merochlorin biosynthetic gene cluster, *mcl22* is the only gene predicted to encode an enzyme with weak homology to *cis*-prenyl diphosphate synthases. Gene deletion of *mcl22* in a heterologous *Streptomyces* host strain expressing the entire merochlorin biosynthetic gene cluster resulted in a complete abolishment of merochlorin production, underlining its pivotal role in biosynthesis (Teufel et al., 2014). Based on the chemical structure of the prenyl moiety of the merochlorins, we predicted GPP and DMAPP as the most likely substrates. Accordingly, Mcl22 activity was tested in enzymatic assays using GPP and DMAPP as substrates, as described later. Indeed, a single product peak was observed by HPLC, which was identified and named as ISLPP by extensive 1D and 2D NMR analyses, HR-MS, as well as comparison with chemically synthesized ISLPP. Enzymatic activity of Mcl22 was furthermore shown to be boosted fourfold by addition of Mg^{2+} to a k_{cat} of $\sim$20 min^{-1}. In contrast, the metal chelation by addition of ethylenediaminetetraacetate (EDTA) completely abolished ISLPP formation, suggesting that Mg^{2+} is crucial for PP_i elimination during catalysis by masking its negative charges and activating it as leaving group similar to other prenyl diphosphate synthases.

Notably, Mcl22 utilizes DMAPP instead of the common IPP for reaction with an allylic acceptor molecule (GPP), which implies an unusual mechanism, i.e., nucleophilic attack on GPP by C2 of DMAPP and subsequent proton elimination from the same carbon, which can be rationalized via a cationic intermediate (Fig. 1D). To further confirm this proposal, deuterated DMAPP analogues were chemically prepared (according to Thulasiram, Phan, Rivera, & Poulter, 2006) and used in enzymatic assays together with GPP. Indeed, the deuterium label was retained in ISLPP for all proffered deuterated substrates apart from the C2-labeled DMAPP that afforded unlabeled ISLPP as verified by HR-MS, consistent with the proposed mechanism.

To further investigate the catalytic potential of Mcl22, assays were conducted with alternative prenyl diphosphate substrates and analyzed by HPLC and HR-MS. For that, GPP was replaced with the alternative electrophilic acceptor molecules farnesyl diphosphate (FPP; C15) or geranylgeranyl diphosphate (GGPP; C20). In further assays, the nucleophile DMAPP was replaced with IPP or GPP. Finally, DMAPP or GPP were tested as sole substrates serving as both nucleophile and electrophilic acceptor.

Indeed, Mcl22 was able to catalyze the in vitro condensation of DMAPP, FPP (C15), or GGPP (C20) with one molecule of DMAPP, as determined by HR-MS data, suggesting formation of the corresponding branched C10, C20, and C25 prenyl diphosphates. IPP was furthermore condensed with GPP, whereas incubation with GPP alone did not afford a branched C20 terpene diphosphate but rather led to partial dephosphorylation of GPP to geraniol. All relative activities for nonnative substrates were $<1\%$.

4. HETEROLOGOUS PRODUCTION AND PURIFICATION OF Mcl22

The cloning of *mcl22*, heterologous production and purification of the enzyme, was conducted according to published procedures (Teufel et al., 2014):

- The *mcl22* gene (amplified by PCR using fosmid 6B10 as DNA template; Kaysser et al., 2012) was inserted between the *Nco*I and *Bam*HI restriction sites of the expression vector pHIS8 (Jez, Ferrer, Bowman, Dixon, & Noel, 2000), derived from pET28a.
- *Escherichia coli* BL21 (DE3) transformed with the pHIS8-EncM plasmid was grown at 28°C in terrific broth (TB) medium containing 50 mg/mL kanamycin until an optical density of about 0.5 was reached. 40–200 μ*M* isopropyl-β-D-thiogalactoside (IPTG) was then added to induce recombinant protein expression under the control of T7 RNA polymerase and a modified lac promoter.
- After induction, cells were additionally grown for 24 h at 28°C and harvested by centrifugation. Cell pellets were resuspended in lysis buffer (20 m*M* HEPES-Na^+ pH 7.8, 300 m*M* NaCl, 10% (v/v) glycerol, and 30 m*M* imidazole), and then lysed by sonication (10 cycles for 15 s) on ice.
- After centrifugation, the supernatant was passed over a Ni^{2+}-nitrilotriacetate column using a FPLC system. Following a wash step with 100 m*M* imidazole to remove unspecifically bound protein, the N-terminal octahistidine-tagged Mcl22 was eluted with lysis buffer supplemented with 500 m*M* imidazole. Mcl22 proved to be soluble and stable, allowing to obtain high yields of the enzyme without further adjustment of the conditions. Mcl22 was then desalted and concentrated using PD-10 and Vivaspin 6 (10 or 30 kDa exclusion size) columns (both GE Healthcare), respectively.

- The protein concentration of purified Mcl22 was determined by a Bradford assay using bovine serum albumin as standard and the purity verified by sodium dodecyl sulfate 4%–12% polyacrylamide gel electrophoresis using Coomassie brilliant blue R-250 staining. Typically, from 1 L of TB media, ~20–40 mg of His-tagged Mcl22 was isolated and then stored at −20°C or −80°C in 20 m*M* Hepes pH 7.8, 300 m*M* NaCl, and 10%–50% glycerol until use. Mcl22 remained stable over several months under these conditions and no significant loss of activity could be observed.

For crystallization, Mcl22 was prepared from a synthetic gene. For that, *mcl22* from *Streptomyces* sp. CNH-189 was chemically synthesized, amplified by PCR, and ligated into the plasmid pET46Ek/LIC. Expression was conducted with *E. coli* BL21 (DE3). Purification of Mcl22 was performed using Ni^{2+}-nitrilotriacetate and DEAE sepharose columns (Gao et al., 2018).

5. ENZYMATIC AND CHEMICAL SYNTHESIS OF ISOSESQUILAVANDULYL DIPHOSPHATE

To investigate the catalytic features, substrate specificity, and role of Mcl22 in merochlorin biosynthesis, the purified enzyme was incubated with various commercially available prenyl diphosphate substrates (purchased from Sigma-Aldrich) in small-scale in vitro assays (0.2 mL) containing:

- 10% (v/v) glycerol, 150 m*M* NaCl, and 20 m*M* HEPES-Na^+ (pH 7.8)
- 0.5–1 m*M* GPP, 0.5–1 m*M* DMAPP
- 5 m*M* $MgCl_2$
- 10 m*M* EDTA (optional, as control)
- alternative substrates in separate assays (2 m*M* each):
  (a) instead of DMAPP: GPP or IPP
  (b) instead of GPP: DMAPP, FPP, or GGPP
- Mcl22 concentration of 0.62 μm (for assays using GPP and DMAPP) or 220 μ*M* (for nonnative substrates IPP, FPP, GGPP, or when either only DMAPP or only GPP were used)

Assays were conducted at 22°C for 1 h with authentic substrates or for 2 h with nonnative substrates. Reactions were quenched with MeOH, centrifuged, and the supernatants analyzed by HPLC/HRMS as discussed later in greater detail.

To confirm the nature of the Mcl22 product (in addition to HR-MS and NMR analysis), ISLPP was chemically synthesized via intermediates **1** and **2** (see Fig. 3) according to the following protocol (as published in Teufel et al., 2014):

- Synthesis of **1**: $CuBr{\cdot}SMe_2$ (1.9 g, 9.4 mmol) was suspended in THF (60 mL) and then cooled to 0°C, before methyl lithium (12 mL, 18.8 mmol; 1.6 *M* in Et_2O) was added slowly. The solution was cooled to −78°C, ethyl 2-butynoate (1.1 mL, 9.4 mmol) was added and then stirred at −78°C for 1 h before addition of commercially available geranyl bromide (4.1 g, 18.8 mmol). The mixture was warmed to 0°C and stirred for another 2 h before quenching by addition of saturated NH_4Cl and dilution with EtOAc. After phase separation, the aqueous phase was 3 × extracted with EtOAc. The combined organic phases were washed with brine and dried over $MgSO_4$, the solvent evaporated, and the residue purified via flash column chromatography (hexanes/EtOAc 40:1 → 9:1) to give **1** as a colorless liquid (2.28 g, 8.6 mmol, 91%).
- Synthesis of **2**: DIBAL-H (26 mL, 26 mmol; 1 *M* in hexanes) was added to a solution of ester **1** (2.28 g, 8.6 mmol) in CH_2Cl_2 (50 mL) at −78°C and the reaction stirred for 2 h at −78°C before quenching with EtOAc and saturated NH_4Cl. After phase separation, the aqueous phase was extracted 3 × with EtOAc and the combined organic phases washed with brine and dried over $MgSO_4$. The solvent was evaporated and the residue purified via flash column chromatography (hexanes/EtOAc 9:1), affording **2** as colorless liquid (1.44 g, 6.5 mmol, 75%).
- Synthesis of ISLPP: PBr_3 (25 μL, 0.27 mmol) was added to a solution of **2** (50 mg, 0.22 mmol) in THF (2 mL) at 0°C and the mixture stirred at 0°C for 45 min before quenching with ice water. The aqueous phase was extracted 3 × with hexane, the combined organic phases washed with brine, dried over $MgSO_4$, the solvent evaporated, and the crude residue dissolved in MeCN (2 mL). Tris(tetrabutylammonium) hydrogen pyrophosphate (400 mg, 0.44 mmol) was added and the mixture was stirred over night before removal of the solvent and purification of the crude product via flash column chromatography (i-PrOH/aq. NH_3/H_2O 6:2:0 → 6:3:1) and preparative HPLC (eluent: 25 m*M* NH_4HCO_3/MeCN 95:5 → 5:95) affording ISLPP as white solid (35 mg, 0.09 mmol, 42%), as confirmed by proton-NMR and ^{13}C NMR (see Teufel et al., 2014 for full MS, NMR, and IR datasets).

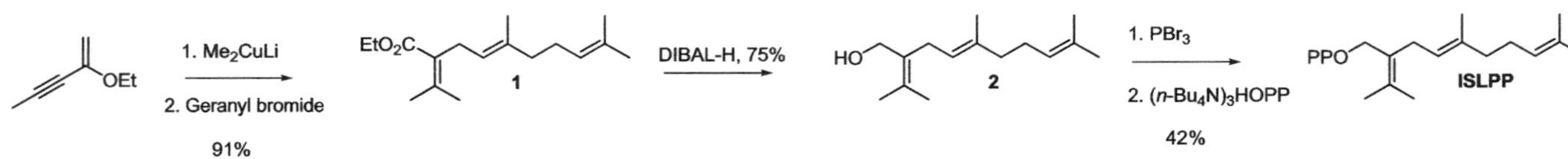

Fig. 3 Chemical synthesis of ISLPP as reported in Teufel et al. (2014). Steps are described in more detail in the main text.

6. HPLC ANALYSIS OF ENZYMATICALLY AND CHEMICALLY PRODUCED ISOSESQUILAVANDULYL DIPHOSPHATE AND OTHER PRENYL DIPHOSPHATES

For analysis of the Mcl22 assays, analytical reversed-phase HPLC analysis was applied using an Agilent 1200 HPLC with a photodiode array detector. Conditions were as follows:

- As column, a 3.5-μm ZORBAX Extend-C18 (150 × 4.6 mm, Agilent Technologies, USA) was employed.
- Water was used as liquid phase buffered with 25 m*M* NH_4HCO_3 (pH 8.0) at a flow rate of 0.7 mL min^{-1}.
- After 3 min, the buffer was gradually exchanged for MeCN using a linear gradient from 0% to 100% (v/v) MeCN over 15 min.
- Semipreparative reversed-phase HPLC for the purification of enzymatically produced ISLPP (for structural elucidation by NMR) was similarly performed using an Agilent 1260 Infinity HPLC with a 5-μ*M* C18(2) column (250 × 10.00 mm, Phenomenex, USA) and a NH_4HCO_3-buffered MeCN gradient.
- Prenyl diphosphates were quantified based on their absorption through peak integration at 214 nm using a standard curve.

Notes

(1) In contrast to long-chain prenyl diphosphates (≥C10), the C5 prenyl diphosphates IPP and DMAPP did not bind sufficiently well to the C18 columns and thus could not be quantified with the described methods. Other columns or methods (e.g., GC-MS) should be applied when analysis and quantification of these compounds are required.

(2) For peak shape and separation, it was critical to inject and analyze small amounts of the prenyl diphosphates (ideally ~1 nmol) to prevent extensive peak tailing and splitting. For the same reason, NH_4HCO_3 was added, which improves the peak shape of prenyl diphosphates (Zhang & Poulter, 1993).

7. CYCLOLAVANDULYL DIPHOSPHATE SYNTHASE (CLDS) SYNTHESIZES CYCLOLAVANDULYL DIPHOSPHATE (CLPP) BY CATALYZING BOTH TERPENE BRANCHING AND CYCLIZATION

Aside from the merochlorins, there are various other bacterial and plant secondary metabolites that feature monoterpene moieties with similar branching patterns such as lavanducyanin, lavanduquinocin, or seselinonol among others. Presumably, two DMAPP molecules undergo head-to-torso [C(1′)–C(2)] condensation and subsequent cyclization. In this context, the enzyme CLDS from *Streptomyces* sp. CL190 was recently identified, which is homologous to Mcl22 but not only catalyzes C(1′)–C(2) branching, but also the subsequent cyclization ultimately producing cyclolavandulyl, which is then incorporated into various bacterial natural products (e.g., lavanducyanin) (Ozaki, Zhao, Shinada, Nishiyama, & Kuzuyama, 2014; Fig. 2B). Usually, such a cyclization requires a separate terpene cyclase or fusion proteins with distinct C-terminal and N-terminal domains for terpene condensation and cyclization, respectively. Hence, CLDS is the first prenyl diphosphate synthase that catalyzes both condensation as well as the subsequent cyclization. A hitherto unidentified prenyltransferase then fuses the cyclolavandulyl moiety with a phenazine backbone to afford lavanducyanin. Interestingly, Mcl22 and CLDS share 67% sequence identity. Similar to Mcl22, CLDS also depends on the presence of divalent cations showing the highest activity with Mg^{2+}.

8. Mcl22 (ISLPPS), CLDS, AND LPPS ARE STRUCTURALLY RELATED TO *CIS*-PRENYL DIPHOSPHATE SYNTHASES

In addition to Mcl22 and CLDS, the recently characterized plant enzyme LPPS also catalyzes the head-to-torso C(1′)–C(2) condensation of two DMAPPs, thus affording LPP (Demissie, Erland, Rheault, & Mahmoud, 2013; Fig. 1B). All three enzymes show weak homology to *cis*-prenyl diphosphate synthases. Recently, their protein structures were elucidated via X-ray crystallography, showing that indeed all three enzymes adopt the typical butterfly-like homodimer structure of *cis*-prenyl diphosphate synthases (Fig. 4; Gao et al., 2018; Liu et al., 2016; Tomita et al., 2017). CLDS and LPPS share active sites and two putative DMAPP-binding sites (S1 and S2) with highly similar architecture. Both enzymes have distinct hydrophobic active site pockets, which, however, slightly differ in size and shape; e.g., CLDS has a narrower active site pocket caused by the bulky side-chain residues of Tyr26 and Phe68 (corresponding to His93 and Leu135, respectively, in LPPS) that protrude into the active site. Modeling of DMAPP into the active site of CLDS suggested a suitable distance of ~3.1 Å between the carbons from the two DMAPPs at S1 and S2 that undergo C–C bond formation via head-to-torso condensation. Notably, the isoleucyl side chain of Ile75

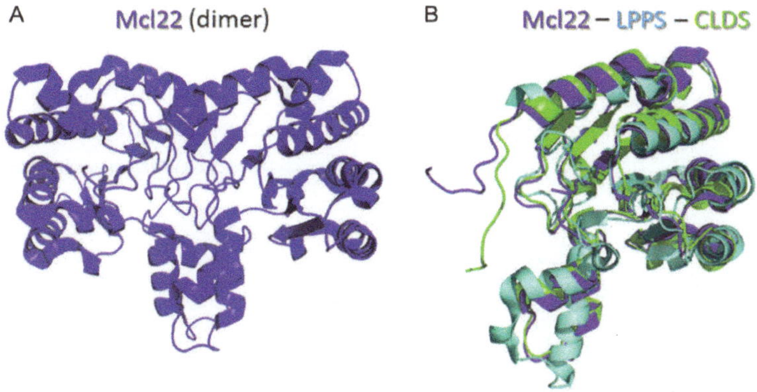

Fig. 4 Structures of Mcl22 (*blue*; pdb code: 5xk3), CLDS (*green*; pdb code: 5guk), and LPPS (*teal*; pdb code: 5hc6). All shown enzymes are homologous to *cis*-prenyl diphosphate synthases and form homodimers. (A) Butterfly-shaped homodimer of Mcl22. (B) Alignment of monomers of Mcl22, CLDS, and LPPS showing the same fold and high structural similarity.

in LPPS is inserted between the two DMAPPs, whereas the corresponding Pro8 of CLDS lies next to the two DMAPP molecules and therefore allows an intramolecular proton transfer between the dimethyl moieties of the two DMAPPs required for the unique cyclization reaction of CLDS (Tomita et al., 2017).

Mcl22 sets itself apart from CLDS and LPPS by catalyzing the head-to-torso coupling of DMAPP with GPP instead of two DMAPPs. The GPP molecule is furthermore not bound in the usual S1 site, as observed in other *cis*-prenyl diphosphate synthases, which process lengthier substrates (e.g., the archetypical *cis*-undecaprenyl diphosphate synthase). Normally, two α-helices form a hydrophobic tunnel in undecaprenyl diphosphate synthase and related enzymes to accommodate the lengthy isoprene moieties of their substrates (Gao et al., 2018). Mcl22, in contrast, binds GPP in a separate hydrophobic pocket at an angle of ~90 degrees compared to the substrates of these structurally related enzymes. This substrate-binding mode is caused by the markedly shorter distance between the two corresponding α-helices in Mcl22 that sterically preclude GPP binding. The substrate-binding mode observed for Mcl22 furthermore suggests that the proposed transient C1 carbocation intermediate (arising from the initial nucleophilic attack of DMAPP on GPP) is quenched by the P1 phosphate-oxygen via deprotonation at C2 (distance of 3.5 Å). The proton is then further relayed to N57 and S54 and thus removed from the active site, consistent with results from mutagenesis of these residues. Notably, in LPPS, one of the methyl groups of DMAPP (rather than the C2) is closest to the P1 phosphate-oxygen, which presumably forms a similar proton relay with the corresponding N124 and S121 residues. This may provide a rationale for the different double-bond arrangement observed in the final branched prenyl diphosphate products (Gao et al., 2018).

9. SUMMARY AND CONCLUSION

In this chapter, I described three recently discovered noncanonical prenyl diphosphate synthases with a focus on the discovery and characterization of the ISLPPS Mcl22 from merochlorin biosynthesis. While terpenoids exhibit a large structural diversity mostly arising from dedicated pathway-specific terpene cyclases, the backbone is commonly synthesized via head-to-tail coupling of IPP with one or several molecules of DMAPP. ISLPPS, CLDS, and LPPS in contrast catalyze noncanonical "head-to-torso" couplings of prenyl diphosphates as a new strategy for the cyclase-independent structural diversification of terpene natural products. Recent structural studies of these enzymes suggest that minor modifications in the active sites alter the substrate binding and positioning of the two respective diprenyl diphosphate substrates and thus account for their unusual terpene coupling chemistry. These insights ultimately provide a more substantial basis for future rational design of these intriguing enzymes and possibly for the generation of novel branched terpenes.

ACKNOWLEDGMENTS

I want to thank all colleagues and collaborators who participated in the discovery and elucidation of the merochlorin biosynthetic pathway and the DFG for funding (Grant: TE 931/2-1).

REFERENCES

Demissie, A., Erland, L. A., Rheault, M. R., & Mahmoud, S. S. (2013). The biosynthetic origin of irregular monoterpenes in Lavandula: Isolation and biochemical characterization of a novel *cis*-prenyl diphosphate synthase gene, lavandulyl diphosphate synthase. *Journal of Biological Chemistry*, *288*, 6333–6341.

Dickschat, J. S. (2016). Bacterial terpene cyclases. *Natural Product Reports*, *33*, 87–110.

Diethelm, S., Teufel, R., Kaysser, L., & Moore, B. S. (2014). A multitasking vanadium-dependent chloroperoxidase as an inspiration for the chemical synthesis of the merochlorins. *Angewandte Chemie International Edition*, *53*, 11023–11026.

Fujihashi, M., Zhang, Y.-W., Higuchi, Y., Li, X.-Y., Koyama, T., & Miki, K. (2001). Crystal structure of cis-prenyl chain elongating enzyme, undecaprenyl diphosphate synthase. *Proceedings of the National Academy of Sciences of the United States of America*, *98*, 4337–4342.

Gao, J., Ko, T.-P., Chen, L., Malwal, S. R., Zhang, J., Hu, X., et al. (2018). "Head-to-middle" and "head-to-tail" *cis*-prenyl transferases: Structure of isosesquilavandulyl diphosphate synthase. *Angewandte Chemie International Edition*, *57*, 683–687.

Jez, J. M., Ferrer, J. L., Bowman, M. E., Dixon, R. A., & Noel, J. P. (2000). Dissection of malonyl-CoA decarboxylation from polyketide formation in the reaction mechanism of a plant polyketide synthase. *Biochemistry*, *39*, 890–902.

Kaysser, L., Bernhardt, P., Namm, S. J., Loesgen, S., Ruby, J. G., Skewes-Cox, P., et al. (2012). Merochlorins A-D, cyclic meroterpenoid antibiotics biosynthesized in divergent pathways with vanadium-dependent chloroperoxidases. *Journal of the American Chemical Society*, *134*, 11988–11991.

Liu, M., Chen, C. C., Chen, L., Xiao, X., Zheng, Y., Huang, J. W., et al. (2016). Structure and function of a "head-to-middle" prenyltransferase: Lavandulyl diphosphate synthase. *Angewandte Chemie International Edition*, *55*, 4721–4724.

Miles, Z. D., Diethelm, S., Pepper, H. P., Huang, D. M., George, J. H., & Moore, B. S. (2017). A unifying paradigm for naphthoquinone-based meroterpenoid (bio)synthesis. *Nature Chemistry*, *9*, 1235–1242.

Ozaki, T., Zhao, P., Shinada, T., Nishiyama, M., & Kuzuyama, T. (2014). Cyclolavandulyl skeleton biosynthesis via both condensation and cyclization catalyzed by an unprecedented member of the *cis*-isoprenyl diphosphate synthase superfamily. *Journal of the American Chemical Society*, *136*, 4837–4840.

Rivera, B., Swedlund, B. D., King, G. J., Bell, R. N., Hussey, C. E., Jr., Shattuck-Eidens,-D. M., et al. (2001). Chrysanthemyl diphosphate synthase: Isolation of the gene and characterization of the recombinant non-head-to-tail monoterpene synthase from *Chrysanthemum cinerariaefolium*. *Proceedings of the National Academy of Sciences of the United States of America*, *98*, 4373–4378.

Sakoulas, G., Nam, S. J., Loesgen, S., Fenical, W., Jensen, P. R., Nizet, V., et al. (2012). Novel bacterial metabolite merochlorin A demonstrates in vitro activity against multi-drug resistant methicillin-resistant *Staphylococcus aureus*. *PLoS One*, 7(1), e29439.

Teufel, R., Kaysser, L., Villaume, M. T., Diethelm, S., Carbullido, M. K., Baran, P. S., et al. (2014). One-pot enzymatic synthesis of merochlorin A and B. *Angewandte Chemie International Edition*, *53*, 11019–11022.

Thulasiram, H. V., Erickson, H. K., & Poulter, C. D. (2008). A common mechanism for branching, cyclopropanation, and cyclobutanation reactions in the isoprenoid biosynthetic pathway. *Journal of the American Chemical Society*, *130*, 1966–1971.

Thulasiram, H. V., Phan, R. M., Rivera, S. B., & Poulter, C. D. (2006). Synthesis of deuterium-labeled derivatives of dimethylallyl diphosphate. *Journal of Organic Chemistry*, *71*, 1739–1741.

Tomita, T., Kobayashi, M., Karita, Y., Yasuno, Y., Shinada, T., Nishiyama, M., et al. (2017). Structure and mechanism of the monoterpene cyclolavandulyl diphosphate synthase that catalyzes consecutive condensation and cyclization. *Angewandte Chemie International Edition*, *129*, 15109–15113.

Zhang, D., & Poulter, C. D. (1993). Analysis and purification of phosphorylated isoprenoids by reversed-phase HPLC. *Analytical Biochemistry*, *213*, 356–361.

CHAPTER FIFTEEN

Identification of Enzymes Involved in Sesterterpene Biosynthesis in Marine Fungi

Jingjing Yan, Jingjing Guo, Wei Yuan, Wanying Mai, Kui Hong[1]
Key Laboratory of Combinatorial Biosynthesis and Drug Discovery, Ministry of Education, and Wuhan University School of Pharmaceutical Sciences, Wuhan University, Wuhan, China
[1]Corresponding author: e-mail address: kuihong31@whu.edu.cn

Contents

Abstract

Diverse and bioactive terpene compounds are produced by marine fungi. The biosynthesis of sesterterpene molecules containing a C25 skeleton is catalyzed by chimeric enzymes that carry out both chain elongation and terpene cyclization reactions. These bifunctional characteristics facilitate using different chain length polyisoprenoid

Methods in Enzymology, Volume 604
ISSN 0076-6879
https://doi.org/10.1016/bs.mie.2018.04.023

diphosphates as precursors to yield the geranylfarnesyl diphosphate intermediate, which is then converted to a sesterterpene in one step. In this chapter, we describe the identification of sesterterpene synthase enzymes, together with other related enzymes such as diterpene and farnesyl diphosphate synthases, in a single fungal strain. The processes are based on genome sequencing, in silico analysis of terpene synthase, in vivo gene (cluster) deletion and complementation, and in vitro protein function verification, together with the methods of detecting terpenes using high-performance liquid chromatography (HPLC) and gas chromatography-mass spectrometry (GC-MS).

1. INTRODUCTION

Sesterterpenes are terpene molecules containing a C25 skeleton, which are rare among terpene compounds (Wang, Yang, Lin, Zhou, & Liu, 2013). Many of them are reported from marine fungi, especially those from mangroves, which include neomangicols A–C and mangicols A–G from the Bahamas mangrove fungus *Fusarium* sp. (Renner, Jensen, & Fenical, 1998; Renner, Jensen, & Fenical, 2000), asperterpenoid A from a south China mangrove endophytic fungus *Aspergillus* sp. (Huang et al., 2013), ophiobolin O produced by *Aspergillus ustus* 094102 from south China mangrove rhizosphere (Yang et al., 2012), and ophiobolins G and K also from marine fungi (Tian, Deng, & Hong, 2017) (Fig. 1).

The biosynthesis of terpenes results from the condensation of five-carbon isoprene building blocks of dimethylallyl diphosphate (DMAPP) and isopentenyl diphosphate (IPP). Prenyltransferases (PTs) catalyze chain elongation from these building blocks to yield polyisoprenoid diphosphates such as geranyl diphosphate (GPP, C10), farnesyl diphosphate (FPP, C15), geranylgeranyl diphosphate (GGPP, C20), geranylfarnesyl diphosphate (GFPP, C25), and hexaprenyl diphosphate (HexPP, C30). Terpene synthases (TS) (or terpene cyclases (TC)) catalyze the cyclization or rearrangement of these polyisoprenoid diphosphates to yield monoterpenes, sesquiterpenes, diterpenes, sesterterpenes and triterpenes, respectively (Christianson, 2007; Quin, Flynn, & Schmidt-Dannert, 2014). Terpene diversity is facilitated by TS (or TC), the catalysis mechanisms of which are very complex (Gao, Honzatko, & Peters, 2012). In the past decade, a new type of bifunctional TC attracted attention of scientists. They possess two catalytically independent domains of PT and TC, and carry out both chain elongation and terpene cyclization reactions. The reported enzymes catalyzed biosynthesis of diterpenes (Minami et al., 2009; Qin et al., 2016; Toyomasu et al., 2007) and sesterterpenes (Chai et al., 2016; Chiba, Minami, Gomi, & Oikawa, 2013; Matsuda, Mitsuhashi, Quan, &

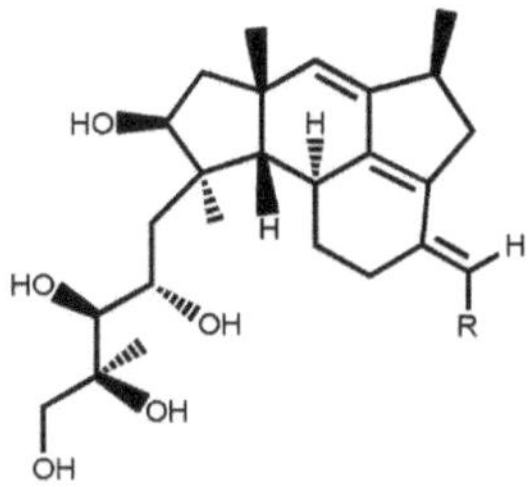

Neomangicols A R=Cl
Neomangicols B R=Br

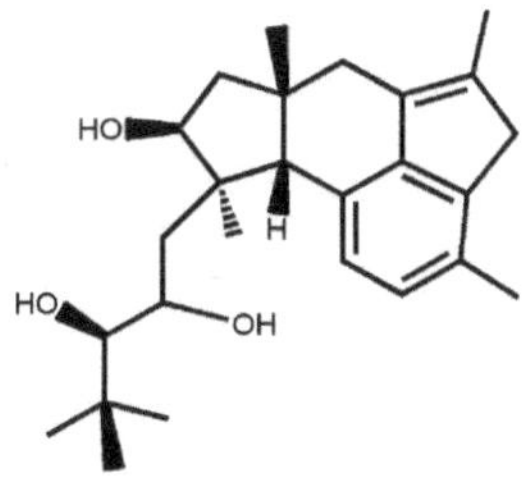

Neomangicols C

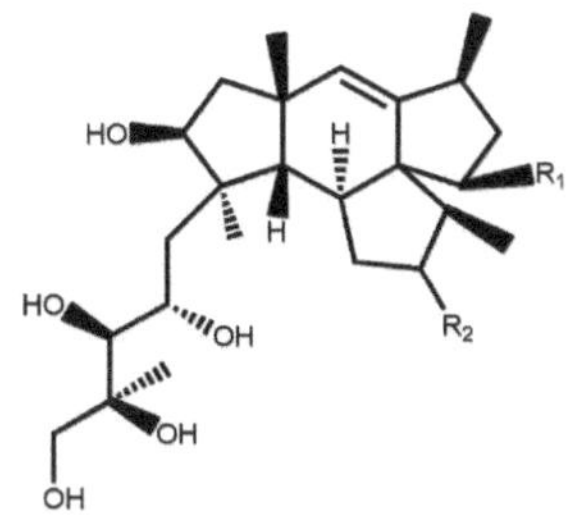

Mangicols A R_1=OH, R_2=H
Mangicols B R_1=H, R_2=OH
Mangicols D R_1=H, R_2=H

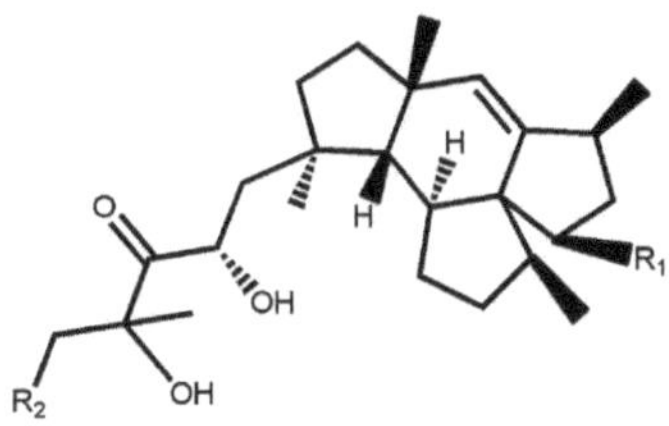

Mangicols C R_1=OH, R_2=H
Mangicols E R_1=H, R_2=OH
Mangicols F R_1=H, R_2=H

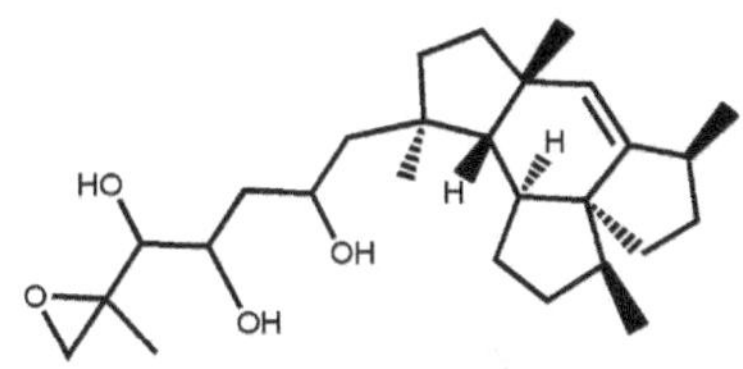

Mangicols G

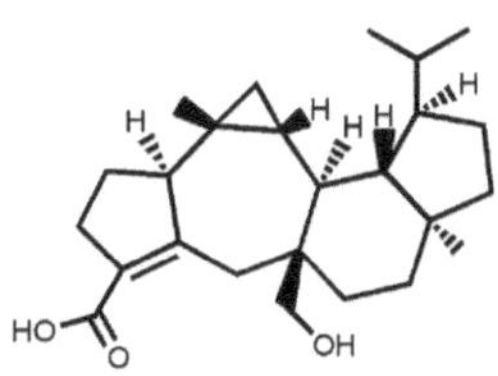

Asperterpenoid A

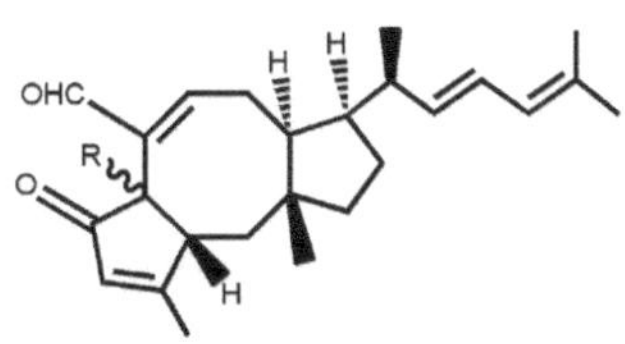

Ophiobolin G R=β-H ^{16cis}
6-Epi ophiobolin G R=α-H^{16cis}

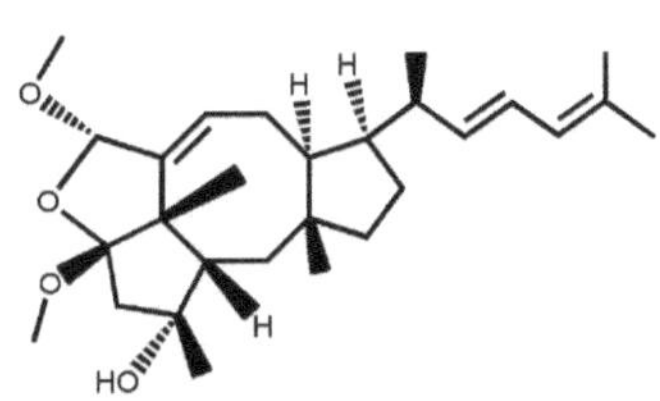

Ophiobolin O

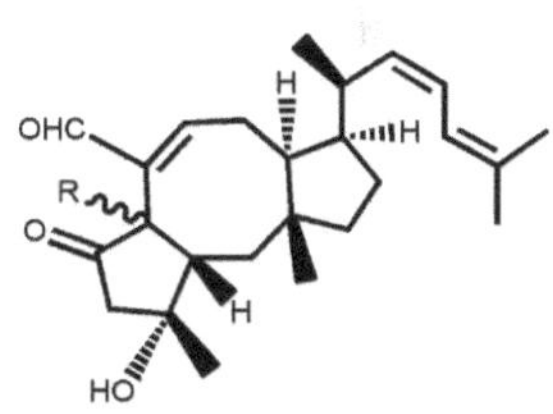

Ophiobolin K R=β-H
6-Epi ophiobolin K R=α-H

Fig. 1 Sesterterpenes produced by marine fungi.

Abe, 2015; Mitsuhashi, Rinkel, Okada, Abe, & Dickschat, 2017; Ye et al., 2015).

In filamentous fungi, genes coding for the enzymes that catalyze secondary metabolites (SM) synthesis, together with those coding for specific regulatory functions and resistance proteins, are usually contiguously aligned in the genome (Hoffmeister & Keller, 2007; Keller, Turner, & Bennett, 2005; Quin et al., 2014). Biosynthetic gene clusters responsible for the production of terpenes can be rapidly identified by genome sequencing and bioinformatics analysis (Quin et al., 2014). Using the program BLASTP or Profile Hidden Markov models, conserved protein sequences, which are a characteristic set of universally conserved functional domains typical of all terpene synthases, can be used to mine and identify terpene synthases. The two universally conserved domains are the aspartate-rich DDXX(D/E) motif and the NSE/DTE triad, (N/D)DXX(S/T)XX(K/R)(D/E) (Cane & Ikeda, 2012). The true biochemical function of a presumed terpene synthase can then be affirmed by experimentation. The approaches are (i) in vivo gene (gene cluster) knockout and complementation (Chai et al., 2016; Gao et al., 2011); (ii) incubation of the expressed recombinant protein with acyclic, allylic diphosphate substrates and identification of the resultant terpene hydrocarbon or alcohol (Chai et al., 2016); and (iii) in vivo expression in heterologous hosts, i.e., *Escherichia coli* (Minami et al., 2009), or other fungi such as *Aspergillus oryzae* (Chiba et al., 2013; Qin et al., 2016; Ye et al., 2015), that can support the production of terpene metabolites.

C25 sesterterpene synthases were discovered only in recent years. Ophiobolin F synthase (AcOS) was found by accident during genome mining for diterpene synthase from *Aspergillus clavatus* (Chiba et al., 2013). Four other sesterterpene synthases were reported thereafter by genome mining, including NfSS for sesterfisherol biosynthesis from *Neosartorya fischeri* (Ye et al., 2015), EvSS for stellatic acid biosynthesis from *Emericella variecolor* (Matsuda et al., 2015), PbSS of sesterbrasiliatriene synthase from *Penicillium brasilianum*, and PvPS of preasperterpenoid A synthase from *Penicillium verruculosum* synthase (Mitsuhashi et al., 2017).

A. ustus 094102 was found to produce sesterterpene ophiobolin and three other kinds of terpenoids, including the sesquiterpenoids drimane, isochromane, and sterols and dipeptides (Lu et al., 2009; Yang et al., 2012). The biosynthesis of sesterterpene ophiobolins was found involving five gene clusters related to C15, C20, C25, and C30 terpenoid biosynthesis (Chai et al., 2016). In this chapter, we describe how to identify sesterterpene biosynthesis enzymes, together with other enzymes such as diterpene and FPP synthases. The processes are based on our former work (Chai et al., 2016) with updated information on genome sequencing, in silico analysis of terpene synthesis, in vivo gene (cluster) deletion and complementation, and in vitro protein function verification, together with the methods of how to detect terpenes using HPLC and GC-MS methods.

2. GENOME SEQUENCING AND BIOINFORMATICS ANALYSIS OF TERPENE-RELATED GENES IN *A. USTUS* 094102

Nowadays whole-genome sequence is becoming more and more important to investigate the biosynthesis pathways of SM (Cane & Ikeda, 2012; Chiba et al., 2013; Matsuda et al., 2015; Ye et al., 2015). After preparing high quality genomic DNA (gDNA) and submitting for sequencing, complete genome information can be readily obtained by gene annotation and assembly.

2.1 Preparation of gDNA

The methods of fungal gDNA extraction include the cetyltrimethylammonium bromide (CTAB) method (Lipp et al., 1999; Zhang, Yang, Castlebury, & Cerniglia, 1996), sodium dodecyl sulfate (SDS) (Goldenberger, Perschil, Ritzler, & Altwegg, 1995), alkaline guanidine thiocyanate boiling (Sandhu, Kline, Stockman, & Roberts, 1995), Triton X-100 lysis (Sinzger, Knapp, Schmidt, Kahl, & Jahn, 1999), microwave (Orsini & Romano-Spica, 2001), and commercial kit methods (Wang, Teng, Tian, Guan, & Wang, 2012).

The gDNA of *A. ustus* 094102 was extracted via the improved CTAB method. The resulting gDNA (Fig. 2) for whole-genome sequencing was submitted to the Beijing Genomics Institution (BGI) in Shenzhen, China.

2.1.1 Buffers and Reagents

- Sodium chloride–EDTA–Tris (SET) buffer: 75 m*M* NaCl, 25 m*M* EDTA (pH 8.0), 20 m*M* Tris–HCl (pH 7.5).
- Liquid nitrogen.

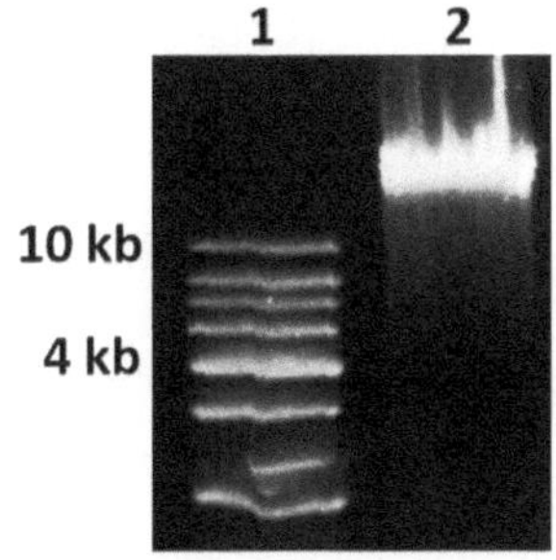

Fig. 2 The genomic DNA of *A. ustus* 094102 with 1.0% agarose gel electrophoresis. Lane 1: 1 kb plus DNA ladder; lane 2: genomic DNA extracted with CTAB method.

- 2% CTAB: 1 *M* Tris–HCl (pH 8.0) 50 mL; 0.5 *M* EDTA (pH 8.0) 20 mL; NaCl 40.908 g; CTAB 10 g; add ddH_2O to 500 mL, store in brown flask, autoclave, add 10 mL beta-mercaptoethanol (β-ME) before use.
- Phenol/chloroform/isoamyl alcohol (v/v/v: 25/24/1), isopropanol, 70% ethyl alcohol.
- 1 kb DNA ladder (Fermentas).
- RNase A (Thermo Scientific, 10 mg/mL).
- Tris–EDTA (TE) buffer: 25 m*M* EDTA (pH 8.0); 25 m*M* Tris–HCl (pH 7.5).
- 3 *M* NaAc (pH 5.2): dissolve 40.8 g $CH_3COONa{\cdot}3H_2O$ in 50 mL ddH_2O, adjust pH to 5.2 with acetic acid, add ddH_2O to 100 mL, autoclave.

All materials are used for extraction of gDNA of *A. ustus* 094102.

2.1.2 Procedure

1. Inoculate fresh mycelium in 100 mL fungal II medium (mannitol 20 g; yeast extract 3 g; $MgSO_4{\cdot}7H_2O$ 0.3 g; sodium glutamate 10 g; KH_2PO_4 0.5 g; maltose 20 g; corn steep liquor 1 g; glucose 10 g; add ddH_2O to 1 L, and adjust pH to 6.5.) at 28°C, 220 rpm for 2–3 days.
2. Collect the mycelium at 18,000 × *g* for 15 min, and wash it with sterile ddH_2O, then resuspend with SET buffer (the volume is to cover the mycelium). Sop up the water with a filter paper and store at −40°C.
3. Grind 20 mg mycelium with liquid nitrogen, add 400 μL 2% CTAB, 65°C for 30 min.
4. Add the same volume of phenol/chloroform/isoamyl alcohol (v/v/v: 25/24/1), shake well to make the protein fully denatured for 30 min. Centrifuge at 13,000 × *g* for 10 min.
5. Remove the supernatant, add RNase to 20 μg/mL, keep at 37°C for 30 min.
6. Repeat steps 3–5 for 1–2 times.
7. Add two volumes of precooled isopropanol, shake softly, and then separate filamentous DNA out.
8. Remove filamentous DNA, wash twice with 70% ethyl alcohol, and then dry.
9. Add TE buffer to dissolve DNA.
10. Use agarose gel electrophoresis to detect the gDNA with 1% agarose.
11. Store the gDNA at −40°C.

2.1.3 Notes

1. Preheat CTAB to 65°C before use to avoid CTAB precipitation.
2. Sop up the water before grinding and precool the mortar and pestle. Keep cool while grinding to avoid DNase renaturation and grind the mycelium to a powder.
3. Remove the gDNA with cut tips softly to avoid gDNA breakage.
4. Dissolve the gDNA with RNA-free water and add about 0.8 μL 10 mg/mL RNase.

2.2 In Silico Analysis for Terpene Synthesis Genes

Submit *A. ustus* 094102 gDNA for whole-genome sequencing to BGI in Shenzhen, China. Genome assembly can be carried using the short oligonucleotide alignment program denovo 1.05 (Li et al., 2010). The genome size of strain 094102 is 40.09 Mb, assembled as 301 contigs and 174 scaffolds (Table 1).

Software and databases for gene prediction and annotation are generally used as following: gene prediction by defining putative open reading frames (ORFs) with GeneMark-ES 2.3e (Ter–Hovhannisyan, Lomsadze, Chernoff, & Borodovsky, 2008); protein-encoding gene annotation through BLASTP searches against the SwissProt, GO, COG, KEGG, and NR databases. Two bioinformatics tools, SMURF (Khaldi et al., 2010) and antiSMASH (Weber et al., 2015), have been developed based on the common features of SM biosynthetic gene clusters and are most frequently used to predict genes and gene clusters involved in biosynthesis of SM.

Based on bioinformatics analysis, 13,982 protein-coding genes from *A. ustus* 094102 were predicted by BLASTP searches against the SwissProt, GO, COG, KEGG, and NR databases. Eleven terpene gene clusters were identified using antiSMASH analysis. Further manually scanning key words of "terpene synthesis" or "terpene cyclase," using the typical conserved protein sequences aspartate-rich DDXX (D/E) motif and the NSE/DTE triad, (N/D)DXX(S/T)XX(K/R)(D/E), resulted in 27 putative terpene synthesis genes which classified to four groups, including sesquiterpene, diterpene, triterpene synthases, and a "function unknown" group of hypothetical proteins (Table 2).

2.2.1 Software and Database

- GeneMark-ES 2.3e for Gene prediction.
- SwissProt (2012-04), GO (release: 1.419), COG (release: 20090331), KEGG (release: 59), and NR (2012-04) databases for protein-encoding genes annotation.

Table 1 *Aspergillus ustus* 094102 Genome Assembly and Main Feature

Statistic	Value
Coverage	>68 ×
Genome assembly size (Mb)	40.09
Number of scaffolds	174
Scaffold N50 (Kb)	1814
Number of contigs	301
Contig N50 (kb)	675
CDS total length (Mb)	20.44 (50.98% of genome)
CDS average size (bp)	1461.62
Predicted protein-coding genes (Number)	13,982
Predicted protein-coding genes, total length (Mb)	23.19
Predicted protein-coding genes, mean length (kb)	1658.33
Exon total length (Mb)	20.44
Exon total number	43,400
Exon average length (bp)	470.88
Intron total length (Mb)	2.75
Intron total number	29,418
Intron average length (bp)	93.50
GC content (%)	51.22

- GenBank.
- MEGA (version 5.0) for amino acid phylogenetic analysis.
- BioEdit 7.0 for multiple sequence alignments.

2.2.2 Procedure

1. Determine putative ORFs with GeneMark-ES 2.3e.
2. Annotate protein-coding genes through BLASTP searches against the databases of SwissProt (2012-04), GO (release: 1.419), COG (release: 20090331), KEGG (release: 59), and NR (2012-04)
3. Search manually terpene synthesis genes using the key words of "terpenoid synthase" or "terpenoid cyclase" from the local genome database. Then check the typical conserved protein sequences

Table 2 BLASTP of the Presumed 27 Terpene Synthases in *A. ustus* 094102 Genome (done in Oct, 2012)

Presumed AUPTS Type	AUPTS ID (Au)	AA No.	Intron No.	RTS	Max Score	Query Coverage (%)	Max Identity (%)
Di AUPTS	6061	343	0	|BAF45925.1| fusicoccadiene synthase [*Phomopsis amygdali*]	207	97	37
	6241	744	7		288	89	29
	7240	762	8		258	94	28
	8003	725	3		517	100	39
	11189	710	5		292	94	30
	13192	708	8		425	98	36
	13606	755	7		307	85	31
	13676	717	9		294	85	30
Unknown AUPTS	606	387	1	|EAA65430.1| hypothetical protein AN0654.2 [*Aspergillus nidulans* FGSC A4]	696	98	86
	7669	362	2	gb|EAA59165.1| hypothetical protein AN8143.2 [*Aspergillus nidulans* FGSC A4]	593	100	83
	11565	692	7	|EAA64518.1| hypothetical protein AN2407.2 [*Aspergillus nidulans* FGSC A4]	759	99	53
	13624	689	7	|EAU32979.1| conserved hypothetical protein [*Aspergillus terreus* NIH2624]	1036	100	72
	329	336	2	|CAK48316.1| unnamed protein product [*Aspergillus niger*]	529	99	74
	3446	456	1	|EAA63201.1| hypothetical protein AN2767.2 [*Aspergillus nidulans* FGSC A4]	832	98	90
	6064	404	3	|CCD54927.1| BcSTC5, similar to sesquiterpene cyclase [*Botryotinia fuckeliana*]	38.1	18	31
	6298	345	1	|EAA59634.1| hypothetical protein AN8012.2 [*Aspergillus nidulans* FGSC A4]	643	99	88

Note: AUPTS: *Aspergillus ustus* presumed terpene synthases; Prefix "Au" for all the ID of the AUPTS.

aspartate-rich DDXX(D/E) motif and the NSE/DTE triad, (N/D)DXX(S/T)XX(K/R)(D/E).
4. Confirm the search result using BLASTX analysis against the GenBank database.
5. Use the BioEdit 7.0 software to perform multiple sequence alignments.
6. Use MEGA (version 5.0) to construct the phylogenetic analysis-based amino acid sequence.

2.2.3 Notes
1. The updated software and database are recommended for further use, as now the above corresponding software and database have been updated to SwissProt (2012-06), GO (release: 1.419), COG (release: 20090331), KEGG (release: 76), NR (20160219), and MEGA (version 7.0)

3. IN VIVO IDENTIFICATION OF GENE (CLUSTERS) INVOLVED IN SESTERTERPENE OPHIOBOLIN BIOSYNTHESIS

Up to now, the tools for in vivo identifying gene function in filamentous fungi include overexpressing (Leger, Joshi, Bidochka, & Roberts, 1996), transposon arrayed gene knockouts (TAGKO) (Hamer et al., 2001), T-DNA tagging (Combier, Melayah, Raffier, Gay, & Marmeisse, 2003), gene trapping (Aichinger et al., 2003), transposon tagging (Firon, Villalba, Beffa, & D'Enfert, 2003), RNA interference (RNAi) (Nakayashiki, 2005), and yeast one/two hybrid (Duyvesteijn et al., 2005). Gene editing, including knockout and complementation experiments, is one of the most direct and effective ways to study gene function. It is based on the principle of DNA homologous recombination to attenuate or terminate the expression of related genes and determine the gene function according to the change of product yield. In addition to random integration, gene targeting based on homologous recombination, RNAi technologies, and emerging gene-editing technologies such as Zinc-finger nucleases (Wang et al., 2013), transcription activator-like effector nucleases (TALENs) (Ménoret et al., 2014), and clustered regulatory interspaced short palindromic repeat/associated systems (CRISPR/Cas) technology can accomplish gene knockout or gene replacement in vivo (Liu, Chen, Jiang, Zhou, & Zou, 2015; Nielsen et al., 2017; Wang et al., 2017).

Generally, the in vivo gene deletion and complementation for fungi is based on the following procedures: protoplast preparation, gene-targeting cassette construction, transformation, and transformants identification. In this section, we describe the in vivo gene (cluster) deletion and complementation of terpene synthases in the marine *Aspergillus* strain 094102.

3.1 Preparation of Fungal Protoplast

The principle of protoplast preparation may be found in the manual described in *Fungal Biology* (Berg & Maruthachalam, 2015), in which the fungal growth phase and conditions, the type of enzyme combinations, the osmotic stabilizer, the incubation time and temperature for cell wall lysing are included.

3.1.1 Equipment

- Microscope (Axio Scope. A1, Zeiss), hemocytometer (10 mm × 10 mm).
- High-speed freezing centrifuge (Centrifuge 5810R, Eppendorf).

3.1.2 Buffers and Reagents

- Glucose minimum medium (GMM): glucose 5 g; 20 × nitrate salts 25 mL; trace elements 500 μL; agar 20 g; in 500 mL dH_2O, at pH 6.5, to obtain a large amount of fungal spores.
- Liquid minimal medium (LMM): GMM without agar, for spore germination.
- 20 × Nitrate salts: $NaNO_3$ 60 g; KCl 5.2 g; $MgSO_4{\cdot}7H_2O$ 5.2 g; KH_2PO_4 15.2 g; add ddH_2O to 500 mL, for GMM preparation.
- Trace elements: $ZnSO_4{\cdot}7H_2O$ 2.2 g; H_3BO_3 1.1 g; $MnCl_2{\cdot}4H_2O$ 0.5 g; $FeSO_4{\cdot}7H_2O$ 0.16 g; $CoCl{\cdot}5H_2O$ 0.16 g; $CuSO_4{\cdot}5H_2O$ 0.16 g; $(NH_4)6Mo_7O_{24}{\cdot}4H_2O$ 0.11 g; Na_4EDTA 5 g; add ddH_2O to 100 mL, for GMM preparation.
- Mycelium wash: $MgSO_4$ 73.95 g; add ddH_2O to 500 mL, for washing spores.
- Trapping buffer: sorbitol 54.65 g; 1 *M* Tris–Cl (pH 7.0) 50 mL, pH 7.0; add ddH_2O to 500 mL, for protoplast preparation.
- Osmotic medium: $MgSO_4$ (147.9 g) and 2 *M* NaPB (2.5 mL), in 1 L dH_2O; the pH was adjusted to 5.8 with 1 *M* Na_2HPO_4, for protoplast preparation.
- Driselase from *Basidiomycetes* sp., purchased from Sigma.
- Sorbitol/Tris/$CaCl_2$ (STC) buffer: Sorbitol 109.3 g; $CaCl_2$ 0.235 g; 1 *M* Tris–Cl (pH 7.5) 5 mL; add ddH_2O to 500 mL, for protoplast transformation.

3.1.3 Procedure

1. Grow spores of *A. ustus* 094102 on at least five GMM plates for 3 days at 28°C.
2. Collect spores with 10–15 mL ddH_2O containing 0.1% Tween-80.
3. Determine spore concentration with a hemocytometer under microscope.
4. Inoculate 10^9 *A. ustus* 094102 spores into 100 mL sterile LMM and shake at 28°C and 200 rpm for 10–11 h. Check the young germlings under the microscope (Fig. 3A).
5. Harvest the germlings in 50-mL tube at 8000 × *g* and 4°C for 15 min, decant the supernatant, and resuspend the cultures in 50 mL mycelium wash buffer. Centrifuge at 8000 × *g* and 4°C for 15 min and decant the supernatant.
6. Transfer the aggregates of germlings to a sterile 50-mL Erlenmeyer flask containing 10 mL osmotic medium with 1% drislase. Shake at 80 rpm and 28°C for 3.5–4 h (Fig. 3B). Observe the formation of protoplast under the microscope at intervals of 0.5 h.
7. Transfer the protoplast into a new 50-mL tube and carefully add 10 mL trapping buffer to cover the mixture and centrifuge at 3000 × *g* and 4°C for 15 min.
8. Remove the protoplast in the interface with a pipette and transfer to a sterile 15-mL falcon tube. Add at least two volumes of STC buffer to dilute the protoplast. Centrifuge at 5000 × *g* and 4°C for 8 min.
9. Decant the supernatant. Resuspend the protoplast at the concentration of 10^8–10^9 with about 1 mL STC buffer. Count with a hemocytometer under the microscope.

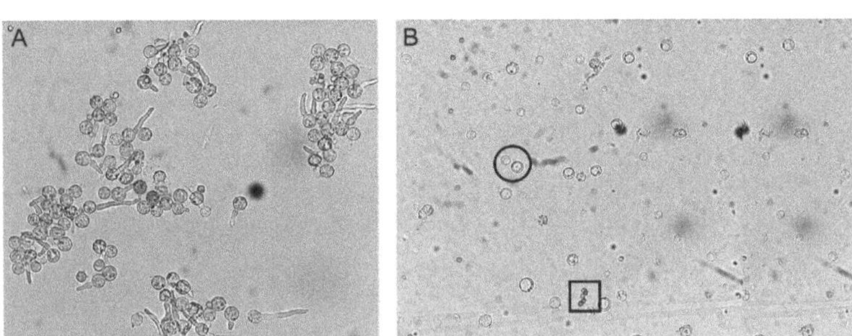

Fig. 3 Protoplast preparation of *A. ustus* 094102. (A) Spore germination in LMM at 28°C and 220 rpm for 11 h. (B) Protoplasts formation in Osmotic medium containing 10 mg mL^{-1} driselase at 28°C and 80 rpm for 4 h. The spores and protoplasts were displayed in the *circle* and *square*, respectively.

3.1.4 Notes

1. Do not adjust the pH of osmotic medium with NaOH, as it will cause precipitation.
2. Do not heat up the osmotic medium in microwave or autoclave, or it will form a precipitate.
3. Time of spore germination must be controlled. If the culture has grown too fast, do ot try to make protoplast.
4. If the protoplast forms rapidly, it will need less than 3–4 h. Protoplasts are thin-walled and about twice as large as spores.
5. Trapping buffer should be kept on the surface of mixture before centrifugation.

3.2 Construction of Gene-Targeting Cassette

The gene-targeting cassette contains upstream homology arm (UHA), downstream homology arm (DHA) and a selection marker. In general, homologous recombination in filamentous fungi (Fig. 4A) requires that the length of homology arms be greater than 1 kb (Asch & Kinsey, 1990; Hynes, 1996). The cassette can be constructed using restriction enzyme digestion and ligation, fusion PCR (Szewczyk et al., 20066), double-joint PCR (Yu et al., 2004), *E. coli* RecE/T system (Chaveroche, Ghigo, & D'Enfert, 2000), and TAGKO (Hamer et al., 2001).

Selective markers commonly used in filamentous fungi include auxotrophic complementary genes such as *pyrG* (orotidine-5′-phosphate decarboxylase gene), *ura5* (uracil auxotrophic gene) (Ruiz-Díez, 2002), and *argB* (encoding ornithine carbamoyltransferase) (Jadoun, Shadkchan, & Osherov, 2004). Carbon, nitrogen, sulfur, and other nutrients such as *amdS* (encoding acetamidase) (Hynes, 1994), *niaD* (nitrate reductase gene) (Malardier et al., 1989), *MPI* (mannose-6-phosphate isomerase) (Gui, Li, Liu, Han, & Li, 2014), and the like can be used. Antibiotic resistance genes such as hygromycin, bialaphos, zeocin, chlorimuronethyl (Chang et al., 2010), benomyl, and oligomycin (Wang et al., 2017) are employed.

We describe construction of the gene-targeting cassette here (Fig. 4B), which involves steps of restriction enzyme digestion, ligation, and subcloning. The pMD18T vector was used as a backbone; the *Sh ble* resistance gene is cloned into pMD18T by TA cloning. UHA and DHA of the *Au3446* gene of the strain 094102 are loaded to the two ends of the resistance gene of pMD18T vector through restriction enzyme digestion and ligation.

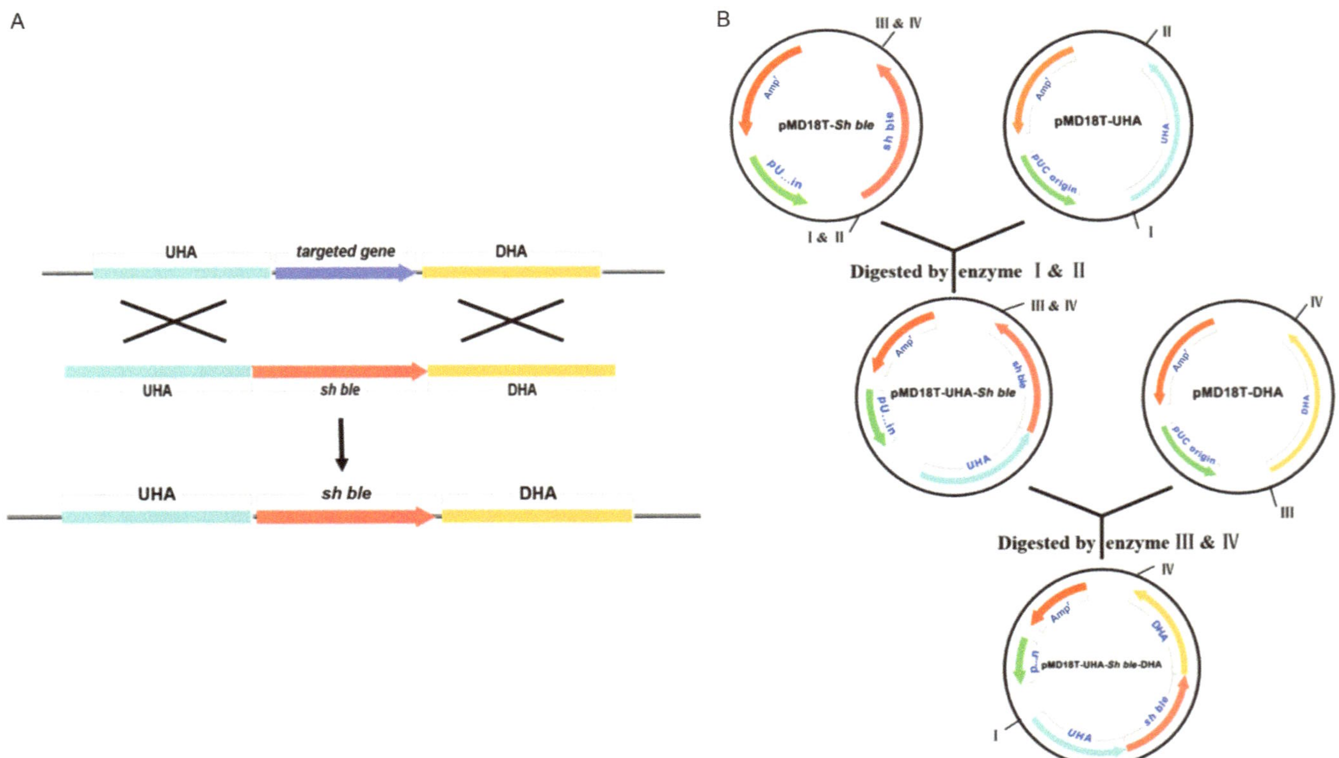

Fig. 4 Gene (cluster) knockout in fungi with UHA-Sh ble-DHA cassette. (A) The schematic diagram of homologous recombination. (B) Construction of the gene-targeting cassette. pMD18T-*Sh ble*, pMD18T-UHA, and pMD18T-DHA were obtained by TA clone. pMD18T-UHA-*Sh ble* and pMD18T-UHA-*Sh ble*-DHA were obtained by restriction enzymes I & II, III & IV digestion, respectively, and ligation of T4 DNA ligase. The gene-targeting cassette was obtained from pMD18T-UHA-*Sh ble*-DHA digested by enzyme I & IV.

3.2.1 Equipment

- PCR instrumentation (C-1000, BIO-RAD)
- Agarose gel electrophoresis equipment (G560E, BIO-RAD)
- Ultramicro spectrophotometer (Nanodrop 2000, Thermo)

3.2.2 Buffers and Reagents

- 2 × ES Taq MasterMix/2 × Pfu MasterMix (CWBio)
- I-5™ 2 × High-Fidelity Master Mix (Molecular Cloning Laboratories)
- Gel/PCR extraction kit/plasmid miniprep kit (50 ×) (OMEGA Bio-tek)
- T Vector pMD18 (Simple) Kit (TaKaRa)
- Restrictive endonuclease/T4 DNA ligase (New England BioLabs)

3.2.3 Procedure

1. Obtain each of *Au3446L′* (1224bp) and *Au3446R′* (1657bp) gene (Fig. 5A) using PCR amplified from *A. ustus* 094102 gDNA with primer *Au3446L′-s/a* and *Au3446R′-s/a* (Table 3) and clone these fragments into pMD18-T-Simple vector (Table 4).
2. Amplify the zeocin-resistant gene *Streptoalloteichus hindustanus ble* (*Sh ble*) (1172bp) (Fig. 5A) that selected as the selection marker from pGPAαA with *sh ble-s′/a′* (Table 4) and cloned into pMD18T vector by TA clone (Table 4).
3. Transform the TA clones to *E. coli* TOP 10 competent cell and incubate at 37°C overnight. Confirm all fragments by DNA sequencing.
4. Digest pMD18-T-Simple-*Au3446L′* and pMD18-T-*Sh ble* with restriction *Eco*RI and *Kpn*I and purify using gel extraction kit, clone *Au3446L′* into linearized pMD18-T-*Sh ble* with T4 DNA ligase to obtain pMD18-T-*Au3446L′-Sh ble* (Table 4).
5. Digest pMD18-T-Simple-*Au3446R′* and pMD18-T-*Au3446L′-Sh ble* with restriction *Pst*I and *Hin*dIII and purify the target DNA fragments. Clone *Au3446R′* into linearized pMD18-T-*Au3446L′-Sh ble* to construct pMD18-T-*Au3446L′-Sh ble-Au3446R′* (Table 4).
6. Digest pMD18-T-*Au3446L′-Sh ble-Au3446R′* with restriction *Eco*RI and *Hin*dIII to obtain linearized *Au3446L′-Sh ble-Au3446R′* target cassette (Fig. 5B).

3.2.4 Notes

1. The size of homology arms is about 1–2kb.
2. Using the above procedure, gene cassette containing UHA, DHA, and the antibiotic-resistant gene can be used in the deletion of other gene or gene clusters by changing different primers of UHA and DHA.

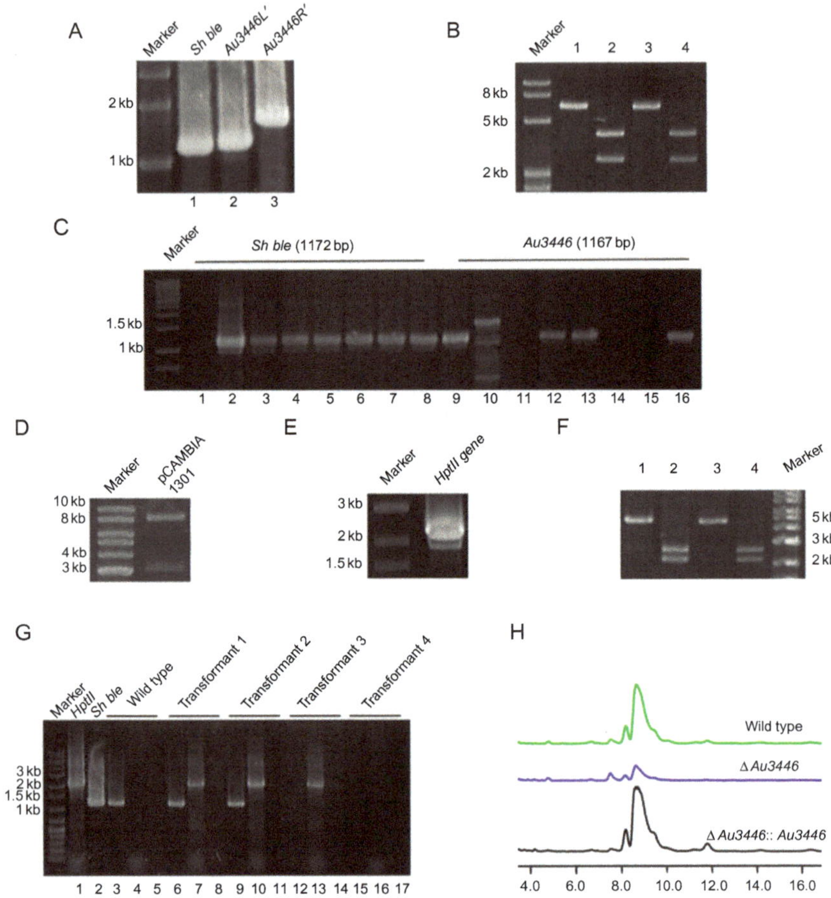

Fig. 5 *Au*3446 gene knockout and complementation in *A. ustus* 094102. (A) *Sh ble, Au3446L′*, and *Au3446R′* gene amplification. Lanes 1–3: *Sh ble* (1171 bp), *Au3446L′* (1224 bp), and *Au3446R′* (1657 bp) gene amplified with primers *Sh ble-s/a, Au3446L′-s/a*, and *Au3446R′-s/a*, respectively. (B) Digestion verification of pMD18T-*Au3446L′-Sh ble-Au3446R′*. Lanes 1 and 3: pMD18T-*Au3446L′-Sh ble-Au3446R′* digested by *Eco*RI (6720 bp); lanes 2 and 4: pMD18T-*Au3446L′-Sh ble-Au3446R′* digested by *Eco*RI and *Hin*dIII (4085 and 2635 bp). (C) Colony PCR of *A. ustus* 094102::Δ*Au3446.* Lanes 1 and 9: *A. ustus* 094102 genomic DNA amplified with primers *Sh ble-s/a* and *Au3446-s/a*; lanes 2 and 10: pGAPZαA amplified with primers *Sh ble-s/a* and *Au3446-s/a*; lanes 3–8, 11–16: six transformants of *A. ustus* 094102::Δ*Au3446 genomic* DNA amplified with primers *Sh ble-s/a* and *Au3446-s/a*, respectively. Six transformants were picked for PCR identification. Among them, three transformants (in lanes 4/12, 5/13, and 8/16) were positive strains, and ectopic insertion of targeting cassette may have occurred in the others. (D) *hptII* gene amplification. The plasmid pCAMBIA 1301 digested by *Sph*I (8619 and 3230 bp). (E) *hptII* gene amplified from digested pCAMBIA 1301 with primers *hptII-s/a* (2146 bp). (F) Digestion verification of pUC18-*CaMV35S-hptII-CaMV35S polyA*.

3.3 Transformation of Gene-Editing Cassette to Fungal Protoplast

Methods of genetic transformation in filamentous fungi include protoplast-mediated transformation, *Agrobacterium tumefaciens*-mediated transformation, biolistics transformation, electroporation and restriction endonuclease-mediated integration (Wang et al., 2017), and lithium acetate transformation (Olmedo-Monfil, Cortés-Penagos, & Herrera-Estrella, 2004).

After the protoplast preparation, the linearized *Au3446L′-Sh ble-Au3446R′* target cassette obtained in Section 3.2 is transformed into *A. ustus* 094102 protoplast by PEG–$CaCl_2$-mediated protoplast transformation (Jiang, Shen, Liu, & Lu, 2014; Zhong et al., 2012; Zhou, Li, Wang, & Zhuge, 2005).

3.3.1 Equipment

- High-speed freezing centrifuge (Centrifuge 5810R, Eppendorf).

3.3.2 Buffers and Reagents

- PEG solution: PEG 4000 (BDH) 60%; $CaCl_2$ 0.235 g; 1 *M* Tris–Cl (pH 7.5) 5 mL; add ddH_2O to 100 mL, for protoplast transformation.
- Stabilized minimal medium (SMM): GMM with sorbitol 218.6 g/L, for protoplast regeneration and screening of transformants.

3.3.3 Procedure

1. Dilute 10 μg linearized *Au3446L′-Sh ble-Au3446R′* target DNA fragment with STC buffer in a sterile falcon tube and keep the volume at 100 μL.
2. Add 100 μL of protoplast and incubate for 50 min on ice.

CaMV35S-hptII-CaMV35S polyA cassette was cloned into pUC18 vector. Lanes 1 and 3: plasmid digested by *Eco*RI (4840 bp); lanes 2 and 4: plasmid digested by *Eco*RI and *Sph*I (2195 and 2645 bp). (G) Colony PCR of *A. ustus* 094102::Δ*Au3446* complementary strains. Lanes 1, 4, 7, 10, 13, and 16: *hptII* gene amplified with primers *hptII-s/a* (2146 bp); lanes 2, 5, 8, 11, 14, and 17: *Sh ble* gene amplified with primers *Sh ble-s/a* (1172 bp); lanes 3, 6, 9, 12, and 15: *Au3446* gene amplified with primers *Au3446-s/a* (1167 bp). Among these transformants, 1 and 2 were positive mutant strains. (H) HPLC analysis of ophiobolin production in *A. ustus* 094102 wild-type, Δ*Au3446* mutant and the *Au3446Δ::Au3446* complementary strains. The production of ophiobolins decreased in the *ΔAu3446* mutant and recovered almost the same to the wild-type strain in the complementary mutant.

Table 3 Primers Used for *Au*3446 Gene and Gene Clusters Deletion and Complementation of *Au*3446

Type	Full Name	Sequence: 5′–3′ (Restriction Endonuclease)	Product Size (bp)
pUC18	M13-47	CGCCAGGGTTTTCCCAGTCACGAC	156
	RV-M	GAGCGGATAACAATTTCACACAGG	
Selective marker	*Sh ble*-s	CCCACACACCATAGCTTCAAAAT	1172
	Sh ble-a	AGCTTGCAAATTAAAGCCTTCG	
	hptII-s	AAATTGACGCTTAGACAACTTAA	2146
	hptII-a	GCAGCTTGCCAACATGGTG	
Genetic deletion and complementation system (gene *Au_3446*)	*Au_3446*-P1	GTGTGGGTGATCTTGACGGATTC	1245
	Au_3446-P3	CTATGGTGTGTGGGGGATCGGGTGGTATTTGCGTGTTTGG	
	Au_3446-P4	GCAAGCTGGAGACCAACATGCGGCCTAATCGAGATGTGC	1672
	Au_3446-P6	AACGGGTTCCTTCAAAATGG	
	Au_3446-P2	ACATACACTCTCATCCTACGCCTG	3402
	Au_3446-P5	CAGCCCAACTGAACTTTTCTAGC	
	Au_3446L′-s	CGGAATTCATCTTGACGGATTC (*Eco*RI)	1224
	Au_3446L′-a	GGGGTACCTTGCGTGTTTGG (*Kpn*I)	
	Au_3446R′-s	AACTGCAGCTAATCGAGATGTGC (*Pst*I)	1657
	Au_3446R′-a	CCCAAGCTTCCTTCAAAATGG (*Hin*dIII)	
	Au_3446-s	AAAGATGTTTCTCCCTTTGGCG	1167
	Au_3446-a	GAGGTCGCGGGCCTGATA	

	Au_3446L′-Au_3446-s	GGGGTACCATACACTCTCATCCTAC (*Kpn*I)	2502
	Au_3446L′-Au_3446-a	CTGCAGAACCAATGCATTGGCCGGTTCATGGCC (*Bst*XI)	
	Au_3446R′-s	CTGCAGAAAAAAAGCTCGCTTTTGCAC (*Pst*I)	1012
	Au_3446R′-a	AAGCTTTGAACTTTTCTAGCCCGCCC (*Hin*dIII)	
POC*3446* deletion	POC*3446L*-s	GAATTCCCACAGATAGTATGCGAGTC (*Eco*RI)	1419
	POC*3446L*-a	TCTAGACAAACGCCAGAACATCACG (*Xba*I)	
	POC*3446R*-s	GTCGACTAGGGTGGTATTTGCGT (*Sal*I)	1128
	POC*3446R*-a	AAGCTTGGGCATCCGTATGATGTCG (*Hin*dIII)	
POC*8003* deletion	POC*8003L*-s	GAATTCGTCGTACAGTCAGGTAGGGA (*Eco*RI)	1085
	POC*8003L*-a	GGTACCAAGGAGGAGTGCAAGGAA (*Kpn*I)	
	POC*8003R*-s	CCTGCAGGTTGCCGAGGACAAGG (*Sbf*I)	1061
	POC*8003R*-a	AAGCTTCATCTGTGGACGGAGTAAAG (*Hin*dIII)	
	Au_8003-s	CGCCCACCACTCATTCTTTTAG	1063
	Au_8003-a	CGACCTGATCCGATGCTGTC	
POC*13192* deletion	POC*13192L*-s	GAATTCTAACTTCAAATCAGCGAGGA (*Eco*RI)	1224
	POC*13192L*-a	TCTAGAATACCCACGGACCCAAC (*Xba*I)	
	POC*13192R*-s	CCTGCAGGAGCAGCATCTTCGATTGG (*Sbf*I)	1077
	POC*13192R*-a	AAGCTTATTCTGGGACAGGAACTAG (*Hin*dIII)	
	Au_13192-s	AAGAACAAGCACGGACTGCC	1096
	Au_13192-a	CCTGCCTCGTCCAATACAGAT	

Continued

Table 3 Primers Used for *Au*3446 Gene and Gene Clusters Deletion and Complementation of *Au*3446—cont'd

Type	Full Name	Sequence: 5′–3′ (Restriction Endonuclease)	Product Size (bp)
POC*11565* deletion	POC*11565L*-s	GAATTCTTGTTGAGGCGTTTTGTG (*Eco*RI)	1058
	POC*11565L*-a	TCTAGAGTCATCGTGCGAGTAGG (*Xba*I)	
	POC*11565R*-s	CCTGCAGGCATCCGCAGTCATAGTAGT (*Sbf*I)	1018
	POC*11565R*-a	AAGCTTCGCCTCCACGTTG (*Hin*dIII)	
	Au_11564-s	CTCGTATTCGCTTTGGGTTGA	1106
	Au_11564-a	CCAGCATTGTCGGGTTCG	
POC*6298* deletion	POC*6298L*-s	GAATTCGCAGCCAGTTCCTTCCCTA (*Eco*RI)	1508
	POC*6298L*-a	GGATCCTTTTTACCCAAGCCCATC (*Bam*HI)	
	POC*6298R*-s	CCTGCAGGAGGGAAATACCGCCTCAAAAG (*Sbf*I)	1208
	POC*6298R*-a	AAGCTTAGCCTCCTCCTCGTTC (*Hin*dIII)	
	Au_6298-s	TTTGAGGCGGTATTTCCC	1038
	Au_6298-a	ACCTCCTTCTTCAGTCCCT	

Table 4 Plasmids Used for *Au*3446 Gene and Gene Clusters Deletion and Complementation of *Au*3446

Type	Serial Number/Full Name	Characteristics
Expression vectors	pGAPZαA	P_{GAP}, α-factor signal, myc epitope tag, C-terminal polyhistidine tag, T_{Aox1}, P_{Tef1}, P_{Em7}, *Sh ble* (Zeocin^r), T_{Cyc1}, pUC ori (ColE1 ori)
Cloning vectors	pMD18T	Derived from pUC18, pUC ori (ColE1 ori), O_{LacZ}, *bla* (Amp^r)
	pMD18-T-Simple	Derived from pMD18T, without multiple cloning site
	pCAMBIA1301	$P_{CAMV35S}$, *Gus*, $T_{Nos\ polyA}$, $P_{CAMV35S}$, *hptII* (HmB^r), $T_{CAMV35S\ polyA}$, pVS1 rep, pVS1 sta, pBR322 ori, pBR322 born, Kan^r
Deletion and complementation vectors	pUC18/P_{Tef1}-P_{Em7}-*Sh ble*-T_{Cyc1}	Derived from pMD18T and pGAPZαA, pUC ori (ColE1 ori), O_{LacZ}, *bla* (Amp^r), P_{Tef1}, P_{Em7}, *Sh ble* (Zeocin^r), T_{Cyc1}
	pUC18/$P_{CAMV35S}$-*hptII*-$T_{CAMV35S\ polyA}$	Derived from pMD18T and pCAMBIA1301, pUC ori (ColE1 ori), O_{LacZ}, *bla* (Amp^r), $P_{CAMV35S}$, *hptII* (HmB^r), $T_{CAMV35S\ polyA}$
Genetic deletion and complementation system (gene *Au_3446*)	pUC18/*Au_3446L*	Derived from pMD18-T-Simple
	pUC18/*Au_3446R*	Derived from pMD18-T-Simple
	pUC18/*Au_3446L′*	Derived from pMD18-T-Simple
	pUC18/*Au_3446R′*	Derived from pMD18-T-Simple

Continued

Table 4 Plasmids Used for *Au*3446 Gene and Gene Clusters Deletion and Complementation of *Au*3446—cont'd

Type	Serial Number/Full Name	Characteristics
	pUC18/*Au_3446L′*-P_{Tef1}-P_{Em7}-*Sh ble*-T_{Cyc1}	Derived from pUC18/P_{Tef1}-P_{Em7}-*Sh ble*-T_{Cyc1}
	pUC18/*Au_3446L′*-P_{Tef1}-P_{Em7}-*Sh ble*-T_{Cyc1}-*Au_3446R′*	Derived from pUC18/*Au_3446L′*-P_{Tef1}-P_{Em7}-*Sh ble*-T_{Cyc1}
	pUC18/*Au_3446*-*Au_3446L′*-$P_{CAMV35S}$-*hptII*-$T_{CAMV35S\ polyA}$	Derive from pUC18/$P_{CAMV35S}$-*hptII*-$T_{CAMV35S\ polyA}$
	pUC18/*Au_3446*-*Au_3446L′*-$P_{CAMV35S}$-*hptII*-$T_{CAMV35S\ polyA}$-*Au_3446R′*	Derive from pUC18/*Au_3446*-*Au_3446L′*-$P_{CAMV35S}$-*hptII*-$T_{CAMV35S\ polyA}$
POC*3446* deletion	pUC18/POC*3446L*	Derived from pMD18-T-Simple
	pUC18/POC*3446R*	Derived from pMD18-T-Simple
	pUC18/POC*3446L*-P_{Tef1}-P_{Em7}-*Sh ble*-T_{Cyc1}	Derived from pUC18/P_{Tef1}-P_{Em7}-*Sh ble*-T_{Cyc1}
	pUC18/POC*3446L*-P_{Tef1}-P_{Em7}-*Sh ble*-T_{Cyc1}-POC*3446R*	Derived from pUC18/POC*3446L*-P_{Tef1}-P_{Em7}-*Sh ble*-T_{Cyc1}
POC*8003* deletion	pUC18/*POC8003L*	Derived from pMD18-T-Simple
	pUC18/*POC8003R*	Derived from pMD18-T-Simple
	pUC18/POC*8003L*-P_{Tef1}-P_{Em7}-*Sh ble*-T_{Cyc1}	Derived from pUC18/P_{Tef1}-P_{Em7}-*Sh ble*-T_{Cyc1}
	pUC18/POC*8003L*-P_{Tef1}-P_{Em7}-*Sh ble*-T_{Cyc1}-POC*8003R*	Derived from pUC18/POC*8003L*-P_{Tef1}-P_{Em7}-*Sh ble*-T_{Cyc1}

POC*13192* deletion	pUC18/POC*13192L*	Derived from pMD18-T-Simple
	pUC18/POC*13192R*	Derived from pMD18-T-Simple
	pUC18/POC*13192L*-P_{Tef1}-P_{Em7}-*Sh ble*-T_{Cyc1}	Derived from pUC18/P_{Tef1}-P_{Em7}-*Sh ble*-T_{Cyc1}
	pUC18/POC*13192L*-P_{Tef1}-P_{Em7}-*Sh ble*-T_{Cyc1}-POC*13192R*	Derived from pUC18/POC*13192L*-P_{Tef1}-P_{Em7}-*Sh ble*-T_{Cyc1}
POC*11565* deletion	pUC18/POC*11565L*	Derived from pMD18-T-Simple
	pUC18/POC*11565R*	Derived from pMD18-T-Simple
	pUC18/POC*11565L*-P_{Tef1}-P_{Em7}-*Sh ble*-T_{Cyc1}	Derived from pUC18/P_{Tef1}-P_{Em7}-*Sh ble*-T_{Cyc1}
	pUC18/POC*11565L*-P_{Tef1}-P_{Em7}-*Sh ble*-T_{Cyc1}-POC*11565R*	Derived from pUC18/POC*11565L*-P_{Tef1}-P_{Em7}-*Sh ble*-T_{Cyc1}
POC*6298* deletion	pUC18/POC*6298L*	Derive from pMD18-T-Simple
	pUC18/POC*6298R*	Derive from pMD18-T-Simple
	pUC18/POC*6298L*-P_{Tef1}-P_{Em7}-*Sh ble*-T_{Cyc1}	Derived from pUC18/P_{Tef1}-P_{Em7}-*Sh ble*-T_{Cyc1}
	pUC18/POC*6298L*-P_{Tef1}-P_{Em7}-*Sh ble*-T_{Cyc1}-POC*6298R*	Derived from pUC18/POC*6298L*-P_{Tef1}-P_{Em7}-*Sh ble*-T_{Cyc1}

3. Add 1.25 mL PEG solution and mix gently by turning on its side and rotating it. Incubate at room temperature for 30 min.
4. Add 2 mL STC buffer, 6.5 mL melt SMM (about 45°C) and proper antibiotic into the mixture, and mix gently.
5. Quickly pour the 10 mL transformed mixture onto a 10-mL solid SMM (Zeocin$^+$, 1000 mg/L) in Petri dish and culture at 28°C for 3–5 days.

3.3.4 Notes

1. The protoplast should be prepared freshly for transformation.

3.4 Identification of Positive Transformants

After culturing the tranformants on the SMM medium containing Zeocin (1000 mg/L), the grown out colonies will be picked out to check whether the target gene (*Au3446*) has been successfully deleted using PCR amplification. If the *Au3446* gene cannot be amplified but the *Sh ble-s/a* gene can, a positive transformant is confirmed.

3.4.1 Procedure

1. Pick out the single colonies and purify if needed.
2. Extract gDNA of the single colony.
3. Confirm positive transformants by PCR using primers *Au3446*-s/a and *Sh ble*-s/a (Fig. 5C).

3.4.2 Notes

1. Pick at least six transformants for production of ophiobolin and drimane to make sure the yield of these compounds is stable.
2. Isolate the positive transformants quickly to avoid crosscontamination because of spores dispersal.

3.5 Gene Complementation

To confirm the role of the deleted gene, the same gene is complemented back to the original location of the genome and its recovered phenotype is confirmed (Mohd-Zin et al., 2016; Penwell, Arivett, & Actis, 2012). We complemented *Au*3446 back into the mutant strain Δ*Au*3446 to get a mutant strain Δ*Au*3446::*Au*3446. In order that the complementation mutant can be easily observed, another resistant gene *hptII* of hygromycin is used. The yield of obtained complementary mutant, together with the wild strain, and the deletion mutant of *Au*3446 were obtained by HPLC analysis as described in Section 3.7. Ophiobolin production of the complementary mutant recovered to near wild-type levels. The yield of sesterterpene decreased but was not abolished in the deletion mutant, which means that the *Au3446* gene is only partially involved in the biosynthesis of the sesterterpene ophiobolin and is not a key gene.

3.5.1 Procedure

1. Digest the pCAMBIA 1301 by *Sph*I (Fig. 5D) and amplify the hygromycin resistance gene *hptII* (2146 bp, together with the promoter pCaMV35S, shown in Fig. 5E) used as a selective marker for complementation with linear pCAMBIA 1301 with primers *hptII*-s/a (Table 3).
2. Clone *hptII* gene into pMD18T to produce pUC18-CaMV35S-*hptII*-CaMV35S polyA (Table 4) and checked by sequencing (Fig. 5F).
3. Amplify upstream fragment of *Au3446-Au3446L′* and downstream *Au3446R′* from *A. ustus* 094102 gDNA using primers *Au3446L′-Au3446*-s/a and *Au3446R′*-s/a, and clone them into pMD18-T-Simple by TA clone (Table 4) and check them by DNA sequencing.
4. Subclone *Au3446-Au3446L′* and *Au3446R′* to either side of *hptII* gene, respectively, to obtain pUC18-*Au3446L′-Au3446-CaMV35S-hptII-CaMV35S polyA-Au3446R′* (Table 4).
5. Digest pUC18-*Au3446L′-Au3446-CaMV35S-hptII-CaMV35S polyA-Au3446R′* by *Kpn*I and *Hin*dIII to obtain the DNA cassette *Au3446L′-Au3446-CaMV35S-hptII-CaMV35S.*
6. Transform the fragment into the protoplast of strain *A. ustus* 094102::Δ*Au3446* to obtain the complementary mutant strain by homologous recombination.
7. Select transformants on SMM with hygromycin B (150 g L^{-1}) and verify using PCR (Fig. 5G) using primers *Au3446*-s/a, *hptII*-s/a, and *Sh ble*-s/a (Table 3). *hptII* and *Au3446* gene can be amplified but *Sh ble* gene cannot be amplified among the positive mutants. Pick out the positive transformants 1 and 2 in Fig. 5G and detect the production of ophiobolin using the HPLC method as described in Section 3.7.

3.6 Gene Cluster Inactivation

In the biosynthetic pathway of filamentous fungal SM, many enzymes are involved in the process of catalysis, and the genes that encode these enzymes usually appear in clusters on the chromosomes (Keller & Hohn, 1997). A gene cluster deletion can be carried while the single gene deletion is not a sufficient evidence to confirm the function of a gene cluster. Such as the incomplete abolishment of ophiobolin production in *A. ustus* 094102::Δ*Au3446*

(Fig. 5H), we speculated that *Au3446* might not be the only gene involved in ophiobolin synthesis.

Methods of gene cluster localization include the in silico prediction and verification by experiment. The gene prediction may be carried by alignment of homologous gene protein sequence (Altschul, Gish, Miller, Myers, & Lipmen, 1990; Bromann et al., 2012) and the antiSMASH bioinformatic pipeline (http://antismash.secondarymetabolites.org) (Weber et al., 2015).

Since terpene biosynthesis includes steps such as isoprenyl chain elongation, cyclization, and postmodification (oxidation, hydroxylation, hydrolysis, etc.), a terpene biosynthesis gene cluster can be predicted based on the terpene synthase/cyclase or prenyl transferase genes, together with flanking genes encoding cytochrome P450 (CYP450), oxidoreductase, hydroxylase, hydrolase, and/or transcription regulators. Based on the structure of the ophiobolin (Fig. 1), the putative gene clusters named POC (Presumed Ophiobolin biosynthesis gene Cluster) can be predicted by in silico analysis of the regions flanking the terpene synthases (including the diterpene and "function unknown" group shown in Table 2). Based on this method, 15 POCs related to ophiobolins synthesis in strain 094102 were obtained. The method of gene cluster knockout is similar to that for single gene deletion. Primers and plasmids are shown in Tables 3 and 4, respectively.

3.7 HPLC Detection of Ophiobolins and Drimanes

Terpenoid extraction methods include steam distillation (Muzika, Campbell, Hanover, & Smith, 1990), solvent extraction (Bowman et al., 1997), headspace analysis (Tholl et al., 2006), supercritical fluid extraction (Pourmortazavi & Hajimirsadeghi, 2007), microwave method (Delazar, Nahar, Hamedeyazdan, & Sarker, 2012), ultrasound extraction (Liao, Zheng, & Qu, 2016), and the like. Commonly used extractants in the solvent extraction method are methanol, ethanol, *n*-hexane, ethyl ether, chloroform, and benzene.

Analytical methods for terpenoids include UV spectroscopy, gas chromatography-based analytical methods such as gas chromatography (Kappers et al., 2005), GC-MS (Fojtová, Lojková, & Kubán, 2008), two-dimensional capillary gas chromatography (Strunk & Bechtold, 1996), HPLC (Keinänen, Oldham, & Baldwin, 2001), liquid chromatography and tandem mass spectrometry, proton transfer reaction-mass spectrometry (Steeghs et al., 2004), and Fourier transform ion cyclotron resonance mass spectrometry (Vahur et al., 2012).

After obtaining the gene and gene cluster deletion mutant of *A. ustus* 094102, the yield of sesterterpene ophiobolins and sesquiterpene drimanes from *A. ustus* 094102 wild-type strain and its mutants can be checked by solvent extraction and the following HPLC analysis.

Five out of 15 POCs knockout mutants changed their ophiobolin yield (Fig. 6), which indicated their involvement in ophiobolin biosynthesis. Among those, the POC 8003 mutation abolished ophiobolin production, and the POC 6298 mutation increased, but POC 13192 and POC 11565 mutation decreased (Fig. 6B). These results indicated that sesterterpene ophiobolin biosynthesis involved multiple gene clusters (Chai et al., 2016).

3.7.1 Equipment

- Rotary evaporator (Vacuum controller V-850, BUCHI)
- HPLC (2998, Waters)
- Phenomenex Gemini column (C18, 250 × 4.6 mm, 5 μm)
- Photodiode Array detector

3.7.2 Materials and Reagents

- Medium: corn niblet 100 g, yeast extract 1.7 g, $MgSO_4$ 0.85 g, KH_2PO_4 0.5 g, sea salt 1.6 g, and tap water 40 g, for solid fermentation of fungus
- Methanol/acetone/ethyl acetate
- Microporous membrane (0.2 μm)
- Disposable sterile syringes (1-mL)

3.7.3 Procedure

1. Inoculate the hypha or spores of wild-type strain 094102 and its mutants in corn solid medium at 20°C for 18 days.
2. Add 30 mL 80% acetone into 20 g fermentation products and stay for 12 h.
3. Break it further by ultrasonication for 20 min.
4. Centrifuge at 8000 × *g* for10 min.
5. Remove acetone from the supernatant with rotary evaporator.
6. Extract the residua with a double volume of ethyl acetate and remove ethyl acetate using rotary evaporator.
7. Dissolve the crude extraction with 15 mL methanol for HPLC detection.
8. Filter fermentation crude extract dissolved with methanol using 0.22 μm microporous membrane and inject the filtrate into the chromatography bottle for HPLC detection.

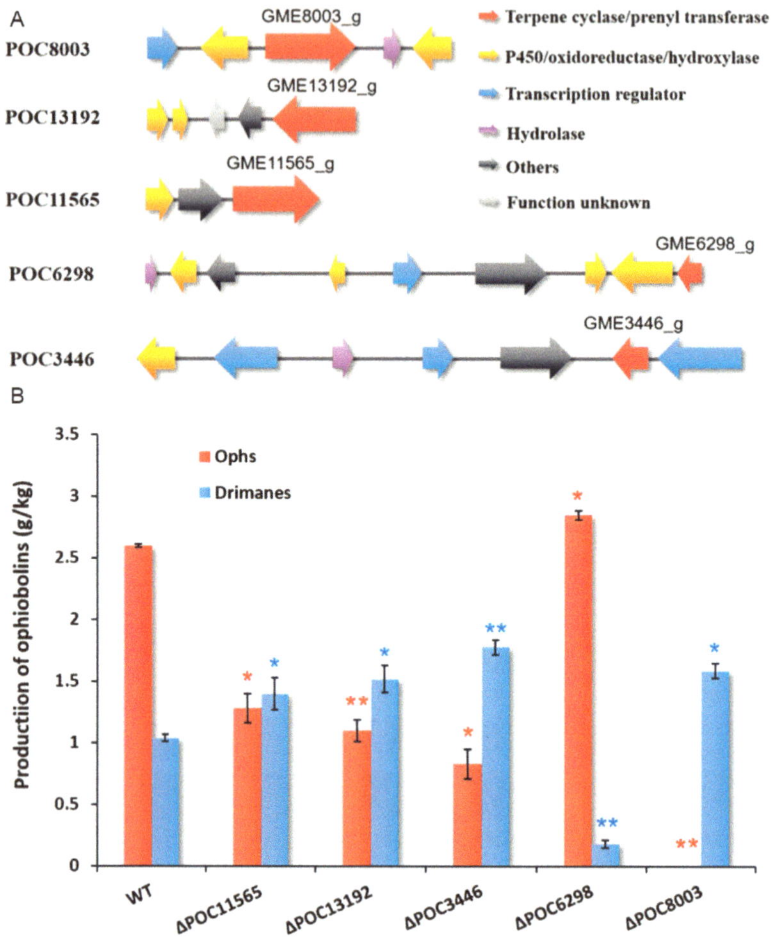

Fig. 6 The inactivation of gene cluster involved in the biosynthesis of ophiobolins. (A) Five gene clusters of *A. ustus* 094102 involved in ophiobolin synthesis. The clusters included the genes of terpene cyclase (or prenyl transferase) (in *red*), cytochrome P450 or oxidoreductase or hydroxylase (in *yellow*), transcription regulator (in *blue*), hydrolase (in *purple*), other function (in *black*), and function unknown (in *gray*). Numbers on or near the open reading frame are gene codes. (B) HPLC detection of five gene cluster mutants and wild-type strain in ophiobolin and drimane. Data are means ± SE of at least five independent measurements. $*P<0.05$, $**P<0.01$, vs wild-type strain. The production of ophiobolin was completely abolished in the ΔPOC8003 mutant. The yield of ophiobolin partially but significantly decreased in the ΔPOC3446 ($P<0.01$), ΔPOC13192 ($P<0.05$), and ΔPOC11565 ($P<0.05$) mutants, but increased significantly in the ΔPOC6298 mutant ($P<0.05$) compared with the wild-type strain. The yield of drimane significantly decreased in the ΔPOC6298 ($P<0.01$) mutants but increased in ΔPOC3446 ($P<0.01$), ΔPOC8003 ($P<0.05$), ΔPOC11565 ($P<0.05$), and ΔPOC13192 ($P<0.05$) mutants.

9. Set the program of HPLC: full wavelength (210–400 nm) scanning; isocratic elution with the mobile phase consisting of methanol and water (85:15, v/v); the injection volume, elution time, and flow rate are 10 μL, 30 min, and 1.0 mL/min, respectively.
10. Prepare ophiobolin G (>96% purity) in methanol at concentration of 0, 0.005, 0.010, 0.015, and 0.020 g L^{-1} for standard curve.
11. Peak areas of ophiobolins and drimanes were summed at 234 and 306 nm, respectively.
12. Determine the yield of ophiobolins and drimanes according to the standard curve.

3.7.4 Notes

1. Dilute the crude extract to a suitable fold if its concentration is too high.
2. If any precipitate appears in the filtrate, it should be filtered again.

4. IN VITRO FUNCTIONAL CHARACTERIZATION OF AU8003, AU13192, AND AU6298

Terpene synthases gene can be identified either using the in vivo gene (cluster) deletion and complementation in native host, or the in vitro biochemical reaction using the heterologously expressed proteins. While using in vitro approach, the prerequisite is to obtain the proteins of the presumed terpene synthases. *E. coli* is often used as a heterologous expression host, due to its pure background and accessibility to manipulate. Using *E. coli* heterologous expression system, the target proteins can be obtained (Ueda et al., 2015; Ye et al., 2015), and furthermore, SM such as terpenes can be produced directly since their precursors are available in *E. coli* (Bian et al., 2017; Jia, Potter, & Peters, 2016).

In this section, we describe the function verification of terpene synthase genes from *A. ustus* 094102 using the in vitro biochemical reaction. The process includes in silico analysis, target gene cloning, protein expression using *E. coli*, biochemical reaction, and product detection.

4.1 In Silico Analysis

Five terpene biosynthesis genes of *Au8003, Au13192, Au11565, Au6298,* and *Au3446*, in the five POCs which are selected by the in vivo gene cluster deletion (in Section 3.6), are further analyzed in silico. Phylogenetic analysis is performed using the amino acid sequences of the five genes' coding proteins, and their corresponding homologous proteins extracted from the

GenBank database. The neighbor-joining phylogenetic tree shows that the five proteins distributed in three clusters corresponding to three enzyme types: di(sester)terpene synthase, FPPS, and HexPPS, which demonstrate the functional diversity of these enzymes (Fig. 7A). Au8003 clustered closely with AcOS, an ophiobolin F synthase (Chiba et al., 2013). Au13192 clustered within the same di (sester)terpene synthase branch but not as closely as Au8003 (Fig. 7A). A multiple sequence alignment using BioEdit software also shows that Au8003, Au13192, and Au11565 each containing both TC and PT domains (Fig. 7B). The closest identified protein matching to Au6298 is the plant EpFPPS (*Euphorbia pekinensis* farnesyl diphosphate synthase), at an identity of 49%; and to Au3446 is the bacterial MlHexPPS (*Micrococcus luteus* heterodimeric hexaprenyl diphosphate synthase), at an identity of 32%. We therefore assume that proteins Au8003 and Au13192 are bifunctional terpene synthases, which are sesterterpene synthase and diterpene synthase, respectively, and that Au6298 may be a FPPS. Functions of genes *Au8003, Au13192,* and *Au6298* are verified further in vitro.

4.1.1 Software and Database

- GenBank database
- MEGA (version 5.0) for amino acid phylogenetic analysis
- BioEdit 7.0 for multiple sequence alignments

4.1.2 Procedure

1. Search the GenBank database by BLASTP (basic local alignment tool—protein) using amino acid sequences of Au8003, Au13192, Au11565, Au6298, and Au3446 as query sequences.
2. Select reference sequences of query proteins from BLASTP search results. Only those sequences of function confirmed by experiment were selected.
3. Use the BioEdit 7.0 software to perform multiple sequence alignments.
4. Use MEGA (version 5.0) to construct the phylogenetic analysis based on amino acid sequences.

4.1.3 Notes

1. The updated software and database are recommended for further use.

4.2 Target Gene Cloning

To study the function of target genes by heterologous expression, obtaining target genes from original strain is one of the most important steps.

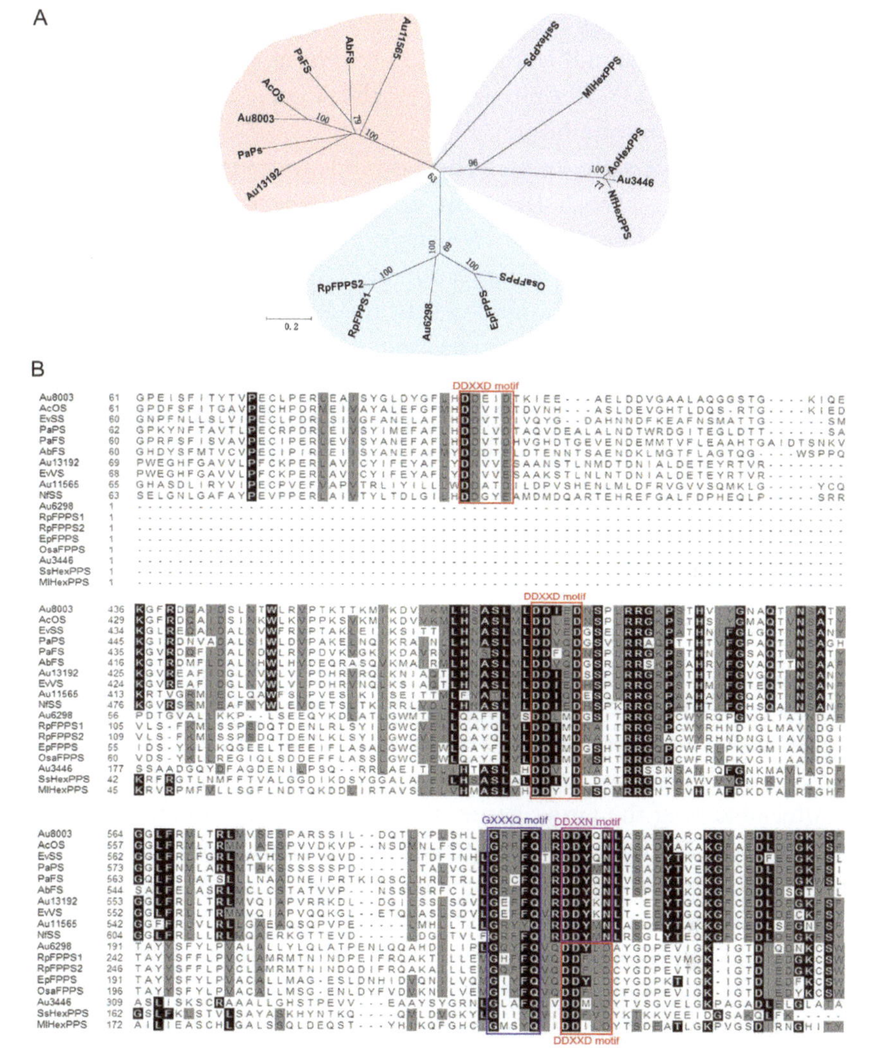

Fig. 7 In silico analysis of Au8003, Au13192, and Au6298. (A) Neighbor-joining phylogenetic tree based on amino acid sequences of bifunctional diterpene and sesterterpene synthases, farnesyl pyrophosphate synthases, and hexaprenyl pyrophosphate synthases. (B) Alignment of amino acid sequences of the five terpene synthases related to ophiobolin biosynthesis with function identified terpene synthases. The three DDXXD/E (terpene cyclase (TC) domain), DDXXD (prenyltransferases (PT) domain), GQ (PT), DDXXN (bifunction PT only) DDXXD (single function PT only) are showed in *colored frame*. Only sequence parts contain characteristic catalytic domain sequence are shown; other part of amino sequence is omitted.

The gDNA of eukaryotic fungi contains both exons and introns, therefore cDNA (complementary DNA from mRNA without introns) is employed as template while using PCR approach to get the target genes for heterologous expressing in prokaryotic hosts.

RT-PCR (reversible transcription PCR) technique is an effective method to acquire cDNA (Keilholz et al., 1998; Zheng, Qiu, & Han, 2015). This method consists of two steps: cDNA synthesis from mRNA by reverse transcription and amplification of PCR. It has been proved useful in amplifying very small amount of specific mRNA to obtain cDNA (Kusec, 1998). However, this approach can be limited by the availability of particular tissues or mRNA, making gDNA a desirable source for target gene cloning (Mitani, Nakayama, Harbers, & Hayashizaki, 2004). On this occasion, overlap PCR using gDNA as template is developed for individual exon amplification, using tagged primers with overlapping regions and their concatenations in additional rounds of PCR amplification. With these methods, most of the target genes can be obtained efficiently.

When the target gene sequences are derived from eukaryotic organisms, rare codons can be a problem on successful expression of these genes, especially for those have diad or triad rare codons structures (Kane, 1995). Under these circumstances, codon optimization of target sequences should be employed to increase their expression level and solubility (Utashima, Yamashita, Arima, & Masaki, 2017).

After obtaining the target genes, they are linked to cloning vectors for high-efficient cloning of DNA fragments. Cloning methods can be divided into two classes, depending on whether or not ligase is required (Zhou, Clark, & Gomezsanchez, 1995). TA cloning is a method that ligase is not required and often used when constructing cloning vectors. It is a convenient and quick way to clone PCR product, taking advantage of Taq DNA polymerase's terminal transferase activity. For PCR product of Pfu DNA polymerase, vector kit based on DNA topoisomerase I may be a good choice (Shuman, 1994).

In case of strain 094102, genes of *Au8003* and *Au6298* can be obtained by mRNA extraction and RT-PCR, but for gene *Au13192*, overlap PCR using gDNA as template is used. After that, target genes are cloned to pEASY-blunt vector and transformed to *E. coli* by heat shock for further cloning.

4.2.1 Equipment

- Honor HNY-1 incubator shaker
- Mortar and pestle

- BIO-RAD C1000TM Thermal Cycler
- Thermo Nano Drop 2000C
- BIO-RAD gel electrophoresis system
- SYNGENE gel image analyzing system

4.2.2 Buffers and Reagents

- 0.1% DEPC (diethyl pyrocarbonate) treated H_2O: 1 mL high-purity (more than 97%) DEPC, dilute with ddH_2O to 1 L, stir until it is fully dissolved.
- QIAGEN RNeasy Mini Kit includes Buffer RLT, Buffer RW1, and Buffer RPE.
- Thermo Scientific RevertAid First Strand cDNA Synthesis Kit, includes oligo (dT) 18, 5 × reaction buffer, Ribo Lock RNase Inhibitor, RevertAid RT, and dNTP Mix.
- NEB pfu DNA polymerase kit, includes Pfu DNA polymerase and 5 × NF buffer.
- OMEGA Gel Extraction Kit, includes Binding Buffer (XP2), SPW Wash Buffer and Elution Buffer.
- Transgen pEASY-Blunt vector kit.
- Maikelaibo plasmid extraction kit, includes Buffer 1, Buffer 2, Buffer 3, Buffer D, Buffer W, and Buffer E.

4.2.3 Procedure

4.2.3.1 RNA Extraction From Strain *Aspergillus* sp. 094102

Extraction of the total RNA from mycelia of strain 094102 is carried out with the German company QIAGEN RNeasy Mini Kit (50) Kit (stock # 74104), do as the following:

1. Culture strain 094102 in fungal II liquid medium at 28°C, 220 rpm for 2–3 days.
2. Take 50 mg of hyphae at one time for microextraction.
3. Grind the sample into a powder with liquid nitrogen, transfer the powder into a 1.5-mL plastic centrifuge tube.
4. Add 450 μL Buffer RLT (lysis buffer) in the centrifugal tube (add 10 μL β-ME per 1 mL Buffer RLT before use), oscillate on the oscillator for 1–3 min, make sure that the solution at the bottom of the tube is fully oscillated.
5. Centrifuge at 13,000 × *g* for 2 min at 4°C, transfer the supernatant to a clean 1.5-mL plastic centrifugal tube. Avoid touching or drawing the middle and lower organic phase, which contain DNA, proteins, and other impurities.

6. Add 0.5 volume of anhydrous ethanol, fully reverse the mix till it is thoroughly mixed.
7. Transfer the mixed solution to a centrifugal adsorption column and centrifuge at 4°C, 8000 × *g* for 15 s, discard the filtrate.
8. Add 700 μL Buffer RW1 (wash buffer) into centrifugal adsorption column and centrifuge at 4°C, 8000 × *g* for 15 s, discard the filtrate.
9. Move the centrifugal column to a new collection tube (no RNA enzymes), add 500 μL Buffer RPE (wash buffer), and centrifuge at 4° C, 8000 × *g* for 15 s, discard the filtrate.
10. Add 500 μL Buffer RPE (wash buffer, add 4 times volume of ethanol before use). Centrifuge for 2 min at 4°C, 13,000 × *g*.
11. Move the centrifugal column to a new collection tube (no RNA enzymes). Centrifuge for 2 min at 4°C, 13,000 × *g*.
12. Move the centrifugal column to a new 1.5-mL plastic centrifugal tube (no RNA enzymes), add 50 μL RNase-free water, centrifuge at 4°C, 8000 × *g* for 15 s.
13. Transfer 40 μL RNA sample to a new 1.5-mL plastic centrifugal tube (no RNA enzymes), add 5 μL each of DNase I and DNase I buffer, incubated at 37°C for 2 h.
14. Add 5 μL of 50 m*M* EDTA, incubated at 65°C for 10 min, preserved at −80°C.

4.2.3.2 Reverse Transcription

America's Revert of Thermo Fisher Scientific Company Aid First strand cDNA synthesis kit (model K1621) was used for reverse transcription, the steps are as follows:

1. Load 2 μg of the prepared RNA samples into a PCR tube, add 1 μL of 10 μ*M* oligo (dT) 18.
2. Add RNase-free water up to 12 μL, incubate at 65°C for 5 min.
3. Cool for 5 min on ice, centrifuge at 4°C, 13,000 × *g* for 2 min.
4. Add 4 μL of 5 × reaction buffer, 1 μL of Ribo Lock RNase Inhibitor, 1 μL of RevertAid RT, 2 μL of 10 m*M* dNTP Mix, and RNase-free water up to 20 μL.
5. Incubate at 42°C for 60 min, then at 70°C for 5 min.
6. Store the cDNA product at −20°C.

4.2.3.3 Amplification of Au8003 and Au6298

1. Design primer pairs of *Au8003* and *Au6298* according to the genomic sequence of strain 094102 (Table 5).

Table 5 Primers Used for Target Gene Cloning

Primer Name	Primer Sequence (5′–3′)
Au8003-F	**CATATG**ATGGAGTATAAGTACTCGACC (*Nde*I)
Au8003-R	**AAGCTT**TCAAACCTTCAGCAGCTCCA (*Hin*dIII)
Au6298-F	CC**CATATG**ATGTCTTCACCGCGTGCC (*Nde*I)
Au6298-R	CC**AAGCTT**TTATTTGGTGCGCTTGTAAA(*Hin*dIII)
Au13192-E1-S	**GCGGCCGC**ATGCCACAAACACAATCC (*Not*I)
Au13192-E1-A	GCCGACTCGACAACATTATCATACAGAAACGCATACTCAAAGATAT AACAAATAA
Au13192-E3-A	CATGAGCATGTCGACGAAG GGGGCGCCGGTATCGAT
Au13192-E4-S	CTTCGTCGACATGCTCATG
Au13192-E4-A	GCCGGGGCGAGGAGGATCTCGTCTCCTAGCTGCAC
Au13192-E5-S	ATCCTCCTCGCCCCGGC
Au13192-E5-A	AGTCTTCGATGTCGTCGAGCATGAGCGAGGCATTGTGTA
Au13192-E6-S	CTCGACGACATCGAAGACT
Au13192-E6-A	CTGCACCATCAACCTCGTAAGCAACCGGAACAACCCACCGGTCTT CTGCCTAACCATCTCCAAAT
Au13192-E8-S	TTACGAGGTTGATGGTGCAGATTGCGCCGGTGCGGCGGAAAGATC TCGACGGCATCTTATCAT
Au13192-E8-A	CCCTTTTGGCCGGTGTACTCTTCAGTTAGGTTCTTGTAGT
Au13192-E9-S	AGTACACCGGCCAAAAGGG
Au13192-E9-A	**TCTAGA**TACACCTTCAACCTCTGCAC (Xba I)

2. Prepare reaction system: 1 μL of Pfu DNA polymerase, 21 μL of sterilized deionized water, 1 μL of Pfu DNA polymerase 5 × NF buffer, 2.5 μL (10 μ*M*) each of forward and reverse primer, and 2 μL of cDNA template.
3. Start PCR, the reaction conditions: initial denaturation at 95°C for 3 min; denaturation at 95°C for 30 s; anneal at 53°C for 20 s; elongation at 72°C for 70 s; cycle for 35 times; elongation at 72°C for 10 min.
4. Verify PCR products by agarose gel electrophoresis (Fig. 8A and B).

4.2.3.4 Amplification of Au13192

Since there are seven introns in *Au13192* and the gene cannot be amplified from cDNA, we carried out overlap PCR using gDNA of *A. ustus* 094102 as template (Fig. 8C). The primers are designed as shown in Table 5 and Fig. 8. The primers for short fragments are designed with partially overlap of the

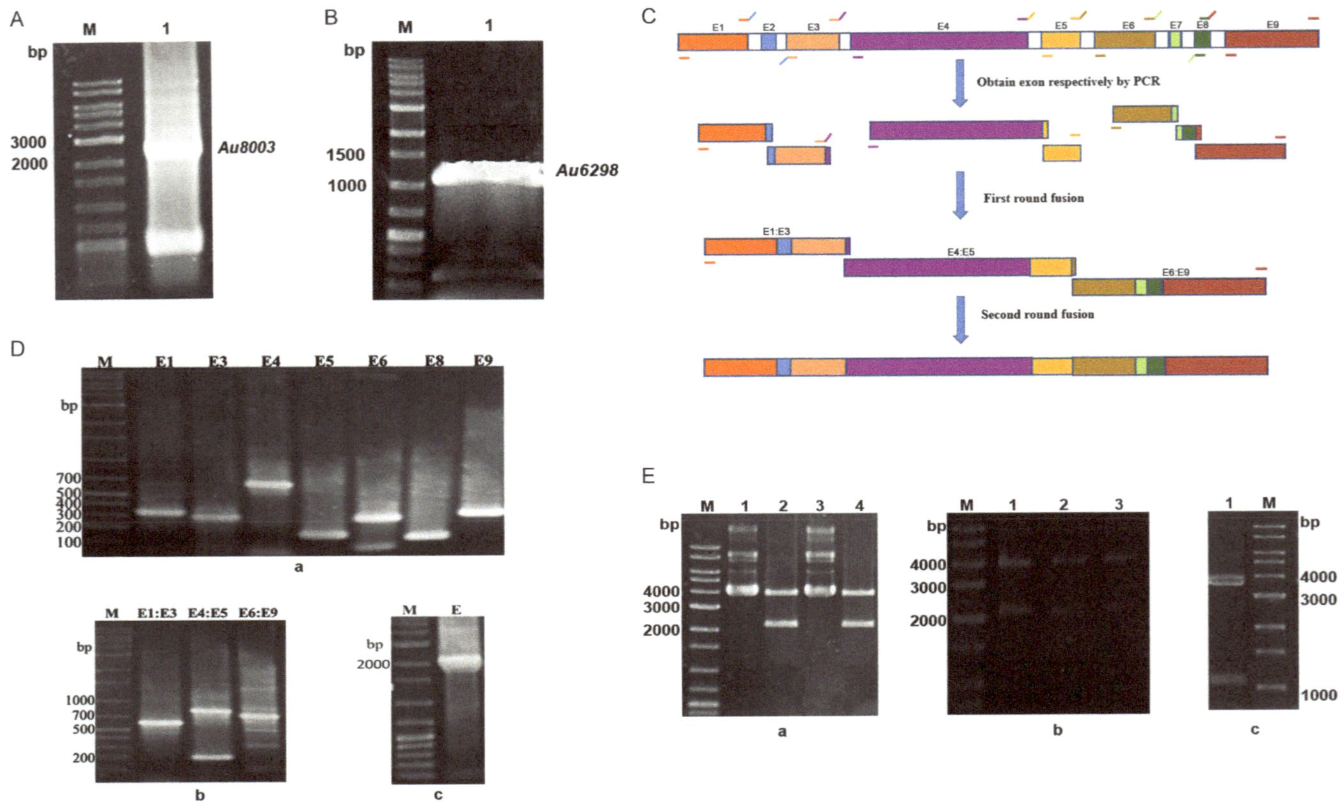

Fig. 8 See figure legend on opposite page.

long fragments, such as fragments of E2 were designed in primers *Au13192*-E1-A and *Au13192*-E3-A. Detailed steps are listed below:

1. First round PCR, obtain seven exon fragments: E1, E3, E4, E5, E6, E8, and E9 (Fig. 8D(a))
2. First round fusion PCR to get E1:E3, E4:E5, and E6:E9 fusion fragments, verify PCR products using agarose gel electrophoresis (Fig. 8D(b)).
3. Second round fusion PCR to fuse the three fragments obtained from step 2 and get full-length *Au13192* cDNA.
4. Verify PCR product by agarose gel electrophoresis (Fig. 8D(c)).

4.2.3.5 Recovery of Target DNA

After the electrophoresis of the PCR product, the target DNA was recovered from the agarose gel according to the Gel Extraction Kit (100) (Series No. D2500-01) supplier (OMEGA, Germany) as following:

1. Cut the target DNA fragment under UV light from the agarose gel to minimize gel size and improve DNA recovery. Absorb liquid on the gel surface using paper towel.
2. Chop the gel block to decrease the gel melting time and improve DNA recovery.
3. Weigh the gel block. Add binding buffer at the ratio of 400 μL per 100 mg agarose gel, incubate at 50–60°C for 10 min (mix every 2 min till all the gel melt).
4. Insert a HiBind DNA Mini Column in a 2-mL collection tube. Transfer less than 700 μL DNA/agarose melt solution to the column, centrifuge at 10,000 × *g* for 1 min at room temperature. Discard the filtrate.

Fig. 8 Target gene cloning of *Au8003*, *Au13192*, and *Au6298*. (A) Electrophoretic analysis of *Au8003* (2350 bp) PCR product. Lane M: marker; lane 1: *Au8003* PCR product. (B) Analysis of *Au6298* (1102 bp) PCR product. Lane M: marker; lane 1: *Au6298* PCR product. (C) Strategy of *Au13192* multiple exon fusion by overlap extension PCR. (D) Electrophoretic analysis of fusion fragments in the process of *Au13192* cDNA (2608 bp) fusion, respectively. (a) PCR product of each exon fragments. M: marker; E1, E3, E4, E5, E6, E8, and E9 represent for corresponding exon fragment. (b) Fusion PCR product after first round fusion. M: marker; E1:E3, E4:E5, and E6:E9 represent for fusion exon fragment E1 & E2, E4 & E5, and E6 & E8–9. (c) *Au13192* cDNA after second round fusion PCR. M: marker; E: *Au13192* cDNA. (E) Double digestion of cloning vectors pEASY-Blunt-*Au8003*, pEASY-Blunt-*Au13192*, pEASY-Blunt-*Au6298*. (a) Double digestion of pEASY-Blunt-*Au8003*. M: marker; lanes 1 and 3: pEASY-Blunt-*Au8003* plasmid; lanes 2 and 4: pEASY-Blunt-*Au8003* double digestion product. (b) Double digestion of pEASY-Blunt-*Au13192*. M: marker; lanes 1–3: pEASY-Blunt-*Au13192* double digestion product. (c) Double digestion of pEASY-Blunt-*Au6298*. M: marker; lane 1: double digestion product of pEASY-Blunt-*Au8003*.

5. Reuse the column, repeat step 4 till all the melt gel are transferred to the column.
6. Add 300 μL XP2 to the adsorption column, centrifuge at room temperature, 10,000 × *g* for 1 min, discard the filtrate, and reuse the collection tube.
7. Add 700 μL SPW wash buffer to the adsorption column, centrifuge at room temperature, 10,000 × *g* for 1 min, discard the filtrate, and reuse collection tube.
8. Add 300 μL SPW wash buffer to the adsorption column, centrifuge at room temperature, 10,000 × *g* for 1 min, discard the filtrate.
9. Centrifuge the HiBind DNA Mini Column at room temperature, maximum speed for 2 min to dry the column matrix.
10. Transfer the adsorption column to a new 1.5-mL tube. Add 30–50 μL elution buffer or sterilized water on the center of the column membrane.
11. Keep at room temperature for 2–5 min, centrifuge at maximum speed at room temperature for 1 min.
12. Check the recovered product with 1% agarose gel electrophoresis.

4.2.3.6 Target DNA Cloning

The recovered DNA is ligated to the pEASY-Blunt vector at 16°C for 1 h. The resultants are transformed to *E. coli* TOP 10 cells and cultured overnight. The positive colonies are selected and checked using rapid testing methods. The procedures are as following:

1. Pick out about 20 clones from the transformation agar plate with a sterile toothpick, streak onto the Luria-Bertani (LB) agar plates containing benzyl ammonia (100 μg/mL), incubate at 37°C for 8 h.
2. Add 30 μL sodium chloride–Tris–EDTA (STE) solution (20 m*M* Tris–HCl, 25 m*M* EDTA, 75 m*M* NaCl) in 1.5-mL plastic centrifugal tube, transfer the single colony individually to the plastic centrifugal tubes with toothpick, vortex 2 min till mixed well.
3. Add equal volume of chloroform:phenol:isoamyl alcohol (v:v:v = 24:25:1), shake the mix, centrifuge at 10,000 × *g* for 5 min at room temperature.
4. Check the supernatant using 1% agarose gel electrophoresis, to determine initially whether the plasmid containing the inserted exogenous fragments.

5. Extract the plasmid containing small foreign fragment using plasmid extraction kits (Wuhan Maikelaibo Technology, Ltd). The procedure is as following:
   (1) Transfer the single colony to ampicillin (100 μg/mL)-resistant LB medium incubated overnight at 37°C, 220 rpm.
   (2) Transfer 2 mL culture to a 2-mL plastic centrifugal tube, centrifuge at 10,000 × *g* for 1 min at room temperature. Discard liquid.
   (3) Add 250 μL Buffer 1, vortex to make the cells suspend thoroughly.
   (4) Add 250 μL Buffer 2, invert the centrifuge tube 5–6 times mildly, till the solution turn to clear.
   (5) Add 350 μL Buffer 3, invert the centrifugal tube 3–5 times mildly to mix thoroughly.
   (6) Centrifuge 12,000 × *g* for 10 min at room temperature.
   (7) Transfer the supernatant carefully into the collection tube adsorption column.
   (8) Centrifuge at 12,000 × *g* for 1 min at room temperature. Pour the liquid out of the collection tube. Put the column back into the collection tube.
   (9) Add 500 μL Buffer D, centrifuge at 12,000 × *g* for 1 min at room temperature. Pour the liquid out of the collection tube, put the column back into the collection tube.
   (10) Add 600 μL Buffer W, centrifuge at 12,000 × *g*, for 1 min at room temperature. Pour the liquid out of the collection tube, put the column back into the collection tube.
   (11) Add 630 μL Buffer W, centrifuge at 12,000 × *g* for 1 min at room temperature.
   (12) Transfer the column to a clean 1.5-mL plastic centrifuge tube, add 50–100 μL Buffer E to the adsorb membrane at the center, standing for 1 min at room temperature, centrifuge at 12,000 × g, for 1 min at room temperature to wash off the cloning plasmid.
   (13) Store the obtained plasmid at −20°C.

4.2.3.7 Verification of Cloning Vectors

1. Digest Plasmid pEASY-Blunt-*Au8003* and pEASY-Blunt-*Au6298* with *Hin*dIII and *Nde*I, in reaction conditions of following: 2 μL cutsmart buffer, 0.5 μL *Hin*dIII, 0.5 μL *Nde*I, 1 μL each plasmid, and 16 μL ddH_2O, incubate at 37°C for 1 h.

2. Digest pEASY-Blunt-*Au13192* with *Hin*dIII and *Not*I, in reaction conditions of following: 2 μL cutsmart buffer, 0.5 μL *Hin*dIII, 0.5 μL *Nde*I, 1 μL each plasmid, and 16 μL ddH_2O, incubate at 37°C for 1 h.
3. Check the digested products with 1% of agarose gel electrophoresis to determine whether the DNA fragment has cloned in plasmid (Fig. 8E(a–c)).
4. Submit 1 mL of culture containing the cloned plasmid of target DNA for sequencing to confirm the target DNA has been successfully cloned.

4.2.4 Notes

1. RNA extraction is supposed to be carried out in an enclosure space without RNase to prevent degradation. The pipettor, pipette tip, and EP tubes being used should have no RNase, and the mortars and pestles should be soaked overnight with 0.1% DEPC ddH_2O and then autoclaved for 40 min.
2. DEPC is toxic, 0.1% DEPC ddH_2O should be prepared in fuming cupboard, and avoid skin and respiratory exposure.
3. In Table 5, bases showed in bold are restriction enzyme cutting sites. Take care that the enzyme cutting sites should not exist in target gene sequence.

4.3 Protein Expression and Purification

For heterologous expression of target genes in *E. coli*, the pET expression system developed by Novagen is a common choice (Mierendorf, Morris, Hammer, & Novy, 1998). Target genes cloned in pET plasmids are controlled by strong bacteriophage T7 transcription and translation signals. T7 promoter is so efficient that target protein can comprise more than 50% of the total cell protein after a few hours of induction (Studier & Moffatt, 1986). Besides, this system is able to maintain target genes transcriptionally silent in the uninduced state, only active when transferred into expression hosts containing a chromosomal copy of the T7 RNA polymerase gene under lacUV5 control, and induced by the addition of IPTG. In addition, the His-T7-tag fusion proteins in pET system facilitate the target protein purification by using affinity nickel resin, as the Ni^{2+} can combine with 6 × His-tag.

A most used host of pET expression system is *E. coli* BL21 (DE3). It is a type B strain which is deficient in the *lon* protease *omp*T and lacks the *omp*T outer membrane protease that can degrade proteins during purification (Grodberg & Dunn, 1988). Taking advantage of these properties, target protein can stay more stably in strain BL21 (DE3).

E. coli BL21 (DE3) containing pET28a (+)-*Au8003*, pET28a (+)-*Au6298*, and pET28a (+)-*Au13192* are constructed to produce the His-T7-tag fusion protein using each gene cloned in Section 4.2, respectively. After elution with imidazole, desalination, and concentration, the target proteins are obtained.

4.3.1 Equipment

- Honour HNY-1 incubator shaker
- Beckman J2-MI centrifuge
- Scientz JY92-IIN Ultrasonic Cell Disruptor
- Novagen Ni-NTA His Bind Resin Nickel column
- Standard SDS-PAGE equipment
- Ultra-15 Centrifugal Filter Devices
- GE Healthcare PD-10 desalting column

4.3.2 Buffers and Reagents

- NEB restriction enzymes
- Thermo T4 DNA Ligase
- HEPES lysis buffer: 20 m*M* HEPES, 300 m*M* NaCl, 10% glycerol, 5 m*M* iminazole
- Elution buffer: 20 m*M* HEPES, 300 m*M* NaCl, 10% glycerol, 50–500 m*M* iminazole
- SDS-PAGE electrophoresis buffer: Tris 3 g, glycine 14.4 g, SDS 1 g, dilute with ddH_2O to 1 L, adjust pH value to 8.3
- 5 × SDS-PAGE loading buffer: glycerol 5 mL, SDS 1 g, 2-mercaptoethanol 0.5 mL, 1 *M* Tris–HCl (pH 6.8) 3.15 mL, 1% bromophenol blue 1 mL
- Coomassie brilliant blue staining solution: isopropanol 250 mL, Coomassie brilliant blue R-250 1 g, glacial acetic acid 100 mL, ddH_2O 650 mL
- Coomassie brilliant blue destaining solution: ethyl alcohol 50 mL, glacial acetic acid 100 mL, ddH_2O 850 mL
- HEPES desalination buffer: 20 m*M* HEPES, 150 m*M* NaCl, 25% glycerol

4.3.3 Procedure

4.3.3.1 Recovery of Target DNA Fragments and pET28a Fragment

1. Digest recombinant vectors pEASY-Blunt-*Au8003*, pEASY-Blunt-*Au6298*, and pET28a with *Hin*dIII and *Nde*I, in reaction conditions of

following: 5 μL cutsmart buffer, 2.5 μL *Hind*III, 2.5 μL *Nde*I, 2 μg each plasmid, add ddH_2O up to 50 μL, incubate at 37°C for 3 h.

2. Digest pEASY-Blunt-*Au13192* and pET28a with *Hind*III and *Not*I, in reaction conditions of following: 5 μL cutsmart buffer, 2.5 μL *Hind*III, 2.5 μL *Nde*I, 2 μg of each plasmid, add ddH_2O up to 50 μL, incubate at 37°C for 3 h.
3. Recover target DNA fragments and pET28a fragment after 1% agarose gel electrophoresis, using OMEGA Gel Extraction Kit.
4. Store the DNA fragments at −20°C.

4.3.3.2 Ligation of Target DNA and pET28a Fragments

1. Ligate target DNA fragments and pET28 fragment using T4 DNA ligase overnight, at a temperature of 4°C.
2. Transform ligation resultants into *E. coli* TOP 10 cells.
3. Positive colonies were selected using rapid testing methods after overnight culture, as described in Section 4.2.3.6.

4.3.3.3 Verification of Expressing Vectors

1. Extract plasmids pET28a-*Au8003*, pET28a-*Au13192*, and pET28a-*Au6298* using plasmid extraction kit (Wuhan Maikelaibo Technology, Ltd).
2. Digest pET28a-*Au8003* and pET28a-*Au6298* with *Hind*III and *Nde*I, respectively; reaction conditions are as described in Section 4.2.3.7.
3. Digest pET28a-*Au8003* and pET28a-*Au6298* with *Hind*III and *Not*I; reaction conditions are as described in Section 4.2.3.7.
4. Check digested products with 1% of agarose gel electrophoresis to determine whether the DNA fragments have been cloned in expressing plasmid successfully (Fig. 9).
5. Submit 1 mL of culture containing the expressing plasmid of target DNA to Qingke Company, China, for sequencing to confirm whether the vectors have been successfully constructed.

4.3.3.4 Expression Quality Analysis

Heterogeneous expression of eukaryotic fungi genes in prokaryotic *E. coli* may come across some problems, such as no gene expression and produce inclusion body. On this account, before purification of target proteins, an assay to detect the expression quality is necessary. Steps are listed below:

1. Add 10 μL of the recombinant plasmids pET28a-*Au8003*, pET28a-*Au13192*, pET28a-*Au6298*, and pET28a to 100 μL *E. coli* BL21

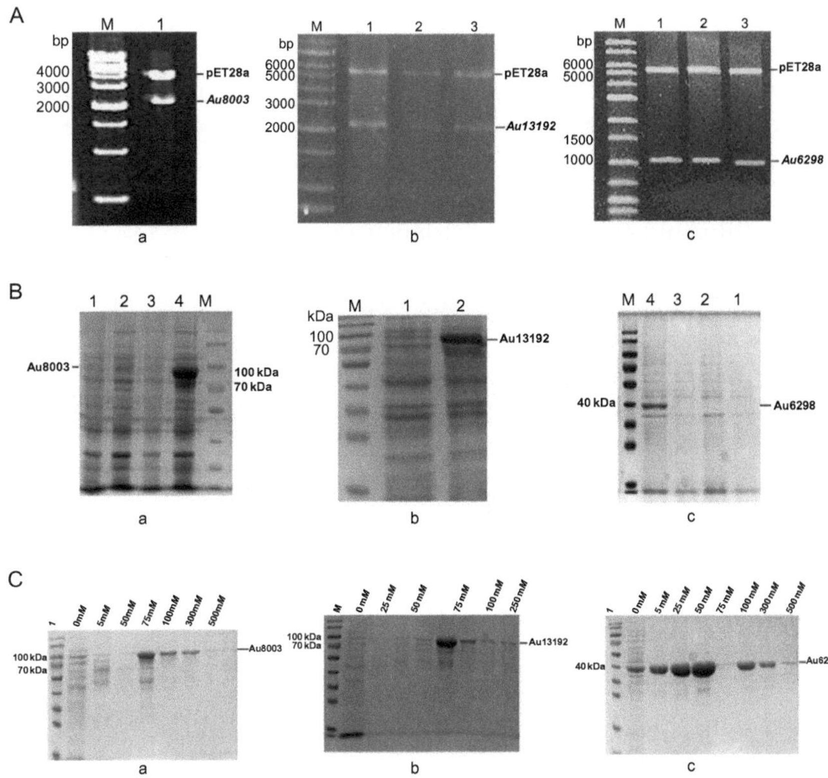

Fig. 9 Expression and purification of Au8003, Au13192, and Au6298. (A) Double digestion of cloning vectors pET-*Au8003*, pET-*Au13192*, pET-*Au6298*. (a) Double digestion of pET-*Au8003*. M: marker; lane 1: pET-*Au8003* double digestion product. (b) Double digestion of pET-*Au13192*. M: marker; lanes 1–3: pET-*Au13192* double digestion product. (c) Double digestion of pET-*Au6298*. M: marker; lanes 1–3: double digestion product of pET-*Au6298*. (B) SDS-PAGE electrophoresis result of expression detection of Au8003, Au13192, and Au6298. (a) Overexpression of *Au8003* gene in *E. coli* BL21 (DE3). Lane M: marker; lane 1: uninduced (without IPTG); lane 2: induced (with IPTG) bacterial protein carrying plasmid pET28a; lane 3: uninduced bacterial protein carrying *Au8003* gene; lane 4: induced bacterial protein carrying *Au8003* gene. (b) Overexpression of *Au13192* gene in *E. coli* BL21 (DE3). Lane M: marker; lane 1: uninduced bacterial protein carrying the 2.1 kb *Au13192* gene; lane 2: induced bacterial protein carrying the 2.1 kb *Au13192* gene. (c) Overexpression of *Au6298* gene in *E. coli* BL21 (DE3). Lane M: marker; lane 1: uninduced bacterial protein carrying plasmid pET28a; lane 2: induced bacterial protein carrying plasmid pET28a; lane 3: uninduced bacterial protein carrying *Au6298* gene; lane 4: induced bacterial protein carrying *Au6298* gene. (C) SDS-PAGE electrophoresis result of different imidazole concentration elutions of Au8003, Au13192, and Au6298. (a) Au8003 purification with different concentration of imidazole. Lane 1: marker; 0, 5, 50, 75, 100, 300 and 500 m*M* represent for corresponding imidazole concentration elutions. (b) Au13192 purification with different concentration of imidazole. Lane M: marker; 0, 25, 50, 75, 100, and 250 m*M* represent for corresponding imidazole concentration elutions. (c) Au13192 purification with different concentration of imidazole. Lane 1: marker; 0, 5, 25, 50, 75, 100, 300, and 500 m*M* represent for corresponding imidazole concentration elutions, respectively.

(DE3) broth on ice for 30 min, respectively. Heat at 42°C for 45–60 s. Keep on ice for 2 min.

2. Inoculate each of the engineered *E. coli* BL21 (DE3) strains containing recombinant plasmid in 5 mL kanamycin (50 μg/mL)-resistant LB medium, culture at 37°C, 220 rpm, overnight.
3. Inoculate each of the fresh culture to another 5 mL LB broth of kanamycin (50 μg/mL) resistant with 1% inoculation. Culture at 37°C for 2–3 h, till the OD_{600} reaches to 0.4–0.6 at 37°C.
4. Take out 1 mL each cultures, then add IPTG to the rest of culture to concentration of 1 m*M*. Continue the inducted culture for 3–5 h.
5. Take 200 μL of each culture of before and after induction, add 5 × SDS-PAGE buffer and boil for 10 min, centrifuge at 12,000 × *g* for 10 min, at room temperature check the supernatant with SDS-PAGE (dodecyl sulfate, sodium salt-polyacrylamide gel electrophoresis).

The results show that, comparing with the control, new protein bands appear after induction (Au8003: 82 kDa; Au13192: 80 kDa; Au6298: 39 kDa), which agree with the target proteins' size, indicating that the target proteins are successfully expressed in *E. coli* BL21 (DE3) (Fig. 9).

4.3.3.5 Purification of Expressed Proteins

In order to get enough quantity of target protein for in vitro analysis, the qualified engineering *E coli* strain is cultured in 500 mL LB broth to accumulate cells and then for protein extraction and purification. The process is as following:

1. Inoculate engineered *E. coli* BL21 (DE3) strains containing recombinant plasmid individually in 5 mL LB broth of kanamycin (50 μg/mL) resistance. Cultivate at 37°C, 220 rpm, overnight.
2. Inoculate each fresh culture to 500 mL LB medium containing kanamycin (50 μg/mL), cultivate at 37°C, 220 rpm. Culture at 37°C for 2–3 h, till the OD_{600} reaches to 0.4–0.6 at 37°C.
3. Add IPTG to a final concentration of 0.1 m*M* to induce the protein expression at 18°C for 16–18 h.
4. Centrifuge at room temperature, 5000 × *g* for 5 min to collect *E. coli* cells, wash the cells with ddH_2O twice.
5. Resuspend the *E. coli* cells with 10 mL HEPES buffer and ultrasonicate on ice (ultrasonicate 5S, 5S interval, power 300 W) for 30 min.
6. Centrifuge at 4°C, 10,000 × *g* for 20 min, filer the supernatant with 0.45 μm filter membrane to remove insoluble substances.
7. Add 2 mL of nickel beads onto the column, let the anhydrous ethanol fully flow out of the column under gravity.

8. Wash column with 20 mL aseptic ddH_2O and balance the column with 20 mL HEPES buffer.
9. Add supernatant containing each target protein to the nickel column, collect effluent liquid, detect with SDS-PAGE.
10. Elute nickel column with iminazole elution at concentrations of 0, 5, 25, 50, 75, 100, 250, 300, 500 m*M*, collect the elution and detect with SDS-PAGE (Fig. 9).
11. Combine iminazole fractions containing target proteins together. Concentrate to approximately 2.5 mL with Ultra-15 Centrifugal Filter Devices.
12. Balance GE Healthcare PD-10 desalting column with 10 mL balance buffer, discard the filtrate. Repeat 4 times.
13. Load the 2.5 mL concentrated protein solution onto the desalting column till it is fully soaked in the column. Discard the filtrate.
14. Add 3.5 mL HEPES desalination buffer. Collect the elution which contain the purified protein. Deposit at −80°C for further use.

4.3.4 Notes

1. To improve the ligation efficiency, the total quantity of target DNA fragments and pET28a fragment should be calculated when carrying out ligation; the ratio of target DNA fragment and plasmid fragment should be 6:1 to 7:1.
2. The pH value of HEPES lysis buffer, HEPES desalination buffer, and elution buffer should be adjusted to an appropriate threshold far away from the isoelectric point of the protein, otherwise it will precipitate. Also, according to the properties of different proteins, protease inhibitor (phenylmethylsulfonyl fluoride) or reductive agents (β-ME and dithiothreitol (DTT)) can be added to buffer to ensure protein activity.
3. Low temperature (4°C) is required during the purification process to keep the protein stable.

4.4 In Vitro Enzymatic Reaction

In vitro enzymatic reaction is a valid method to study functions of terpene synthetase (Anthony et al., 2009; Bian et al., 2017; Chiba et al., 2013). This method provides a simple background, and many influence factors are eliminated, which is helpful to identify the function of target protein.

An in vitro reaction system for terpene synthetase catalization including substances as following: DMAPP, GPP, FPP, GGPP, and IPP are substrates (Chiba et al., 2013; Toyomasu et al., 2007); $MgCl_2$ is needed to provide Mg^{2+} for reaction as a cofactor (Laskovics & Poulter, 1981).

Low concentration (<10 m*M*) of DTT is added to protect target proteins against inactivation by oxidation on disulfide bond (Aros et al., 2012; Lin et al., 2014).

We describe the in vitro biochemical reaction of the purified proteins Au8003, Au13192, and Au6298 in this section, which are examples of the sesterterpene, diterpene, and FPP synthases, respectively.

4.4.1 Equipment

- Grant GD120 water bath

4.4.2 Buffers and Reagents

- Substrates: IPP, DMAPP, GPP, FPP, GGPP, from Sigma-Aldrich.
- HEPES desalination buffer: 20 m*M* HEPES, 150 m*M* NaCl, 25% glycerol.

4.4.3 Procedure

4.4.3.1 Au8003 in vitro Reaction

Au8003 is predicted to be a chimeric sesterterpene synthetase. It is supposed that four kinds of substrates, including DMAPP, GPP, FPP, and GGPP, can be catalized by Au8003 on adding IPP to produce GFPP, and then cyclized by the same protein to produce ophiobolin F. The in vitro biochemical reactions are carried out as following:

1. Take five 1.5-mL tubes as reaction vessels, and label them with numbers 1–5.
2. Add four groups of substrates: 50 μ*M* DMAPP with 50 μ*M* IPP, 50 μ*M* GPP with 50 μ*M* IPP, 50 μ*M* FPP with 50 μ*M* IPP, and 50 μ*M* GGPP with 50 μ*M* IPP, into tubes Nos. 1–4, respectively.
3. Add 50 μ*M* DMAPP and 100 μ*M* IPP into the tube No. 5 as a control.
4. Add His-T7-tag-Au8003 recombinant protein (6.1 μ*M*) into tubes Nos. 1–4, add the same volume boiled protein into tube No. 5.
5. Add HEPES reaction buffer, which contains 20 m*M* HEPES, 10% glycerol, 5 m*M* DTT, and 5 m*M* $MgCl_2$ into all of the five tubes, to the final volume of 200 μL.
6. Mix the components by vortex oscillation. Centrifuge at 3500 × *g* for 1 min at room temperature.
7. Incubate at 30°C for 3 h. Add equal volume of *n*-hexane to terminate the reaction.

4.4.3.2 Au13192 in vitro Reaction

Au13192 is a hypothetical diterpene synthetase; in this reaction, three groups of substrates (DMAPP with IPP, GPP with IPP, FPP with IPP) are subjected to test its catalytic activity. Steps are listed below:

1. Take four 1.5-mL tubes as reaction vessels and label them with numbers 1–4.
2. Add three groups of substrates, 50 μ*M* DMAPP with 50 μ*M* IPP, 50 μ*M* GPP with 50 μ*M* IPP, and 50 μ*M* FPP with 50 μ*M* IPP into tubes Nos. 1–3, respectively.
3. Add 50 μ*M* DMAPP and 50 μ*M* IPP into tube No. 4 as a control.
4. Add His-T7-tag-Au13192 recombinant protein (6.1 μ*M*) into tubes Nos. 1–3, add the same volume boiled protein into tube No. 4.
5. Add HEPES reaction buffer, which contains 20 m*M* HEPES, 10% glycerol, 5 m*M* DTT, and 5 m*M* $MgCl_2$ into all of the four tubes, to the final volume of 200 μL.
6. Mix the components by vortex oscillation. Centrifuge at 3500 × *g* for 1 min at room temperature.
7. Incubate at 37°C for 3 h, add equal volume of *n*-hexane to terminate the reaction.

4.4.3.3 Au6298 in vitro Reaction

Au6298 is predicted as a FPP synthetase, and using DMAPP and IPP as substrates, the reaction steps are as following:

1. Take two 1.5-mL tubes, label them with Nos. 1 and 2.
2. Add 50 μ*M* DMAPP and 50 μ*M* IPP into both of the tubes.
3. Add purified His-T7-tag-Au6298 recombinant protein (4.2 μ*M*) into tube No. 1, and add the same volume boiled protein into tube No. 2 as a control.
4. Mix the components by vortex oscillation. Centrifuge at 3500 × *g* for 1 min at room temperature.
5. Add HEPES reaction buffer which contains 20 m*M* HEPES, 10% glycerol, 5 m*M* DTT, and 5 m*M* $MgCl_2$ up to 600 μL into both of the tubes.
6. Incubate at 30°C for 5 h and add equal volume of H_2O saturated *n*-butanol to end the reaction.

4.4.4 Note

1. After adding protein to the reaction system, be careful to observe whether the protein is precipitate to avoid negative result caused by protein denaturation.

4.5 GC-MS Analysis

GC-MS is a rapid method that provides rich information on natural products analysis and is used to analyze complex organic and biochemical mixtures (Hussain & Maqbool, 2014). Taking advantage of this method, in vitro products of purified terpene synthetase and in vivo products by heterologous expression can be analyzed rapidly and conveniently (Chai et al., 2016; Chiba et al., 2013; Matsuda et al., 2015; Okada et al., 2016).

Cyclic terpenes products of the in vitro reaction can be detected directly by GC-MS. But for the polyprenyl pyrophosphate end product of the farnesol synthase, it should be dephosphorized to the corresponding alcohol, then detected by GC-MS. Acid phosphatase is recommended in this case (Fujii, Koyama, & Ogura, 1982).

In this section, we show the process of GC-MS analysis of each reaction product of Au8003, Au13192, and Au6298. The peaks show on the gas chromatograph of those molecules that their corresponding MS data are in agreement with the prediction of a sesterterpene ophiobolin, a diterpene variediene, and a farnesol (Fig. 10).

4.5.1 Equipment

- Miulab NK200-1B digital dry bath
- Varian 450-GC-320TQ GC-MS
- Varian FactorFour VF-5 MS capillary column (30 m × 0.25 mm, 0.25 μm film thickness)

4.5.2 Buffers and Reagents

- *n*-Hexane for extraction
- *n*-Butanol saturated by ddH_2O
- Acetate buffer (pH 6.0): sodium acetate 54.6 g, 1 *M* acetic acid 20 mL, add H_2O to 500 mL
- Methanol for washing syringes
- Potato acid phosphatase, purchased from Sigma-Aldrich

4.5.3 Procedure

4.5.3.1 Sample Preparation of Au8003 and Au13192 in vitro Reaction Products

1. For Au8003 and Au13192 in vitro reaction products, extract the resultants using *n*-hexane at the same volume for 3 times after the reaction, respectively.
2. Evaporate the organic solvent by dry bath.
3. Add 50 μL *n*-hexane to dissolve the products and filter the solution.
4. Load the filtered *n*-hexane solutions into chromatography bottles.

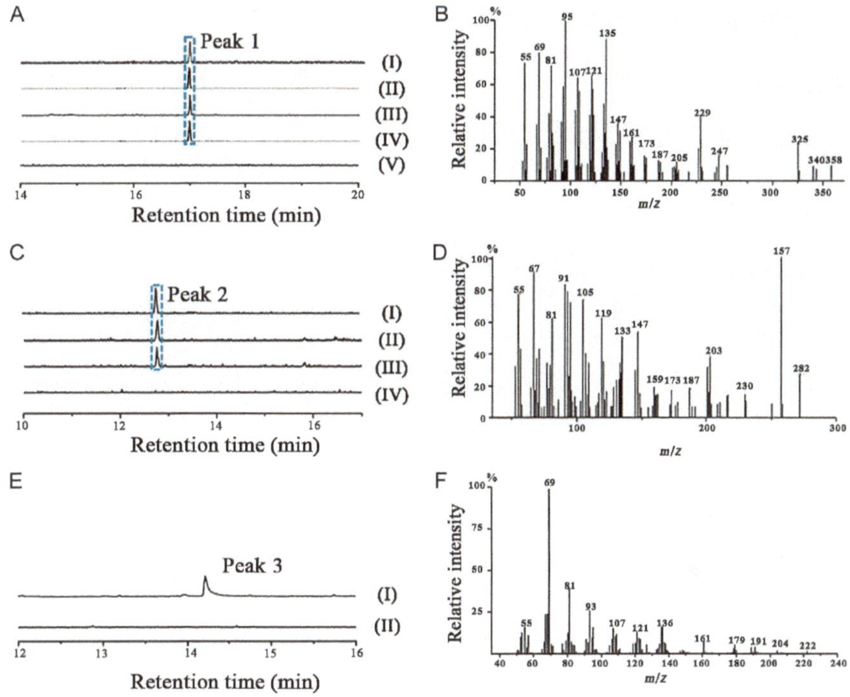

Fig. 10 GC-MS profile of Au8003, Au13192, and Au6298 reaction products. (A) The chromatograms of Au8003 reaction product ophiobolin F extracted at *m/z* 358: (I) GGPP+IPP, (II) FPP+IPP, (III) GPP+IPP, (IV) DMAPP+IPP, and (V) Control, and MS Spectra of peak 1 (B); (C) Chromatograms of Au13192 reaction product variediene extracted at *m/z* 272: (I) FPP+IPP, (II) GPP+IPP, (III) DMAPP+IPP, (IV) Control, and MS Spectra of peak 2 (D); (E) Chromatograms of phosphatase hydrolysate of Au6298 reaction product FPP extracted at *m/z* 222: (I) DMAPP+IPP, (II) Control, and MS Spectra of peak 3 (F).

4.5.3.2 Sample Preparation of Au6298

1. For Au6298 reaction product, after the synthesis reaction, extract the resultant using water saturated *n*-butanol at the same volume for 3 times.
2. Evaporate the organic solvent by dry bath.
3. Add 5 μL acetate buffer, 0.05 μL Triton X-100, 30 μL methanol, 2 μL potato acid phosphatase to a 1.5-mL tube, then add ddH_2O to 50 μL.
4. Mix the components by vortex oscillation. Centrifuge at 3500 × *g* for 1 min at room temperature.
5. Incubate at 37°C for 3 h.
6. Extract the resultant using *n*-hexane at the same volume for 3 times after the synthesis.
7. Evaporate the organic solvent by dry bath.
8. Dissolve the products with 50 μL ethyl acetate and then filtered to chromatography bottle.

4.5.3.3 Preparation for Running GC-MS

1. Before sample injection, the microsyringe should be washed by methanol.
2. Inject 1 μL *n*-hexane to check baseline of GC-MS.

4.5.3.4 GC-MS Conditions

1. For in vitro reaction products of Au8003 and Au13192, inject each sample into the column at 60°C in the splitless mode. After a 2-min isothermal hold at 60°C, increase the column temperature to 150°C at 30°C min^{-1}, from 150°C to 180°C at 10°C min^{-1}, from 180°C to 210°C at 2°C min^{-1} with a 5-min isothermal hold at 210°C. Set the helium carrier gas flow rate at 0.66 mL min^{-1}.
2. For farnesol analysis (dephosphorized product of Au6298), inject the sample into the column at 80°C in the splitless mode. After a 2-min isothermal hold at 80°C, increase the column temperature to 290°C at 10°C min^{-1}, with a 5-min isothermal hold at 290°C. Set the helium carrier gas flow rate at 0.66 mL min^{-1}.

4.5.4 Note

1. Samples for GC-MS analysis should have no high concentration substances, no sediment, or no water; otherwise GC-MS instrument will be damaged.
2. Organic solvents used to extract products can be adjusted according to the polarity of terpenes.
3. GC temperature programming will influence the retention time of sample, so the temperature programming should keep the same while running one specific product.

5. DISCUSSION AND CONCLUSION

With the development of genome sequencing, more and more SM-related unknown enzyme will be identified. The first step is to predict the function of an interested gene or gene clusters. A high quality genome sequence is required, and the software that helps to identify the SM gene clusters from the genome sequence is important. The bioinformatics analysis for terpene synthases is relatively immature, particularly for chimeric fungal terpene synthases, but can be carried out through local alignment methods, for instance, the extensively used BLAST algorithm (Altschul et al., 1990; Yamada et al., 2015). AntiSMASH, a powerful software for SM searching

(Weber et al., 2015), in our case gave only 11 terpene gene clusters. But using "terpene synthase" or "terpene cyclase" as the key word to search against the local gene annotation data base, and confirmed by using the typical conserved sequences of aspartate-rich DDXX(D/E) motif and the NSE/DTE triad, (N/D)DXX(S/T)XX(K/R)(D/E), 27 terpene synthase genes were selected. During the past decade, more and more chimeric terpene synthases were discovered, with fantastic catalytic characteristics (Bian et al., 2017; Chai et al., 2016; Chiba et al., 2013; Minami et al., 2009; Qin et al., 2016; Toyomasu et al., 2007; Ye et al., 2015). Hence, the combination of software and manually local database searching is recommended on thoroughly searching of terpene synthases from a fungal genome. Furthermore, a thorough analysis of terpene synthases on genome scale, resulted in the discovery of multiple gene clusters involved in the biosynthesis of sesterterpene ophiobolin in *A. ustus* 094102 (Chai et al., 2016). This network biosynthesis may be explained by the characteristics of chimeric terpene synthase enzymes (i.e., Au8003 and Au13192) whose PT domain can catalyze chain elongation from each substrate of DMAPP, GPP, FPP, or GGPP. These precursors could be a pool for these enzymes in the fungal strain and affect the end terpene product synthesis.

The method of in vivo gene functional identification for fungi may encounter the challenge of constructing mutant strains owing to protoplast preparation and choice of resistance genes for yielding correct transformations (Pudake, Kumari, Sahu, & Sultan, 2017). The other methods employed in fungi for disrupting a gene/cluster or blocking its expression are TALENs and CRISPR/Cas, RNAi, ribozymes, etc. (Wang et al., 2017). For *Aspergillus* species, the methods for mutant alleles can be constructed by PEG-mediated protoplast fusion, electroporation, shock waves, and biolistic transformation fusion such as *A. tumefaciens* (Pudake et al., 2017). If the object of interest is a nonmodel organism, the process of targeting specific gene/cluster is possibly difficult. In this case, a CRISPR–Cas9-based system facilitates in the genetically manipulating of filamentous fungi (Nødvig, Nielsen, Kogle, & Mortensen, 2015).

The in vitro method of enzyme function verification needs to obtain the purified enzyme of interest first and carry the corresponding reaction subsequently. The common hosts for protein expression are *Saccharomyces cerevisiae* (Cacho & Tang, 2016; Liu, Martínez, Liu, Petranovic, & Nielsen, 2014) and *E. coli* (Rosano & Ceccarelli, 2014). In this chapter, the function of the enzymes Au8003, Au13192, Au6298 were revealed by this means via expression and purification from *E. coli* BL21 (DE3) and verified their function by the in vitro reactions. It should be kept in mind that the real function and the physiological activity of the enzyme in the native host may be different from the in vitro biochemical reaction. Another method for functional characterization is cloning the corresponding coding gene in proper heterologous hosts, such as Aspergilli (Anyaogu & Mortensen, 2015; Lv et al., 2017), and thus revealing the function via comparing product profiles.

In conclusion, this chapter systematically introduces the methodology for the elucidation of terpene synthases and hopes to boost the development of unknown terpene synthase, in particular, chimeric synthase, since there are many novel terpene compounds discovered from marine fungi (Ebel, 2010; Elissawy, El-Shazly, Ebada, Singab, & Proksch, 2015). Major breakthroughs in other fields, such as the characterized crystalline structure of a chimeric synthase PaFS (Chen, Chou, Toyomasu, Cane, & Christianson, 2016) and maturing synthetic biology platform (Gruchattka, Hädicke, Kayser, Klamt, & Schütz, 2013), weighed well in broadening the applications of this chapter.

REFERENCES

Anthony, J. R., Anthony, L. C., Nowroozi, F., Kwon, G., Newman, J. D., & Keasling, J. D. (2009). Optimization of the mevalonate-based isoprenoid biosynthetic pathway in *Escherichia coli* for production of the anti-malarial drug precursor amorpha-4, 11-diene. *Metabolic Engineering*, *11*, 13–19.

Aichinger, C., Hansson, K., Eichhorn, H., Lessing, F., Mannhaupt, G., Mewes, W., et al. (2003). Identification of plant-regulated genes in *Ustilago maydis* by enhancer-trapping mutagenesis. *Molecular Genetics and Genomics: MGG*, *270*(4), 303–314.

Altschul, S. F., Gish, W., Miller, W., Myers, E. W., & Lipmen, D. J. (1990). Basic local alignment search tool. *Journal of Molecular Biology*, *215*, 403–410.

Anyaogu, D. C., & Mortensen, U. H. (2015). Heterologous production of fungal secondary metabolites in aspergilli. *Frontiers in Microbiology*, *6*, 77.

Aros, D., Gonzalez, V., Allemann, R. K., Müller, C. T., Rosati, C., & Rogers, H. J. (2012). Volatile emissions of scented alstroemeria genotypes are dominated by terpenes, and a myrcene synthase gene is highly expressed in scented alstroemeria flowers. *Journal of Experimental Botany*, *63*(7), 2739.

Asch, D. K., & Kinsey, J. A. (1990). Relationship of vector inserts size to homologous integration during transformation of Neurospora crassa with cloned am (GDH) gene. *Molecular Genetics and Genomics: MGG*, *221*, 37–43.

Berg, M., & Maruthachalam, K. (2015). Genetic transformation systems in fungi. *Fungal biology: Vol. 1. Springer International Publishing.*

Bian, G., Han, Y., Hou, A., Yuan, Y., Liu, X., Deng, Z., et al. (2017). Releasing the potential power of terpene synthases by a robust precursor supply platform. *Metabolic Engineering*, *42*, 1–8.

Bowman, J. M., Braxton, M. S., Churchill, M. A., Hellie, J. D., Starrett, S. J., Causby, G. Y., et al. (1997). Extraction method for the isolation of terpenes from plant tissue and subsequent determination by gas chromatography. *Microchemical Journal*, *56*(1), 10–18.

Bromann, K., Toivari, M., Viljanen, K., Vuoristo, A., Ruohonen, L., & Nakarisetälä, T. (2012). Identification and characterization of a novel diterpene gene cluster in *Aspergillus nidulans*. *PLoS One*, 7(4), 350–354.

Cacho, R. A., & Tang, Y. (2016). Reconstitution of fungal nonribosomal peptide synthetases in yeast and in vitro. In B. Evans (Ed.), *Methods in molecular biology, Vol. 1401. Nonribosomal peptide and polyketide biosynthesis*. New York, NY: Humana Press.

Cane, D. E., & Ikeda, H. (2012). Exploration and mining of the bacterial terpenome. *Accounts of Chemical Research*, *45*(3), 463–472. https://doi.org/10.1021/ar200198d. (Epub 2011 Oct 31).

Chai, H., Ru, Y., Liu, Y., Meng, H., Zhou, X., Zhou, G., et al. (2016). Sesterterpene ophiobolin biosynthesis involving multiple gene clusters in *Aspergillus ustus*. *Scientific Reports*, *6*, 27181. https://doi.org/10.1038/srep27181.

Chang, Q., Yang, J., Liu, L., Su, Y., Li, J., Zhu, Y., et al. (2010). Chlorimuronethyl resistance selectable marker unsuited for the transformation of rice blast fungus. *IFIP Advances in Information and Communication Technology*, *344*, 335–342.

Chaveroche, M. K., Ghigo, J. M., & D'Enfert, C. (2000). A rapid method for efficient gene replacement in the filamentous fungus *Aspergillus nidulans*. *Nucleic Acids Research*, *28*(22), E97.

Chen, M., Chou, W. K., Toyomasu, T., Cane, D. E., & Christianson, D. W. (2016). Structure and function of fusicoccadiene synthase, a hexameric bifunctional diterpene synthase. *ACS Chemical Biology*, *11*, 889–899.

Chiba, R., Minami, A., Gomi, K., & Oikawa, H. (2013). Identification of ophiobolin F synthase by a genome mining approach: A sesterterpene synthase from Aspergillus clavatus. *Organic Letters*, *15*(3), 594–597.

Christianson, D. W. (2007). Roots of biosynthetic diversity. *Science*, *316*, 60–61.

Combier, J. P., Melayah, D., Raffier, C., Gay, G., & Marmeisse, R. (2003). Agrobacterium tumefaciens-mediated transformation as a tool for insertional mutagenesis in the symbiotic ectomycorrhizal fungus *Hebeloma cylindrosporum*. *FEMS Microbiology Letters*, *220*(1), 141.

Delazar, A., Nahar, L., Hamedeyazdan, S., & Sarker, S. D. (2012). Microwave-assisted extraction in natural products isolation. *Methods in Molecular Biology*, *864*, 89.

Duyvesteijn, R. G., van Wijk, R., Boer, Y., Rep, M., Cornelissen, B. J., & Haring, M. A. (2005). Frp1 is a *Fusarium oxysporum* F-box protein required for pathogenicity on tomato. *Molecular Microbiology*, *57*(4), 1051–1063.

Ebel, R. (2010). Terpenes from marine-derived fungi. *Marine Drugs*, *8*, 2340–2368.

Elissawy, A. M., El-Shazly, M., Ebada, S. S., Singab, A. B., & Proksch, P. (2015). Bioactive terpenes from marine-derived fungi. *Marine Drugs*, *13*(4), 1966–1992.

Firon, A., Villalba, F., Beffa, R., & D'Enfert, C. (2003). Identification of essential genes in the human fungal pathogen Aspergillus fumigatus by transposon mutagenesis. *Eukaryotic Cell*, *2*(2), 247.

Fojtová, J., Lojková, L., & Kubán, V. (2008). GC/MS of terpenes in walnut-tree leaves after accelerated solvent extraction. *Journal of Separation Science*, *31*(1), 162.

Fujii, H., Koyama, T., & Ogura, K. (1982). Efficient enzymatic hydrolysis of polyprenyl pyrophosphates. *Biochimica et Biophysica Acta*, *712*(3), 716.

Gao, X., Chooi, Y. H., Ames, B. D., Wang, P., Walsh, C. T., & Tang, Y. (2011). Fungal indole alkaloid biosynthesis: Genetic and biochemical investigation of the tryptoquialanine pathway in *Penicillium aethiopicum*. *Journal of the American Chemical Society*, *133*(8), 2729–2741.

Gao, Y., Honzatko, R. B., & Peters, R. J. (2012). Terpenoid synthase structures: A so far incomplete view of complex catalysis. *Natural Product Reports*, *29*, 1153–1175.

Goldenberger, D., Perschil, I., Ritzler, M., & Altwegg, M. (1995). A simple "universal" DNA extraction procedure using SDS and proteinase k is compatible with direct PCR amplification. *PCR Methods and Applications*, *4*(6), 368–370.

Grodberg, J., & Dunn, J. J. (1988). ompT encodes the Escherichia coli outer membrane protease that cleaves t7 RNA polymerase during purification. *Journal of Bacteriology*, *17*(3), 1245–1253.

Gruchattka, E., Hädicke, O., Kayser, O., Klamt, S., & Schütz, V. (2013). In silico profiling of *Escherichia coli* and *Saccharomyces cerevisiae* as terpenoid factories. *Microbial Cell Factories*, *12*, 84.

Gui, H., Li, X., Liu, Y., Han, K., & Li, X. (2014). The relationship between PMI (manA) gene expression and optimal selection pressure in indica rice transformation. *Plant Cell Reports*, *33*(7), 1081–1090.

Hamer, L., Adachi, K., Montenegro-Chamorro, M. V., Tanzer, M. M., Mahanty, S. K., Lo, C., et al. (2001). Gene discovery and gene function assignment in filamentous fungi. *Proceedings of the National Academy of Sciences of the United States of America*, *98*(9), 5110–5115.

Hussain, & Maqbool. (2014). GC-MS: Principle, technique and its application in food science. *International Journal of Current Science*, *13*(E), 116–126.

Hoffmeister, D., & Keller, N. P. (2007). Natural products of filamentous fungi: Enzymes, genes, and their regulation. *Natural Product Reports*, *24*, 393–416.

Huang, X., Huang, H., Li, H., Sun, X., Huang, H., Lu, Y., et al. (2013). Asperterpenoid A, a new sesterterpenoid as an inhibitor of Mycobacterium tuberculosis protein tyrosine phosphatase B from the culture of Aspergillus sp.16-5c. *Organic Letters*, *15*, 721–723.

Hynes, M. J. (1994). Regulatory circuits of the amdS gene of Aspergillus nidulans. *Antonie Van Leeuwenhoek*, *65*(3), 179–182.

Hynes, M. J. (1996). Genetic transformation of filamentous fungi. *Journal of Genetics*, *75*, 297–311.

Jadoun, J., Shadkchan, Y., & Osherov, N. (2004). Disruption of the *Aspergillus fumigatus* argB gene using a novel in vitro transposon-based mutagenesis approach. *Current Genetics*, *45*(4), 235–241.

Jia, M., Potter, K. C., & Peters, R. J. (2016). Extreme promiscuity of a bacterial and a plant diterpene synthase enables combinatorial biosynthesis. *Metabolic Engineering*, *37*, 24.

Jiang, H., Shen, Y., Liu, W., & Lu, L. (2014). Deletion of the putative stretch-activated ion channel Mid1 is hyper virulent in *Aspergillus fumigatus*. *Fungal Genetics and Biology*, *62*(1), 62.

Kane, J. F. (1995). Effects of rare codon clusters on high-level expression of heterologous proteins in *Escherichia coli*. *Current Opinion in Biotechnology*, *6*(5), 494–500.

Kappers, I. F., Aharoni, A., van Herpen, T. W., Luckerhoff, L. L., Dicke, M., & Bouwmeester, H. J. (2005). Genetic engineering of terpenoid metabolism attracts bodyguards to *Arabidopsis*. *Science*, *309*(5743), 2070–2072.

Keilholz, U., Willhauck, M., Rimoldi, D., Brasseur, F., Dummer, W., Rass, K., et al. (1998). Reliability of reverse transcription-polymerase chain reaction (RT-PCR)-based assays for the detection of circulating tumour cells: A quality-assurance initiative of the eortc melanoma cooperative group. *European Journal of Cancer*, *34*(5), 750–753.

Keinänen, M., Oldham, N. J., & Baldwin, I. T. (2001). Rapid HPLC screening of jasmonate-induced increases in tobacco alkaloids, phenolics, and diterpene glycosides in *Nicotiana attenuata*. *Journal of Agricultural and Food Chemistry*, *49*(8), 3553–3558.

Keller, N. P., & Hohn, T. M. (1997). Metabolic pathway gene clusters in filamentous fungi. *Fungal Genetics and Biology*, *21*(1), 17–29.

Keller, N. P., Turner, G., & Bennett, J. W. (2005). Fungal secondary metabolism—From biochemistry to genomics. *Nature Reviews. Microbiology*, *3*, 937–947.

Khaldi, N., Seifuddin, F. T., Turner, G., Haft, D., Nierman, W. C., Wolfe, K. H., et al. (2010). SMURF: Genomic mapping of fungal secondary metabolite clusters. *Fungal Genetics and Biology*, *47*(9), 736–741.

Kusec, R. (1998). Reverse transcriptase-polymerase chain reaction (RT-PCR). *Methods in Molecular Medicine*, *16*(2), 149.

Laskovics, F. M., & Poulter, C. D. (1981). Prenyltransferase: Determination of the binding mechanism and individual kinetic constants for farnesylpyrophosphate synthetase by rapid quench and isotope partitioning experiments. *Biochemistry*, *20*(7), 1893–1901.

Leger, R. J. S., Joshi, L., Bidochka, M. J., & Roberts, D. W. (1996). Construction of an improved mycoinsecticide overexpressing a toxic protease. *Proceedings of the National Academy of Sciences of the United States of America*, *93*(13), 6349–6354.

Li, R., Zhu, H., Ruan, J., Qian, W., Fang, X., Shi, Z., et al. (2010). De novo assembly of human genomes with massively parallel short read sequencing. *Genome Research*, *20*(2), 265–272.

Liao, J., Zheng, N., & Qu, B. (2016). An improved ultrasonic-assisted extraction method by optimizing the ultrasonic frequency for enhancing the extraction efficiency of lycopene from tomatoes. *Food Analytical Methods*, *9*(8), 2288–2298.

Lin, H. C., Tsunematsu, Y., Dhingra, S., Xu, W., Fukutomi, M., Chooi, Y. H., et al. (2014). Generation of complexity in fungal terpene biosynthesis: Discovery of a multifunctional cytochrome p450 in the fumagillin pathway. *Journal of the American Chemical Society*, *136*(11), 4426–4436.

Lipp, M., Brodmann, P., Pietsch, K., Pauwels, J., Anklam, E., Börchers, T., et al. (1999). IUPAC collaborative trial study of a method to detect genetically modified soybeans and maize in dried powder. *Journal of AOAC International*, *82*, 923–928.

Liu, R., Chen, L., Jiang, Y. P., Zhou, Z. H., & Zou, G. (2015). Efficient genome editing in filamentous fungus *Trichoderma reesei* using the CRISPR/Cas9 system. *Cell Discovery*, *1*, 15007.

Liu, L., Martínez, J. L., Liu, Z., Petranovic, D., & Nielsen, J. (2014). Balanced globin protein expression and heme biosynthesis improve production of human hemoglobin in Saccharomyces cerevisiae. *Metabolic Engineering*, *21*, 9–16.

Lu, Z., Wang, Y., Miao, C., Liu, P., Hong, K., & Zhu, W. (2009). Sesquiterpenoids and benzofuranoids from the marine-derived fungus *Aspergillus ustus* 094102. *Journal of Natural Products*, *72*, 1761–1767.

Lv, J. M., Hu, D., Gao, H., Kushiro, T., Awakawa, T., Chen, G. D., et al. (2017). Biosynthesis of helvolic acid and identification of an unusual C-4-demethylation process distinct from sterol biosynthesis. *Nature Communications*, *8*, 1644.

Malardier, L., Daboussi, M. J., Julien, J., Roussel, F., Scazzocchio, C., & Brygoo, Y. (1989). Cloning of the nitrate reductase gene (niaD) of *Aspergillus nidulans* and its use for transformation of *Fusarium oxysporum*. *Gene*, *78*(1), 147–156.

Matsuda, Y., Mitsuhashi, T., Quan, Z., & Abe, I. (2015). Molecular basis for stellatic acid biosynthesis: A genome mining approach for discovery of sesterterpene synthases. *Organic Letters*, *17*(18), 4644.

Ménoret, S., Tesson, L., Rémy, S., Usal, C., Thépenier, V., Thinard, R., et al. (2014). Gene targeting in rats using transcription activator-like effector nucleases. *Methods*, *69*(1), 102–107.

Mierendorf, R. C., Morris, B. B., Hammer, B., & Novy, R. E. (1998). Expression and purification of recombinant proteins using the pet system. *Methods in Molecular Medicine*, *13*, 257.

Minami, A., Tajima, N., Higuchi, Y., Toyomasu, T., Sassa, T., Kato, N., et al. (2009). Identification and functional analysis of brassicicene C biosynthetic gene cluster in *Alternaria brassicicola*. *Bioorganic & Medicinal Chemistry Letters*, *19*, 870–874.

Mitani, Y., Nakayama, T., Harbers, M., & Hayashizaki, Y. (2004). Aptamer-dependent full-length cDNA synthesis by overlap extension PCR. *BioTechniques*, *37*(1), 124. 126, 128.

Mitsuhashi, T., Rinkel, J., Okada, M., Abe, I., & Dickschat, J. S. (2017). Mechanistic characterization of two chimeric sesterterpene synthases from *Penicillium*. *Chemistry*, *23*(42), 10053–10057.

Mohd-Zin, S. W., Abdullah, N. L., Abdullah, A., Greene, N. D., Cheah, P. S., Ling, K. H., et al. (2016). Identification of the genomic mutation in Epha4 (rb-2J/rb-2J) mice. *Genome*, *59*(7), 439.

Muzika, R. M., Campbell, C. L., Hanover, J. W., & Smith, A. L. (1990). Comparison of techniques for extracting volatile compounds from conifer needles. *Journal of Chemical Ecology*, *16*(9), 2713–2722.

Nakayashiki, H. (2005). RNA silencing in fungi: Mechanisms and applications. *FEBS Letters*, *579*(26), 5950–5957.

Nielsen, M. L., Isbrandt, T., Rasmussen, K. B., Thrane, U., Hoof, J. B., Larsen, T. O., et al. (2017). Genes linked to production of secondary metabolites in *Talaromyces atroroseus* revealed using CRISPR-Cas9. *PLoS One*, *12*(1), e0169712.

Nødvig, C. S., Nielsen, J. B., Kogle, M. E., & Mortensen, U. H. (2015). A CRISPR-Cas9 system for genetic engineering of filamentous fungi. *PLoS One*, *10*, e0133085.

Okada, M., Matsuda, Y., Mitsuhashi, T., Hoshino, S., Mori, T., Nakagawa, K., et al. (2016). Genome-based discovery of an unprecedented cyclization mode in fungal sesterterpenoids biosynthesis. *Journal of the American Chemical Society*, *138*(31), 10011–10018.

Olmedo-Monfil, V., Cortés-Penagos, C., & Herrera-Estrella, A. (2004). Three decades of fungal transformation: Key concepts and applications. *Methods in Molecular Biology*, *267*, 297.

Orsini, M., & Romano-Spica, V. (2001). A microwave-based method for nucleic acid isolation from environmental samples. *Letters in Applied Microbiology*, *33*(1), 17.

Penwell, W. F., Arivett, B. A., & Actis, L. A. (2012). The acinetobacter baumannii entA gene located outside the acinetobactin cluster is critical for siderophore production, iron acquisition and virulence. *PLoS One*, 7(5), e36493.

Pourmortazavi, S. M., & Hajimirsadeghi, S. S. (2007). Supercritical fluid extraction in plant essential and volatile oil analysis. *Journal of Chromatography A*, *1163*(1–2), 2–24.

Pudake, R. N., Kumari, M., Sahu, B. B., & Sultan, E. (2017). *Targeted gene disruption tools for fungal genomics in modern tools and techniques to understand microbes*. Springer International Publishing.

Qin, B., Matsuda, Y., Mori, T., Okada, M., Quan, Z., Mitsuhashi, T., et al. (2016). An unusual chineric diterpene synthase form Emericella variecolor and its functional conversion into a sesterterpene synthase by domain swapping. *Angewandte Chemie International Edition*, *55*, 1658–1661.

Quin, M. B., Flynn, C. M., & Schmidt-Dannert, C. (2014). Traversing the fungal terpenome. *Natural Product Reports*, *31*(10), 1449–1473.

Renner, M. K., Jensen, P. R., & Fenical, W. (1998). Neomangicols: Structures and absolute stereochemistries of unprecedented halogenated sesterterpenes from a marine fungus of the genus *Fusarium*. *The Journal of Organic Chemistry*, *63*, 8346–8354.

Renner, M. K., Jensen, P. R., & Fenical, W. (2000). Mangicols: Structures and biosynthesis of a new class of sesterterpene polyols from a marine fungus of the genus *Fusarium*. *The Journal of Organic Chemistry*, *65*, 4843–4852.

Rosano, G. L., & Ceccarelli, E. A. (2014). Recombinant protein expression in Escherichia coli: Advances and challenges. *Frontiers in Microbiology*, *5*, 172.

Ruiz-Díez, B. (2002). Strategies for the transformation of filamentous fungi. *Journal of Applied Microbiology*, *92*(2), 189.

Shuman, S. (1994). Novel approach to molecular cloning and polynucleotide synthesis using vaccinia DNA topoisomerase. *The Journal of Biological Chemistry*, *269*(51), 32678–32684.

Sinzger, C., Knapp, J., Schmidt, K., Kahl, M., & Jahn, G. (1999). A simple and rapid method for preparation of viral DNA from cell associated cytomegalovirus. *Journal of Virological Methods*, *81*(1–2), 115.

Steeghs, M., Bais, H. P., de Gouw, J., Goldan, P., Kuster, W., Northway, M., et al. (2004). Proton-transfer-reaction mass spectrometry as a new tool for real time analysis of root-secreted volatile organic compounds in *Arabidopsis*. *Plant Physiology*, *135*(1), 47–58.

Strunk, M. R., & Bechtold, W. E. (1996). Automated two-dimensional interface for capillary gas chromatography. US, US5492555.

Studier, F. W., & Moffatt, B. A. (1986). Use of bacteriophage T7 RNA polymerase to direct selective high-level expression of cloned genes. *Journal of Molecular Biology*, *189*(1), 113.

Szewczyk, E., Nayak, T., Oakley, C. E., Edgerton, H., Xiong, Y., Taheritalesh, N., et al. (2006). Fusion PCR and gene targeting in *Aspergillus nidulans*. *Nature Protocols*, *1*(6), 3111–3120.

Ter–Hovhannisyan, V., Lomsadze, A., Chernoff, Y. O., & Borodovsky, M. (2008). Gene prediction in novel fungal genomes using an ab initio algorithm with unsupervised training. *Genome Research*, *18*, 1979–1990.

Tholl, D., Boland, W., Hansel, A., Loreto, F., Rose, U., & Schnitzler, J. (2006). Practical approaches to plant volatile analysis. *Plant Journal*, *45*(4), 540–560.

Tian, W., Deng, Z., & Hong, K. (2017). The biological activities of sesterterpenoid-type ophiobolins. *Marine Drugs*, *15*, 229. https://doi.org/10.3390/md15070229.

Toyomasu, T., Tsukahara, M., Kaneko, A., Niida, R., Mitsuhashi, W., Dairi, T., et al. (2007). Fusicoccins are biosynthesized by an unusual chimera diterpene synthase in fungi. *Proceedings of the National Academy of Sciences of the United States of America*, *104*(9), 3084–3088.

Utashima, Y., Yamashita, S., Arima, T. H., & Masaki, K. (2017). Codon optimization enables the zeocin resistance marker's use in the ascomycete yeast *Debaryomyces occidentalis*. *The Journal of General and Applied Microbiology*, *63*(4), 254–257.

Ueda, D., Yamaga, H., Murakami, M., Totsuka, Y., Shinada, T., & Sato, T. (2015). Biosynthesis of sesterterpene, head-to-tail triterpene, and sesquarterpene in Bacillus clausii: Identification of multifunctional enzymes and analysis of isoprenoid metabolites. *Chembiochem*, *16*(9), 1371–1377.

Vahur, S., Teearu, A., Haljasorg, T., Burk, P., Leito, I., & Kaljurand, I. (2012). Analysis of dammar resin with MALDI-FT-ICR-MS and APCI-FT-ICR-MS. *Journal of Mass Spectrometry*, *47*(3), 392–409.

Wang, L., Lin, J., Zhang, T., Xu, K., Ren, C., & Zhang, Z. (2013). Simultaneous screening and validation of effective zinc finger nucleases in yeast. *PLoS One*, *8*(5), e64687.

Wang, L., Yang, B., Lin, X., Zhou, X., & Liu, Y. (2013). Sesterterpenoids. *Natural Product Reports*, *30*, 455–473.

Wang, S., Chen, H., Tang, X., Zhang, H., Chen, W., & Chen, Y. Q. (2017). Molecular tools for gene manipulation in filamentous fungi. *Applied Microbiology and Biotechnology*, *101*, 8063–8075.

Wang, X., Teng, D., Tian, F., Guan, Q., & Wang, J. (2012). Comparison of three DNA extraction methods for feed products and four amplification methods for the 5′-junction fragment of roundup ready soybean. *Journal of Agricultural and Food Chemistry*, *60*(18), 4586–4595.

Weber, T., Blin, K., Duddela, S., Krug, D., Kim, H. U., Bruccoleri, R., et al. (2015). antiSMASH 3.0—A comprehensive resource for the genome mining of biosynthetic gene clusters. *Nucleic Acids Research*, *43*(W1), W237–W243.

Sandhu, G. S., Kline, B. C., Stockman, L., & Roberts, G. D. (1995). Molecular probes for diagnosis of fungal infections. *Journal of Clinical Microbiology*, *33*(11), 2913–2919.

Yamada, Y., Kuzuyama, T., Komatsu, M., Shin-ya, K., Omura, S., Cane, D. E., et al. (2015). Terpene synthases are widely distributed in bacteria. *Proceedings of the National Academy of Sciences of the United States of America*, *112*, 857–862.

Yang, T., Lu, Z., Meng, L., Wei, S., Hong, K., Zhu, W., et al. (2012). The novel agent ophiobolin O induces apoptosis and cell cycle arrest of MCF-7 cells through activation of MAPK signaling pathways. *Bioorganic & Medicinal Chemistry Letters*, *22*, 579–585.

Ye, Y., Minami, A., Mandi, A., Liu, C., Taniguchi, T., Kuzuyama, T., et al. (2015). Genome mining for sesterterpenes using bifunctional terpene synthases reveals a unified intermediated of di/sesterterenes. *Journal of the American Chemical Society*, *137*(36), 11846–11853.

Yu, J. H., Hamari, Z., Han, K. H., Seo, J. A., Reyes-Domínguez, Y., & Scazzocchio, C. (2004). Double-joint PCR: A PCR-based molecular tool for gene manipulations in filamentous fungi. *Fungal Genetics and Biology*, *41*(11), 973–981.

Zhang, D., Yang, Y., Castlebury, L. A., & Cerniglia, C. E. (1996). A method for the large scale isolation of high transformation efficiency fungal genomic DNA. *FEMS Microbiology Letters*, *145*(2), 261–265.

Zheng, Z. L., Qiu, X. H., & Han, R. C. (2015). Identification of the genes involved in the fruiting body production and cordycepin formation of *Cordyceps militaris* fungus. *Mycobiology*, *43*(1), 37–42.

Zhong, G., Wei, W., Guan, Q., Ma, Z., Wei, H., Xu, X., et al. (2012). Phosphoribosyl pyrophosphate synthetase, as a suppressor of the sepH mutation in *Aspergillus nidulans*, is required for the proper timing of septation. *Molecular Microbiology*, *86*(4), 894–907.

Zhou, L. H., Li, G. Q., Wang, Z. X., & Zhuge, J. (2005). Preparation and regeneration of protoplasts from *Monascus purpureus* and genetic transformation system. *Hereditas*, *27*(3), 423–428.

Zhou, M. Y., Clark, S. E., & Gomezsanchez, C. E. (1995). Universal cloning method by TA strategy. *BioTechniques*, *19*(1), 34–35.

FURTHER READING

Toyomasu, T., Kaneko, A., Tokiwano, T., Kanno, Y., Kanno, Y., Niida, R., et al. (2009). Biosynthetic gene-based secondary metabolite screening: A new diterpene, methyl phomosenonate, from the fungus *Phomopsis amygdali*. *The Journal of Organic Chemistry*, *74*, 1541–1548.

CHAPTER SIXTEEN

Assaying Oxidative Coupling Activity of CYP450 Enzymes

Vinayak Agarwal[*,†,1]
[*]School of Chemistry and Biochemistry, Georgia Institute of Technology, Atlanta, GA, United States
[†]School of Biological Sciences, Georgia Institute of Technology, Atlanta, GA, United States
[1]Corresponding author: e-mail address: vagarwal@gatech.edu

Contents

Abstract

Cytochrome P450 (CYP450) enzymes are ubiquitous catalysts in natural product biosynthetic schemes where they catalyze numerous different transformations using radical intermediates. In this protocol, we describe procedures to assay the activity of a marine bacterial CYP450 enzyme Bmp7 which catalyzes the oxidative radical coupling of polyhalogenated aromatic substrates. The broad substrate tolerance of Bmp7, together with rearrangements of the aryl radical intermediates leads to a large number of products to be generated by the enzymatic action of Bmp7. The complexity of the product pool generated by Bmp7 thus presents an analytical challenge for structural elucidation. To address this challenge, we describe mass spectrometry-based procedures to provide structural insights into aryl crosslinked products generated by Bmp7, which can complement subsequent spectroscopic experiments. Using the procedures described here, for the first time, we show that Bmp7 can efficiently accept polychlorinated aryl substrates, in addition to the physiological polybrominated substrates for the biosynthesis of polyhalogenated marine natural products.

Methods in Enzymology, Volume 604
ISSN 0076-6879
https://doi.org/10.1016/bs.mie.2018.01.029

1. INTRODUCTION

Cytochrome P450 (CYP450) enzymes catalyze an unparalleled diversity of primary and secondary metabolic reactions across all domains of life. Two single electron transfers and activation by molecular oxygen of a resting Fe^{III} bound to a porphyrin scaffold in the CYP450 enzyme active site leads to the formation of an electrophilic $Fe^{IV}=O+^{\bullet}$ oxoferryl species that can homolytically cleave X—H bonds (X = C, N, O, and S) in organic substrates to generate a radical intermediate. The substrate radical can then participate in myriad downstream reactions with regio- and stereospecific control exercised by the CYP450 enzyme active site. CYP450 enzymes have evolved to take full advantage of the redox potential of the oxoferryl species by accommodating a wide diversity of substrates to match the diversity of reactions that they catalyze. Several excellent reviews that describe the underlying iron-porphyrin activation scheme for CYP450 enzymes and the ensuing reactions that they catalyze are available (Denisov, Makris, Sligar, & Schlichting, 2005; Podust & Sherman, 2012; Poulos, 2014; Rudolf, Chang, Ma, & Shen, 2017; Tang, Zou, Watanabe, Walsh, & Tang, 2017). Efforts to engineer CYP450 enzymes to catalyze reactions that fall outside the current scope of these enzymes further highlight their tremendous catalytic prowess (Girvan & Munro, 2016; McIntosh, Farwell, & Arnold, 2014). CYP450 enzymes require two partner enzymes—a ferredoxin-$NAD(P)^+$ oxidoreductase and a ferredoxin—for activity. These reaction partners channel two electrons from a NAD(P)H donor via a single electron adapter (flavin), finally delivering two single electrons sequentially to the iron-porphyrin reaction center in the CYP450 enzyme active site via the ferredoxin Fe–S cluster (Poulos, 2014; Tripathi, Li, & Poulos, 2013). While numerous variants to this theme exist, the underlying electron transfer route from NAD(P)H to the iron-porphyrin reaction center is fairly conserved (Guengerich & Munro, 2013).

Aryl crosslinking is a well-represented transformation among the reactions catalyzed by CYP450 enzymes. Perhaps the most celebrated examples for CYP450-mediated aryl crosslinking reactions are the installation of the intra- and intermolecular ether and carbon–carbon bonds in the biosynthesis of glycopeptide antibiotics such as vancomycin (**1**, Fig. 1) and plant alkaloids such as berbamunine (**2**) and magnoflorine (**3**) (Ikezawa, Iwasa, & Sato, 2008; Kraus & Kutchan, 1995; Mizutani & Sato, 2011; Yim, Thaker, Koteva, & Wright, 2014). Furthermore, a dedicated CYP450 enzyme recruitment domain has recently been described to be embedded within the nonribosomal

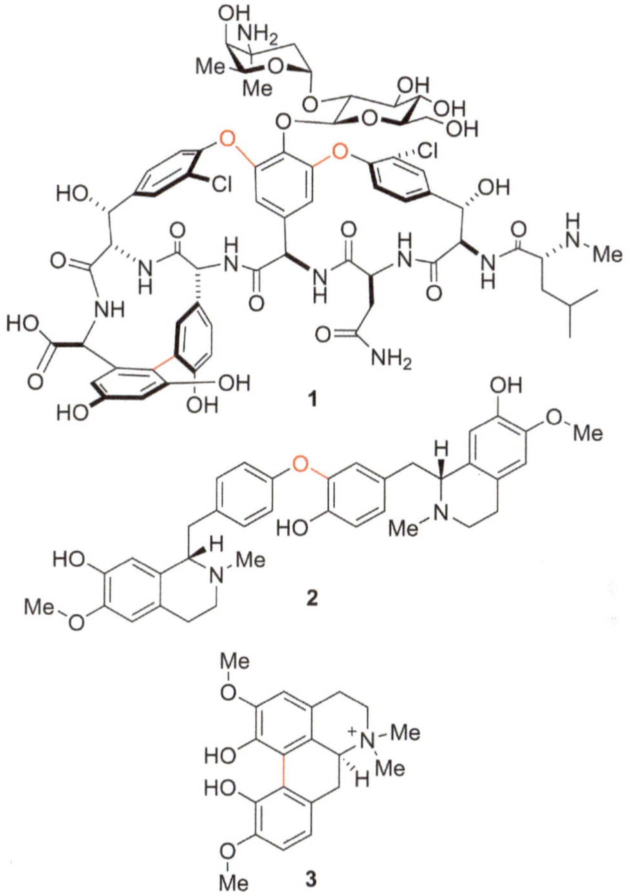

Fig. 1 Chemical structures of **1–3** with the C—C and C—O bonds generated by the respective CYP450 enzyme catalysts highlighted in *red*.

peptide synthetases that catalyze the formation of the glycopeptide antibiotics (Brieke, Peschke, Haslinger, & Cryle, 2015; Haslinger, Peschke, Brieke, Maximowitsch, & Cryle, 2015; Peschke, Gonsior, Sussmuth, & Cryle, 2016).

Each of the CYP450 enzymes alluded to above are highly selective catalysts, in that, the three aryl coupling reactions en route the biosynthesis of **1** are catalyzed by three different CYP450 enzymes in a sequential manner (Fig. 1). In sharp contrast is the recently described marine bacterial CYP450 enzyme Bmp7 that catalyzes the terminal carbon–carbon bond formation en route the biosynthesis of the marine natural product pentabromopseudilin (**4**, Fig. 2) (Agarwal et al., 2014). The total in vitro reconstitution of the biosynthesis of **4** revealed that the physiological substrates for Bmp7 are

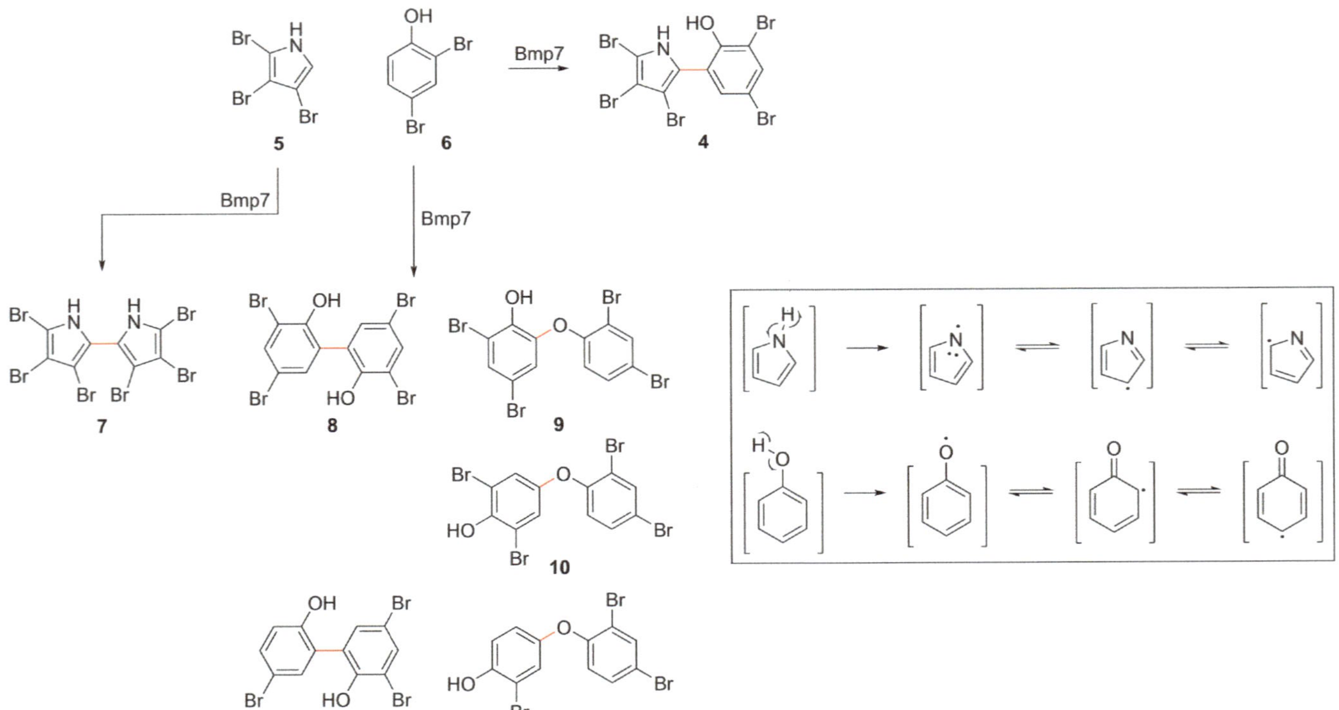

Fig. 2 Products **4**, **7–12** generated by Bmp7 using **5** and **6** as substrates. The coupling reactions catalyzed by Bmp7 are postulated to involve rearranged pyrrolic and phenoxy radical intermediates as shown in the *box*.

2,3,4-tribromopyrrole (**5**) and 2,4-dibromophenol (**6**) (Agarwal et al., 2014; El Gamal, Agarwal, Diethelm, et al., 2016; El Gamal, Agarwal, Rahman, & Moore, 2016). In addition to the communesin core generating CYP450 CnsC, Bmp7 is the only other known CYP450 capable of catalyzing intermolecular crosslinking of two different substrate molecules (Lin et al., 2016). Bmp7 can accept **5** and **6** individually as substrates to catalyze the formation of the marine natural product hexabromo-2,2′-bipyrrole (**7**), and a suite of dimeric phenolic products coupled via ether and carbon–carbon bonds as illustrated in Fig. 2 (Agarwal et al., 2014). The dimeric phenolic products generated by Bmp7 include 3,3′,5,5′-tetrabromobiphenyl-2,2′-diol (**8**) and 2-OH-BDE68 (**9**, BDE68 refers to brominated diphenyl ether congener 68), methoxylated derivatives of which have been shown to accumulate in the blubber of marine mammals (Marsh et al., 2005; Teuten, Xu, & Reddy, 2005). Structures of **4**, **7**–**9** can be rationalized based on the coupling of the rearranged pyrrolic and phenoxy radicals postulated to be generated by Bmp7, followed by rearomatization of the aryl rings (Fig. 2, boxed). In addition to the *ortho* C—O coupled **9**, isomeric *para* C—O coupled molecule **10** was observed as a product of the Bmp7 reaction, in addition to C—C coupled and *para* C—O coupled products **11** and **12**, respectively. Enzymatic synthesis of **11** and **12** involves dehalogenation accompanying oxidative radical coupling, as has been observed previously for halophenols (Dec, Haider, & Bollag, 2003; Osborne, Coggins, Raner, Walla, & Dawson, 2009; Osman, Boeren, Boersma, Veeger, & Rietjens, 1997). Remarkably, the substrate tolerance of Bmp7 extends even further, in that, it can catalyze the formation of dibenzo-*p*-dioxin products starting from (poly)brominated catechols as substrates (Agarwal & Moore, 2014).

Protocols described in this chapter detail the recombinant expression and purification of the CYP450 Bmp7, and its electron transfer partner proteins Bmp9 (ferredoxin) and Bmp10 (ferredoxin-NAD(P)$^+$ oxidoreductase) from the marine gammaproteobacterium *Pseudoalteromonas luteoviolacea* 2ta16, together with the NAD(P)H regenerating phosphite dehydrogenase enzyme PtdH (Agarwal et al., 2014; Johannes, Woodyer, & Zhao, 2007). We then describe procedures to assay for the activity of Bmp7 at analytical as well as preparative scales. Owing to the large number of products that Bmp7 generates, which depends on the possible intermediate aryl-radical rearrangements, characterization of its reaction product profile presents an analytical challenge as the chromatographic separation of products and their structure determination using spectroscopic techniques is untenable. To address this challenge, mass spectrometric methods to guide the characterization of Bmp7 products are presented that utilize the unique fragmentation patterns of C—C coupled, *ortho* C—O coupled, and *para* C—O coupled halophenols. Using the procedures described here, we demonstrate that Bmp7 can accommodate chlorinated aromatic molecules as substrates, in addition to brominated and iodinated substrates (Agarwal et al., 2014). The broad substrate tolerance of Bmp7 coupled with its ability to generate diverse products makes it an attractive candidate for future engineering efforts to develop a broadly applicable aryl coupling enzyme catalysts.

2. EQUIPMENTS AND CONSUMABLES

Equipments

1. Temperature-controlled incubators and shakers for bacterial growth
2. Refrigerated centrifuges
3. Sonicator
4. SDS-PAGE gel casting accessories and electrophoresis chamber
5. GE Healthcare Amersham Imager 600 or equivalent
6. White light transilluminator
7. Spectrophotometer
8. GE Healthcare AKTA Prime chromatography system or equivalent
9. Agilent 1260 Infinity Quaternary LC coupled to Agilent 6530 QToF mass spectrometer or equivalent

Consumables

1. Chromatography: His-Trap FF 5 mL (GE Healthcare); C18 $100 \times 4.6\,mm^2$ (Phenomenex).
2. Antibiotics: Kanamycin monosulfate and ampicillin trihydrate (Fisher).
3. Chemical and reagents: All chemicals and reagents were purchased from Fisher Scientific or Sigma Aldrich.
4. Solvents: Chromatography grade acetonitrile, water, and formic acid (Fisher).
5. Media: Luria-Bertani (LB) broth and Terrific broth (TB) (Fisher).
6. Centrifugal filters: Amicon Ultra-15 centrifugal filters with molecular weight cut-off (MWCO) of 3 and 10 kDa (EMD Millipore).

3. PREPARATION OF RECOMBINANT PROTEINS

3.1 Plasmid DNA for Recombinant Protein Expression

Table 1 lists the plasmids used in this chapter. To assist in purification, all plasmids are appended with fusion tags.

Table 1 Plasmids Used in This Study

Construct	Source Organism	Fusion Tag	Plasmid Backbone	Reference
Bmp7	*Pseudoalteromonas luteoviolacea* 2ta16	N-terminal His_6	pET-28b(+)	Agarwal et al. (2014)
Bmp9	*P. luteoviolacea* 2ta16	N-terminal His_6	pET-28b(+)	Agarwal et al. (2014)
Bmp10	*P. luteoviolacea* 2ta16	N-terminal His_6	pET-28b(+)	Agarwal et al. (2014)
PtdH	*Pseudomonas stutzeri*	N-terminal His_6	pET-15b(+)	Johannes et al. (2007)

3.2 Expression and Purification of Recombinant Proteins

Expression and purification of all proteins described in this chapter follows identical procedures unless otherwise noted. The following buffers are used in the procedures described in this section:

- Resuspension buffer: 20 mM Tris–HCl (pH 8.0), 500 mM NaCl, 10% glycerol.
- Wash buffer: 20 mM Tris–HCl (pH 8.0), 1 M NaCl, 30 mM imidazole.
- Elution buffer: 20 mM Tris–HCl (pH 8.0), 1 M NaCl, 250 mM imidazole.
- Storage buffer: 100 mM potassium phosphate (pH 7.6), 10% glycerol.

Experimental protocol

1. Transform plasmid DNA into chemically competent BL21Gold(DE3) cells. Plate the transformants on LB agar media with antibiotic selection and allow growth for 14 h at 37°C.
2. Pick individual colonies from LB agar plates in 10 mL TB media with antibiotic selection and grow at 37°C with vigorous shaking for 14 h.
3. Use 5 mL of the starter culture to innoculate 1 L of TB media with antibiotic selection and grow with vigorous shaking at 37°C till optical density (measured at 600 nm) reaches 0.6.
4. Reduce the incubation temperature to 18°C and induce protein expression by the addition of 0.3 mM IPTG (isopropyl-β-D-1-thiogalactopyranoside).
   a. For cultures expressing Bmp7, add 0.15 mM 5-aminolevulinic acid from a filter-sterilized 1 M stock solution.
   b. For cultures expressing Bmp9, add 10 mg $Fe(II)SO_4 \cdot 7H_2O$.
   c. For cultures expressing Bmp10, add 5 mg riboflavin. The culture will turn intense yellow upon addition of riboflavin.

5. Grow for additional 24 h with vigorous shaking at 18°C.
6. Collect biomass by centrifugation at 5000 × *g* for 15 min at 4°C and resuspend the pellet in 50 mL prechilled resuspension buffer.
7. Lyse bacterial cells by sonication. Care should be exercised to maintain the temperature close to 4°C during sonication.
8. Clarify the lysate by centrifugation at 30,000 × *g* for 1 h at 4°C. Collect the supernatant. The pellet can be discarded. The supernatants for cultures expressing Bmp7 and Bmp10 will be visibly red and yellow, respectively.
9. Apply the supernatant to a 5 mL His-Trap chromatography column using an AKTA Prime chromatography system operating at 4°C.
10. After the supernatant has been applied to the column, wash the column extensively with wash buffer.
11. Elute the bound protein by a continuous linear gradient to 100% elution buffer across 20 column volumes and collect the eluant in 5 mL fractions.
12. Analyze the eluant fractions by SDS-PAGE with Coomassie staining.
13. Pool the fractions with pure protein and concentrate to 2.5 mL final volume using centrifugal filters with 10 kDa MWCO. For Bmp9, use centrifugal filters with 3 kDa MWCO.
14. Desalt the concentrated protein into storage buffer using gravity flow desalting columns.
15. For Bmp7, record an absorbance spectrum from 350 to 700 nm. The absorbance spectra for purified Bmp7 are shown in Fig. 3.
16. Measure protein concentration using Bradford assay. Store purified Bmp7, Bmp10, and PtdH at −80°C in 100 μL aliquots. Store Bmp9 at 4°C.

Note: Procedures described here typically yield greater than 30 mg of Bmp7, Bmp10, and PtdH, and greater than 10 mg of Bmp9. Bmp9, as purified above, retains activity for ~2 days at 4°C. A fresh purification of Bmp9 is recommended after that Bmp7, Bmp10, and PtdH retain activity for an extended period of time at −80°C.

4. ACTIVITY ASSAYS

The following solvents are used in the procedures described in this section:

- HPLC solvent A: Water + 0.1% formic acid
- HPLC solvent B: MeCN + 0.1% formic acid
- Ethyl acetate
- Methanol

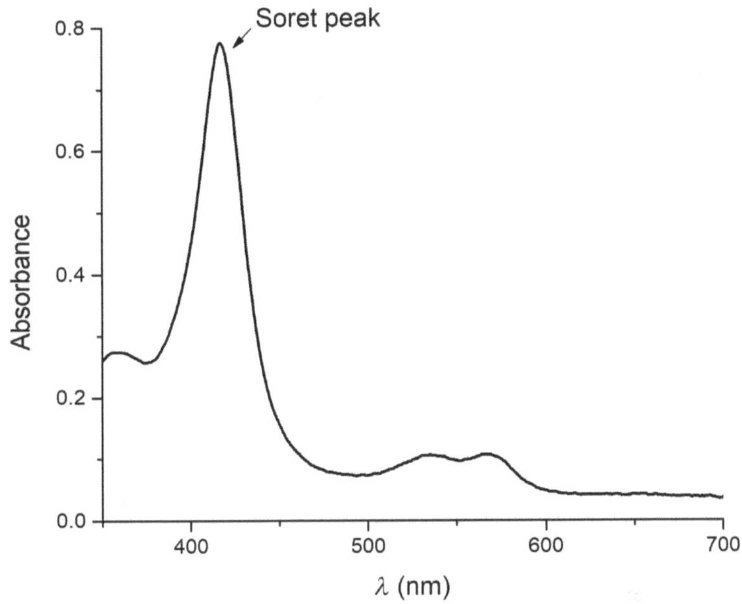

Fig. 3 UV–Vis absorbance profile for purified Bmp7 demonstrating the characteristic Soret peak at 420 nm.

4.1 Analytical Scale Assays

Experimental protocol

1. Thaw aliquots of Bmp7, Bmp10, and PtdH on ice. Centrifuge briefly using a table-top centrifuge.
2. Dissolve the small-molecule substrate(s) at an appropriate concentration in methanol.
3. Add reaction components to the final concentrations as listed in Table 2. For each substrate, the reactions should be set up in duplicate, with one of the reactions serving as the negative control (see below).
4. Initiate one of the two reactions by adding Na phosphite. Adjust the final reaction volume to 1.0 mL with water.
5. For the other reaction, add water in place of Na phosphite. This reaction is henceforth referred to as "negative control."
6. Incubate the reactions at 30°C for 12 h.
7. Centrifuge at 16,000 × *g* for 10 min to remove any aggregated proteins.
8. Extract the reactions twice with equal volume of ethyl acetate. Combine the organic layers.
9. Remove the solvent in vacuo. Resuspend the dried residue in 100 μL methanol.

Table 2 Final Concentration of Assay Components for Coupling of **13** by Bmp7 at the Analytical Scale

Component	Concentration in Assay
Bmp7	2 μ*M*
13	0.2 m*M*
Bmp9	2 μ*M*
Bmp10	2 μ*M*
PtdH	2 μ*M*
$NADP^+$	0.2 m*M*
Na phosphite	5 m*M*
Potassium phosphate (pH 7.6)	100 m*M*
Water	

10. Centrifuge at 16,000 × *g* for 10 min to remove any particulate matter. The solution can also be filtered through a 0.2 μ syringe filter to remove any particulate matter.

11. Inject 10 μL of this sample on a HPLC–MS/MS system and elute using the following solvent gradient at a flow rate of 0.7 mL/min using an C18 100 × 4.6 mm^2 column:

a. 0–5 min: 5% HPLC solvent B
b. 5–10 min: linear gradient to 70% HPLC solvent B
c. 10–35 min: linear gradient to 95% HPLC solvent B
d. 35–40 min: 95% HPLC solvent B
e. 40–41 min: linear gradient to 5% HPLC solvent B
f. 41–43 min: 5% HPLC solvent B
g. 43–44 min: linear gradient to 95% HPLC solvent B
h. 44–47 min: 95% HPLC solvent B
i. 47–48 min: linear gradient to 5% HPLC solvent B
j. 48–53 min: 5% HPLC solvent B

Note: As the polyhalogenated products generated by Bmp7 are very hydrophilic and bind strongly to the C18 stationary phase, the elution gradient described here includes reintroduction of high concentration of the organic solvent in the mobile phase which affords a rigorous regeneration of the stationary phase.

12. Acquire mass spectrometry data between 4 and 53 min in the negative ionization mode.

4.2 Preparative Scale Assays

Experimental protocol

1. Thaw aliquots of Bmp7, Bmp10, and PtdH on ice. Centrifuge briefly using a table-top centrifuge.
2. Dissolve the small-molecule substrate(s) at an appropriate concentration in methanol.
3. Add reaction components to the final concentrations as listed in Table 3. Duplicate reactions are not set up at the preparative scale.
4. Initiate the reactions by adding Na phosphite. Adjust the final reaction volume to 100 mL with water.
5. Incubate the reactions at 30°C.
6. After every 4 h, add a fresh aliquot of Na phosphite.
7. After 12 h, centrifuge at 25,000 × *g* for 10 min to remove any aggregated proteins.
8. Extract the reactions twice with equal volume of ethyl acetate. Combine the organic layers. Addition of saturated brine can be useful to resolve the aqueous and organic layers during extractions.
9. Remove the solvent in vacuo. Resuspend the dried residue in final volume of 1.0 mL methanol.
10. Centrifuge at 16,000 × *g* for 10 min to remove any particulate matter. The solution can also be filtered through a 0.2 μ syringe filter to remove any particulate matter.

Table 3 Final Concentration of Assay Components for Coupling of **13** by Bmp7 at the Preparative Scale

Component	Concentration in Assay
Bmp7	2 μ*M*
13	1.0 m*M*
Bmp9	2 μ*M*
Bmp10	2 μ*M*
PtdH	2 μ*M*
$NADP^+$	0.2 m*M*
Na phosphite	10 m*M*
Potassium phosphate (pH 7.6)	100 m*M*
Water	

11. A diluted aliquot of the sample thus generated can be analyzed by analytical HPLC–MS/MS as described in Section 4.1.
12. Sequentially inject 50 μL of this sample on a HPLC–MS/MS system and elute using the following solvent gradient at a flow rate of 3.0 mL/min using an C18 semipreparative column:
    a. 0–5 min: 5% HPLC solvent B
    b. 5–10 min: linear gradient to 70% HPLC solvent B
    c. 10–35 min: linear gradient to 95% HPLC solvent B
    d. 35–40 min: 95% HPLC solvent B
    e. 40–41 min: linear gradient to 5% HPLC solvent B
    f. 41–43 min: 5% HPLC solvent B
    g. 43–44 min: linear gradient to 95% HPLC solvent B
    h. 44–47 min: 95% HPLC solvent B
    i. 47–48 min: linear gradient to 5% HPLC solvent B
    j. 48–53 min: 5% HPLC solvent B

Note: During fractionation using semipreparative or preparative HPLC columns, eluate corresponding to molecule(s) of interest can be collected for further characterization as needed, such as by NMR spectroscopy. Typical NMR experiments to characterize products generated by Bmp7 have been described in literature (Agarwal et al., 2017, 2014; Agarwal & Moore, 2014; Calcul et al., 2009). Resolution at the semipreparative and at the preparative scale HPLC might be as poorer as compared to the analytical scale. The elution profile should be adjusted to achieve the desired resolution by employing isocratic elution profiles and exploring different stationary phase matrices.

5. DATA ANALYSIS

Aryl radical rearrangements typically result in a large number of products being generated by Bmp7. Approaches combining mass spectrometry and NMR spectroscopy for the characterization of polybrominated products generated by Bmp7 have been described previously (Agarwal et al., 2017, 2014; Agarwal & Moore, 2014). Oxidative coupling of polyhalogenated phenols and catechols can lead to proton deficient ring systems that are challenging to be unambiguously resolved by NMR experiments (Calcul et al., 2009; Choi, Engene, Smith, Preskitt, & Gerwick, 2010; Liu et al., 2016). In these cases, syntheses of descriptive model compounds and comparison of NMR spectra can be employed for structural assignments (Agarwal et al., 2015). Here, we describe a typical mass spectrometric characterization of products generated by Bmp7 using 2,4-dichlorophenol (**13**, Fig. 4) as substrate. Mass spectrometry data are generated as described in Section 4.1. Characterization of halogenated products is aided by the natural isotopic distribution of halides. Isotopic distributions of (poly)chlorinated, (poly)brominated, and mixed halogenated molecules have been reviewed previously (Vetter, 2006).

Data analysis protocol

1. Generate the total ion chromatogram (TIC) for the MS/MS dataset. A typical TIC for a Bmp7 enzymatic reaction using **13** as the substrate is shown in Fig. 4.
2. By manual examination, determine the peaks on the chromatogram that correspond to polyhalogenated products. These peaks are marked as A–D in Fig. 4.
3. For each of the four peaks, determine the MS^1 and the MS^2 spectra. The MS^1 and MS^2 spectra for peaks A, B and C, and D in Fig. 4 are shown in Figs. 5–7, respectively.
4. Guided by the isotopic distribution, determine the molecular formula for the chemical species corresponding to the MS^1 spectra within the error tolerance of the mass spectrometer. As shown in Figs. 5 and 6A–C peaks in Fig. 4 correspond to isomeric tetrachlorinated products generated by Bmp7 using **13** as the substrate. As shown in Fig. 7, peak D in Fig. 4 corresponds to a trichlorinated product generated by Bmp7.
5. For each molecular species, determine the MS^2 spectra. Each of the three different kinds of coupling products generated by Bmp7 generates a distinct MS^2 signature.
   a. Fig. 5A illustrates the MS^1 spectra corresponding to a likely C—C coupled tetrachlorinated product generated by Bmp7. The C—C coupling is supported by the MS^2 spectra (Fig. 5B) which shows successive losses of Cl atoms, as is typical for polyhalogenated biphenyls (Agarwal et al., 2014).
   b. Fig. 6A shows the MS^1 spectra corresponding to a likely C—O—C coupled tetrachlorinated product in which the ether linkage has been installed *para* relative to the phenoxyl residue. This assignment is supported by the characteristic dichlorinated hydroquinone radical MS^2 product that is observed (Fig. 6B), as is supported by prior observations (Agarwal et al., 2014, 2015; Mas et al., 2007).
   c. Fig. 6C shows the MS^1 spectra for a third tetrachlorinated product generated by Bmp7 in which the ether linkage is likely installed *ortho* relative to the phenoxyl. The MS^2 spectra (Fig. 6D) show

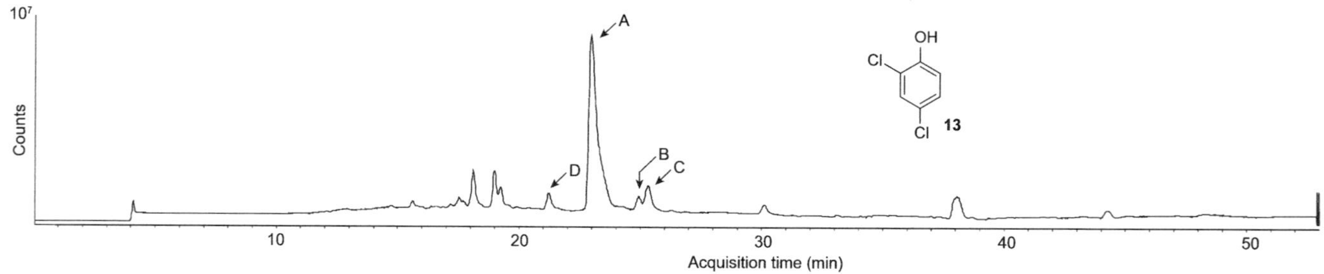

Fig. 4 TIC for Bmp7 reaction as described in Section 4.1 using **13** as the substrate. Peaks on the TIC corresponding to halogenated products are marked A–D.

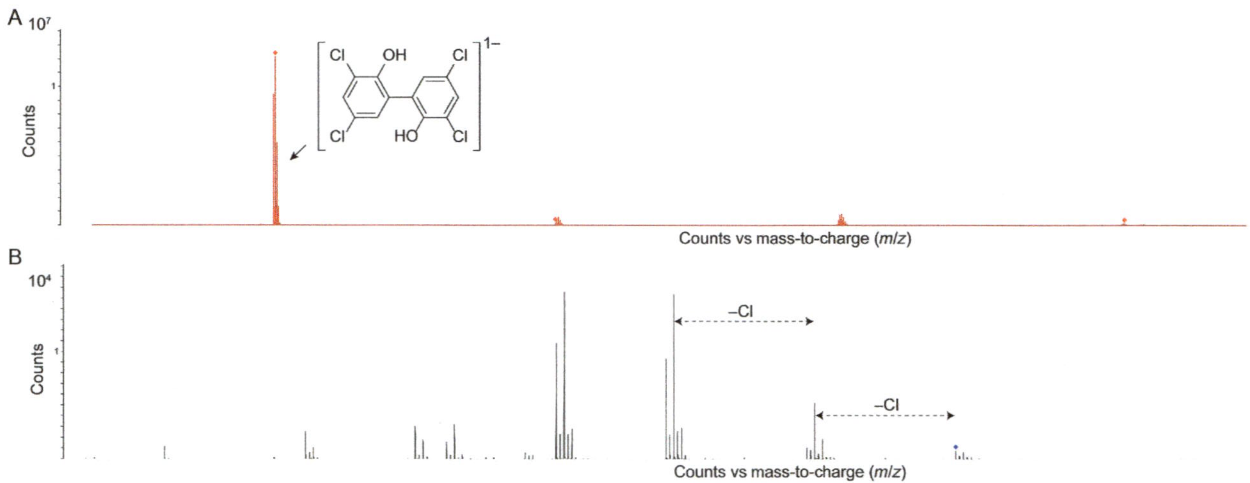

Fig. 5 (A) MS^1 and (B) MS^2 spectra corresponding to the tetrachlorobiphenyl product generated by Bmp7 (peak A, Fig. 4). The proposed structure is shown in panel A and is supported by successive losses of Cl atoms in the MS^2 spectra, as is typical of biphenyl molecules.

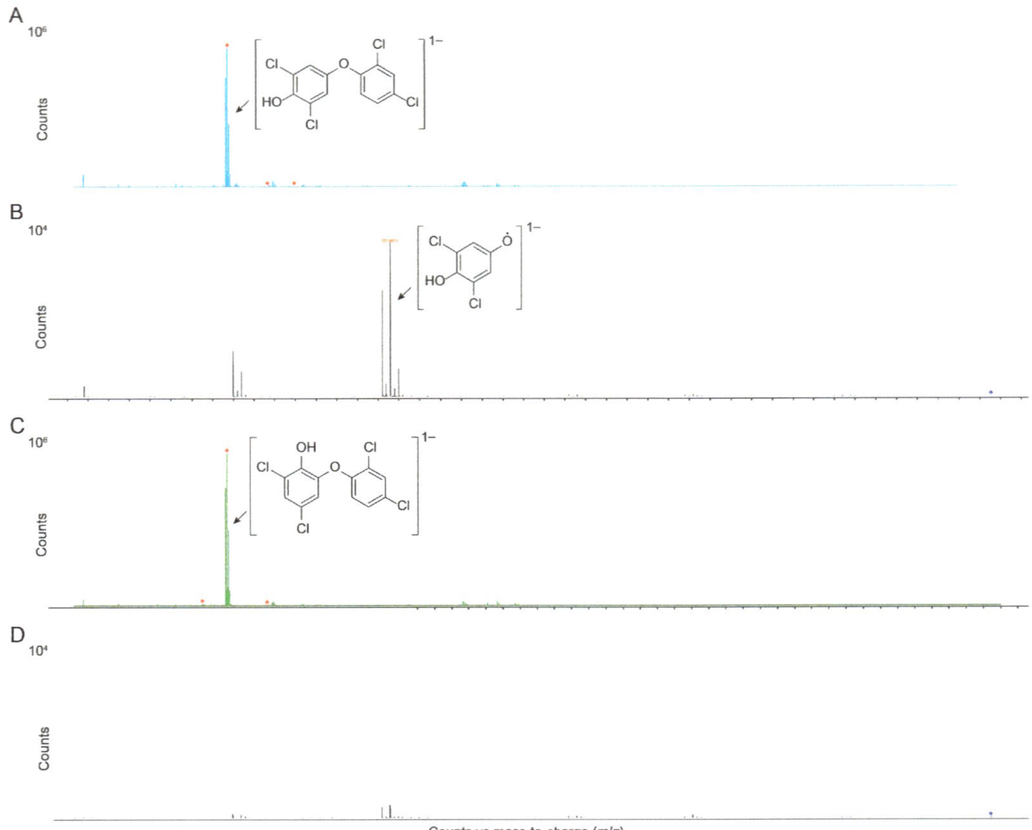

Fig. 6 (A) MS^1 and (B) MS^2 spectra corresponding to the *para* ether coupled tetrachlorobiphenyl product generated by Bmp7 (peak B, Fig. 4). (C) MS^1 and (D) MS^2 spectra corresponding to the *ortho* ether coupled tetrachlorobiphenyl product generated by Bmp7 (peak C, Fig. 4).

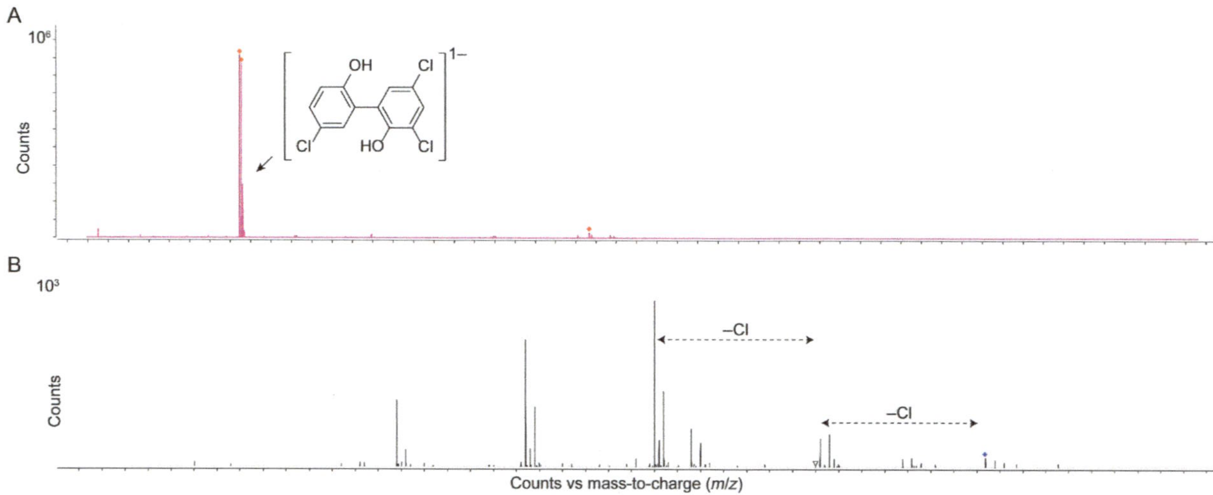

Fig. 7 (A) MS^1 and (B) MS^2 spectra corresponding to the trichlorobiphenyl product generated by Bmp7 (peak D, Fig. 4). The proposed structure is shown in panel A and is supported by successive losses of Cl atoms in the MS^2 spectra, as is typical of biphenyl molecules (see Fig. 5B).

no major fragmentation product ions, as are typical of this structural class of Bmp7 products (Agarwal et al., 2014, 2015; Mas et al., 2007).

d. The MS^1 spectra corresponding to peak D in Fig. 4 is shown in Fig. 7A. The MS^1 spectra leads to its assignment as a trichlorinated biphenyl product as is supported by the successive loss of Cl atoms in the MS^2 spectra (Fig. 7B).

Note: The postulated molecular structures shown in Figs. 5–7 are illustrated to demonstrate the use of MS/MS fragmentation in structural assignment of Bmp7 products. Further verification by NMR spectroscopy and other analytical techniques can be guided by the mass spectrometry-based findings.

6. CONCLUSIONS

Protocols described in this chapter utilize *P. luteoviolacea* 2ta16 derived Bmp9 (ferredoxin) and Bmp10 (ferredoxin-NAD(P)$^+$ oxidoreductase) as the electron transport partners for the CYP450 enzyme Bmp7 (Table 1, Section 1). Bmp7, as for other CYP450 enzymes, can accept noncognate ferredoxins and ferredoxin-NAD(P)$^+$ oxidoreductases as reaction partners as well. Feeding halophenols to *E. coli* cultures expressing Bmp7 (but not Bmp9 and Bmp10) leads to the production of coupled products, demonstrating that *E. coli* ferredoxin(s) and ferredoxin-NAD(P)$^+$ oxidoreductase(s) can substitute Bmp9 and Bmp10 to support the activity of Bmp7 (Agarwal et al., 2014). Furthermore, the Bmp7 active site can accommodate chlorinated, brominated, and iodinated halophenols as substrates. Given the relatively larger steric differences between polybrominated phenols, catechols, and pyrroles, all of which are accepted as substrates by Bmp7, the fact that differentially halogenated phenols such as **6** and **13** are substrates for Bmp7 is perhaps not surprising. A cocrystal structure in complex with polyhalogenated substrate(s) will help describe the molecular bases for the broad substrate tolerance of Bmp7.

Crystal structures of CYP450 enzymes that catalyze aryl oxidative radical coupling reactions in complex with their physiological substrates are already available. Two examples include the *Streptomyces coelicolor* flaviolin dimerizing CYP158A2 and the *Mycobacterium tuberculosis* CYP121 that installs the intramolecular carbon–carbon bond in cyclodityrosine (Belin et al., 2009; Zhao et al., 2005). Both cocrystal structures demonstrate that only one of the two substrate aryl rings is bound proximal to the iron-porphyrin cofactor while the second aryl ring is bound distal to the cofactor (Fig. 8A and B). Given this steric arrangement, how two aryl radicals are generated within

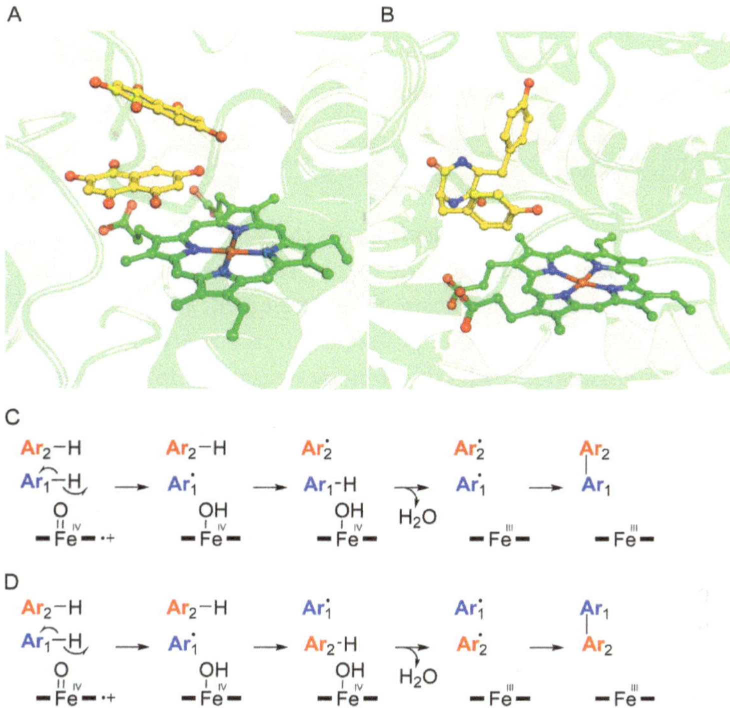

Fig. 8 Actives site of (A) CYP158A2; PDB: 1T93 (Zhao et al., 2005) and (B) CYP121; PDB: 3G5H (Belin et al., 2009) showing the heme cofactor and the substrates bound in stick-ball representation with the carbon atoms colored *green* and *yellow*, respectively. Two possible mechanistic routes (C and D) for biradical aryl coupling activity of CYP450 enzymes (see Section 6 for details). The aryl rings are abbreviated as "Ar."

the same CYP450 enzyme active site is presently not clear. Two models have been proposed to address this conundrum. In the first model, radical generated on the aryl substrate bound proximal to the heme cofactor is electronically transferred to the distal aryl ring. A second radical initiation event on the proximal aryl ring then leads to the biradical coupled product (Fig. 8C). In a second proposed model, following radical generation on the proximal aryl ring, the change in the electronic structure leads to a steric switch in which the second aryl ring now moves proximal to the cofactor. Now, the second radical can be generated followed by biradical coupling (Fig. 8D). As both substrate aryl rings for CYP158A2 and CYP121 are identical, distinguishing between the two scenarios is challenging. However, biradical cross coupling of two different aryl substrates **5** and **6** by Bmp7 could perhaps provide an experimentally tractable system to mechanistically distinguish between the two possibilities. In addition to the two biradical mechanisms described earlier, a third possible mechanism could entail radical formation on the heme proximal substrate followed by addition to the distal substrate leading to a radical formation on the distal species which is subsequently quenched by a single-electron transfer event.

Bmp7 products, such as **7–9** (Fig. 2), are persistent pollutants that accumulate in the marine environment and can potentially be transferred to the human population via our trophic connections to the oceans (Wan et al., 2009). In addition to potentially addressing the mechanistic curiosities described earlier, Bmp7 can also serve as a diagnostic genetic element to mine marine (meta) genomic datasets to query the potential for natural production of polybrominated pollutants. Already, this strategy of using Bmp7 as a query element to mine complex metagenomes has led to the discovery of a biosynthetic gene clusters present in the marine sponge endosymbiotic cyanobacteria that lead to the production of **9**, among other polybrominated diphenyl ethers (Agarwal et al., 2017). The cyanobacterial homolog of Bmp7 was found to be highly selective for the production of **9** using **6** as the substrate, production of other oxidative radical coupling products generated by the proteobacterial Bmp7 were not observed for the cyanobacterial Bmp7 enzyme (Agarwal et al., 2017). The widely represented oxidative radical coupling activity of CYP450 enzymes can thus have profound effects on human health via the biosynthesis of diverse natural products, ranging from lifesaving antibiotics such as **1** to persistent polybrominated environmental pollutants.

ACKNOWLEDGMENTS

We gratefully acknowledge financial support from the NIH (R00ES026620) and Georgia Institute of Technology.

REFERENCES

Agarwal, V., Blanton, J. M., Podell, S., Taton, A., Schorn, M. A., Busch, J., et al. (2017). Metagenomic discovery of polybrominated diphenyl ether biosynthesis by marine sponges. *Nature Chemical Biology*, *13*(5), 537–543. https://doi.org/10.1038/nchembio.2330.

Agarwal, V., El Gamal, A. A., Yamanaka, K., Poth, D., Kersten, R. D., Schorn, M., et al. (2014). Biosynthesis of polybrominated aromatic organic compounds by marine bacteria. *Nature Chemical Biology*, *10*(8), 640–647. https://doi.org/10.1038/nchembio.1564.

Agarwal, V., Li, J., Rahman, I., Borgen, M., Aluwihare, L. I., Biggs, J. S., et al. (2015). Complexity of naturally produced polybrominated diphenyl ethers revealed via mass spectrometry. *Environmental Science & Technology*, *49*(3), 1339–1346. https://doi.org/10.1021/es505440j.

Agarwal, V., & Moore, B. S. (2014). Enzymatic synthesis of polybrominated dioxins from the marine environment. *ACS Chemical Biology*, *9*(9), 1980–1984. https://doi.org/10.1021/cb5004338.

Belin, P., Le Du, M. H., Fielding, A., Lequin, O., Jacquet, M., Charbonnier, J. B., et al. (2009). Identification and structural basis of the reaction catalyzed by CYP121, an essential cytochrome P450 in *Mycobacterium tuberculosis*. *Proceedings of the National Academy of Sciences of the United States of America, 106*(18), 7426–7431. https://doi.org/10.1073/pnas.0812191106.

Brieke, C., Peschke, M., Haslinger, K., & Cryle, M. J. (2015). Sequential in vitro cyclization by cytochrome P450 enzymes of glycopeptide antibiotic precursors bearing the X-domain from nonribosomal peptide biosynthesis. *Angewandte Chemie (International ed. in English), 54*(52), 15715–15719. https://doi.org/10.1002/anie.201507533.

Calcul, L., Chow, R., Oliver, A. G., Tenney, K., White, K. N., Wood, A. W., et al. (2009). NMR strategy for unraveling structures of bioactive sponge-derived oxy-polyhalogenated diphenyl ethers. *Journal of Natural Products, 72*(3), 443–449. https://doi.org/10.1021/np800737z.

Choi, H., Engene, N., Smith, J. E., Preskitt, L. B., & Gerwick, W. H. (2010). Crossbyanols A-D, toxic brominated polyphenyl ethers from the Hawai'ian bloom-forming cyanobacterium *Leptolyngbya crossbyana*. *Journal of Natural Products, 73*(4), 517–522. https://doi.org/10.1021/np900661g.

Dec, J., Haider, K., & Bollag, J. M. (2003). Release of substituents from phenolic compounds during oxidative coupling reactions. *Chemosphere, 52*(3), 549–556. https://doi.org/10.1016/S0045-6535(03)00236-4.

Denisov, I. G., Makris, T. M., Sligar, S. G., & Schlichting, I. (2005). Structure and chemistry of cytochrome P450. *Chemical Reviews, 105*(6), 2253–2277. https://doi.org/10.1021/cr0307143.

El Gamal, A., Agarwal, V., Diethelm, S., Rahman, I., Schorn, M. A., Sneed, J. M., et al. (2016). Biosynthesis of coral settlement cue tetrabromopyrrole in marine bacteria by a uniquely adapted brominase–thioesterase enzyme pair. *Proceedings of the National Academy of Sciences of the United States of America, 113*(14), 3797–3802. https://doi.org/10.1073/pnas.1519695113.

El Gamal, A., Agarwal, V., Rahman, I., & Moore, B. S. (2016). Enzymatic reductive dehalogenation controls the biosynthesis of marine bacterial pyrroles. *Journal of the American Chemical Society, 138*(40), 13167–13170. https://doi.org/10.1021/jacs.6b08512.

Girvan, H. M., & Munro, A. W. (2016). Applications of microbial cytochrome P450 enzymes in biotechnology and synthetic biology. *Current Opinion in Chemical Biology, 31*, 136–145. https://doi.org/10.1016/j.cbpa.2016.02.018.

Guengerich, F. P., & Munro, A. W. (2013). Unusual cytochrome p450 enzymes and reactions. *Journal of Biological Chemistry, 288*(24), 17065–17073. https://doi.org/10.1074/jbc.R113.462275.

Haslinger, K., Peschke, M., Brieke, C., Maximowitsch, E., & Cryle, M. J. (2015). X-domain of peptide synthetases recruits oxygenases crucial for glycopeptide biosynthesis. *Nature, 521*(7550), 105–109. https://doi.org/10.1038/nature14141.

Ikezawa, N., Iwasa, K., & Sato, F. (2008). Molecular cloning and characterization of CYP80G2, a cytochrome P450 that catalyzes an intramolecular C–C phenol coupling of (S)-reticuline in magnoflorine biosynthesis, from cultured *Coptis japonica* cells. *Journal of Biological Chemistry, 283*(14), 8810–8821. https://doi.org/10.1074/jbc.M705082200.

Johannes, T. W., Woodyer, R. D., & Zhao, H. (2007). Efficient regeneration of NADPH using an engineered phosphite dehydrogenase. *Biotechnology and Bioengineering, 96*(1), 18–26. https://doi.org/10.1002/bit.21168.

Kraus, P. F., & Kutchan, T. M. (1995). Molecular cloning and heterologous expression of a cDNA encoding berbamunine synthase, a C–O phenol-coupling cytochrome P450 from the higher plant *Berberis stolonifera*. *Proceedings of the National Academy of Sciences of the United States of America, 92*(6), 2071–2075.

Lin, H. C., McMahon, T. C., Patel, A., Corsello, M., Simon, A., Xu, W., et al. (2016). P450-mediated coupling of indole fragments to forge communesin and unnatural isomers. *Journal of the American Chemical Society, 138*(12), 4002–4005. https://doi.org/10.1021/jacs.6b01413.

Liu, H., Lohith, K., Rosario, M., Pulliam, T. H., O'Connor, R. D., Bell, L. J., et al. (2016). Polybrominated diphenyl ethers: Structure determination and trends in antibacterial activity. *Journal of Natural Products*, *79*(7), 1872–1876. https://doi.org/10.1021/acs.jnatprod.6b00229.

Marsh, G., Athanasiadou, M., Athanassiadis, I., Bergman, A., Endo, T., & Haraguchi, K. (2005). Identification, quantification, and synthesis of a novel dimethoxylated polybrominated biphenyl in marine mammals caught off the coast of Japan. *Environmental Science & Technology*, *39*(22), 8684–8690.

Mas, S., Jauregui, O., Rubio, F., de Juan, A., Tauler, R., & Lacorte, S. (2007). Comprehensive liquid chromatography-ion-spray tandem mass spectrometry method for the identification and quantification of eight hydroxylated brominated diphenyl ethers in environmental matrices. *Journal of Mass Spectrometry*, *42*(7), 890–899. https://doi.org/10.1002/jms.1224.

McIntosh, J. A., Farwell, C. C., & Arnold, F. H. (2014). Expanding P450 catalytic reaction space through evolution and engineering. *Current Opinion in Chemical Biology*, *19*, 126–134. https://doi.org/10.1016/j.cbpa.2014.02.001.

Mizutani, M., & Sato, F. (2011). Unusual P450 reactions in plant secondary metabolism. *Archives of Biochemistry and Biophysics*, *507*(1), 194–203. https://doi.org/10.1016/j.abb.2010.09.026.

Osborne, R. L., Coggins, M. K., Raner, G. M., Walla, M., & Dawson, J. H. (2009). The mechanism of oxidative halophenol dehalogenation by Amphitrite ornata dehaloperoxidase is initiated by H_2O_2 binding and involves two consecutive one-electron steps: Role of ferryl intermediates. *Biochemistry*, *48*(20), 4231–4238. https://doi.org/10.1021/bi900367e.

Osman, A. M., Boeren, S., Boersma, M. G., Veeger, C., & Rietjens, I. M. (1997). Microperoxidase/H_2O_2-mediated alkoxylating dehalogenation of halophenol derivatives in alcoholic media. *Proceedings of the National Academy of Sciences of the United States of America*, *94*(9), 4295–4299.

Peschke, M., Gonsior, M., Sussmuth, R. D., & Cryle, M. J. (2016). Understanding the crucial interactions between cytochrome P450s and non-ribosomal peptide synthetases during glycopeptide antibiotic biosynthesis. *Current Opinion in Structural Biology*, *41*, 46–53. https://doi.org/10.1016/j.sbi.2016.05.018.

Podust, L. M., & Sherman, D. H. (2012). Diversity of P450 enzymes in the biosynthesis of natural products. *Natural Product Reports*, *29*(10), 1251–1266. https://doi.org/10.1039/c2np20020a.

Poulos, T. L. (2014). Heme enzyme structure and function. *Chemical Reviews*, *114*(7), 3919–3962. https://doi.org/10.1021/cr400415k.

Rudolf, J. D., Chang, C. Y., Ma, M., & Shen, B. (2017). Cytochromes P450 for natural product biosynthesis in Streptomyces: Sequence, structure, and function. *Natural Product Reports*, *34*(9), 1141–1172. https://doi.org/10.1039/c7np00034k.

Tang, M. C., Zou, Y., Watanabe, K., Walsh, C. T., & Tang, Y. (2017). Oxidative cyclization in natural product biosynthesis. *Chemical Reviews*, *117*(8), 5226–5333. https://doi.org/10.1021/acs.chemrev.6b00478.

Teuten, E. L., Xu, L., & Reddy, C. M. (2005). Two abundant bioaccumulated halogenated compounds are natural products. *Science*, *307*(5711), 917–920. https://doi.org/10.1126/science.1106882.

Tripathi, S., Li, H., & Poulos, T. L. (2013). Structural basis for effector control and redox partner recognition in cytochrome P450. *Science*, *340*(6137), 1227–1230. https://doi.org/10.1126/science.1235797.

Vetter, W. (2006). Marine halogenated natural products of environmental relevance. *Reviews of Environmental Contamination and Toxicology*, *188*, 1–57.

Wan, Y., Wiseman, S., Chang, H., Zhang, X., Jones, P. D., Hecker, M., et al. (2009). Origin of hydroxylated brominated diphenyl ethers: Natural compounds or man-made flame retardants? *Environmental Science and Technology*, *43*(19), 7536–7542.

Yim, G., Thaker, M. N., Koteva, K., & Wright, G. (2014). Glycopeptide antibiotic biosynthesis. *Journal of Antibiotics*, *67*(1), 31–41. https://doi.org/10.1038/ja.2013.117.

Zhao, B., Guengerich, F. P., Bellamine, A., Lamb, D. C., Izumikawa, M., Lei, L., et al. (2005). Binding of two flaviolin substrate molecules, oxidative coupling, and crystal structure of *Streptomyces coelicolor* A3(2) cytochrome P450 158A2. *Journal of Biological Chemistry*, *280*(12), 11599–11607. https://doi.org/10.1074/jbc.M410933200.

CHAPTER SEVENTEEN

Preparation and Characterization of the Favorskiiase Flavoprotein EncM and Its Distinctive Flavin-N5-Oxide Cofactor

Robin Teufel[1]
ZBSA, Center for Biological Systems Analysis, University of Freiburg, Freiburg, Germany
Faculty of Biology, University of Freiburg, Freiburg, Germany
[1]Corresponding author: e-mail address: robin.teufel@zbsa.uni-freiburg.de

Contents

Abstract

Flavoenzymes often function as oxygenases and have been extensively studied for many decades. Commonly, oxygenation reactions are mediated by a transient C4a-peroxyflavin formed from reaction of reduced flavin with O_2. EncM, however, employs a previously unrecognized flavin-oxygenating species, the flavin-N5-oxide, which is key to a complex oxidative Favorskii-type rearrangement and cyclization cascade in the biosynthesis of the bacterial polyketide antibiotic enterocin produced by the marine bacterium *Streptomyces maritimus*. Here, the methodology and key experiments are described that led to the discovery of this novel flavin redox species and granted insight into one of the most astounding single-enzyme-catalyzed reaction cascades in natural product biosynthesis.

Methods in Enzymology, Volume 604
ISSN 0076-6879
https://doi.org/10.1016/bs.mie.2018.01.036

1. INTRODUCTION

Polyketides represent one of the most prevalent and medically relevant classes of natural products, whose biosynthesis has been extensively studied in the past (Shen, 2003). Commonly, a starter unit (e.g., acetyl-CoA) is elongated with malonyl-CoA extender units via decarboxylative Claisen condensations affording a poly-β-ketone chain, which usually undergoes additional modifications allowing formation of structurally diverse natural products. Three types of polyketide synthases (PKSs) are generally recognized. While type I PKS consists of large, modular, and usually non-iteratively acting megasynthases that use an acyl-carrier protein (ACP) for substrate binding, the chalcone synthase-like type III PKSs directly act on acyl-CoA substrates and generate aromatic products, which may be further modified by additional enzymes to the mature natural product. In contrast, a type II PKS machinery comprises various monofunctional enzymes including the "minimal PKS" consisting of two homologous ketosynthase (KS) subunits and an ACP, in which the catalytically active subunit KS_α produces a linear, reactive poly-β-ketone whose length is determined by the KS_β subunit (also referred to as "chain-length factor"). In addition, the ACP serves as an anchor for the growing polyketide chain. Structural diversity in type II PKS systems is generally mediated by additional tailoring enzymes that may drastically alter the chemical shape and properties of the parent molecule (Wang, Gao, & Tang, 2012). Typically, type II PKS-derived polyketides feature a polycyclic, planar, and aromatic structure as exemplified by tetracycline and are thus often referred to as aromatic polyketides. While the cyclization and aromatization may occur spontaneously, cyclases and aromatases usually assist in this process in order to prevent formation of unwanted side products (Hertweck, Luzhetskyy, Rebets, & Bechthold, 2007).

A remarkable exception is found in enterocin biosynthesis that lacks such cyclases and aromatases and represents one of the most thoroughly investigated type II PKS pathways to date (Fig. 1; Piel et al., 2000). Here, a single flavin-dependent enzyme, EncM, prevents the aromatization of the mature poly-β-ketone chain in form of an ACP-bound C7,O4-dihydrooctaketide by triggering an oxidative Favorskii-type carbon–carbon rearrangement, aldol condensations, as well as two heterocycle-forming reactions (Xiang, Kalaitzis, & Moore, 2004). Ultimately, five chiral centers and four rings are formed by EncM, giving rise to the unique and characteristic tricyclic nonaromatic core of enterocin (Fig. 1; Teufel et al., 2013). This arguably

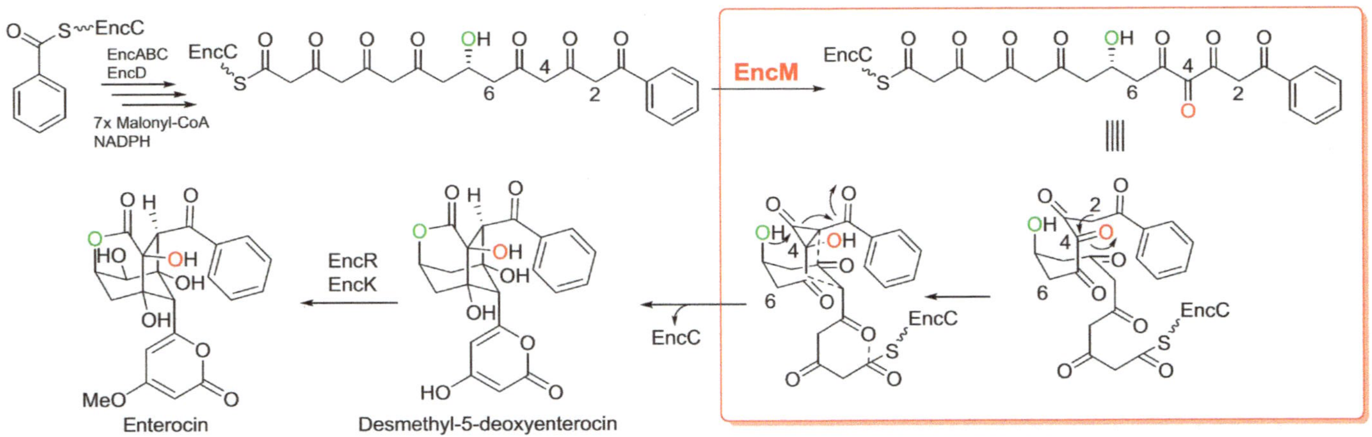

Fig. 1 Enterocin biosynthetic pathway. First, the starter unit benzoic acid is ligated to the holo-ACP EncC (not shown), before benzoyl-EncC undergoes chain elongation through decarboxylative Claisen condensation mediated by the minimal PKS EncAB using seven units of malonyl-CoA. Likely during chain elongation, the ketoreductase EncD reduces the C7-ketone to a hydroxyl group. EncM then catalyzes a dual oxidation at C4 of the substrate. The newly introduced ketone group then triggers the Favorskii rearrangement and subsequent aldol condensations and heterocyclization reactions affording desmethyl-5-deoxyenterocin (*red box*). An additional hydroxylase (EncR) and a methyltransferase (EncK) ultimately produce enterocin.

represents one of the most outstanding series of cascade-like reactions catalyzed by a single enzyme. Surprisingly, when comparing the *encM* gene to data bases such as GenBank, it is predicted to encode an inconspicuous member of the functionally diverse vanillyl alcohol oxidase (VAO)/p-cresol methyl-hydroxylase (PCMH) flavoprotein family, most of which function as classic flavin-dependent oxidoreductases involved in two electron reductions (reductases) or oxidations (oxidases/dehydrogenases) (Ewing, Fraaije, Mattevi, & van Berkel, 2017). However, detailed structural and mechanistic investigation revealed that EncM instead acts as an unusual member of the so-called internal flavin monooxygenases, i.e., flavoenzymes catalyzing a dual oxidation by using their own substrate as electron donor for flavin reduction (instead of more common exogenous reductants such as NADPH), which allows subsequent reaction with dioxygen (O_2). This presumably gives rise to a covalent flavin: oxygen adduct that serves as an oxygenating species and thus enables the subsequent hydroxylation of the substrate. EncM, however, first catalyzes the hydroxylation before dehydrogenation and may thus be considered an "inverted" internal oxygenase (Teufel, 2017; Teufel et al., 2013). As will be described later, this is only possible due to the remarkable flavin-N5-oxide cofactor of EncM (Teufel et al., 2015).

Before the characterization of EncM, it was assumed that all flavin monooxygenases employ a reactive transient flavin-C4a-peroxide as oxygenating species, which has been intensively investigated for many decades (Huijbers, Montersino, Westphal, Tischler, & van Berkel, 2014; Walsh & Wencewicz, 2012). Interestingly, EncM was instead shown to form and stably maintain a flavin-N5-oxide (Fl_{N5ox}); i.e., the resting state of EncM's flavin cofactor under aerobic conditions is the Fl_{N5ox} rather than the typically observed oxidized flavin (Fl_{ox}) (Fig. 2; Teufel et al., 2013, 2015). The Fl_{N5ox} species thus allows direct oxygenation of the poly-β-ketone substrate, followed by dehydrogenation of the oxidized carbon, which converts a 1,3-diketone motif into an even more reactive 1,2,3-triketone. The formation of an additional electrophilic center then triggers the Favorskii-type rearrangement of the carbon backbone that entails additional cyclization reactions and ultimately affords desmethyl-5-deoxyenterocin, which is further converted into enterocin by two additional enzymes. The exact oxidation mechanism of EncM has not been elucidated; however, a hydroxylation reaction involving a N5-nitroxyl radical seems plausible due to chemical precedence for similar oxidation reactions mediated by nitroxyl radicals (e.g., 2,2,6,6-tetramethylpiperidinyloxyl, "TEMPO").

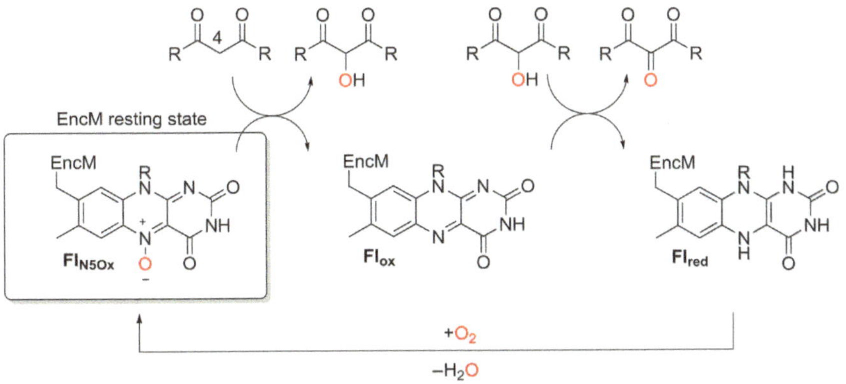

Fig. 2 Simplified reaction scheme and flavin redox states of EncM. The Fl_{N5ox} represents the resting state of EncM, which allows hydroxylation of the substrate. This step also generates Fl_{ox}, which can further oxidize the C4 of the intermediate and thus produce the final ketone that triggers subsequent cyclization reactions (see also Fig. 1).

The hydroxylated poly-β-ketone intermediate is then further oxidized by the now oxidized flavin cofactor (EncM-Fl_{ox}) via dehydrogenation at the same carbon, thus converting the hydroxyl group into a ketone and thereby setting up the following cyclization reactions. Concomitantly, Fl_{red} is formed that reacts with O_2 and allows restoration of the Fl_{N5ox}, presumably via a transient N5-hydroperoxide intermediate (Teufel, 2017; Teufel, Agarwal, & Moore, 2016; Teufel et al., 2013, 2015). The following sections highlight how the enzyme was produced, isolated, and characterized.

2. HETEROLOGOUS PRODUCTION AND PURIFICATION OF EncM

The cloning of *encM*, heterologous production, and purification of the enzyme was conducted according to standard procedures (Teufel et al., 2013):

- The *encM* gene (amplified by PCR) was inserted between the *Nco*I and *Eco*RI restriction sites of the expression vector pHIS8 (Jez, Ferrer, Bowman, Dixon, & Noel, 2000), derived from pET28a.
- *Escherichia coli* BL21 (DE3) transformed with the pHIS8-EncM plasmid was grown at 28°C in terrific broth medium containing 50 mg/mL kanamycin until an optical density of about 0.5 was reached. 25 μ*M* isopropyl-β-D-thiogalactoside (IPTG) was then added to induce recombinant protein expression under the control of T7 RNA polymerase and a modified lac promoter. Addition of IPTG was not obligatory, due to

the "leakiness" of the pHIS8 vector allowing sufficient gene expression. *Note*: IPTG concentrations >0.1 m*M* led to a decrease in soluble enzyme due to formation of inclusion bodies.

- After induction, cells were additionally grown for 24h at 28°C and harvested by centrifugation. Cell pellets were resuspended in lysis buffer (50 m*M* sodium phosphate pH 7.7, 300 m*M* sodium chloride, 10% (v/v) glycerol) with 10 m*M* imidazole, and then lysed by sonication (10 cycles for 15 s) on ice.
- After centrifugation, the supernatant was passed over a Ni^{2+}-nitrilotriacetate column using a FPLC system. Following a wash step with 50 m*M* imidazole to remove unspecifically bound protein, the N-terminal octahistidine-tagged EncM was eluted with lysis buffer supplemented with 500 m*M* imidazole. EncM was desalted and concentrated using PD-10 and Vivaspin 6 (10 or 30 kDa exclusion size) columns (both GE Healthcare), respectively.

Notes

(1) Addition of extra riboflavin in the broth did not significantly improve the yield of EncM, likely because it was not a limiting factor but rather the generally modest stability/solubility of EncM when heterologously produced in *E. coli*. In general, however, riboflavin addition should be considered when overproducing a flavoprotein.

(2) An elongated wash step was critical to acquire highly pure EncM for subsequent UV–vis spectroscopy (see later sections). For that, the column with bound EncM was washed for at least 40 min at a flow of 1 mL/min until the detector reached the baseline level. Alternatively, ResourceQ (GE Healthcare) anion-exchange chromatography was applied as additional purification step (also for protein crystallization) with a linear gradient from 0 to 1 *M* NaCl over 30 min in 10 m*M* TES-Na^+ buffer (pH 7.7) with 10% (v/v) glycerol.

(3) For the discovery of the novel flavin-N5-oxide cofactor of EncM, it proved pivotal to forgo addition of reductants during protein purification (see also discussion later). For example, the presence of the common additive dithiothreitol (DTT) during EncM purification resulted in the artifactual reduction of the Fl_{N5ox} into Fl_{ox}. This can be observed, e.g., in the originally published UV–vis spectrum of EncM (Cheng, Xiang, Izumikawa, Meluzzi, & Moore, 2007, supplemental fig. 2). For comparison with the spectrum of intact EncM-Fl_{N5ox}, see Fig. 3.

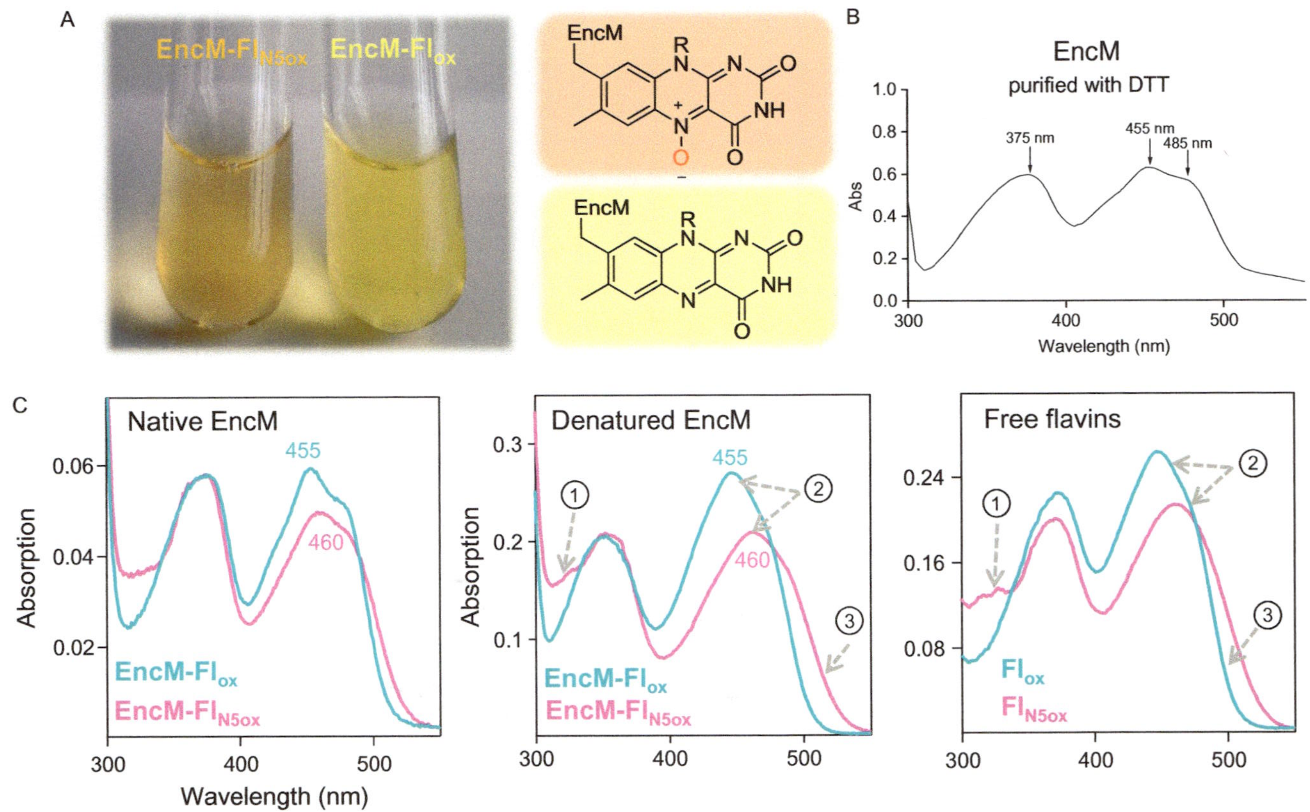

Fig. 3 See legend on next page.

3. CHEMICAL SYNTHESIS OF POLY-β-KETONE SUBSTRATE ANALOGS

One of the main obstacles that impede in-depth studies of enzymes involved in polyketide biosynthesis is the innate reactivity of key intermediates, in particular open-chain poly-β-ketones such as the proposed ACP-bound C7,O4-dihydrooctaketide substrate of EncM. These compounds readily undergo spontaneous cyclization and aromatization and as a result can neither be isolated from bacterial producers nor chemically synthesized. Hence, it was critical for the project to obtain suitable substrate analogs for cocrystallization and enzyme assays. As a minimum requirement for a convertible analog to follow the anticipated EncM-catalyzed Favorskii rearrangement, a partially reduced tetraketide was required mimicking the terminal moiety of the ACP-bound C7,O4-dihydrooctaketide. Accordingly, a C7,O4-dihydrotetraketide was synthesized along with other, nonconvertible analogs. The synthesis of these molecules is described in detail in the supplemental information of Teufel et al. (2013).

Note: C7,O4-dihydrotetraketide was unstable when exposed to air at room temperature and slowly decomposed. It was thus important to aliquot and properly store the substrate analog in the freezer at −20°C and thaw it shortly before each assay. Due to low water solubility, the compound was dissolved in DMSO.

Fig. 3 (A) Heterologously produced and purified EncM shows a distinct *yellow-orangish color* indicative of the Fl_{N5ox} cofactor. When EncM is anaerobically reduced and chemically reoxidized without O_2, the cofactor can be converted to brightly *yellow* Fl_{ox}. (B) UV–vis spectrum of EncM when first described (as published in Cheng et al. (2007). The Fl_{N5ox} cofactor was reduced to Fl_{ox}, most likely due to the presence of DTT in the purification buffer. (C) The depicted UV–vis spectra have been published before in Teufel et al. (2013). *Left panel*: UV–vis spectra of native EncM as isolated containing a covalently bound Fl_{N5ox} (*purple*) and after reduction and anaerobic chemical reoxidation forming Fl_{ox} (*blue*). *Middle panel*: the subtle differences in the UV–vis spectra of the two flavin redox states become more pronounced in denatured EncM. Compared to Fl_{ox}, Fl_{N5ox} absorbs light at ~320 nm (1), exhibits a lower molar absorption coefficient in the ~450–460 nm region (2), and shows a small shift of the absorption maximum to 460 nm, thus absorbing light up to ~530 nm (3). *Right panel*: free flavin cofactors show highly similar spectral features compared to denatured EncM. Fl_{ox} (*blue*) is commercially available, Fl_{N5ox} was chemically synthesized.

4. AEROBIC AND ANAEROBIC EncM ACTIVITY ASSAYS AND HPLC ANALYSIS

EncM enzymatic activity was tested with reversed-phase high-performance liquid chromatography (HPLC) assays using the chemically synthesized substrate analogs. Previously, it could be shown that EncM is active when the natural substrate is provided in a fully in vitro reconstituted pathway. Assays were conducted in HEPES-Na^+ pH 7.5, 300 m*M* NaCl, containing at least 10% (v/v) glycerol, and 0.7 m*M* of C7,O4-dihydrooctaketide (both racemic and enantiomerically pure). EncM was subsequently precipitated using 50% (v/v) of methanol following centrifugation and injection of the supernatant on a Sync Polar RP column 4 μ (150 x 4.6 mm; ES Industries) connected to a reversed-phase HPLC (1200 series; Agilent) using a gradient with 10% (v/v) acetonitrile as liquid phase buffered with 90% (v/v) of 20 m*M* ammonium acetate (pH 5.0). Gratifyingly, the conversion of the substrate analog could indeed be observed, giving rise to a diastereomeric product pair without detectable intermediates. Using extensive NMR and mass spectrometric analyses as well as chemical synthesis, the products were surprisingly identified as ring-opened derivatives of the expected enterocin-core-like lactones. Circular dichroism experiments were used to show that the configuration at C7 of the C7, O4-dihydrooctaketide was maintained during the transformation and that the racemization was most likely caused by the lactonization step in the Favorskii-type rearrangement (producing a new stereocenter at C4), before an unforeseen retro-Claisen lactone cleavage afforded the observed products. A two-step oxidation mechanism as outlined in Fig. 2 was further supported by an anaerobic experiment single-turnover experiment, in which 20 μ*M* EncM-Fl_{N5ox} was incubated with C7,O4-dihydrotetraketide. The substrate-dependent reduction of the Fl_{N5ox} species first to Fl_{ox} (rapid) and then to Fl_{red} (very slow) could be clearly observed (Teufel et al., 2013). Due to a lack of O_2, the Fl_{N5ox} species could not be regenerated and hence only one turnover was possible.

Notes

(1) Product and substrate peaks from the EncM-catalyzed conversion of C7,O4-dihydrooctaketide showed a poor overall shape (including "tailing") when larger amounts were analyzed by HPLC. For clear separation and good peak shape, it was thus crucial to inject and analyze lower amounts.

(2) It is unclear why the expected lactone products from conversion of C7, O4-dihydrotetraketide could not be observed as they should not be overly unstable. However, when the natural substrate is converted, this unwanted retro-Claisen cleavage is likely prevented by the aldol condensation between C5 and C10 in the enzymatic environment.

5. SPECTROSCOPIC CHARACTERIZATION OF EncM

Flavoenzymes are among the most extensively studied enzymes not only due to their general importance and prevalence but also because of wonderful spectroscopic properties that facilitate determining the flavin redox state and formation of covalent flavin adducts (with O_2 or the substrate) and thus in-depth investigation of the enzymatic mechanism. Typically, aerobically isolated holoflavoenzymes adopt a brightly yellow color (when adequately concentrated), indicative of bound Fl_{ox} (Fig. 3). Hence, as a typical first step after flavoenzyme isolation, UV–vis spectroscopic analysis is commonly used to further verify the presence of the flavin cofactor. More detailed investigation to follow formation of short-lived intermediates can be achieved using a stopped-flow spectrophotometer to record spectral changes in the millisecond range. Accordingly, such measurements were also conducted with EncM (Teufel et al., 2013):

- The flavin absorption spectrum of purified EncM was recorded using an Agilent Cary 50 UV–vis spectrophotometer or a Shimadzu UV-2501 PC. When EncM was first isolated from *E. coli*, it showed a seemingly inconspicuous yellow-orangish coloration and a UV–vis spectrum similar to common oxidized flavin cofactors. However, it turned out that the spectrum and color instead derive from EncM-Fl_{N5ox} that differs only marginally from Fl_{ox}, which may thus have prevented earlier discovery of this novel redox state (Fig. 3).
- When EncM (EncM-Fl_{N5ox}) was incubated with a convertible substrate analog (see discussion later), the spectrum underwent minor but distinct permanent changes that could initially not be explained but later turned out to stem from EncM-Fl_{ox} formation, i.e., EncM-Fl_{N5ox} oxidized the substrate analog and was thereby converted into EncM-Fl_{ox}.
- Further confusion arose from determination of the molar absorption coefficients of EncM. For that, a solution of anaerobic dithionite was calibrated in a gastight syringe by titrating a known concentration of flavin mononucleotide (FMN) to full reduction. The dithionite syringe was then transferred to an anaerobic cuvette containing EncM and titrated

with the calibrated dithionite to complete reduction. The amount of dithionite needed to fully reduce EncM was used to determine a molar absorption coefficient ε of 4800 M^{-1} cm^{-1} at 460 nm based on the original absorbance spectrum. This is extremely low for a flavoprotein, as enzymatic flavin cofactors commonly exhibit absorption coefficients of ~12,000 M^{-1} cm^{-1} in the 450 nm region that usually only slightly deviate from unbound flavin. As was found out later, this odd result can be rationalized by the fact that four electrons are required to fully reduce the superoxidized EncM-Fl_{N5ox} into Fl_{red}, whereas the oxidized FMN used for calibration only requires two electrons. Hence, the actual molar absorption coefficient of EncM-Fl_{N5ox} is twice as high with 9600 M^{-1} cm^{-1} at 460 nm. This is still lower than what is observed for the average flavoenzyme and caused by the small, yet significant different spectral properties of Fl_{N5ox} vs Fl_{ox} (Fig. 3).

- To further investigate EncM, EncM-Fl_{N5ox} was converted to EncM-Fl_{ox} by first reducing EncM anaerobically, before reoxidation with the chemical oxidant 2,6-dichlorophenolindophenol (DCIP). The absence of O_2 thereby ensured that EncM-Fl_{red} could only be oxidized into EncM-Fl_{ox}. The molar absorption coefficient of EncM-Fl_{ox} was 11,900 M^{-1} cm^{-1} at 450 nm. Upon titration of EncM-Fl_{ox} with the single-electron reductant dithionite, formation of the anionic (red) flavin semiquinone could be observed. While this is not unique for EncM and also observed in other flavoenzymes (mainly oxidases), we postulated that the stabilization of this flavin radical with high spin density at N5 is critical for formation of the EncM-Fl_{N5ox} via radical coupling with the superoxide (Teufel et al., 2015). Interestingly, a highly similar species can be observed when EncM-Fl_{N5ox} is titrated with dithionite, which can likely be attributed to the respective anionic flavin-N5-nitroxyl radical.

6. CRYSTALLIZATION AND STRUCTURAL ELUCIDATION OF EncM

In order to gain more insight into the reaction mechanism, EncM was crystallized (Teufel et al., 2013, 2015). The solved protein structure of EncM showed a homodimer in crystal form which was confirmed in solution by hydrodynamic analysis using size-exclusion chromatography (Fig. 4). Each monomer features distinct domains for substrate (residues 211–418) and flavin binding (residues 2–210 and 419–461). The flavin-binding domain sequesters the ADP-ribosyl of the cofactor, whereas the reactive

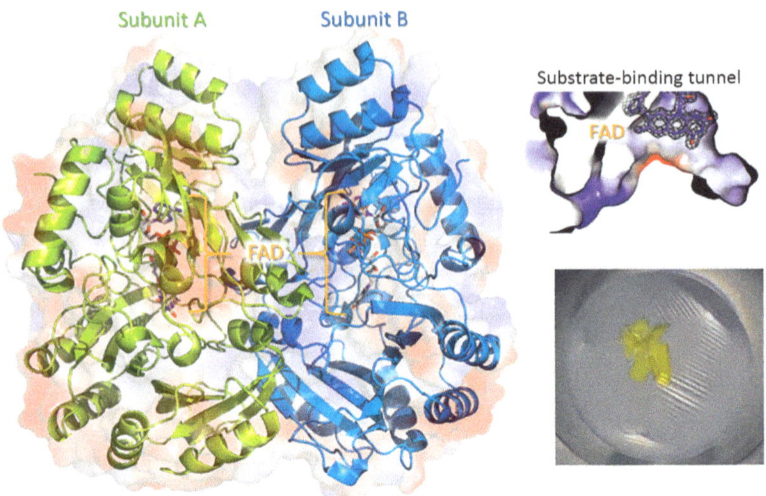

Fig. 4 Crystal structure of homodimeric EncM (*left side*) as published in Teufel et al. (2013). Each subunit contains a covalently bound FAD in a secluded substrate-binding tunnel (*top right*). *Yellow crystals* of EncM typically form after 5–7 days using hanging-drop vapor diffusion (*bottom right*).

isoalloxazine ring is positioned at the interface of the substrate and cofactor domains. The C8-methyl group of the isoalloxazine ring is covalently linked to histidine-78 of EncM, analogous to closely related enzymes such as 6-hydroxy-D-nicotine oxidase (6HDNO). Unlike 6HDNO, each EncM monomer has an elongated L-shaped tunnel extending about 30 Å from the surface past the flavin and ending in a hydrophobic pocket. This tunnel is complementary to the shapes of the phosphopantetheine arm of the ACP and the attached substrate (Fig. 4). A few positively charged amino acid side-chain residues are located adjacent to the tunnel entrance and form a basic patch, which was shown to provide an interaction surface for the acidic surface of the ACP via mutagenesis and computational docking analysis (Teufel et al., 2013). Cocrystal structures furthermore showed that the shortened analogs are bound in a way that positions the site of oxidative attack close to N5 of the flavin, consistent with the structural constraints required for substrate dehydrogenation by N5 of the flavin (Teufel, 2017). Likely, this binding mode also allows the prior hydroxylation by the Fl_{N5ox}.

The crystallization and structural elucidation of EncM was conducted as follows:

- EncM was first digested with thrombin to remove the His8 tag, followed by His-trap purification to remove the cleaved tag and ResourceQ anion chromatography (as described earlier).

- Next, yellow crystals of EncM (Fig. 4) were acquired within 1 week using hanging-drop vapor diffusion at 4°C from a 1:1 mixture of protein solution (5 mg/mL) in 10 m*M* TES-Na^+, pH 7.7 with 10% (v/v) glycerol and reservoir solution (0.1 *M* HEPES-Na^+ pH 7.5, 0.2 *M* calcium acetate, 20% (w/v) PEG3350). The crystals in reservoir solution were frozen in liquid nitrogen (containing 25% (v/v) glycerol as a cryo-protectant) until X-ray data collection. Using the protein structure of 6HDNO (PDB code 2BVG) as a search model, the initial phases could be determined by molecular replacement.
- Additionally, EncM was mixed with various substrate analogs (2 m*M*) before addition of the reservoir solution to acquire cocrystals.

Note: Despite compelling evidence for the EncM-Fl_{N5ox}, this flavin redox state could never be observed by X-ray crystallography. This is most likely due to rapid X-radiation-induced reduction, which is a well-known phenomenon and problematic for proteins containing redox centers such as flavins. Chemically synthesized Fl_{N5ox} has furthermore been reported to be highly susceptible to reduction. The bleaching of the yellowish EncM crystals could also be observed during data collection, indicative of a rapid conversion of the Fl_{N5ox} into Fl_{red}. Conceivably, the Fl_{N5ox} has thus been overlooked in other protein structures in the past.

7. EVIDENCE FOR THE FLAVIN-N5-OXIDE COFACTOR OF EncM

In the beginning of the in-depth investigation of EncM, the presence of the Fl_{N5ox} cofactor was by no means clearly obvious due to several reasons:

1. The highly similar UV–vis spectra of Fl_{N5ox} and Fl_{ox} and the fact that the Fl_{N5ox} spectrum—although having been reported in the late 1970s—was not commonly known.
2. The absence of any visible modifications of the flavin in the EncM crystal structure.
3. The sheer unlikelihood of an unrecognized flavin redox state and oxygenating species after many decades of flavoprotein studies.

At first, we suspected that EncM harbors a second redox center which may interact with the flavin, possibly explaining why four electrons are required to fully reduced EncM and accounting for the slight spectral changes observed for EncM oxidized with O_2 vs DCIP. The only redox active residue close to the flavin was a methionine, which may conceivably form a catalytically active methionine sulfoxide. However, mutation of this residue neither altered the EncM UV–vis spectrum nor abolish activity (Teufel et al., 2015). After ruling out other possibilities, the flavin cofactor itself had to be considered, also because denatured EncM mostly retained its spectrum. The common flavin-C4a-peroxide was ruled out based on spectroscopic features and stability. Several lines of evidence eventually led to the proposal of a Fl_{N5ox} species in EncM. It is generally known that the N5-C4a locus of the flavin serves as entry/exit points for electrons and formation of covalent adducts, and we thus focused our attention on this region. The Fl_{N5ox} seemed the most plausible possibility, as nitroxyl groups are commonly relatively stable and are applied in organic synthesis, e.g., in radical oxidation reactions. Indeed, chemically synthesized Fl_{N5ox} exhibited the same key spectral features as EncM, in particular after denaturation of EncM that minimized the effects of the protein surroundings on the flavin cofactor (Fig. 3). Further evidence came from digestion of EncM with proteinase K, which was required due to the covalent link of the flavin with His78 (Teufel et al., 2015):

- EncM was digested for 4–6 h at 22°C in the dark with 2 mg/mL fungal proteinase K, 5 m*M* $CaCl_2$, and 25 m*M* Tris–HCl (pH 7.8).
- Peptide fragments were separated from undigested protein using Nanosep spin columns (10 kDa cutoff, Pall Corp.).
- Peptides were analyzed by high-resolution electrospray ionization tandem mass spectrometry (HR-ESI-$LCMS^2$) experiments in positive mode using a 1290 Infinity LC system with a Luna 5 μm C18E (2) column (Phenomenex, Torrence, CA) (150 × 4.6 mm) coupled to a 6530 Accurate-Mass Q-TOF MS system (Agilent Technologies, Santa Clara, CA). An acetonitrile gradient of 2%–100% (v/v) was used over 25 min in 0.1% (v/v) formic acid. Collision-induced dissociation for tandem MS spectra was set to 35 with a fragmentor voltage of 175 V. Pseudo-MS^3 measurements were conducted by increasing the in-source fragmentor voltage to 430 V.
- MS, MS^2, and pseudo-MS^3 data analysis revealed an ion of a pentapeptide fragment containing His78 with covalently bound flavin as well as an additional oxygen, consistent with the proposed Fl_{N5ox}.

Notes

(1) Proteins under investigation should not be overdigested, as this will result in partial or complete decay of the Fl_{N5ox}. A few hours at ambient temperature should suffice.

(2) The flavin cofactor significantly facilitates identification of relevant fragments when MS analysis is combined with UV–vis detection of the flavin at ~450 nm, e.g., by using a diode array detector, which allows facile determination of the retention time of flavinylated peptides.

8. STRATEGIES TO IDENTIFY NOVEL ENZYMES EMPLOYING A FLAVIN-N5-OXIDE COFACTOR

Many of the highlighted methods in this chapter may allow to scrutinize other flavoproteins for formation of Fl_{N5ox}. However, depending on the protein, adjustments may be required as recently shown for two other flavoenzymes that form a Fl_{N5ox} but are interestingly not structurally related to EncM (Adak & Begley, 2016; Adak & Begley, 2017a, 2017b; Teufel, 2017). One of the main challenges for detection of the Fl_{N5ox} is the susceptibility of this flavin redox state for reduction. EncM itself shows an ambiguous behavior toward reductants, i.e., the cellular biological reductant NADPH (and to a lesser extent NADH) reduces EncM-Fl_{ox} but not EncM-Fl_{N5ox} and thereby promotes Fl_{N5ox} formation through reaction of Fl_{red} with O_2 (Teufel et al., 2015). Presumably, accessibility of the reductants to the flavin is an important factor and NADPH may only have access to EncM's active site when in the Fl_{ox} redox state, although the molecular basis for such a behavior remains to be elucidated. Given the narrow tunnel in EncM and the absence of an NADPH-binding motif, it is hard to envisage how NADPH can react with one flavin redox state but not with the other. Notably, cocrystallization attempts with NADPH could not shed light on the interaction with EncM (Teufel et al., 2015). In contrast to NADPH, other (nonbiological) reductants such as DTT show a markedly different reactivity toward Fl_{N5ox} and readily reduced it to Fl_{ox} (Teufel et al., 2015).

In contrast to EncM, detection of the Fl_{N5ox} species in the flavoenzymes RutA and DszA (involved in bacterial catabolic pathways) was only possible in the absence of NAD(P)H (Adak & Begley, 2016; Adak & Begley, 2017b), likely because their flavin cofactors were not as secluded and protected compared to EncM. Taking these observations into account, a guideline for the discovery of Fl_{N5ox} in other flavoproteins is suggested:

i. Record the UV–vis spectrum of the flavoprotein of interest (purified without reductants such as DTT) and compare to the typical spectra of Fl_{ox} and Fl_{N5ox} (see Fig. 3). If the spectrum shows key features of the Fl_{N5ox}, proceed with (ii), if it resembles Fl_{ox}, go to (iv).

ii. Test if the flavin cofactor is covalently bound by denaturation of the protein (e.g., with heat or organic solvents), followed by centrifugation. If the yellow color is found in the pellet, the flavin is most likely covalently attached, proceed with (v), if the color remains in the supernatant go to (iii).

iii. Analyze your flavin cofactor first by checking whether FMN or FAD is present, this can be done, e.g., by first reducing the presumed Fl_{N5ox} into Fl_{red} (e.g., using fresh sodium dithionite) that should spontaneously reoxidize to Fl_{ox} in the absence of enzyme. Then, compare the extracted cofactor with commercially available FAD and FMN standards via HPLC. Analyze another sample of the flavin cofactor (not treated with dithionite or other reductants) by LC-MS to verify the presence of an additional oxygen atom bound to the flavin. To confirm the presence of Fl_{N5ox}, a consistent spectrum and correct mass should be adequate. Additionally, the isolated cofactor can be compared to chemically synthesized FAD-N5-oxide or FMN-N5-oxide standard (for synthesis, see Teufel et al., 2013).

iv. If the cofactor is in the Fl_{ox} state, reduce it under anaerobic conditions (e.g., in a cuvette under N_2 atmosphere) and avoid contact with reductants. Flavin photoreduction with EDTA is recommended, as described by Adak & Begley (2016). Then, the enzyme is reoxidized in the dark by exposure to air. Check the flavin redox state before and after each step by UV–vis spectroscopy. If the flavin reoxidizes to Fl_{ox}, the investigated protein most likely does not form a Fl_{N5ox}. If the spectrum resembles Fl_{N5ox}, proceed with (ii). As control experiment and comparison, the Fl_{red} can be reoxidized anaerobically with DCIP, which produces Fl_{ox}.

v. Digest your protein with proteinase K (or other proteases) before separating and analyzing the peptide fragments by LC-MS^2 and UV–vis detection as described in this chapter. Check for UV–vis signals at a wavelength of 450 nm indicative of the flavin cofactor and whether the spectrum is consistent with Fl_{ox} or Fl_{N5ox}. Search for detected ions corresponding to flavinylated peptides (consider both FMN and FAD if the nature of the cofactor is not known) and confirm the presence or absence of an additional oxygen atom. In the MS^2 spectra, one can typically observe the loss of the AMP moiety of the FAD.

9. SUMMARY AND CONCLUSION

This chapter describes the methodology that led to the discovery and characterization of EncM's Fl_{N5ox} oxygenating species catalyzing a key step in the biosynthesis of the polyketide natural product enterocin. While other enzymes forming a Fl_{N5ox} have meanwhile been reported, the full significance and role of this novel flavin redox state remain to be elucidated. As of yet, it is unknown how widespread the Fl_{N5ox} is and what reactions can be catalyzed. Presumably, the Fl_{N5ox} may have gone undetected in enzymes already characterized, due to the inconspicuous UV–vis spectrum and susceptibility to X-ray radiation-induced reduction or chemical reductants such as DTT commonly used for protein purification. The detection of the Fl_{N5ox} in other flavoproteins may be relatively simple, if the Fl_{N5ox} is stably maintained in the resting state as observed for EncM. Likely, this requires a secluded active site that impedes access of reductants to the flavin. Other flavoproteins may require additional analyses as described herein. Ultimately, other flavoenzymes employing the Fl_{N5ox} need to be studied in detail to gain a more comprehensive understanding of this novel flavin redox state and its catalytic repertoire.

ACKNOWLEDGMENTS

I thank all past and current colleagues and collaborators involved in the enterocin and EncM project and the DFG for funding (Grant: TE 931/2-1).

REFERENCES

Adak, S., & Begley, T. P. (2016). Dibenzothiophene catabolism proceeds via a flavin-N5-oxide intermediate. *Journal of the American Chemical Society*, *138*(20), 6424–6426.

Adak, S., & Begley, T. P. (2017a). Flavin-N5-oxide: A new, catalytic motif in flavoenzymology. *Archives of Biochemistry and Biophysics*, *632*, 4–10.

Adak, S., & Begley, T. P. (2017b). RutA-catalyzed oxidative cleavage of the uracil amide involves formation of a flavin-N5-oxide. *Biochemistry*, *56*(29), 3708–3709.

Cheng, Q., Xiang, L., Izumikawa, M., Meluzzi, D., & Moore, B. S. (2007). Enzymatic total synthesis of enterocin polyketides. *Nature Chemical Biology*, *3*(9), 557–558.

Ewing, T. A., Fraaije, M. W., Mattevi, A., & van Berkel, W. J. H. (2017). The VAO/PCMH flavoprotein family. *Archives of Biochemistry and Biophysics*, *632*, 104–117.

Hertweck, C., Luzhetskyy, A., Rebets, Y., & Bechthold, A. (2007). Type II polyketide synthases: Gaining a deeper insight into enzymatic teamwork. *Natural Product Reports*, *24*(1), 162–190.

Huijbers, M. M., Montersino, S., Westphal, A. H., Tischler, D., & van Berkel, W. J. (2014). Flavin dependent monooxygenases. *Archives of Biochemistry and Biophysics*, *544*, 2–17.

Jez, J. M., Ferrer, J. L., Bowman, M. E., Dixon, R. A., & Noel, J. P. (2000). Dissection of malonyl-CoA decarboxylation from polyketide formation in the reaction mechanism of a plant polyketide synthase. *Biochemistry*, *39*, 890–902.

Piel, J., Hertweck, C., Shipley, P. R., Hunt, D. M., Newman, M. S., & Moore, B. S. (2000). Cloning, sequencing and analysis of the enterocin biosynthesis gene cluster from the marine isolate 'Streptomyces maritimus': Evidence for the derailment of an aromatic polyketide synthase. *Chemistry & Biology*, 7, 943–955.

Shen, B. (2003). Polyketide biosynthesis beyond the type I, II and III polyketide synthase paradigms. *Current Opinion in Chemical Biology*, 7(2), 285–295.

Teufel, R. (2017). Flavin-catalyzed redox tailoring reactions in natural product biosynthesis. *Archives of Biochemistry and Biophysics*, *9861*(17), 30270–30279.

Teufel, R., Agarwal, V., & Moore, B. S. (2016). Unusual flavoenzyme catalysis in marine bacteria. *Current Opinion in Chemical Biology*, *31*, 31–39.

Teufel, R., Miyanaga, A., Michaudel, Q., Stull, F., Louie, G., Noel, J. P., et al. (2013). Flavin-mediated dual oxidation controls an enzymatic Favorskii-type rearrangement. *Nature*, *503*, 552–556.

Teufel, R., Stull, F., Meehan, M. J., Michaudel, Q., Dorrestein, P. C., et al. (2015). Biochemical establishment and characterization of EncM's flavin-N5-oxide cofactor. *Journal of the American Chemical Society*, *137*, 8078–8085.

Walsh, C. T., & Wencewicz, T. A. (2012). Flavoenzymes: Versatile catalysts in biosynthetic pathways. *Natural Product Reports*, *30*(1), 175–200.

Wang, P., Gao, X., & Tang, Y. (2012). Complexity generation during natural product biosynthesis using redox enzymes. *Current Opinion in Chemical Biology*, *16*(3–4), 362–369.

Xiang, L., Kalaitzis, J. A., & Moore, B. S. (2004). EncM, a versatile enterocin biosynthetic enzyme involved in Favorskii oxidative rearrangement, aldol condensation, and heterocycle-forming reactions. *Proceedings of the National Academy of Sciences of the United States of America*, *101*, 15609–15614.

FSC
www.fsc.org
MIX
Papier | Fördert gute Waldnutzung
FSC® C083411

Kolibri 360 GmbH
Ferdinand-Jühlke-Straße 7
99095 Erfurt, Deutschland
produktsicherheit@kolibri360.de